Research Problems
in Discrete Geometry

Peter Brass
William Moser
János Pach

Research Problems
in Discrete Geometry

 Springer

Peter Brass
Department of Computer Science
City College, City University of New York
138th Street & Convent Ave.
New York, NY 10031

William O.J. Moser
Department of Mathematics and Statistics
McGill University
805 Sherbrooke St. West
Montreal, QC H3A 2K6

János Pach
Courant Institute
New York University
251 Mercer St.
New York, NY 10012

Mathematics Subject Classification (2000): 52-00

Library of Congress Control Number: 2005924022

ISBN-10: 0-387-23815-8   Printed on acid-free paper.
ISBN-13: 978-0387-23815-8

Printed in the United States of America.   (SB)

9 8 7 6 5 4 3 2 1

springeronline.com

You say you've got a real solution
Well, you know
We'd all love to see the plan
You ask me for a contribution
Well, you know
We are doing what we can

(John Lennon)

Preface

The forerunner of this book had a modest beginning in July 1977 at the Discrete Geometry Week (organized by H.S.M. Coxeter) in Oberwolfach, Germany. There, William Moser distributed a list of 14 problems that he called RPDG (Research Problems in Discrete Geometry). The problems had first appeared in a 1963 mimeographed collection of 50 problems proposed by Leo Moser (1921–1970) with the title "Poorly formulated unsolved problems in combinatorial geometry." Five new editions of RPDG appeared between 1977 and 1981, with hundreds of copies mailed to interested geometers; reviews of RPDG appeared in *Mathematics Magazine* 53 (1980) p. 189; *American Mathematical Monthly* 87 (1980) p. 236; *Zentralblatt für Mathematik* Zbl 528.52001 and *Mathematical Reviews* MR 84c:51003, MR 85h:52002. The 1986 edition of RPDG reported on the solution of several outstanding problems in earlier editions and was prepared with the collaboration of János Pach; the 1993 edition appeared as DIMACS Technical Report 93-32, 131 pp. We had hoped to publish a book soon thereafter. Indeed, Paul Erdős, the great problem proposer and collector, wrote a preface for that book in the expectation that it would soon be published. However, the book-writing project languished until 2000, when Peter Brass joined the project; his hard and careful work was instrumental in bringing the project to a conclusion. The book finally exists.

Many problems had to be left out, for in a subject with an active research community and a tradition of problem proposing it is natural that the number of open problems explodes over time. Our selection of problems is subjective, and many areas, such as art gallery problems, Helly-type questions, stochastic geometry, and problems about convex polytopes, are completely missing. We decided not to delay further, since a published incomplete book is more useful than an unpublished book (which would also be incomplete). Perhaps later in this century we will expand the collection in a second edition and report then that many current problems have been solved. Meanwhile, we invite the readers to submit their comments, corrections, and new problems to the site `http://www.math.nyu.edu/~pach/`. Whenever it was possible, we tried to give proper credit to the original problem proposers and problem solvers, but we have surely made many mistakes. We apologize for them, and we urge our readers to point out any error of this kind that they may discover.

Our aim all along has been to achieve a collection of research problems in discrete geometry containing a statement of each problem, an account of progress, and an up-to-date bibliography. It was meant to be a resource for everyone, but particularly for students and for young mathematicians, to help them in finding an interesting problem for research. Apart from the important open problems in the field, we have included a large number of less well known but beautiful questions whose solutions may not require deep methods. We wish the reader good luck in finding solutions.

We sincerely thank all those who helped us with encouragement, information, and corrections. These include Boris Aronov, Vojtech Bálint, Imre Bárány, András Bezdek, Károly Bezdek, Károly Böröczky Jr., Helmut Brass, Erik Demaine, Adrian Dumitrescu, Herbert Edelsbrunner, György Elekes, Christian Elsholtz, Gábor Fejes Tóth, Eli Goodman, Ronald Graham, Branko Grünbaum, Heiko Harborth, Martin Henk, Aladár Heppes, Ferran Hurtado, Dan Ismailescu, Gyula Károlyi, Arnfried Kemnitz, Wlodzimierz Kuperberg, Endre Makai, Rados Radoičić, Andrej Raĭgorodskiĭ, Imre Z. Ruzsa, Micha Sharir, Alexander Soifer, József Solymosi, Konrad Swanepoel, Gábor Tardos, Csaba D. Tóth, Géza Tóth, Pavel Valtr, Katalin Vesztergombi, Jörg Wills, Chuanming Zong, and two students, Zheng Zhang and Mehrbod Sharifi. We apologize to those whose names have inadvertently been left out. We thank Marion Blake, David Kramer, Ina Lindemann, Paula Moser, and Mark Spencer for valuable editorial assistance, and Danielle Spencer for her help in preparing the cover design. We thank the mathematics libraries at the Free University Berlin, the Technical University Braunschweig, the Mathematische Forschungsinstitut Oberwolfach and at Courant Institute, New York University; our work would not have been possible without access to these excellent libraries. We also thank all our friends who obtained literature for us that we could not get ourselves.

This book is dedicated to Gisela and Helmut Brass and to Heiko Harborth (respectively parents and advisor of Peter Brass); to Beryl Moser and Leo Moser (respectively wife and brother of William Moser); to Klára and Zsigmond Pál Pach (parents of János Pach).

City College New York	Peter Brass
McGill University	William Moser
City College New York, NYU, and Rényi Institute	János Pach

Preface to an Earlier Version of RPDG

My friend Leo Moser (1921–1970) was an avid creator, collector, and solver of problems in number theory and combinatorics. At the 1963 Number Theory Conference in Boulder, Colorado, he distributed mimeographed copies of his list of fifty problems, which he called "Poorly formulated unsolved problems in combinatorial geometry." Although some parts of this collection have been reproduced several times, the entire list in its original form appeared in print only recently (*Discrete Applied Math.* **31** (1991), 201–225).

After Leo Moser's death, his brother Willy put together his *Research Problems in Discrete Geometry (RPDG)*, which was based on some questions proposed by Leo and was first distributed among the participants of the Discrete Geometry week in Oberwolfach, July 1977. This collection has been revised and largely extended by W. Moser and J. Pach. It has become an excellent resource book of fascinating open problems in combinatorial and discrete geometry which had nine different editions circulating in more than a thousand copies. In the last fifteen years it has reached virtually everybody interested in the field, and has generated a lot of research. In addition to the many new questions, a number of important but badly forgotten problems have also been publicized in these collections. They include Heilbronn's (now famous) triangle problem and my old questions about the distribution of distances among n points in the plane, just to mention two areas where much progress has been made recently. The present book is an updated "final" version of a large subset of the problems that appeared in the previous informal editions of *Research Problems in Discrete Geometry*. The authors have adopted a very pleasant style that allows the reader to get not only a feel for the problems but also an overview of the field.

And now let me say a few words about discrete geometry. As a matter of fact, I cannot even give a reasonable definition of the subject. Perhaps it is not inappropriate to recall the following old anecdote. Some years ago, when pornography was still illegal in America, a judge was asked to define pornography. He answered: "I cannot do this, but I sure can recognize it when I see it."

Perhaps discrete geometry started with the feud between Newton and Gregory about the largest number of solid unit ball spheres that can be placed to touch a "central" unit ball sphere. Newton believed this number to be twelve, while Gregory believed it was thirteen. This controversy was settled in Newton's favor only late in the last century. Even today little is known about similar problems in higher dimensions, although these questions were kept alive by the nineteenth century crystallographers and have created a lot of interest among physicists and biologists.

Minkowski's book *Geometrie der Zahlen* (1896) opened a new and important chapter in mathematics. It revealed some surprising connections between number theory and convex geometry, particularly between diophantine approximation and packing problems. This branch of discrete geometry was developed in books by Cassels (*An Introduction to the Geometry of Numbers*), Lekkerkerker (*Geometry of Numbers*), Coxeter (*Regular Polytopes*), and L. Fejes Tóth (*Lagerungen in der Ebene, auf der Kugel und im Raum*). "Alles Konvexe interessiert mich," said Minkowski, and I share his feeling.

Another early source is Sylvester's famous "orchard problem." In 1893 he also raised the following question: Given n points in the plane, not all on a line, can one always find a line passing through exactly two points? This problem remained unsolved and was completely forgotten before I rediscovered it in 1933. I was reading the Hilbert and Cohn-Vossen book (*Anschauliche Geometrie*) when the question occurred to me, and I thought it was new. It looked innocent, but to my surprise and annoyance I was unable to resolve it. However, I immediately realized that an affirmative answer would imply that any set of n noncollinear points in the plane determines at least n connecting lines. A couple of days later, Tibor Gallai came up with an ingenious short proof which turned out to be the first solution of Sylvester's problem. This was the starting point of many fruitful investigations about the incidence structure of sets of points and lines, circles, etc. Recently, these results have attracted a lot of attention, because they proved to be relevant in computational geometry.

In 1931, E. Klein observed that from any five points in the plane in general position one can choose four that determine a convex quadrilateral, and she asked whether the following generalization was true: For any $k \geq 4$ there exists an integer n_k such that any n_k-element set of points in general position in the plane contains the vertex set of a convex k-gon. Szekeres and I managed to establish this result; for the first proof we needed, and Szekeres rediscovered, Ramsey's theorem! Our paper raised many fascinating new questions which, I think, gave a boost to the development of combinatorial geometry and extremal combinatorics. A large variety of problems of this kind is discussed in the books of Hadwiger and Debrunner (*Combinatorial Geometry in the Plane*, translated and extended by Klee), Grünbaum (*Convex Polytopes*), Croft, Falconer, and Guy (*Unsolved Problems in Geometry*), and in the collection of my papers (*The Art of Counting*). I hope that the reader will forgive me that the above sketch of the recent history of combinatorial and discrete geometry is very subjective and, of course, overemphasizes my own contribution to the field.

There are certain areas of mathematics where individual problems are less important. However, I feel that problems play a very important role in elementary number theory and geometry. Hilbert and Hermann Weyl had the same opinion, but many eminent mathematicians disagree. I cannot decide who is right, but I am certainly on the side of Grünbaum in his old controversy with Dieudonné, who claimed that geometry is "dead." We are convinced that if a subject is rich in simple and fascinating unsolved problems, then it has a great future! The present collection of research problems by Moser and Pach proves beyond doubt the richness of discrete geometry.

I wish the reader good luck with the solutions!

Budapest, May 1991

Paul Erdős

Contents

0. Definitions and Notations

In this short chapter, we have collected some definitions and notations that are used in many places in this book. All of the concepts are quite standard; we list them for completeness and to explain the notation.

A set C is *convex* if for any two points $p, q \in C$ the entire line segment pq is also contained in C. A set C is *star-shaped* if for some point $p \in C$ and all points $q \in C$, the entire line segment pq is also contained in C. A set is a *convex body* if it is convex, compact and has nonempty interior. In general, a *body* is a set homeomorphic to a ball. Let $\mathbb{R}^d$ stand for the d-dimensional Euclidean space. In $\mathbb{R}^d$, the d-dimensional ball of radius r around the origin is denoted by $B^d(r)$, and the unit ball $B^d(1)$ by B^d. The two-dimensional ball B^2 is called a *circle* (we try to avoid the word "disk," which is often used in the literature for plane convex bodies).

Two bodies are *nonoverlapping* if they do not have an interior point in common, and they *touch* each other if they are nonoverlapping but have a common boundary point.

Some important functions defined for convex bodies C are the *volume* $\mathrm{Vol}(C)$; the *diameter* $\mathrm{diam}(C)$, which is the maximum distance between two points of C; the *width* $\mathrm{width}(C)$, which is the smallest distance of two parallel hyperplanes such that C lies in the slab between them; and the *inradius* and *circumradius*, which are the radii of the largest ball contained in C and the smallest ball containing C.

The Minkowski sum $X + Y$ of two sets is the set $\{x + y \mid x \in X, y \in Y\}$. Similarly, $\lambda X = \{\lambda x \mid x \in X\}$ denotes a scaled copy of X, and $-X = -1X = \{-x \mid x \in X\}$ denotes a copy of X reflected through the point 0. These operations depend on the choice of the origin 0, but the results are the same up to translation, and the operations should be viewed as acting on translation equivalence classes.

The *Hausdorff distance* of two compact sets $X, Y \subset \mathbb{R}^d$ is defined by

$$d^{\overset{\text{Haus-}}{\text{dorff}}}(X, Y) = \max\left(\sup_{x \in X} \inf_{y \in Y} d_{\mathrm{eucl}}(x, y) \, , \, \sup_{y \in Y} \inf_{x \in X} d_{\mathrm{eucl}}(x, y) \right).$$

An alternative description using Minkowski sums is

$$d^{\overset{\text{Haus-}}{\text{dorff}}}(X, Y) = \min\left\{ \lambda \geq 0 \, \middle| \, X + \lambda B^d \supseteq Y \text{ and } Y + \lambda B^d \supseteq X \right\}.$$

Some important classes of set mappings are *translations, homotheties, congruences, similarities,* and *affine maps*. A translate of a set $X \subset \mathbb{R}^d$ is a set $X + t$, $t \in \mathbb{R}^d$, a homothetic copy is a scaled translate $\lambda X + t$ with $\lambda > 0$. Negative homothetic copies with $\lambda < 0$ are allowed only where it is

explicitly stated. A congruence is an isometry (reflections are allowed), and
a similarity is a scaled congruence. An affinity is a nondegenerate linear
transformation followed by a translation.

A *symmetry* of a set is a congruence that maps the set onto itself. A
set X is called *centrally symmetric* about the origin if $X = -X$. In general,
X is centrally symmetric about the point (vector) t if $X = -X + 2t$.

A *lattice* Λ can be viewed in two ways, as a set of translations or as a
set of points, the *lattice points*.

As a set of translations, Λ is the set of all linear combinations of the
elements of a basis of the space with integer coefficients, which is the group
of translations generated by this basis. For any d linearly independent
vectors $u_1, \ldots, u_d$ in d-dimensional space, let $\Lambda = \Lambda(u_1, \ldots, u_d)$ denote the
lattice generated by them, so

$$\Lambda = \{m_1 u_1 + \cdots + m_d u_d \mid m_1, \ldots, m_d \in \mathbb{Z}\}.$$

A *fundamental domain* of Λ is a closed set whose translates by the elements
of Λ tile the space.

As a set of points, Λ is the orbit of any point p under the above set of
translations, that is, the set $\{p + m_1 u_1 + \cdots + m_d u_d \mid m_1, \ldots, m_d \in \mathbb{Z}\}$.
Thus, the lattice points of Λ are a translation equivalence class of point
sets.

The parallelepiped P induced by the 2^d vertices of the form $m_1 u_1 + \cdots + m_d u_d$, where $m_i \in \{0, 1\}$ for every i, is called a *fundamental parallelepiped* of the lattice. The same lattice can of course be generated in
many different ways and, therefore, has infinitely many different fundamental parallelepipeds. All of them are fundamental domains of the lattice, seen
as a group of translations, and therefore all of them have the same volume

$$\mathrm{Vol}(P) = |\det(u_1, \ldots, u_d)|.$$

The *density* of a lattice is defined as the reciprocal of this determinant
$|\det(u_1, \ldots, u_d)|$. This number is equal to the limit of the number of lattice
points in the ball $B^d(r)$ divided by $\mathrm{Vol}(B^d(r))$, as r tends to infinity.

Any lattice similar to the planar lattice generated by two adjacent sides
of a square or equilateral triangle is called a *square lattice* or a *triangular
lattice*, respectively.

A finite-dimensional *normed space*, also called *Minkowski space*, is a
finite-dimensional linear space X equipped with metric which is translation-
invariant $(d(p, q) = d(p + t, q + t)$ for translations $t)$ and homogeneous
$(d(\lambda p, 0) = \lambda d(p, 0)$ for $\lambda > 0)$. Since $d(p, q) = d(p - q, 0)$, this metric is
completely described by the distance of every point x from the origin, which
is called the *norm* $\|x\| = d(x, 0)$. A normed space can be characterized by

its *unit ball*, which is the set of points whose norm is at most one. The unit ball is a centrally symmetric convex body, and any centrally symmetric convex body can be chosen as the unit ball defining a normed space. Two normed spaces are isometric if their unit balls are affine images of each other.

A normed space is said to be *strictly convex* if the boundary of its unit ball does not contain any line segment, or, equivalently, if the distances between any three noncollinear points satisfy the triangle inequality with strict inequality.

The d-dimensional L_p-*space* is the space defined by the unit ball

$$\{(x_1, \ldots, x_d) \in \mathbb{R}^d \mid |x_1|^p + \cdots + |x_d|^p \le 1\},$$

if $p \in [1, \infty)$, and by the unit ball

$$\{(x_1, \ldots, x_d) \in \mathbb{R}^d \mid \max(|x_1|, \ldots, |x_d|) \le 1\},$$

if $p = \infty$. The d-dimensional L_2 space is the usual Euclidean space. All L_p-spaces, with the exception of L_1 and L_∞, are strictly convex. The unit ball of the L_1-space is the d-dimensional crosspolytope, and the unit ball of the L_∞-space is the d-dimensional cube.

The term *arrangement* is used in two different ways in this book. In the first four chapters, we use it for collections of bodies, including packings, coverings, or tilings. It is also used, especially in Chapter 7, for families $\mathcal{F}$ of lines, pseudolines, curves, hyperplanes, and other hypersurfaces subdividing the plane or space. In these cases, the arrangement $\mathcal{A}(\mathcal{F})$ is the decomposition of space into *cells* of various dimensions induced by the members of $\mathcal{F}$. Each *cell* is a connected component of

$$\bigcap_{G \in \mathcal{G}} G \setminus \bigcup_{F \in \mathcal{F} \setminus \mathcal{G}} F,$$

for a subcollection $\mathcal{G} \subseteq \mathcal{F}$, where the first intersection is taken to be the entire space if $\mathcal{G}$ is empty.

In the latter sense of the term, two arrangements are *isomorphic* if they generate combinatorially isomorphic cell decompositions. A planar arrangement $\mathcal{A}(\mathcal{F})$ is *simple* if each of its vertices (0-dimensional cells) belongs to precisely two members (curves) in $\mathcal{F}$. A d-dimensional arrangement $\mathcal{F}$ is simple if for every $0 \le i < d$ each i-dimensional cell is the intersection of $d - i$ members (hypersurfaces) of $\mathcal{F}$.

Let f and g be two nonnegative functions. We say f is $O(g(n))$, or $f(n) \le O(g(n))$, if $f(n) \le Cg(n)$ for some $C > 0$ and for all sufficiently large n. Similarly, we say $f(n)$ is $\Omega(g(n))$, or $f(n) \ge \Omega(g(n))$, if $f(n) \ge Cg(n)$

for some $C > 0$ and for all sufficiently large n. Finally, $f(n) = \Theta(g(n))$ means that we have $C_1 g(n) \le f(n) \le C_2 g(n)$ for some $C_1, C_2 > 0$ and for all sufficiently large n.

The functions $\log^*(n)$ and $\alpha(n)$ are very slowly growing functions that often occur in combinatorial and computational geometry. The *iterated logarithm function* $\log^*(n)$ is defined by

$$\log^* n = k, \text{ if } \quad \left.2^{2^{2^{\cdot^{\cdot^{\cdot^2}}}}}\right\}k \text{ times} \le n < \left.2^{2^{2^{\cdot^{\cdot^{\cdot^2}}}}}\right\}k+1 \text{ times}.$$

It grows slower than any fixed number of iterations of the logarithm function. The *inverse Ackermann function* $\alpha(n)$ grows even slower: it is defined as $\alpha(n) = \min\{k \mid A(k,1) \ge n\}$, where $A(i,j)$ is given by $A(i,j) = A(i-1, A(i, j-1))$ for $i, j \ge 2$, and $A(i,1) = A(i-1,2)$, $A(1,j) = 2j$.

1. Density Problems for Packings and Coverings

1.1 Basic Questions and Definitions

In his lecture at the Scandinavian Natural Science Congress in 1892, Axel Thue [Th892] claimed that the density of any arrangement of non-overlapping equal circles in the plane is at most $\pi/\sqrt{12}$, the ratio of the area of the circle to the area of the regular hexagon circumscribed about it.

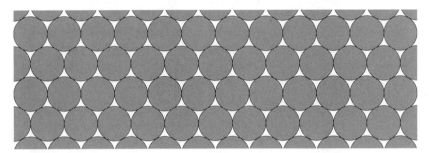

THE DENSEST PACKING OF EQUAL CIRCLES

Thue's first proof contained an error, but later he returned to this problem [Th10] and solved it. His result remained largely unnoticed. However, at about the same time, Minkowski [Mi896] discovered that problems of this kind are closely related to some basic questions in Diophantine approximation, and this connection enabled him to give simple geometric proofs for several number-theoretic results. Hilbert [Hi900] immediately realized the significance of the question. In his famous lecture delivered at the International Congress of Mathematicians in Paris in 1900, he asked, *"How can one arrange most densely in space an infinite number of equal solids of given form, e.g., spheres with given radii or regular tetrahedra with given edges (or in prescribed position), that is, how can one fit them together so that the ratio of the filled to the unfilled space may be as great as possible?"* Still in the context of the same problem, Hilbert raised the question whether one can characterize all polytopes that permit *"... by suitable juxtaposition of congruent copies a complete filling up of all space."* The route Hilbert suggested for the solution of this problem turned out to be impassable, but his group-theoretic viewpoint (strongly influenced by Felix Klein's Erlangen Program and by some spectacular discoveries in crystallography) has passed into common currency.

More than a century has elapsed since Thue's pioneering result and Hilbert's lecture, but we still do not have satisfactory answers to the above questions. Nevertheless, the subject has accumulated a large number of deep results, and has become a vast, separate discipline within mathematics, with its own powerful techniques. There are some valuable monographs on the subject: Cassels [Ca72], L. Fejes Tóth [FeT64], [FeT72], Rogers [Ro64], Gruber and Lekkerkerker [GrL87], Erdős, Gruber, and Hammer [ErGH89], Grünbaum and Shephard [GrS87], Pach and Agarwal [PaA95], Conway and Sloane [CoS03], Zong [Zo99], Böröczky Jr. [Bö04].

For comprehensive surveys of the most recent literature, the reader is referred to G. Fejes Tóth [FeT83], [FeT97], [FeT99], G. Fejes Tóth and W. Kuperberg [FeTK93a], [FeTK93b], L. Fejes Tóth [FeT84], Schulte [Sch93], Gritzmann and Wills [GrW93].

First we introduce the basic definitions and notation that will be used throughout this chapter.

A collection $\mathcal{C} = \{C_1, C_2, \ldots\}$ of compact sets with nonempty interiors is said to form a *packing* in a domain $D \subseteq \mathbb{R}^d$ if $\bigcup_i C_i \subseteq D$ and no two members of $\mathcal{C}$ have an interior point in common. If $\bigcup_i C_i \supseteq D$, then $\mathcal{C}$ is called a *covering* of D. If $\mathcal{C}$ consists of all translates of a particular set C by vectors belonging to a given lattice Λ, i.e.,

$$\mathcal{C} = \{C + \lambda \,|\, \lambda \in \Lambda\},$$

then $\mathcal{C}$ is said to be a *lattice arrangement*. If, in addition, $\mathcal{C}$ is a packing in $\mathbb{R}^d$ (a covering of $\mathbb{R}^d$), then it is called a *lattice packing* (*lattice covering*).

The *density* of a collection $\mathcal{C}$ relative to a bounded domain D is defined as

$$d(\mathcal{C}, D) = \frac{\sum_{C \in \mathcal{C}} \text{Vol}\,(C \cap D)}{\text{Vol}\,D}.$$

If $D = \mathbb{R}^d$ is the whole space, then we define the *upper* and *lower densities* of $\mathcal{C}$ as follows:

$$\overline{d}(\mathcal{C}, \mathbb{R}^d) = \limsup_{r \to \infty} d(\mathcal{C}, B^d(r)), \qquad \underline{d}(\mathcal{C}, \mathbb{R}^d) = \liminf_{r \to \infty} d(\mathcal{C}, B^d(r)),$$

where $B^d(r)$ denotes the ball of radius r centered at the origin. If these two numbers are the same, then it is called the *density* of $\mathcal{C}$ in $\mathbb{R}^d$, denoted by $d(\mathcal{C}, \mathbb{R}^d)$. Throughout this chapter we shall deal only with collections of Jordan-measurable (usually convex) sets, so the above definitions make sense.

Let C be a d-dimensional compact set with nonempty interior. Let the *packing density* of C be defined as the largest density of a packing of congruent copies of C in $\mathbb{R}^d$. That is, let

$$\delta(C) = \sup_{\mathcal{C}\ \text{packing}} d(\mathcal{C}, \mathbb{R}^d),$$

where the supremum is taken over all packings of congruent copies of C which have a density. It is not hard to see that this supremum is reached, and that

$$\delta(C) = \max_{\mathcal{C}\ \text{packing}} d(\mathcal{C}, \mathbb{R}^d) = \sup_{\mathcal{C}\ \text{packing}} \overline{d}(\mathcal{C}, \mathbb{R}^d),$$

where the supremum is taken over all packings with congruent copies of C.

If we restrict our attention to lattice packings, or to packings with translates of C, then we obtain the similarly defined notions of the *lattice packing density* and *translative packing density* of C, denoted by $\delta_L(C)$ and $\delta_T(C)$, respectively. That is,

$$\delta_L(C) = \sup_{\substack{\mathcal{C}\ \text{lattice} \\ \text{packing}}} d(\mathcal{C}, \mathbb{R}^d), \qquad \delta_T(C) = \sup_{\substack{\mathcal{C}\ \text{translative} \\ \text{packing}}} d(\mathcal{C}, \mathbb{R}^d).$$

Again these suprema are reached.

Completely analogously, we can define the *covering density* of C as the least density of a covering of $\mathbb{R}^d$ with congruent copies of C, i.e.,

$$\theta(C) = \inf_{\mathcal{C}\ \text{covering}} d(\mathcal{C}, \mathbb{R}^d),$$

where the infimum is taken over all coverings with C which have a density. Again, it is easy to see that this infimum is reached, and that

$$\theta(C) = \min_{\mathcal{C}\ \text{covering}} d(\mathcal{C}, \mathbb{R}^d) = \inf_{\mathcal{C}\ \text{covering}} \underline{d}(\mathcal{C}, \mathbb{R}^d),$$

where the infimum is taken over all coverings of $\mathbb{R}^d$ with congruent copies of C.

Similarly, let the *lattice covering density* and the *translative covering density* of C be defined as

$$\theta_L(C) = \inf_{\substack{\mathcal{C}\ \text{lattice} \\ \text{covering}}} d(\mathcal{C}, \mathbb{R}^d), \qquad \theta_T(C) = \inf_{\substack{\mathcal{C}\ \text{translative} \\ \text{covering}}} d(\mathcal{C}, \mathbb{R}^d),$$

and again these infima are reached.

It is obvious that for any C,

$$\delta_L(C) \le \delta_T(C) \le \delta(C) \le 1 \le \theta(C) \le \theta_T(C) \le \theta_L(C).$$

If $\delta(C) = \theta(C) = 1$, then one can show that there is an arrangement $\mathcal{C}$, of congruent copies of C, that is a packing and a covering at the same time. In other words, $\mathbb{R}^d$ can be covered by nonoverlapping congruent copies of C. Such an arrangement is usually called a *tiling*, and C is said to be a *tile* (or *space-filler*). Schmidt [Sch61] showed that if the boundary of C is smooth, then $\delta(C) < 1 < \theta(C)$, i.e., C cannot be a tile.

Sometimes we shall consider packings and coverings with unbounded sets (mainly with cylinders) and with incongruent sets, whose densities can be defined in a completely analogous way. The density $\delta(C)$ is often called the *density of the densest* (or *most economical*) packing with congruent copies of C. Similarly, the covering number $\theta(C)$ is often called the density of the *thinnest* (or *most economical*) *covering* of space with congruent copies of C. Of course these terms are somewhat informal, because the most economical arrangements are usually not uniquely determined, and it is not even clear a priori that there exists a most economical arrangement, although in all problems we are going to discuss, this can readily be checked.

Let C be an arrangement (family) of congruent d-dimensional bodies. An isometry of $\mathbb{R}^d$ that takes every member of C onto another one is called a *symmetry* of C, and the collection of all symmetries of C (including the identity) forms the *symmetry group* of C. The symmetry group of C is *nontrivial* if it contains an isometry other than the identity. The arrangement C is called *periodic* if its symmetry group contains d linearly independent translations, and *aperiodic* if its symmetry group contains no translations. (Note that when $d \geq 2$, a nonperiodic arrangement is not necessarily aperiodic!).

[Bö04]   K. BÖRÖCZKY JR.: *Finite Packing and Covering*, Cambridge University Press, 2004.

[Ca72]   J.W.S. CASSELS: *An Introduction to the Geometry of Numbers* (2nd edition), Springer-Verlag, 1972.

[CoS03]   J.H. CONWAY, N.J.A. SLOANE: *Sphere Packings, Lattices and Groups* (3rd edition), Springer-Verlag, 2003.

[ErGH89] P. ERDŐS, P.M. GRUBER, J. HAMMER: *Lattice Points*, Pitman Monographs and Surveys in Pure and Applied Mathematics **39**, Longman Scientific and Technical, Wiley, 1989.

[FeT99]   G. FEJES TÓTH: Recent progress on packing and covering, in: *Advances in Discrete and Computational Geometry* (South Hadley, MA, 1996), *Contemporary Math.* **223**, Amer. Math. Soc., 1999, 145–162.

[FeT97]   G. FEJES TÓTH: Packing and covering, in: *Handbook of Discrete and Computational Geometry*, J.E. Goodman, J. O'Rourke, eds., CRC Press, 1997, 19–41.

[FeT83]   G. FEJES TÓTH: New results in the theory of packing and covering, in: *Convexity and Its Applications*, P.M. Gruber, J.M. Wills, eds., Birkhäuser-Verlag, 1983, 318–359.

[FeTK93a] G. FEJES TÓTH, W. KUPERBERG: A survey of recent results in the theory of packing and covering, in: *New Trends in Discrete and Computational Geometry*, J. Pach, ed., *Algorithms Combin. Series* **10**, Springer-Verlag, 1993, 251–279.

[FeTK93b] G. FEJES TÓTH, W. KUPERBERG: Packing and covering with convex sets, in: *Handbook of Convex Geometry*,

P.M. Gruber, J.M. Wills, eds., North-Holland, 1993, 799–860.

[FeT84]   L. Fejes Tóth: Density bounds for packing and covering with convex discs, *Expositiones Math.* **2** (1984) 131–153.

[FeT72]   L. Fejes Tóth: *Lagerungen in der Ebene, auf der Kugel und im Raum* (2. Auflage), Springer-Verlag, 1972.

[FeT64]   L. Fejes Tóth: *Regular Figures*, Pergamon Press, 1964.

[GrW93]   P. Gritzmann, J.M. Wills: Finite packing and covering, in: *Handbook of Convex Geometry*, P.M. Gruber, J.M. Wills, eds., North-Holland, 1993, 861–897.

[GrL87]   P.M. Gruber, P.M. Lekkerkerker: *Geometry of Numbers*, North-Holland, 1987.

[GrS87]   B. Grünbaum, G.C. Shephard: *Tilings and Patterns*, W.H. Freeman and Co., 1987. The first seven chapters are also available separately as *Tilings and Patterns – An Introduction*.

[Hi900]   D. Hilbert: Mathematische Probleme. (Lecture delivered at the International Congress of Mathematicians, Paris 1900.) *Göttinger Nachrichten* (1900) 253–297, or *Gesammelte Abhandlungen* Band 3, 290–329. English translation in *Bull. Amer. Math. Soc.* (1902) 437–479, and *Bull. Amer. Math. Soc.* New Ser. **37** (2000) 407–436.

[Mi896]   H. Minkowski: *Geometrie der Zahlen*, Teubner-Verlag, Leipzig, 1896.

[PaA95]   J. Pach, P.K. Agarwal: *Combinatorial Geometry*, Wiley-Interscience, 1995.

[Ro64]   C.A. Rogers: *Packing and Covering*, Cambridge University Press, 1964.

[Sch61]   W.M. Schmidt: Zur Lagerung kongruenter Körper im Raum, *Monatshefte Math.* **65** (1961) 154–158.

[Sch93]   E. Schulte: Tilings, in: *Handbook of Convex Geometry*, P.M. Gruber, J.M. Wills, eds., North-Holland, 1993, 899–932.

[Th10]   A. Thue: On the densest packing of congruent circles in the plane (in Norwegian), *Skr. Vidensk-Selsk, Christiania* **1** (1910) 3–9, or in *Selected mathematical papers*, T. Nagell et al., eds., Universitetsforlaget Oslo, 1977.

[Th892]   A. Thue: On some geometric number-theoretic theorems (in Danish), *Forhandlingerne ved de Skandinaviske Naturforskeres* **14** (1892) 352–353, or in: *Selected Mathematical Papers*, T. Nagell et al., eds., Universitetsforlaget Oslo, 1977.

[Zo99]   C. Zong: *Sphere Packings*, Springer-Verlag, 1999.

1.2 The Least Economical Convex Sets for Packing

The closer the packing density of a convex body C is to 1, the larger the number of congruent copies of C that can be packed into a given big container. Therefore, the least economical convex bodies for packing are those whose packing density is minimum.

Problem 1   *Is the least economical convex body for packing in the plane centrally symmetric? More precisely, is it true that for any plane convex body C there exists a centrally symmetric convex body C' such that $\delta(C') \leq \delta(C)$?*

A result of G. Kuperberg and W. Kuperberg [KuK90] suggests that the answer to this question is probably in the negative. It is possible that the worst convex body from the point of view of economical packing in the plane is the regular heptagon (whose densest known packing is of density 0.8926...; see also [Bl83]).

Problem 2   *What is the minimum of $\delta(C)$ over all convex bodies C in the plane?*

Chakerian and Lange [ChL71] proved that every plane convex body C is contained in a quadrilateral Q with area$(Q) \leq \sqrt{2}\,$area(C). Since every quadrilateral tiles the plane, this implies that

$$\delta(C) \geq 1/\sqrt{2} \qquad \text{for every plane convex } C.$$

W. Kuperberg [Ku87] improved this estimate to 25/32. The current best bound in this direction is due to G. Kuperberg and W. Kuperberg [KuK90], who showed that

$$\delta(C) \geq \sqrt{3}/2 = 0.86602540\ldots \qquad \text{for every plane convex body } C.$$

Doheny [Do95] proved by a compactness argument that this inequality can be improved. For centrally symmetric convex bodies, this result had already been proved by Mahler [Ma46, Ma47] and by L. Fejes Tóth [FeT48].

The problem becomes easier if we restrict our attention to *lattice packings*. According to a beautiful theorem of Fáry [Fá50],

$$\delta_L(C) \geq 2/3 \qquad \text{for every convex } C,$$

where equality holds if and only if C is a triangle. Courant [Co65] found an elegant alternative proof of this result.

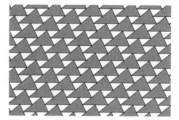

DENSEST LATTICE PACKING
OF TRIANGLES

L. Fejes Tóth [FeT50] proved that $\delta(C) = \delta_L(C)$ for every centrally symmetric convex body C in the plane. That is, the density of a packing with congruent copies of a centrally symmetric convex body can never exceed the density of the densest lattice packing. This partially explains why lattices and number-theoretic concepts play a special role in this field.

Conjecture 3 *(Reinhardt [Re34]) For every centrally symmetric convex body C in the plane,*

$$\delta(C) = \delta_L(C) \geq \frac{8 - 4\sqrt{2} - \ln 2}{2\sqrt{2} - 1} = 0.90241418\ldots,$$

with equality only for the so-called smoothed octagon.

The "smoothed octagon" results from a regular octagon by cutting off each vertex v_i with a hyperbolic arc from the hyperbola that is tangential to $v_i v_{i-1}$ and $v_i v_{i+1}$ and has the lines $v_{i-1} v_{i-2}$ and $v_{i+1} v_{i+2}$ as asymptotes. Nazarov [Na86] proved that Reinhardt's smoothed octagon is, in a certain sense, locally optimal.

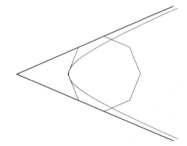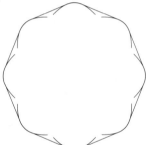

THE SMOOTHED OCTAGON

As we have already mentioned, K. Mahler and L. Fejes Tóth proved that $\delta_L(C) \geq \sqrt{3}/2$ for every centrally symmetric convex C. This bound was improved by Ennola [En61] and later by Tammela [Ta70] to about 0.89265.

For the three-dimensional case, Smith [Sm99] used Tammela's result to establish

$$\delta(C) \geq \delta_L(C) \geq 0.46421 \ldots \qquad \text{if } C \text{ is convex and centrally symmetric.}$$

However, it is possible that this lower bound can be improved to $\pi/\sqrt{18} = 0.74048\ldots$, the packing density of the ball.

The corresponding problems in higher-dimensional spaces seem to be hopelessly difficult. The celebrated Minkowski–Hlawka theorem (see [Hl43]) gives a bound of

$$\delta_L(C) \geq \zeta(d)/2^{d-1} \qquad \text{if } C \text{ is convex and centrally symmetric,}$$

where $\zeta(d) = \sum_{k=1}^{\infty} k^{-d}$ is Riemann's zeta function, so $\frac{\pi^2}{6} = \zeta(2) > \zeta(d) > 1$, for $d \geq 3$.

The asymptotically best known general result is due to Schmidt [Sch63] who established the inequality

$$\delta(C) \geq \delta_L(C) \geq c\frac{d}{2^d} \qquad \text{if } C \text{ is convex and centrally symmetric,}$$

provided that $c < \log 2$ and d is sufficiently large. This bound may well be far from best possible, but even for balls B^d the best known lower estimate, due to Ball [Ba92], is only a constant times better: $\delta(B^d) \geq \delta_L(B^d) \geq (d-1)\zeta(d)/2^{d-1}$. For some special classes of convex bodies, including the so-called superballs, Rush et al. [Ru93], [Ru97], [ElOR91] established essentially better lower bounds.

One of the major difficulties is that in higher-dimensional spaces $\delta(C) \neq \delta_L(C)$ for many centrally symmetric convex bodies C. As a matter of fact, it is generally believed that this is the case for spheres.

Conjecture 4 *(Rogers [Ro64]) Let B^d denote the unit ball in $\mathbb{R}^d$. There exists a dimension $d > 3$ such that*

$$\delta(B^d) \neq \delta_L(B^d).$$

An interesting observation of A. Bezdek and W. Kuperberg [BeK91] seems to support this conjecture: For any $d \geq 3$, there is an ellipsoid E in $\mathbb{R}^d$ such that $\delta(E) > \delta_L(E)$.

[Ba92]   K.M. BALL: A lower bound for the optimal density of lattice packings, *Internat. Math. Res. Notices* **10** (1992) 217–221.

[BeK91]   A. BEZDEK, W. KUPERBERG: Packing Euclidean space with congruent cylinders and with congruent ellipsoids, in: *Applied Geometry and Discrete Mathematics: The Victor Klee Festschrift*, P. Gritzmann et al., eds., *DIMACS Ser. Discrete Math. Comp. Sci.*, AMS and ACM, 1991, 71–80.

[Bl83]   G. BLIND: Research Problem 34, *Periodica Math. Hungar.* **14** (1983) 309–312.

[ChL71]   G.D. CHAKERIAN, L.H. LANGE: Geometric extremum problems, *Math. Mag.* **44** (1971) 57–69.

[Co65]   R. COURANT: The least dense lattice packing of two-dimensional convex bodies, *Comm. Pure. Appl. Math.* **18** (1965) 339–343.

[**Do95**] K.R. DOHENY: On the lower bound of packing density for convex bodies in the plane, *Beiträge Algebra Geom.* **36** (1995) 109–117.

[**ElOR91**] N.D. ELKIES, A.M. ODLYZKO, J.A. RUSH: On the packing densities of superballs and other bodies, *Invent. Math.* **105** (1991) 613–639.

[**En61**] V. ENNOLA: On the lattice constant of a symmetric convex domain, *J. London Math. Soc.* **36** (1961) 135–138.

[**Fá50**] I. FÁRY: Sur la densité des réseaux de domaines convexes, *Bull. Soc. Math. France* **78** (1950) 152–161.

[**FeT50**] L. FEJES TÓTH: Some packing and covering theorems, *Acta Sci. Math. (Szeged)* **12 A** (1950) 62–67.

[**FeT48**] L. FEJES TÓTH: On the densest packing of convex domains, *Proc. Nederl. Akad. Wetensch. (Amsterdam)* **51** (1948) 544–547.

[**Hl43**] E. HLAWKA: Zur Geometrie der Zahlen, *Mathematische Z.* **49** (1943) 285–312.

[**KuK90**] G. KUPERBERG, W. KUPERBERG: Double lattice packings of convex bodies in the plane, *Discrete Comput. Geom.* **5** (1990) 389–397.

[**Ku87**] W. KUPERBERG: On packing the plane with congruent copies of a convex body, in: *Intuitive Geometry* (Siófok, 1985), K. Böröczky et al., eds., *Colloq. Math. Soc. János Bolyai* **48** North-Holland, 1987, 317–329.

[**Ma47**] K. MAHLER: On the minimum determinant and the circumscribed hexagons of a convex domain, *Proc. Nederl. Akad. Wetensch. (Amsterdam)* **50** (1947) 695–703.

[**Ma46**] K. MAHLER: The theorem of Minkowski–Hlawka, *Duke Math. J.* **13** (1946) 611–621.

[**Na86**] F.L. NAZAROV: On the Reinhardt problem of lattice packings of convex regions. Local extremality of the Reinhardt octagon (in Russian), *Zap. Nauchn. Sem. Leningrad. Otdel. Mat. Inst. Steklov. (LOMI)* **151** (1986) Issled. Teor. Chisel. **9** 104–114, 197–198; translation in *J. Soviet Math.* **43** (1988) 2687–2693.

[**Re34**] K. REINHARDT: Über die dichteste gitterförmige Lagerung kongruenter Bereiche in der Ebene und eine besondere Art konvexer Kurven, *Abh. Math. Sem. Hamburg* **10** (1934) 216–230.

[**Ro64**]   C.A. ROGERS: *Packing and Covering*. Cambridge University Press, 1964.

[**Ru97**]   J.A. RUSH: Sphere packing and coding theory, in: *Handbook of Discrete and Computational Geometry*, J.E. Goodman, J. O'Rourke, eds., CRC Press, 1997, 917–932.

[**Ru93**]   J.A. RUSH: A bound, and a conjecture on the maximum lattice-packing density of a superball, *Mathematika* **40** (1993) 137–143.

[**Sch63**]   W.M. SCHMIDT: On the Minkowski–Hlawka theorem, *Illinois J. Math.* **7** (1963) 18–23.

[**Sm99**]   E.H. SMITH: A density bound for efficient packings of 3-space with centrally symmetric convex bodies, *Mathematika* **46** (1999) 137–144.

[**Ta70**]   P. TAMMELA: An estimate of the critical determinant of a two-dimensional convex symmetric domain (in Russian), *Izv. Vyss. Ucebn. Zaved. Math.* **12** (1970) 103–107.

1.3 The Least Economical Convex Sets for Covering

The closer the covering density of a convex body C is to 1, the fewer the number of congruent copies of C needed to cover a given big domain. Therefore, the most economical convex bodies for covering are those that admit a tiling, and the least economical ones are the convex bodies with maximum covering density. According to an old result of Kershner [Ke39], the covering density of the circle is $2\pi/\sqrt{27}$.

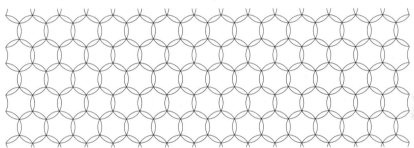

THE THINNEST COVERING OF THE PLANE BY CIRCLES

Problem 1   *What is the maximum of $\theta(C)$ over all convex bodies C in the plane? In particular, does there exist a C with $\theta(C) > \frac{2\pi}{\sqrt{27}} \approx 1.209199$?*

It follows from a result of Sas [Sa39] that every centrally symmetric plane convex body C contains a centrally symmetric convex hexagon H with $\mathrm{area}(H) \geq \frac{\sqrt{27}}{2\pi} \mathrm{area}(C)$. Since H tiles the plane in a lattice-like manner, it follows that

$$\theta(C) \leq \theta_L(C) \leq \frac{2\pi}{\sqrt{27}} \qquad \text{for every centrally symmetric convex } C.$$

Moreover, the only case in which both inequalities here become equalities is that of C a circle or an ellipse sufficiently close to a circle. (In fact, for very long ellipses C, it is not known whether $\theta(C) = \theta_L(C)$. See Section 1.9.)

The currently best known upper bound was obtained by Ismailescu [Is98]

$$\theta(C) \leq 1.2281772\ldots \qquad \text{for every convex body } C \text{ in the plane,}$$

improving a previous result of $\frac{8}{3}(2\sqrt{3} - 3) = 1.23760430\ldots$ due to W. Kuperberg [Ku89] (see also Smith [Sm94]).

The least economical convex sets for *lattice covering* are the triangles. According to Fáry's theorem [Fá50], we have $\theta_L(C) \leq 3/2$ for every convex body C in the plane, where equality holds if and only if C is a triangle. It is very likely that the same is true for *translative coverings*. Of course, the problem is invariant under affine transformations of the plane, so it does not really matter which triangle we consider.

THE THINNEST LATTICE COVERING OF THE PLANE BY TRIANGLES

Problem 2   *Is it true that the minimum density of a covering of the plane with translates of a triangle Δ is at least $3/2$? In other words, does $\theta_T(\Delta) = 3/2$ hold?*

Here, in contrast to packings, the analogous questions for higher-dimensional spaces $\mathbb{R}^d$ do not seem to be completely hopeless. Rogers [Ro57] (see also Erdős and Rogers [ErR62], Füredi and Kang [FüK05]) showed that

$$\theta(C) \leq \theta_T(C) \leq d \ln d + d \ln \ln d + 5d \qquad \text{for every convex body } C \subseteq \mathbb{R}^d,$$

while Coxeter, Few, and Rogers [CoFR59] established the inequality

$$\theta(B^d) \geq \tau_d \ \sim \ \frac{d}{e^{3/2}},$$

where τ_d is defined as follows. Let T^d be a regular simplex inscribed in the unit ball $B^d \subseteq \mathbb{R}^d$, and draw a unit ball around each of its vertices. Then τ_d is the ratio between the total volume of the intersections of these $d+1$ balls with T^d and the volume of T^d.

Problem 3   *Does there exist a constant γ such that for any $d \geq 2$,*

$$\theta(C) \leq \gamma d \qquad \text{for every convex body } C \subseteq \mathbb{R}^d?$$

It is perhaps interesting to note that, again in contrast to packings, the most economical known translative coverings in higher dimensions are not lattice-like. In particular, the best known general upper bound on $\theta_L(C)$, due to Rogers [Ro59], is $d^{\log_2 \log_2 d + O(1)}$, where C is an arbitrary convex body in $\mathbb{R}^d$. Better bounds have been established for balls and other convex bodies with many symmetries (see Gritzmann [Gr85]).

We cannot settle the following attractive problem of Heppes and Lenz [Le56a], [Le56b] even in the plane.

Conjecture 4 *(Heppes [He98], Lenz) For any covering C of $\mathbb{R}^d$ with convex bodies of (at most) unit diameter and for any $r > 0$, let $f(C, r)$ denote the number of elements in C lying in a ball of radius r around the origin, divided by the volume of this ball. Then*

$$\liminf_{r \to \infty} f(C, r)$$

attains its minimum when C is the thinnest covering of space with congruent balls of diameter one.

Paradoxically, in some (very different) sense, this means that from the point of view of covering space with bodies of unit diameter, balls are the *most economical* figures. In fact, the elegant formulation of Heppes's conjecture hides a statement on tilings rather than coverings. In the plane, for instance, the relevant figures are regular *hexagons*: the hexagons inscribed in the elements of a thinnest circle covering form a tiling, in which, according to the above conjecture, the number of cells per area is as small as possible.

[**CoFR59**] H.S.M. COXETER, L. FEW, C.A. ROGERS: Covering space with equal spheres, *Mathematika* **6** (1959) 147–157.

[**ErR62**] P. ERDŐS, C.A. ROGERS: Covering space with convex bodies, *Acta Arith.* **7** (1962) 281–285.

[**Fá50**] I. FÁRY: Sur la densité des réseaux de domaines convexes, *Bull. Soc. Math. France* **178** (1950) 152–161.

[**FüK05**] Z. FÜREDI, J.-H. KANG: Covering Euclidean n-space by translates of a convex body, manuscript.

[**Gr85**] P. GRITZMANN: Lattice covering of space with symmetric convex bodies, *Mathematika* **32** (1985) 311–315.

[**He98**] A. HEPPES: Research problem, *Period. Math. Hungar.* **36** (1998) 181–182.

[**Is98**] D. ISMAILESCU: Covering the plane with copies of a convex disk, *Discrete Comput. Geom.* **20** (1998) 251–263.

[Ke39]   R.B. Kershner: The number of circles covering a set, *Amer. J. Math.* **61** (1939) 665–671.

[Ku89]   W. Kuperberg: Covering the plane with congruent copies of a convex body, *Bull. London Math. Soc.* **21** (1989) 82–86.

[Le56a]   H. Lenz: Über die Bedeckung ebener Punktmengen durch solche kleineren Durchmessers, *Archiv Math.* **6** (1956) 34–40.

[Le56b]   H. Lenz: Zerlegung ebener Bereiche in konvexe Zellen von möglichst kleinem Durchmesser, *Jahresbericht Deutsch. Math. Ver.* **58** (1956) 87–97.

[Ro59]   C.A. Rogers: Lattice coverings of space, *Mathematika* **6** (1959) 33–39.

[Ro57]   C.A. Rogers: A note on coverings, *Mathematika* **4** (1957) 1–6.

[Sa39]   E. Sas: Über eine Extremumeigenschaft der Ellipsen, *Compositio Math.* **6** (1939) 468–470.

[Sm94]   E.H. Smith: Covering the plane with congruent copies of a convex disk, *Beiträge Algebra Geom.* **35** (1994) 101–108.

1.4 How Economical Are the Lattice Arrangements?

For any convex body C in the plane, let h_C (and H_C) denote a hexagon inscribed in C (circumscribed about C) with maximum (minimum) area.

Conjecture 1   *(L. Fejes Tóth [FeTL72]) For any convex body C in the plane, the density $\theta(C)$ of the thinnest covering with congruent copies of C satisfies*

$$\theta(C) \geq \frac{\text{area}(C)}{\text{area}(h_C)}.$$

    L. Fejes Tóth [FeT50] was able to establish the following weaker result. The density of any covering of the plane with *noncrossing* congruent copies of C is at least $\frac{\text{area}(C)}{\text{area}(h_C)}$. (Two sets C_1 and C_2 are said to *cross* if both $C_1 \setminus C_2$ and $C_2 \setminus C_1$ are disconnected.) Since two translates of the same convex body can never cross each other, it follows that

$$\theta_T(C) \geq \frac{\text{area}(C)}{\text{area}(h_C)}.$$

    Furthermore, it follows from a theorem of Dowker [Do44] that if C is centrally symmetric, then h_C (and H_C) can also be chosen to be centrally symmetric. Since any centrally symmetric hexagon tiles the plane in a lattice-like manner, we obtain

$$\theta_L(C) = \theta_T(C) = \frac{\text{area}(C)}{\text{area}(h_C)} \qquad \text{if } C \text{ is convex and centrally symmetric.}$$

    Thus, if the above conjecture of L. Fejes Tóth is true, then we also get an affirmative answer to the following question.

Conjecture 2   *(Bambah–Rogers [BaR52]; L. Fejes Tóth [FeT84]) For any centrally symmetric convex body C in the plane, we have* $\theta(C) = \theta_L(C).$

    Dropping the condition of central symmetry, we can expect only that the following is true.

Conjecture 3   *(Bambah–Rogers [BaR52])*

$$\theta_T(C) = \theta_L(C) \qquad \text{for any convex body } C \text{ in the plane.}$$

The basic difficulty lurking behind the above problems is that we cannot exclude the possibility that the most economical coverings necessarily contain crossing pairs of sets. In fact, Wegner [We80] exhibited a whole series of examples of coverings of a bounded convex domain D whose members cannot be rearranged so as to cover D without crossing. (For a simple example of Heppes, see the figure below.)

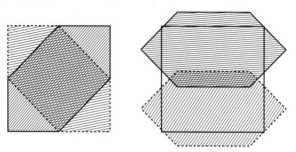

Only a crossing arrangement of hexagons covers the square

Moreover, as G. Fejes Tóth and W. Kuperberg [FeTK95] pointed out, it is very likely that in higher-dimensional spaces the most economical coverings with, for example, sufficiently long ellipsoids is not crossing-free. However, as was shown by Heppes [He03], the density of any covering of the plane with congruent copies of a "fat" ellipse is at least $2\pi/\sqrt{27}$, the density of the thinnest lattice packing. G. Fejes Tóth [FeT05] strengthened and generalized Heppes's result by proving the above three conjectures for any r-fat plane convex body C (inscribed in a unit disk and containing a concentric disk of radius r) with $r \geq 0.933$ (and $r \geq 0.741$, in case C is an ellipse).

It would be sufficient for the solution of the above problems to give a positive answer to the following question, at least for coverings with congruent copies C_i of the same convex body C in the plane.

Problem 4   (L. Fejes Tóth [FeT84]) Let $\mathcal{C} = \{C_1, \ldots, C_n\}$ be a system of plane convex bodies forming a covering of a bounded convex domain D. Does there exist a noncrossing covering $\mathcal{C}' = \{C_1', \ldots, C_n'\}$ of D such that for every i,
(1) $\mathrm{area}(C_i') = \mathrm{area}(C_i)$,
(2) C_i' is the image of C_i under some affine transformation of the plane?

The situation for packings is much simpler. L. Fejes Tóth [FeT50] proved that

$$\delta(C) \leq \frac{\mathrm{area}(C)}{\mathrm{area}(H_C)},$$

which implies that $\delta(C) = \delta_L(C)$ for every centrally symmetric convex C. Rogers [Ro51] noticed that by central symmetrization we also obtain that

$\delta_T(C) = \delta_L(C)$ for every convex C, and he also gave an independent proof of this fact. Another very elegant alternative proof of the last statement was found by L. Fejes Tóth [FeT83]. It was pointed out by A. Bezdek and Kertész [BeK87] that $\delta_T(C)$ and $\delta_L(C)$ do not necessarily coincide for *pseudoconvex* bodies, i.e., for compact sets C in the plane, having two parallel supporting lines whose points of contact with C divide the boundary into two arcs so that at least one of them is convex.

Of course, for a triangle T, for example, we have $1 = \delta(T) \neq \delta_L(T) = 2/3$. Moreover, G. Fejes Tóth [FeT95] proved that in the Baire category sense, for *most* plane convex bodies C, we have $\delta(C) \neq \delta_L(C)$. That is, the densest packing of a "typical" convex body is not lattice-like. The dual statement for coverings was established in [FeTZ94]. Some other related results can be found in Bambah, Rogers, and Zassenhaus [BaRZ64] and G. Fejes Tóth [FeTG72], [FeT77], [FeT88].

Problem 5   *Does there exist a d-dimensional convex body C for any $d \geq 3$ for which*

$$\delta_T(C) \neq \delta_L(C) \quad \text{or} \quad \theta_T(C) \neq \theta_L(C)?$$

[BaR52]   R.P. Bambah, C.A. Rogers: Covering the plane with convex sets, *J. London Math. Soc.* **27** (1952) 304–314.

[BaRZ64]   R.P. Bambah, C.A. Rogers, H. Zassenhaus: On coverings with convex domains, *Acta Arith.* **9** (1964) 191–207.

[BeK87]   A. Bezdek, G. Kertész: Counterexamples to a packing problem of L. Fejes Tóth, in: *Intuitive Geometry* (Siófok, 1985), K. Böröczky et al., eds., *Colloq. Math. Soc. János Bolyai* **48**, North-Holland, 1987, 29–36.

[Do44]   C.H. Dowker: On minimum circumscribed polygons, *Bull. Amer. Math. Soc.* **50** (1944) 120–122.

[FeT05]   G. Fejes Tóth: Covering with fat convex discs, *Discrete Comput. Geom.*, to appear.

[FeT95]   G. Fejes Tóth: Densest packings of typical convex sets are not lattice-like, *Discrete Comput. Geom.* **14** (1995) 1–8.

[FeT88]   G. Fejes Tóth: Note to a paper of Bambah, Rogers and Zassenhaus, *Acta Arith.* **50** (1988) 119–122.

[FeT77]   G. Fejes Tóth: On the intersection of a convex disc and a polygon, *Acta Math. Acad. Sci. Hungar.* **29** (1977) 149–153.

[FeTG72]   G. Fejes Tóth: Covering the plane by convex discs, *Acta Math. Acad. Sci. Hungar.* **23** (1972) 263–270.

[FeTK95]   G. Fejes Tóth, W. Kuperberg: Thin non-lattice covering
           with an affine image of a strictly convex body, *Mathematika*
           42 (1995) 239–250.

[FeTZ94]   G. Fejes Tóth, T. Zamfirescu: For most convex discs
           thinnest covering is not lattice-like, in: *Intuitive Geometry*
           (Szeged, 1991) K. Böröczky et al., eds., *Colloq. Math. Soc.
           János Bolyai* **63**, North-Holland, 1994, 105–108.

[FeT84]   L. Fejes Tóth: Research Problems, *Periodica Math. Hun-
           gar.* **15** (1984) 249–250.

[FeT83]   L. Fejes Tóth: On the densest packing of convex discs,
           Mathematika **30** (1983) 1–3.

[FeTL72]   L. Fejes Tóth: *Lagerungen in der Ebene, auf der Kugel
           und im Raum* (2. Auflage). Springer-Verlag, 1972.

[FeT50]   L. Fejes Tóth: Some packing and covering theorems, *Acta
           Sci. Math. (Szeged)* **12 A** (1950) 62–67.

[He03]   A. Heppes: Covering the plane with fat ellipses without
           non-crossing assumption, *Discrete Comput. Geom.* **29** (2003)
           477–481.

[Ro51]   C.A. Rogers: The closest packing of convex two-dimen-
           sional domains, *Acta Math.* **86** (1951) 309–321.

[We80]   G. Wegner: Zu einem ebenen Überdeckungsproblem, *Stu-
           dia Sci. Math. Hung.* **15** (1980) 287–297.

1.5 Packing with Semidisks, and the Role of Symmetry

One of the basic themes in the theory of packings and coverings is the role of symmetry. Is it true, for instance, that under fairly general circumstances the density of a packing with congruent copies of a convex body C in the plane cannot exceed the maximum density of a periodic packing? As we have seen before, $\delta(C) = \delta_L(C)$ for any centrally symmetric plane convex body C (L. Fejes Tóth [FeT50]). Of course, we cannot expect the same statement to be true without the assumption of central symmetry, as is shown by examples of tilings with congruent triangles. However, even in this case we can hope that the maximum packing density can be attained by a periodic packing, or perhaps even by an arrangement that is the union of two lattice packings.

Conjecture 1 (*L. Fejes Tóth, personal communication*) *Let C be any plane convex body with k-fold rotational symmetry for some odd integer k. Then the maximum packing density of C can be attained by an arrangement that is the union of two lattice packings.*

Unfortunately, we have no general methods for computing $\delta(C)$ if C is not centrally symmetric. Even the following frustratingly simple question is open.

Problem 2 (*L. Fejes Tóth [FeT71]*) *Determine the packing density of the semidisk.*

At first glance one would think that the densest packing can be obtained by putting two semicircles together so as to form a circle, and then arranging the resulting disks in the densest lattice packing. However, Groemer and Heppes [GrH75] have exhibited a packing of semidisks with density

$$\frac{\pi}{\sqrt{3} + 5\tan\frac{\pi}{10}} = 0.935\ldots > \frac{\pi}{\sqrt{12}} \approx 0.906,$$

which is conjectured to be optimal. (See also [Mö91], [Rö92].)

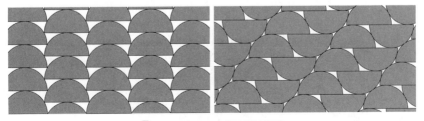

PACKINGS OF SEMIDISKS

Moreover, they proved the following general theorem: Every centrally symmetric plane convex body C other than a hexagon or a parallelogram can be split into two (congruent) parts C_1 and C_2 by a line through its center so that $\delta(C_1) > \delta(C)$. They have also established the analogous result for covering densities.

Conjecture 3   *(Groemer–Heppes [GrH75]) Let C be a strictly convex centrally symmetric body in $\mathbb{R}^d$, $d \geq 3$. Then there are two partitions of C, $C_1 \cup C_2$ and $C_1' \cup C_2'$, determined by two hyperplanes passing through the center of C, such that*

$$\delta(C_1) > \delta(C) \quad \text{and} \quad \theta(C_1') < \theta(C).$$

Returning to our original question, if we drop the condition that C is convex, then we can no longer expect that the maximum density of a packing with congruent copies of C can be realized by a periodic arrangement. By a clever modification of the so-called Wang tiles (see, e.g., Gardner [Ga77]), Schmitt [Sch91] constructed a connected set C in the plane that is the union of two convex bodies, and no periodic packing with congruent copies of C has density $\delta(C)$.

Restricting our investigations to translative packings, we have more hope to obtain some positive results.

Problem 4   *(Grünbaum–Shephard [GrS87]) Let C be a nonconvex body in the plane such that $\delta_T(C) = 1$, i.e., the plane can be tiled by translates of C. Is it true that $\delta_L(C) = 1$?*

As a matter of fact, we do not even know whether $\delta_T(C) = 1$ implies that there is a periodic tiling of the plane with translates of C.

According to a result of Venkov [Ve54], rediscovered by McMullen [McM80], for *convex* bodies C the answer to the question in Problem 4 is affirmative in every dimension.

Conjecture 5   *(L. Fejes Tóth [FeT85]) If C is a connected set that is the union of two plane convex bodies, then*

$$\delta_T(C) = \delta_L(C).$$

This conjecture is known to be true if C is convex (L. Fejes Tóth [FeT50]; Rogers [Ro51]) or if C is "limited semiconvex" (a slightly weaker property, introduced in L. Fejes Tóth [FeT85]). The conjecture has been confirmed by L. Fejes Tóth [FeT86] in the special case of C the union of two equal circular disks. This result was generalized by Heppes [He01] to more complicated regions bounded by circular arcs and by Kertész [Ke87] to the union of two translates of any plane convex body. On the other hand, modifying a construction of A. Bezdek and Kertész [BeK87], Heppes [He87], [He90] found a star-shaped set C in the plane that is the union of three convex bodies and for which $\delta_T(C) \neq \delta_L(C)$. In this sense, the above conjecture, if true, would be best possible.

 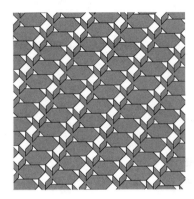

EXAMPLE OF SET WITH HIGHER TRANSLATIVE PACKING DENSITY
THAN LATTICE PACKING DENSITY

A very simple centrally symmetric star-shaped set C that is the union of two convex bodies and satisfies $\theta_T(C) \neq \theta_L(C)$ was found by Loomis [Lo83]: it is a cross obtained from a central unit square by attaching a 2×1 rectangular arm at each of its four sides. In fact, the translative *covering* density of this body satisfies $\theta_T(C) = 9/8$, while $\theta_L(C) = 9/7$. Similar examples were constructed by Bambah et al. [BaDH77], and in higher dimensions by Szabó [Sz83] and Stein [St86].

We close this section with two further problems about sets that can be obtained as the union of finitely many equal circles (i.e., circular disks of the same radius).

Conjecture 6   (*L. Fejes Tóth [FeT86]*) *Let C be the union of n unit circles whose centers lie on a line, equally spaced at a distance at most 2. Then*
$$\delta_T(C) = \delta_L(C).$$

L. Fejes Tóth [FeT86] proved the last conjecture for $n = 2$. He has also raised the following interesting problem: Determine $d_T(n) := \sup \delta_T(C)$, where the supremum is taken over all regions C that can be obtained as the union of n unit circles. He conjectured that for $n \leq 5$, the optimum is attained for a configuration in which the centers of the circles are equally spaced at a properly chosen distance. However, as with the "sausage catastrophe" discussed in Section 1.11, the situation drastically changes as we pass from $n = 5$ to 6. It appears that at this point some more complicated constructions take over, in which most of the circles form an economical covering of a regular hexagon, with some extra circles around the boundary that have a "smoothing" effect as n increases.

Conjecture 7 *(L. Fejes Tóth [FeT86]; Pach) Let $d(n) := \sup \delta(C)$, where the supremum is taken over all sets C that can be obtained as the union of n unit circles. Then*

$$d(n) = 1 - \frac{4\sqrt{3} - 2\pi}{3\sqrt{3}} \frac{1}{n} + o\left(\frac{1}{n}\right) \qquad \text{as } n \to \infty.$$

If this conjecture is true, we obviously have

$$\lim_{n \to \infty} n\left(1 - d(n)\right) = \lim_{n \to \infty} n\left(1 - d_T(n)\right).$$

[BaDH77] R.P. Bambah, V.C. Dumir, R.J. Hans-Gill: Covering by star domains, *Indian J. Pure Appl. Math.* **8** (1977) 344–350.

[BeK87] A. Bezdek, G. Kertész: Counter-examples to a packing problem of L. Fejes Tóth, in: *Intuitive Geometry* (Siófok, 1985), K. Böröczky et al., eds., *Colloq. Math. Soc. János Bolyai* **48** North-Holland, 1987, 29–36.

[FeT86] L. Fejes Tóth: Densest packing of translates of the union of two circles, *Discrete Comput. Geom.* **1** (1986) 307–314.

[FeT85] L. Fejes Tóth: Densest packing of translates of a domain, *Acta Math. Hungar.* **45** (1985) 437–440.

[FeT71] L. Fejes Tóth: The densest packing of lenses in the plane (in Hungarian), *Mat. Lapok* **22** (1971) 209–213.

[FeT50] L. Fejes Tóth: Some packing and covering theorems, *Acta Sci. Math. Szeged* **12 A** (1950) 62–67.

[Ga77] M. Gardner: Mathematical Games: Extraordinary nonperiodic tiling that enriches the theory of tiles, *Scientific Amer.* **236/1** (1977) 110–121.

[GrH75] H. Groemer, A. Heppes: Packing and covering properties of split disks, *Studia Sci. Math. Hungar.* **10** (1975) 185–189.

[GrS87] B. Grünbaum, G.C. Shephard: *Tilings and Patterns.* W.H. Freeman and Co., 1987.

[He01] A. Heppes: Packing of rounded domains on a sphere of constant curvature, *Acta Math. Hungar.* **91** (2001) 245–252.

[He90] A. Heppes: On the packing density of translates of a domain, *Studia Sci. Math. Hungar.* **25** (1990) 117–120.

[He87]   A. HEPPES: On the density of translates of a domain, Közl.
           MTA Számítástech. Automat. Kutató Int. Budapest 36 (1987)
           93–97.

[Ke87]   G. KERTÉSZ: Packing with translates of a special domain,
           Tagungsberichte Math. Forschungsinstitut Oberwolfach, Koll.
           Diskrete Geometrie, 1987.

[Lo83]   P. LOOMIS: The covering constant for a certain symmetric
           star body, Sitzungsberichte Österreich. Akad. Wiss., Math.-
           Naturw. Kl. II 192 (1983) 295–308.

[McM80]   P. MCMULLEN: Convex bodies which tile space by transla-
           tion, Mathematika 27 (1980) 113–121.

[Mö91]   W. MÖGLING: Über Packungen und Überdeckungen mit
           Halbellipsen, in: 4. Kolloquium Geometrie und Kombinatorik
           (Chemnitz, 1991), 1991, 65–70.

[Ro51]   C.A. ROGERS: The closest packing of convex two-dimensional
           domains, Acta Math. 86 (1951) 309–321.

[Rö92]   S. RÖDIGER: Untersuchungen zur Packungskonstante von
           Halbellipsen, Wiss. Z. Pädagogische Hochschule Erfurt/Mühl-
           hausen Math.-Natur. Reihe 28 (1992) 101–115.

[Sch91]   P. SCHMITT: Disks with special properties of densest pack-
           ings, Discrete Comput. Geom. 6 (1991) 181–190.

[St86]   S.K. STEIN: Tiling, packing, and covering by clusters, Rocky
           Mountain J. Math. 16 (1986) 277–321.

[Sz83]   S. SZABÓ: A symmetric star polyhedron that tiles but not
           as a fundamental domain, Proc. Amer. Math. Soc. 89 (1983)
           563–566.

[Ve54]   B.A. VENKOV: On a class of Euclidean polyhedra (in Rus-
           sian), Vestnik Leningrad. Univ. Ser. Mat. Fiz. Him. 9 (1954)
           no. 2, 11–31.

1.6 Packing Equal Circles into Squares, Circles, Spheres

The problems collected in this section have attracted the interest of an army of mathematicians and amateurs having a special taste for small, sporadic point configurations with some remarkable properties. The literature concerning these questions is so large that we cannot even attempt to give a full survey.

Problem 1   (L. Moser [Mo60]) Let $f_\square(n)$ denote the maximum number $f_\square$ with the property that one can place n points in the closed unit square so that the minimum distance between them is at least $f_\square$. Determine the exact values of $f_\square(n)$ at least for small integers $n \geq 2$.

Of course, the asymptotic behavior of $f_\square(n)$ is well known. According to Thue's theorem cited in the introduction ([Th892], [Th10]; see also [FeT49], [SeM44]), the maximal density of a packing of congruent circles in the plane is $\pi/\sqrt{12}$. This immediately implies that

$$f_\square(n) = \frac{1}{\sqrt{n}} \left(\frac{\sqrt{2}}{\sqrt[4]{3}} + o(1) \right)$$

and if n is large, then the extremal configuration(s) must be close to the vertex set of a regular triangular lattice [Gr00]. However, this has little relevance to the case in which n is small.

The first few values of $f_\square(n)$ are listed below:

n	2	3	4	5	6	7	8	9
$f_\square(n)$	$\sqrt{2}$	$\sqrt{6}-\sqrt{2}$	1	$\frac{1}{2}\sqrt{2}$	$\frac{1}{6}\sqrt{13}$	$2(2-\sqrt{3})$	$\frac{1}{2}(\sqrt{6}-\sqrt{2})$	$\frac{1}{2}$

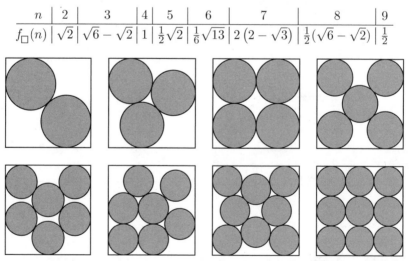

THE DENSEST PACKINGS OF 2 TO 9 CONGRUENT CIRCLES IN A SQUARE

The cases $n = 2$, 3, 4, 5 are easy; the solution for $n = 6$ was found by Graham and Schwartz (see [Me94a] for a short argument), for $n = 7$ by Schaer, for $n = 8$ by Schaer and Meir [SchM65], for $n = 9$ by Schaer [Sch65]. For all values of n up to 27, Peikert et al. [PeW*92] and Nurmela and Östergård [NuÖ97], [NuÖ99] provided computer-aided solutions, while Markót [Ma04] settled the case $n = 28$. See [MaFP95] for a different approach, based on nonlinear programming. Boll, Donovan, Graham, Lubachevsky [BoD*00] and Lubachevsky, Graham, and Stillinger [LuGS97] describe efficient new numerical procedures, the "billiard algorithm," for generating dense packings. (See also [AmB00] for a survey.)

Note that the constructions for $n \leq 9$, depicted in the figure, are "normalized" in the sense that we place n nonoverlapping circles of maximum radius $r(n)$ in a *unit* square. Then the centers of the circles fit in a square of side length $1 - 2r(n)$, so that we have

$$f_\square(n) = \frac{2r(n)}{1 - 2r(n)}.$$

It follows from the above asymptotic formula for $f(n)$ that if k is sufficiently large, $f_\square(k^2) > 1/(k - 1)$; i.e., the grid arrangement is not optimal. It is known that $f_\square(k^2) = 1/(k - 1)$ holds for $k \leq 5$ [We83], [We87]. Moreover, it was announced in [KiW87] that $f_\square(36) = 1/5$, but most experts believe that the proof has serious gaps. On the other hand, a construction in [NuÖ97] shows that $f_\square(49) > 1/6$.

Problem 2   *(G. Wengerodt) Is it true that $f_\square(36) = 1/5$?*

Nurmela, Östergård, and aus dem Spring [NuÖ*99] came up with the following construction. For any positive integers a and b satisfying $a \leq b < \sqrt{3}a$, let

$$P_{ab} = \left\{ \left(\frac{i}{a}, \frac{j}{b} \right) \,\middle|\, \begin{array}{l} 0 \leq i \leq a, \\ 0 \leq j \leq b, \end{array} i + j \equiv 0 \bmod 2 \right\}.$$

Clearly, we have $|P_{ab}| = \left\lceil \frac{(a+1)(b+1)}{2} \right\rceil$, and the minimum distance between two points in P_{ab} is $\sqrt{\frac{1}{a^2} + \frac{1}{b^2}}$. In particular, the minimum distance is never attained between a pair of points having the same x-coordinate.

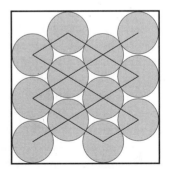

CONJECURED DENSEST PACKING OF 15 CIRCLES IN A SQUARE

Conjecture 3   *[NuÖ*99] For infinitely many pairs of integers (a, b), the configuration P_{ab} is optimal; i.e., we have*
$$f_\square \left(\left\lceil \frac{(a+1)(b+1)}{2} \right\rceil \right) = \sqrt{\frac{1}{a^2} + \frac{1}{b^2}}.$$

To obtain a conjecturally optimal packing, the ratio $a/b > 1/\sqrt{3}$ must be close to $1/\sqrt{3}$. Consider the continued fraction expansion of $1/\sqrt{3}$; its partial fractions are

$$\frac{1}{1}, \frac{1}{2}, \frac{3}{5}, \frac{4}{7}, \frac{11}{19}, \frac{15}{26}, \frac{41}{71}, \ldots$$

Every other element in this sequence is larger than $1/\sqrt{3}$. It is conjectured in [NuÖ*99] that if we make a and b equal the numerator and denominator of such a term, respectively, we always obtain an optimal configuration. The case $a = 3$, $b = 5$ is depicted in the figure.

Similar questions can be asked about dense circle packings, for example, in the unit circle or in the equilateral triangle of side length 1.

Problem 4   *(Kravitz [Kr67], Pirl [Pi69]) Let $f_{\bigcirc}(n)$ denote the maximum number $f_{\bigcirc}$ with the property that one can place n points in the closed unit circle so that the minimum distance between them is at least $f_{\bigcirc}$. Determine the exact values of $f_{\bigcirc}(n)$, at least for small integers $n \geq 2$.*

Pirl proved that

$$f_{\bigcirc}(n) = \begin{cases} 2\sin\frac{\pi}{n}, & \text{for } 2 \leq n \leq 6, \\ 2\sin\frac{\pi}{n-1}, & \text{for } 7 \leq n \leq 9. \end{cases}$$

The exact values of $f_{\bigcirc}(10), f_{\bigcirc}(11), f_{\bigcirc}(12), f_{\bigcirc}(13), f_{\bigcirc}(14)$, and $f_{\bigcirc}(19)$ were determined in [Pi69], [Me94b], [Fo00], [Fo03b], [Fo03a], and [Fo99], respectively. For a survey, consult [GrL*98], where the best known constructions are presented for all $n \leq 65$. For $15 \leq n \leq 18$, they are conjectured to be optimal.

The analogous question about placements of n points in an equilateral triangle maximizing the minimum distance appears to be slightly easier in the sense that the optimal configurations are known [Ol61] for infinitely many values of n: for all positive integers of the form $k(k+1)/2$. Denote this maximum by $f_{\triangle}(n)$. Erdős and Oler asked whether it is true that if an equilateral triangle contains $n - 1$ nonoverlapping unit disks for some $n = k(k+1)/2$, then it can also hold n nonoverlapping unit disks. This is equivalent to the following question.

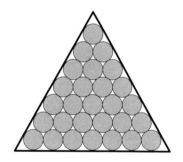

DENSEST PACKING OF 28 CIRCLES IN AN EQUILATERAL TRIANGLE

Problem 5   (Erdős and Oler [Ol61]) Is it true that $f_\triangle(n-1) = f_\triangle(n)$
for all values of n of the form $k(k+1)/2$?

Actually, it is possible that a much stronger statement is true: for every
positive integer i, there may exist a k_i such that if an equilateral triangle
contains $n - i$ nonoverlapping unit disks for some $n = k(k+1)/2$, $k \geq k_i$,
then it can also hold n nonoverlapping unit disks.

It is also very likely that a similar phenomenon occurs for containers
of many other shapes, including the regular hexagon. Heppes believes (personal communication) that answering the above question may lead to the
solution of L. Fejes Tóth's famous conjecture about the "solidity" of the
arrangement obtained from the densest lattice packing of circles by deleting
one of its elements (see Section 2.2).

Melissen [Me93], [Me94c] determined all values of $f_\triangle(n)$ for $n \leq 12$.
Payan [Pa97] verified that $f_\triangle(14) = f_\triangle(15)$. For $n = 16, 17, 18$, some constructions are presented in [MeS95]. Using computer experiments, Graham
and Lubachevsky [GrL95] produced many other examples up to $n = 260$,
and they also described several infinite classes of constructions that are
conjectured to be optimal.

It is interesting to note that the problem of determining $f_\triangle(n)$ was
first studied by the Hungarian mathematician Farkas Bolyai in 1833. In his
famous monograph, *Tentamen juventutem studiosam in elementa mathe-
seos purae, elementaris ac sublimioris, methodo intuitiva, evidentiaque huic
propria, introducendi* (whose appendix, written by his son, János Bolyai,
became a classic, as the first treatise ever written on noneuclidean geometry), he properly defined the notion of the density of a circle packing in
an equilateral triangle. He also constructed some economical packings with
equal circles that turned out to be suboptimal, and correctly computed the
limit of their densities as the number of circles tends to infinity.

Perhaps the best known variant of the above questions was raised by
the botanist Tammes [Ta30], who studied the distribution of pores on pollen
grains.

Problem 6   (Tammes [Ta30]) Let $f_\ominus(n)$ denote the maximum number
$f_\ominus$ with the property that one can place n points on the
surface of the unit sphere $\mathbb{S}^2$ (in three-dimensional space)
so that the minimum distance between them is at least
$f_\ominus$. Determine the exact values of $f_\ominus(n)$, at least for small
integers $n \geq 2$.

As far as we know, the exact value of $f_\ominus(n)$ is known only for $n \leq$
12 and $n = 24$. The cases $n = 3, 4, 6$, and 12 were settled by L. Fejes
Tóth [FeT43], [FeT72], the cases $n = 5, 7, 8$ and 9 by Schütte and van der
Waerden [SchW51], the cases $n = 10, 11$ by Danzer [Da86], Hárs [Há86],

and Böröczky [Bö83], the case $n = 24$ by Robinson [Ro61]. In particular, we have $f_\ominus(5) = f_\ominus(6)$, $f_\ominus(11) = f_\ominus(12)$, and, as was shown in [TaG91a], $f_\ominus(23) > f_\ominus(24)$.

Conjecture 7   *(Robinson [Ro69], Tarnai–Gáspár [TaG91a]) We have*
$$f_\ominus(n-1) \neq f_\ominus(n) \text{ for all } n \geq 13.$$

The best known constructions providing lower bounds on $f_\ominus(n)$ for all $n \leq 90$ are described in Kottwitz [Ko91] and Clare and Kepert [ClK86], [ClK91]. For further details and long lists of additional references, the interested reader should consult Tarnai [Ta83], Tarnai and Gáspár [TaG83], Lazić, Šenk, Šeškar [LaŠŠ87], Böröczky and Szabó [BöS03]. Some axially symmetric and multisymmetric point configurations were considered by Tarnai [Ta84], Tarnai and Gáspár [TaG87], Gáspár [Gá89], Teshima and Ogawa [TeO00]. See also the homepage of Neil Sloane.

The above problems raise three very general questions. For any compact region C (in the plane, on the sphere, or in higher dimensions), let $f_C(n)$ denote the largest number f with the property that one can place n points in C so that the minimum distance between them is at least f.

Problem 8   *Given a compact region C in the plane, bounded by a piecewise algebraic closed curve, design an efficient algorithm for determining the exact value of $f_C(n)$.*

An $n^{O(n)}$-time algorithm can be obtained by quantifier elimination [Ba99], [BaPR96].

Given a set P of n points in a compact region C such that the minimum distance between its elements is $f_C(n)$, a point $p \in P$ is said to be *freely movable* if p can be continuously moved without decreasing the minimum distance.

Problem 9   *(G. Fejes Tóth [FeT05]) Given a compact region C, can one find for infinitely many values of n a set of n points with minimum distance $f_C(n)$ that has at least one freely movable element? Does there exist a finite upper bound for the number of freely movable elements, independent of n? Can one find infinitely many values of n for which there exist extremal configurations with no freely movable elements?*

Problem 10   *(G. Fejes Tóth [FeT05]) Given a compact region C, does there exist a number $k = k(C)$ such that $f_C(n + k) \neq f_C(n)$ for every n?*

We can also formulate the dual counterparts of the above problems for coverings.

Problem 11   *Let $g_\square(n)$, $g_\bigcirc(n)$, $g_\triangle(n)$, and $g_\ominus(n)$ denote the least number r with the property that the unit square, the unit circle, the unit equilateral triangle, and the unit sphere, respectively, can be covered by n circles of radius r. Determine the exact values of $g_\square(n)$, $g_\bigcirc(n)$, $g_\triangle(n)$, and $g_\ominus(n)$, at least for small integers $n \geq 2$.*

By a well-known result of Kershner [Ke39], the minimum density of a covering of the plane with equal circles is $2\pi/\sqrt{27}$, which yields asymptotically tight estimates for each of these functions, as n tends to infinity.

The function $g_\square(n)$ was first studied by Verblunsky [Ve49], while the first computer-aided results on $g_\bigcirc(n)$ were obtained by Zahn [Za62]. For $n \leq 5$ and $n = 7$, the exact values of $g_\square(n)$ were determined by Heppes and Melissen [HeM97], and many efficient constructions for $n \leq 20$ were exhibited in [TaG95], [MeS96], [Nu99]. In fact, in [HeM97] and [MeS00] the more general problem of covering rectangles was also considered. Obviously, the next (perhaps not too hopeless) challenge is to determine the value of $g_\square(6)$.

Molnár [Mo43] showed that $g_\bigcirc(3) = \sqrt{3}/2$, and $g_\bigcirc(4) = 1/\sqrt{2}$. Correcting a proof in [Ne15], K. Bezdek [Be83] obtained that $g_\bigcirc(5) = 0.6098\ldots$ and $g_\bigcirc(6) = 0.5559\ldots$. The easy proof of $g_\bigcirc(7) = 1/2$ can be found in Škljarskiĭ, Čencov, and Jaglom [ŠkČJ74]. For $n = 8, 9$, and 10, the values of $g_\bigcirc(n)$ were determined by G. Fejes Tóth [FeT05]. For further constructions, conjectures, and results on $g_\bigcirc(n)$ up to $n = 30$, see [Nu99], [NuÖ00].

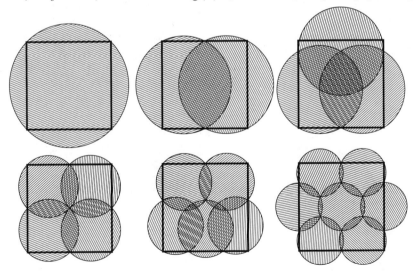

THE THINNEST COVERINGS OF A SQUARE BY 1–5 AND 7 CIRCLES

The problem of thinnest covering of a unit equilateral triangle with n equal circles was solved by Melissen [Me97] for $n \leq 6$ and for $n = 9, 10$.

This work as well as [Nu00] presented many conjecturally optimal coverings for $n \leq 36$.

Conjecture 12 *[BeB85] For any positive integer of the form*
$$n = k(k+1)/2, \text{ we have } g_\triangle(n) = k\sqrt{3}.$$

A. and K. Bezdek proved this conjecture for $k \leq 4$.

Melnyk, Knop, and Smith [MeKS77] and Tarnai and Gáspár [TaG91b] give extensive surveys of all known results about $g_\ominus(n)$. The exact values of $g_\ominus(n)$ for $n \leq 7$ and $n = 12$ were determined by L. Fejes Tóth [FeT43], [FeT72] and Schütte [Sch55]. The only other known values, $g_\ominus(10)$ and $g_\ominus(14)$, were conjectured by Jucovič [Ju60] and proved by G. Fejes Tóth [FeT69]. G. and R. Blind [BlB94], [BlB95] studied the following related question: how should one arrange n congruent circular caps on the unit sphere so as to maximize the total area of those regions that are covered by precisely one of the caps? They solved this problem for $n = 3, 4, 6$, and 12.

Conjecture 13 *(L. Fejes Tóth)* $g_\ominus(n-1) \neq g_\ominus(n)$ *for all* $n \geq 4$.

Very little is known about the generalizations of the above problems to higher-dimensional spaces.

The problem of arranging n points in the three-dimensional unit cube so as to maximize their minimum distance was solved for $n \leq 10$ by Schaer [Sch66a], [Sch66b], [Sch66c], [Sch71], [Sch94]. Some partial results for $n \geq 11$ were obtained by Goldberg [Go71b]. The same problem in higher dimensions was studied by Chepanov, Ryškov, and Yakovlev [ChRY91] and by G. and W. Kuperberg, who solved a few cases in four and five dimensions (personal communication). The only other constructions known to be optimal depend on the existence of certain Hadamard matrices.

The following variant of this question was raised by L. Moser [Mo66] (see also [Mo91], pp. 208–221, Problem LM 41).

Problem 14   *Let $k(d)$ denote the maximum number of points that can be arranged in the d-dimensional unit cube so that all mutual distances are at least 1. Determine the exact values of $k(d)$ at least for small d.*

Obviously, $k(d) = 2^d$ for $d = 1, 2, 3$. A. Meir (personal communication) has shown that $k(4) = 17$, and several others have proved that

$$\log k(d) = \left(\frac{1}{2} + o(1) \right) d \log d, \qquad \text{as } d \to \infty.$$

V. Bálint and V. Bálint Jr. [BáB03] have supplied the first written proofs for these results and made the conjectures $k(5) = 34$, $k(6) = 76$, $k(7) = 152$, and $k(8) = 353$. Makai Jr. (personal communication) established the nontrivial upper bound $k(d) \leq d^{d/2} 0.63901^{(1+o(1))d}$.

There are very few nontrivial results concerning packings and coverings of higher-dimensional unit spheres $\mathbb{S}^d \subset \mathbb{R}^{d+1}$ ($d \geq 3$) with spherical balls. The densest packing with spherical balls of radius φ is known only for very special values. Letting $n_d(\varphi)$ denote the largest number of nonoverlapping d-dimensional spherical balls of radius $0 < \varphi \leq \pi/2$ that can be packed into $\mathbb{S}^d$, we have $n_d(\pi/4) = 2d$. Böröczky [Bö78], Levenšteĭn, Odlyzko, and Sloane [Le79], [OdS79] proved that $n_3(\pi/10) = 120$, $n_7(\pi/6) = 240$, and $n_{23}(\pi/6) = 196560$. Recently, Musin [Mu05], [Mu03] has shown that $n_3(\pi/6) = 24$, which had been claimed by W.-Y. Hsiang for a while. Consider a four-dimensional cube centered at the origin $0 \in \mathbb{R}^4$. Let X denote the 24-element set obtained by projecting the vertices of the cube and the centroids of its eight facets from 0 onto $\mathbb{S}^3$. It is easy to see that the spherical balls of radius $\pi/6$ centered at the elements of X form a packing, showing that $n_3(\pi/6) \geq 24$.

Conjecture 15 *[Le79], [OdS79] For any packing of 24 spherical balls of radius $\pi/6$ in $\mathbb{S}^3$, the set of centers of the balls is congruent to X.*

The problem of arranging $n = 840$ balls on the three-dimensional unit sphere $\mathbb{S}^3$ so as to maximize their minimum distance was addressed by Coxeter [Co87].

The dual problem of covering $\mathbb{S}^d$ with n equal spherical balls of radius φ is intimately related to polyhedral approximations of the ball B^{d+1}. The optimal configurations are known only for $n \leq d + 3$ and for $d = 3$ and $n = 8$. In the latter case, the centers of the spherical balls form the vertex set of a regular four-dimensional crosspolytope [DaL*00].

Problem 16   *(Larman, Zong) Consider a covering of $\mathbb{S}^d$ with $n = 2d+2$ equal spherical balls of minimum radius, $d > 3$. Do the centers of the balls form the vertex set of a regular d-dimensional crosspolytope?*

Actually, it is possible that for every $d \geq 3$ and $d + 2 \leq n \leq 2d + 2$, the set of centers of the balls in an optimal configuration can always be obtained as the union of the vertex sets of $n - d - 1$ mutually orthogonal regular simplices of circumradius one, whose dimensions are as equal as possible. This is true for $n = d + 2$ and $d + 3$, see [BöW03].

Problem 17   *[BöW03] Does there exist an absolute constant $c > 0$ such that the density of any covering of the d-dimensional sphere $\mathbb{S}^d$ by equal spherical balls of an acute radius is at least cd?*

For any $0 < \varphi < \pi/2$, Böröczky Jr. and Wintsche [BöW03] constructed a covering of $\mathbb{S}^d$ with spherical balls of radius φ such that every point is covered at most $400d \ln d$ times.

In an equivalent formulation, the problem of Tammes asks for the smallest diameter of a two-dimensional sphere that contains n points whose mutual distances are at least one. Schütte [Sch63] defined $I(n, d)$ to be the smallest diameter of a set of n points in d-space with minimum distance at least one. Among other results, he proved that $I(6, 3) = \sqrt{2}$. Define $J(n, d) \leq I(n, d)$ as the smallest diameter of a convex d-polytope, all of whose edges are of length at least one. Some simple properties of this function, including the equation $J(6, 3) = I(6, 3) = \sqrt{2}$, were established in [BeB*01].

[AmB00]   P. AMENT, G. BLIND: Packing equal circles in a square, *Studia Sci. Math. Hungar.* **36** (2000) 313–316.

[BáB03]   V. BÁLINT, V. BÁLINT JR.:   On the number of points at distance at least one in the unit cube, *Geombinatorics* **12** (2003) 157–166.

[Ba99]   S. BASU: New results on quantifier elimination over real closed fields and applications to constraint databases, *J. ACM* **46** (1999) 537–555.

[BaPR96]   S. BASU, R. POLLACK, M.-F. ROY: On the combinatorial and algebraic complexity of quantifier elimination, *J. ACM* **43** (1996) 1002-1046.

[BeB85]   A. BEZDEK, K. BEZDEK: Über einige dünnste Kreisüberdeckungen konvexer Bereiche durch endliche Anzahl von kongruenten Kreisen, *Beiträge Algebra Geom.* **19** (1985) 159–168.

[Be83]   K. BEZDEK: Über einige Kreisüberdeckungen, *Beiträge Algebra Geom.* **14** (1983) 7–13.

[BeB*01]   K. BEZDEK, G. BLEKHERMAN, R. CONNELLY, B. CSIKÓS: The polyhedral Tammes problem, *Arch. Math.* **76** (2001) 314–320.

[BlB95]   G. BLIND, R. BLIND: Über ein Kreisüberdeckungsproblem auf der Sphäre, *Studia Sci. Math. Hungar.* **30** (1995) 197–203.

[BlB94]   G. BLIND, R. BLIND: Ein Kreisüberdeckungsproblem auf der Sphäre, *Studia Sci. Math. Hungar.* **29** (1994) 107–164.

[BoD*00]   D.W. BOLL, J. DONOVAN, R.L. GRAHAM, B.D. LUBACHEVSKY: Improving dense packings of equal disks in a square, *Electronic J. Combin.* **7** (2000), # R46.

[Bö83]   K. BÖRÖCZKY: The problem of Tammes for $n = 11$, *Studia Sci. Math. Hungar.* **18** (1983) 165–171.

[Bö78]   K. BÖRÖCZKY: Packing of spheres in spaces of constant curvature, *Acta Math. Acad. Sci. Hungar.* **32** (1978) 243–261.

[BöS03]   K. BÖRÖCZKY, L. SZABÓ: Arrangements of thirteen points on a sphere, in: *Discrete Geometry: In Honor of W. Kuperberg's 60th Birthday*, A. Bezdek, ed., Marcel Dekker, 2003, 111–184.

[BöW03]   K. BÖRÖCZKY JR., G. WINTSCHE: Covering the sphere by equal spherical balls, in: *Discrete and Computational Geometry — The Goodman–Pollack Festschrift*, B. Aronov et al., eds., Springer-Verlag, 2003, 237–253.

[ChRY91] S.A. CHEPANOV, S.S. RYŠKOV, N.N. YAKOVLEV: On the disjointness of point systems (in Russian), *Trudy Mat. Inst. Steklov* **196** (1991) 147–155.

[ClK91]   B.W. CLARE, D.L. KEPERT: The optimal packing of circles on a sphere, *J. Math. Chem.* **6** (1991) 325–349.

[ClK86]   B.W. CLARE, D.L. KEPERT: The closest packing of equal circles in a sphere, *Proc. Royal Soc. London Ser. A* **405** (1986) 329–344.

[Co87]   H.S.M. COXETER: A packing of 840 balls of radius $9°0'19''$ on the 3-sphere, in: *Intuitive Geometry* (Siófok, 1985), K. Böröczky et al., eds., *Colloq. Math. Soc. János Bolyai* **48**, North-Holland, 1987, 127–137.

[DaL*00] L. DALLA, D.G. LARMAN, P. MANI-LEVITSKA, C. ZONG: The blocking numbers of convex bodies, *Discrete Comput. Geom.* **24** (2000) 267–277.

[Da86]   L. DANZER: Finite point-sets on S^2 with minimum distance as large as possible, *Discrete Math.* **60** (1986) 3–66.

[FeT05]   G. FEJES TÓTH: Thinnest covering of a circle by eight, nine, or ten congruent circles, in: *Combinatorial and Computational Geometry*, J.E. Goodman et al., eds., Cambridge Univ. Press, *MSRI Publications* **52** (2005), to appear.

[FeT69]   G. FEJES TÓTH: Kreisüberdeckungen der Sphäre, *Studia Sci. Math. Hung.* **4** (1969) 225–247.

[FeT72]   L. FEJES TÓTH: *Lagerungen in der Ebene, auf der Kugel und im Raum* (2. Aufl.), Springer-Verlag, 1972.

[FeT49]   L. FEJES TÓTH: Über dichteste Kreislagerung und dünnste Kreisüberdeckung, *Comm. Math. Helv.* **23** (1949) 342–349.

[FeT43]   L. Fejes Tóth: On covering a spherical surface with equal spherical caps (in Hungarian), *Mat. Fiz. Lapok* **50** (1943) 40–46.

[Fo03a]   F. Fodor: Packing 14 congruent circles in a circle, *Stud. Univ. Žilina, Math. Ser.* **16** (2003) 25–34.

[Fo03b]   F. Fodor: The densest packing of 13 congruent circles in a circle, *Beiträge Algebra Geom.* **44** (2003) 431–440.

[Fo00]   F. Fodor: The densest packing of 12 congruent circles in a circle, *Beiträge Algebra Geom.* **21** (2000) 401–409.

[Fo99]   F. Fodor: The densest packing of 19 congruent circles in a circle, *Geometriae Dedicata* **74** (1999) 139–145.

[Gá89]   Z. Gáspár: Some new multisymmetric packings of equal circles on a 2-sphere, *Acta Crystal. Ser. B* **45** (1989) 452–453.

[Go71a]   M. Goldberg: Packing 14, 16, 17 and 20 circles in a circle, *Math. Mag.* **44** (1971) 134–139.

[Go71b]   M. Goldberg: On the densest packing of equal spheres in a cube, *Math. Mag.* **44** (1971) 199–208.

[Go70]   M. Goldberg: The packing of equal circles in a square, *Math. Mag.* **43** (1970) 24–30.

[GrL95]   R.L. Graham, B.D. Lubachevsky: Dense packings of equal disks in an equilateral triangle: from 22 to 34 and beyond, *Electron. J. Combin.* **2** (1995) #A1.

[GrL*98]   R.L. Graham, B.D. Lubachevsky, K.J. Nurmela, P.R.J. Östergård: Dense packings of congruent circles in a circle, *Discrete Math.* **181** (1998) 139–154.

[Gr00]   P.M. Gruber: In many cases optimal configurations are almost regular hexagonal (*3rd Int. Conf. Stochastic Geometry, Convex Bodies and Empirical Measures, Part II* (Mazara del Vallo, 1999), *Rend. Circ. Mat. Palermo (2), Suppl.* **65/II** (2000), 121–145.

[Há86]   L. Hárs: The Tammes problem for $n = 10$, *Studia Sci. Math. Hungar.* **21** (1986) 439–451.

[HeM97]   A. Heppes, J.B.M. Melissen: Covering a rectangle with equal circles, *Period. Math. Hungar.* **34** (1997) 65–81.

[Ju60]   E. Jucovič: Some coverings of a spherical surface with equal circles (in Slovakian), *Mat.-Fyz. Časopis Slovensk. Akad. Vied.* **10** (1960) 99–104.

[Ke39]   R.B. Kershner: The number of circles covering a set, *Amer. J. Math.* **61** (1939) 665–671.

[KiW87]   K. KIRCHNER, G. WENGERODT: Die dichteste Packung von
          36 Kreisen in einem Quadrat, *Beiträge Algebra Geom.* **25**
          (1987) 147–159.

[Ko91]   D.A. KOTTWITZ: The densest packing of equal circles on a
          sphere, *Acta Cryst. Sect. A* **47** (1991) 158–165 (Erratum, p.
          851).

[Kr67]   S. KRAVITZ: Packing cylinders into cylindrical containers,
          Math. Mag. **40** (1967) 65–71.

[LaŠŠ87] D.E. LAZIĆ, V. ŠENK, I. ŠEŠKAR: Arrangements of points
          on a sphere which maximize the least distance, *Bull. Appl.
          Math. Techn. Univ. Budapest* **47** (1987) no. 479, 5–21.

[Le79]   V.I. LEVENŠTEĬN: On bounds for packings in n-dimensional
          Euclidean space, *Soviet Math. Dokl.* **20** (1979) 417–421.

[LuGS97] B.D. LUBACHEVSKY, R.L. GRAHAM, F.H. STILLINGER:
          Patterns and structures in disk packings, *Period. Math. Hun-
          gar.* **34** (1997) 123–142.

[Ma04]   M.CS. MARKÓT: Optimal packing of 28 equal circles in a
          unit square – the first reliable solution, *Numerical Algorithms*
          37 (2004) 253–261.

[MaFP95] C.D. MARANAS, CH.A. FLOUDAS, P.M. PARDALOS: New
          results in the packing of equal circles in a square, *Discrete
          Math.* **142** (1995) 287–293.

[Me97]   J.B.M. MELISSEN: *Packing and Covering with Circles*
          (Ph.D. dissertation), University of Utrecht, 1997.

[Me94a]   J.B.M. MELISSEN: Densest packing of six equal circles in a
          square, *Elemente Math.* **49** (1994) 27–31.

[Me94b]   J.B.M. MELISSEN: Densest packings of eleven congruent
          circles in a circle, *Geometriae Dedicata* **50** (1994) 15–25.

[Me94c]   J.B.M. MELISSEN: Optimal packings of eleven equal circles
          in an equilateral triangle, *Acta Math. Hungar.* **65** (1994)
          389–393.

[Me93]   J.B.M. MELISSEN: Densest packings of congruent circles in
          an equilateral triangle, *Amer. Math. Monthly* **100** (1993)
          916–925.

[MeS00]   J.B.M. MELISSEN, P.C. SCHUUR: Covering a rectangle with
          six and seven circles, *Discrete Appl. Math.* **99** (2000) 149–156.

[MeS96]   J.B.M. MELISSEN, P.C. SCHUUR: Improved coverings of a
          square with six and eight equal circles, *Electron. J. Combin.*
          3 (1996) # R31.

[MeS95] J.B.M. Melissen, P.C. Schuur: Packing 16, 17 or 18 circles in an equilateral triangle, *Discrete Math.* **145** (1995) 333–342.

[MeKS77] T.W. Melnyk, O. Knop, W.R. Smith: Extremal arrangement of points and unit charges on a sphere: equilibrium configurations revisited, *Canad. J. Chem.* **55** (1977) 1745-1761.

[Mo43] J. Molnár: On an extremal problem in elementary geometry (in Hungarian), *Mat. Fiz. Lapok* **49** (1943) 249–253.

[Mo66] L. Moser: *Poorly Formulated Unsolved Problems of Combinatorial Geometry.* Mimeographed, 1966. Reprinted in [Mo91].

[Mo60] L. Moser: Problem 24 (corrected), *Canad. Math. Bull.* **3** (1960) 78.

[Mo91] W.O.J. Moser: Problems, problems, problems, *Discrete Appl. Math.* **31** (1991) 201–225.

[Mu05] O.R. Musin: The kissing number in four dimensions, arXiv: math.MG/0309430, manuscript.

[Mu03] O.R. Musin: The problem of the twenty-five spheres, *Russ. Math. Surv.* **58** (2003) 794–795.

[Ne15] E.H. Neville: On the solution of numerical functional equations, *Proc. London Math. Soc.* 2. Ser. **14** (1915) 308–326.

[Nu00] K.J. Nurmela: Conjecturally optimal coverings of an equilateral triangle with up to 36 equal circles, *Experimental Math.* **9** (2000) 241–250.

[Nu99] K.J. Nurmela: Circle coverings in the plane, *Proceedings of the Seventh Nordic Combinatorial Conference (Turku, 1999)*, TUCS Gen. Publ. **15**, Turku Cent. Comput. Sci., Turku, 1999, 71–78.

[NuÖ00] K.J. Nurmela, P.R.J. Östergård: Covering a square with up to 30 equal circles, *Helsinki University of Technology Laboratory for Theoretical Computer Science Research Reports* **62** (2000) 1–14.

[NuÖ99] K.J. Nurmela, P.R.J. Östergård: More optimal packings of equal circles in a square, *Discrete Comput. Geom.* **22** (1999) 439–457.

[NuÖ97] K.J. Nurmela, P.R.J. Östergård: Packing up to 50 equal circles in a square, *Discrete Comput. Geom.* **18** (1997) 111–120.

[NuÖ*99] K.J. Nurmela, P.R.J. Östergård, R. aus dem Spring: Asymptotic behavior of optimal circle packings in a square, *Canad. Math. Bull.* **42** (1999) 380–385.

[OdS79]   A.M. Odlyzko, N.J.A. Sloane: New bounds on the num-
          ber of unit spheres that can touch a unit sphere in n dimen-
          sions, *J. Combinatorial Theory Ser. A* **26** (1979) 210–214.

[Ol61]   N. Oler: A finite packing problem, *Canad. Math. Bull.* **4**
          (1961) 153–155.

[Pa97]   Ch. Payan: Empilement de cercles égaux dans un triangle
          équilatéral. À propos d'une conjecture d'Erdős-Oler, *Discrete
          Math.* **165-166** (1997) 555–565.

[PeW*92] R. Peikert, D. Würtz, M. Monagan, C. de Groot:
          Packing circles in a square: A review and new results, in: *Sys-
          tems Modelling and Optimization 1991*, P. Kall, ed., Springer
          Lecture Notes Control Inf. Sci. **180** (1992) 45–54.

[Pi69]   U. Pirl: Der Mindestabstand von n in der Einheitskreis-
          scheibe gelegenen Punkten, *Math. Nachr.* **40** (1969) 111–124.

[Ro69]   R.M. Robinson: Finite sets of points on a sphere with each
          nearest to five others, *Math. Ann.* **179** (1969) 296–318.

[Ro61]   R.M. Robinson: Arrangement of 24 points on a sphere,
          Math. Ann. **144** (1961) 17–48.

[Sch94]   J. Schaer: The densest packing of ten congruent spheres in
          a cube, in: *Intuitive Geometry* (Szeged, 1991), K. Böröczky et
          al., eds., *Colloq. Math. Soc. János Bolyai* **63**, North-Holland,
          1994, 403–424.

[Sch71]   J. Schaer: On the packing of ten equal circles in a square,
          Math. Mag. **44** (1971) 139–140.

[Sch66a] J. Schaer: On the densest packing of spheres in a cube,
          Canad. Math. Bull. **9** (1966) 265–270.

[Sch66b] J. Schaer: On the densest packing of five spheres in a cube,
          Canad. Math. Bull. **9** (1966) 271–274.

[Sch66c] J. Schaer: On the densest packing of six spheres in a cube,
          Canad. Math. Bull. **9** (1966) 275–280.

[Sch65]   J. Schaer: The densest packing of 9 circles in a square,
          Canad. Math. Bull. **8** (1965) 273–277.

[SchM65] J. Schaer, A. Meir: On a geometric extremum problem,
          Canad. Math. Bull. **8** (1965) 21–27.

[Sch79]   K. Schlüter: Kreispackung in Quadraten, *Elemente Math.*
          34 (1979) 12–14.

[Sch63]   K. Schütte: Minimale Durchmesser endlicher Punktmen-
          gen mit vorgeschriebenem Mindestabstand, *Math. Ann.* **150**
          (1963) 91–98.

[Sch55]   K. SCHÜTTE: Überdeckung der Kugel mit höchstens acht
            Kreisen, Math. Ann. **129** (1955) 181–186.

[SchW51]   K. SCHÜTTE, B.L. VAN DER WAERDEN: Auf welcher Kugel
            haben 5, 6, 7, 8 oder 9 Punkte mit Mindestabstand eins Platz?
            Math. Ann. **123** (1951) 96–124.

[SeM44]   B. SEGRE, K. MAHLER: On the densest packing of circles,
            Amer. Math. Monthly **51** (1944) 261–270.

[ŠkČJ74]   D.O. ŠKLJARSKIĬ, N.N. ČENCOV, I.M. JAGLOM: *Geomet-*
            rical estimates and problems from combinatorial geometry (in
            Russian). Library of the Mathematical Circle, No. 13, Nauka,
            Moscow 1974.

[Ta30]   R.M.L. TAMMES: On the origin of number and arrangement
            of the places of exit on the surface of pollen grains, Rec. Trav.
            Bot. Nederl. **27** (1930) 1–84.

[Ta84]   T. TARNAI: Multisymmetric packing of equal circles on a
            sphere, Ann. Univ. Sci. Budapest Eötvös, Sect. Math. **27**
            (1984) 199–204.

[Ta83]   T. TARNAI: Packing of 180 equal circles on a sphere, Ele-
            mente Math. **38** (1983) 119–122.

[TaG95]   T. TARNAI, Z. GÁSPÁR: Covering a square by equal circles,
            Elemente Math. **50** (1995) 167–170.

[TaG91a]   T. TARNAI, Z. GÁSPÁR: Arrangement of 23 points on a
            sphere (on a conjecture of R.M. Robinson), Proc. Royal Soc.
            London A **433** (1991) 257–267.

[TaG91b]   T. TARNAI, Z. GÁSPÁR: Covering a sphere by equal circles,
            and the rigidity of its graph, Math. Proc. Camb. Phil. Soc.
            110 (1991) 71–89.

[TaG87]   T. TARNAI, Z. GÁSPÁR: Multisymmetric close packings of
            equal spheres on the spherical surface, Acta Cryst. Sect. A
            43 (1987) 612–616.

[TaG86]   T. TARNAI, Z. GÁSPÁR: Covering the sphere with 11 equal
            circles, Elemente Math. **41** (1986) 35–38.

[TaG83]   T. TARNAI, Z. GÁSPÁR: Improved packing of equal circles
            on a sphere and rigidity of its graph, Math. Proc. Cambridge
            Philos. Soc. **93** (1983) 191–218.

[TeO00]   Y. TESHIMA, T. OGAWA: Dense packing of equal circles on a
            sphere by the minimum-zenith method: symmetrical arrange-
            ment, Forma **15** (2000) 347–364.

[Th10]   A. THUE: On the densest packing of congruent circles in the plane (in Norwegian), *Skr. Vidensk-Selsk, Christiania* **1** (1910) 3–9. Also in: *Selected Mathematical Papers*, T. Nagell et al., eds., Universitetsforlaget Oslo, 1977, 257–263.

[Th892]   A. THUE: On some geometric–number-theoretic theorems (in Danish), *Forhandlingerne ved de Skandinaviske Naturforskeres* **14** (1892) 352–353. Also in: *Selected Mathematical Papers*, T. Nagell et al., eds., Universitetsforlaget Oslo, 1977, 5–6.

[Va89]   G. VALETTE: A better packing of ten equal circles in a square, *Discrete Math.* **76** (1989) 57–59.

[Ve49]   S. VERBLUNSKY: On the least number of unit circles which can cover a square, *J. London Math. Soc.* **24** (1949) 164–170.

[We87]   G. WENGERODT: Die dichteste Packung von 25 Kreisen in einem Quadrat, *Ann. Univ. Sci. Budapest. Eötvös. Sect. Math.* **30** (1987) 3–15.

[We83]   G. WENGERODT: Die dichteste Packung von 16 Kreisen in einem Quadrat, *Beiträge Algebra Geom.* **16** (1983) 173–190.

[Za62]   C.T. ZAHN: Black box maximization of circular coverage, *J. Res. Nat. Bureau Standards* **66B** (1962) 181–216.

1.7 Packing Equal Circles or Squares in a Strip

Let $\mathcal{C}$ be a collection of plane convex bodies lying in a parallel strip S_w of width w, and let $B(r)$ denote a circle of radius r around the (arbitrarily fixed) origin. The *density of $\mathcal{C}$ relative to the strip* can be defined as

$$d(\mathcal{C}, S_w) = \lim_{r \to \infty} \frac{\sum_{C \in \mathcal{C}} \text{area}(C \cap B(r))}{\text{area}(S_w \cap B(r))},$$

if this limit exists.

Conjecture 1 *(Molnár [Mo78]) The maximum density of a packing of unit circles in a parallel strip of width $w \geq 2$ is*

$$\frac{(n+1)(n+2)\pi}{2w \left(n + \sqrt{4 - \left(w - 2 - n\sqrt{3}\right)^2}\right)},$$

where $n = \lfloor (w-2)/\sqrt{3} \rfloor$.

This bound, if valid, is sharp, as is shown by the following construction. Let T denote a set of $(n+1)(n+2)/2$ unit circles such that the convex hull of their centers is an equilateral triangle of side $2n$, and they form a lattice-like packing of $n+1$ rows with i circles in the ith row $(1 \leq i \leq n+1)$. Let us arrange a sequence $\{T_i : -\infty < i < +\infty\}$ of nonoverlapping copies of T in a strip of width w such that $n+1$ circles of T_i are tangent to the upper boundary line of the strip if i is even and to the lower one if i is odd, and T_i is touching T_{i+1} for every i.

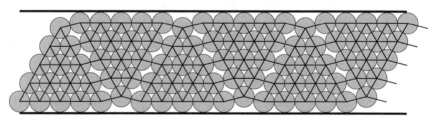

It is easy to see that Molnár's conjecture is valid for $w \leq 2 + \sqrt{3}$. Kertész [Ke82] was able to verify this for every $w \leq 2 + 2\sqrt{2}$, and his argument was extended by Füredi [Fü91] to all $w \leq 2 + 2\sqrt{3}$. On the other hand, it follows from a result of Groemer [Gr60] that the conjecture is true for every $w = 2 + m\sqrt{3}$, where m is a positive integer.

We can raise similar questions about packings of congruent copies of any convex body in a strip.

Problem 2 *Determine (or estimate) the maximum density $f(w)$ of a packing of unit squares in a parallel strip of width $w \geq 1$.*

Evidently, $f(w) = 1$ for any integer w. If w is not an integer, then the density of every packing of squares with sides parallel (or perpendicular) to the direction of the strip is at most $\lfloor w \rfloor / w$. However, we can achieve larger densities by tilting the squares. The optimal solution is known only for $1 \le w < 2$. For $1 \le w \le 2\sqrt{2} - 1$ we have $f(w) = 1/w$. If $2\sqrt{2} - 1 \le w < 2$, then $f(w) = \frac{1}{w(2\sqrt{2}-w)}$.

The following slightly different form of this problem was considered by Erdős and Graham [ErG75]. What is the largest number of unit squares that can be packed into a square of side w? It follows from their results that the density satisfies

$$f(w) \ge 1 - c\left(\frac{1}{w}\right)^{15/11},$$

where c is a suitable constant. This is a dramatic improvement of the trivial lower bound $\left(\frac{\lfloor w \rfloor}{w}\right)^2 > 1 - \frac{2}{w}$, provided that $w - \lfloor w \rfloor$ is not very small. Montgomery somewhat increased the exponent $15/11$, but his result remained unpublished.

The best upper bound so far follows from the arguments of Roth and Vaughan [RoV78]:

$$f(w) \le 1 - c'\frac{|w|^{1/2}}{w^{3/2}},$$

where c' is a suitable constant and $|w|$ denotes the distance of w from the nearest integer. Which of the two bounds is closer to the truth if, say, $|w| = 1/2$?

Another almost equivalent form of the Erdős–Graham problem is the following.

Problem 3   *Let $w(n)$ denote the side length of the smallest square into which n unit squares can be packed. Determine the exact value of $w(n)$, at least for small n.*

Obviously, we have for every n,

$$\sqrt{n} \le w(n) \le \lceil \sqrt{n} \rceil ,$$

which is sharp if n is a perfect square. We have that $w(2) = w(3) = 2$, $w(5) = 2 + 1/\sqrt{2}$, $w(6) = w(7) = w(8) = 3$, and the values $w(15), w(24)$, and $w(35)$ are also known. For more detailed information on this problem, see El Moumni [El99], Kearney and Shiu [KeS02], and the survey papers of Göbel [Gö79] and Friedman [Fr00]. The latter contains efficient constructions for many values of n up to 100.

For any k, let $n(k)$ denote the smallest integer n such that $w(n^2+1) \leq n + \frac{1}{k}$, i.e., a square of side length $n + \frac{1}{k}$ can accommodate $n^2 + 1$ nonoverlapping unit squares. It is shown in [KeS02] by an explicit construction that $n(2) \leq 43$ and that $n(k) \leq 27k^3/2 + O(k^2)$.

As far as we know, the dual questions for coverings have been largely ignored. We propose the following problem.

Problem 4   *Determine (or estimate) the minimum density of a covering of a parallel strip of width w with unit squares (or with unit circles).*

L. Fejes Tóth [FeT89] stated some conjectures concerning the most economical configurations in the circular case. In an unpublished manuscript, Heppes and Melissen showed that if $w \leq \sqrt{3}$, the optimal configuration consists of a string of circles whose centers are equally spaced along a line. For $w > \sqrt{3}$ the situation changes: it is very likely that for slightly larger values of w, there is an optimal covering that can be extended to a lattice arrangement whose fundamental parallelogram has a diagonal parallel to the strip.

Similar questions in higher-dimensional spaces have been addressed by Horváth and Molnár [HoM67]. Molnár [Mo78] determined the maximum density of a packing of unit spheres in a three-dimensional strip (slab) of width $w \leq 2 + \sqrt{2}$. Surprisingly, he found that there are infinitely many essentially different extremal configurations. Horváth [Ho72], [Ho74], [Ho75] studied the same question in four-dimensional space, and he also considered sphere packings in d-dimensional cylinders ($d \geq 3$). In [Sz95], [HoTZ01], dense packings of unit balls touching an infinite unit cylinder were considered.

[El99]   S. El Moumni: Optimal packings of unit squares in a square, *Studia Sci. Math. Hungar.* **35** (1999) 281–290.

[El97]   S. El Moumni: Rangements optimaux de carrés unité dans une bande infinie, *Ann. Fac. Sci. Toulouse Math. (6)* **6** (1997) 121–125.

[ErG75]   P. Erdős, R.L. Graham: On packing squares with equal squares, *J. Combinatorial Theory Ser. A* **19** (1975) 119–123.

[FeT89]   L. Fejes Tóth: Research Problem 45, *Period. Math. Hungar.* **20** (1989) 169-171.

[Fr00]   E. Friedman: Packing unit squares in squares: a survey and new results, *Electron. J. Combin.* **7** (2000) *Dynamic Survey* 7, 24 pp.

[Fü91]   Z. FÜREDI: The densest packing of equal circles into a parallel strip, *Discrete Comput. Geom.* **6** (1991) 95–106.

[Gö79]   F. GÖBEL: Geometrical packing and covering problems, in: *Packing and Covering in Combinatorics*, A. Schrijver, ed., *Mathematical Centre Tracts* **106**, Mathematisch Centrum Amsterdam, 1979, 179–199.

[Gr60]   H. GROEMER: Über die Einlagerung von Kreisen in einen konvexen Bereich, *Math. Zeitschrift* **73** (1960) 285–294.

[Ho75]   J. HORVÁTH: The densest packing of unit balls in 3-dimensional and 4-dimensional layers (in Russian), *Ann. Univ. Sci. Budapest. Eötvös Sect. Math.* **18** (1975) 171–176.

[Ho74]   J. HORVÁTH: Die Dichte einer Kugelpackung in einer 4-dimensionalen Schicht, *Periodica Math. Hungar.* **5** (1974) 195–199.

[Ho72]   J. HORVÁTH: The densest packing of an n-dimensional cylinder by unit spheres (in Russian), *Annales Univ. Sci. Budapest. Eötvös Sect. Math.* **15** (1972) 139–143.

[HoM67]   J. HORVÁTH, J. MOLNÁR: On the density of non-overlapping unit spheres lying in a strip, *Ann. Univ. Sci. Budapest. Eötvös Sect. Math.* **10** (1967) 193–201.

[HoTZ01]   J. HORVÁTH, Á.H. TEMESVÁRI, J. ZÁVOTI: Über Kugelpackungen um einen Zylinder, *Publ. Math. Debrecen* **58** (2001) 411–422.

[KeS02]   M. KEARNEY, P. SHIU: Efficient packing of unit squares in a square, *Electron. J. Combin.* **9** (2002) # R14.

[Ke82]   G. KERTÉSZ: *On a Problem of Parasites* (in Hungarian), master's thesis, Eötvös University, Budapest, 1982.

[Mo78]   J. MOLNÁR: Packing of congruent spheres in a strip, *Acta. Math. Hungar.* **31** (1978) 173–183.

[RoV78]   K.F. ROTH, R.C. VAUGHAN: Inefficiency in packing squares with unit squares, *J. Combinatorial Theory Ser. A* **24** (1978) 170–186.

[Sz95]   L. SZABÓ: On the density of unit balls touching a unit cylinder, *Arch. Math.* **64** (1995) 459–464.

1.8 The Densest Packing of Spheres

Draw a unit ball around each point of $\mathbb{R}^3$ of the form $(x\sqrt{2}, y\sqrt{2}, z\sqrt{2})$, where x, y, z are integers and their sum is even. It is easy to check that we obtain a lattice packing with density $\pi/\sqrt{18} = 0.74048\ldots$. Gauss [Ga831] proved (in a book review) that this is the densest lattice packing with balls, i.e., $\delta_L(B^3) = \pi/\sqrt{18}$.

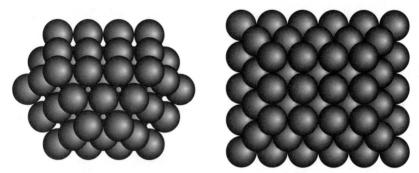

THE DENSEST PACKING OF CONGRUENT BALLS: TWO VIEWS

Over four hundred years ago, Sir Walter Raleigh asked his assistant to find a formula for the number of cannonballs in regularly stacked piles. The question raised the curiosity of Johannes Kepler, as illustrated in [Ke611]. Therefore, the following long-standing conjecture, which is one of the most well known problems in discrete geometry, is often attributed to Kepler. Actually, it was not precisely formulated even by Hilbert, who considered it in connection with his famous 18th problem.

Conjecture 1 *(Kepler) The (upper) density of any packing of equal balls in 3-dimensional space is at most $\pi/\sqrt{18}$, i.e., we have $\delta(B^3) = \delta_L(B^3)$.*

Using an ingenious argument that works in any dimension, Rogers [Ro58] obtained the upper bound

$$\delta(B^3) \le \sqrt{18}\left(\arccos\frac{1}{3} - \frac{\pi}{3}\right) = 0.77963\ldots .$$

This bound has been subsequently improved by Lindsey [Li86] and Muder [Mu88], [Mu93] to $0.773055\ldots$. Hsiang [Hs93a], [Hs93b], [Hs01] proposed an elaborate line of attack (along the ideas of L. Fejes Tóth suggested 40 years earlier), but his claim that he settled Kepler's conjecture seems exaggerated [Ha94]. However, so far no one has found any serious gap in the approach of Hales [Ha97a], [Ha97b], [Ha00], [Ha05a], [Ha05b], [Ha05c],

[Ha05d] and Ferguson [FeH05], [Fe05], although no one has been able to fully verify it either. This is not too surprising, given that the argument is described in (at least) seven separate papers and relies on long computer-aided calculations of more than 5000 subproblems. Whether or not we welcome this new development, it is very likely that in the future many other geometric optimization problems will be settled by similar arguments, whose correctness cannot be checked without computers. (The recent proof of Berge's perfect graph conjecture shows that this may also happen in other areas of mathematics.) Hales's new project is to develop software for geometric proof verification. Meanwhile, the search continues for simpler proofs for at least some special cases of the above conjecture.

A very promising approach was suggested by Böröczky [Bö75]. An infinite series of nonoverlapping unit balls is called a *string* if all of their centers are on the same line (the so-called *axis*) and each ball is tangent to exactly two others.

Conjecture 2 (*Böröczky*) *If a packing of unit balls in* $\mathbb{R}^3$ *consists of strings, then its density cannot exceed* $\pi/\sqrt{18}$. *In other words, denoting by* S_∞ *the union of all balls belonging to a string,* $\delta(S_\infty) = \pi/\sqrt{18}$.

Bezdek, Kuperberg, and Makai [BeKM91] proved the following weaker version of this conjecture. Any packing of unit balls in $\mathbb{R}^3$ that consists of strings with *parallel axes* has density at most $\pi/\sqrt{18}$, so $\delta_T(S_\infty) = \pi/\sqrt{18}$. In fact, Heppes and Makai [HeM05] have also determined the maximum density of a packing of "generalized strings" with parallel axes, that is, sequences of balls of radius 1, whose centers form a collinear equidistant set with distance d, where $2 \leq d \leq 2\sqrt{2}$. Their theorem states that no packing of translates of a generalized string is denser than the densest lattice packing of these generalized strings. It is worth mentioning that for $d \neq 2, 2\sqrt{2}$, this result does not follow from Kepler's conjecture.

A related two-dimensional question was studied in [BlF64].

A subset of a string consisting of k consecutive balls is called a *k-string*. Another possible weakening of the original question is the following:

Conjecture 3 (*Bezdek, Kuperberg, Makai [BeKM91]*) *Let* k *be a fixed number, and let* S_k *denote the union of all balls belonging to a k-string. Then*
(1) $\delta_T(S_k) = \pi/\sqrt{18}$,
(2) $\delta(S_k) = \pi/\sqrt{18}$.

There is no value of k for which any of the above statements has been verified. In fact, it would be of interest to find any small subcollection $\mathcal{C}$ of the densest lattice packing of balls for which it can be relatively easily shown that the maximum density of a packing that can be decomposed into congruent copies of $\mathcal{C}$ is equal to $\pi/\sqrt{18}$.

Another important special case of Kepler's conjecture was formulated by L. Fejes Tóth [FeT69]: The density of any packing of unit balls in $\mathbb{R}^3$, in which each ball is tangent to 12 others, is at most $\pi/\sqrt{18}$. This is an excellent test case for Hsiang's and Hales's analyses, and both of them have graduate students working on clean proofs under this assumption [Le02]. Another plan was announced by Böröczky (see [BöS03] for its first completed step). In fact, L. Fejes Tóth's conjecture and the proposed proofs are more ambitious: it is very likely that *every* packing of unit balls with the above property can be decomposed into parallel "hexagonal" layers such that in each layer the centers of the balls are coplanar and form a regular triangular lattice.

It is interesting to note that a fairly natural variant of Kepler's conjecture, due to L. Fejes Tóth [FeT76], has a relatively simple complete solution. The *gap size* of a packing is the supremum of the radius of a ball disjoint from all of its members. Confirming Fejes Tóth's conjecture, Böröczky [Bö86] proved that the gap size of any packing of unit balls in $\mathbb{R}^3$ is at least as large as in the packing obtained by placing the centers of the balls at the vertices and at the centers of the cubes of a cubic lattice of side length $4/\sqrt{3}$. Another very similar result can be found in [Bö01]. See more about this topic in Section 1.10.

In recent years, the construction of dense sphere packings in higher- (but still low-) dimensional spaces has become a rapidly developing area in discrete geometry. Much research in this field has been motivated by applications of economical sphere packings and coverings in coding (information transmission) and in pattern recognition (artificial intelligence); see, e.g., [Va95], [CoS96]. Detailed accounts of estimates and constructions are given in Conway and Sloane [CoS95], [CoS99]. For small dimensions d, the best upper bounds for $\delta(B^d)$ were found by Rogers [Ro58], K. Bezdek [Be02], and recently by Cohn et al. [CoE03],[Co02], [CoK05], who developed linear programming techniques to determine some of these values with large precision. For example, in 8 dimensions their bound differs from the conjectured optimum only by a factor of 1.000001, and in 24 dimensions they determined the exact value of $\delta_L(B^{24})$.

As $d \to \infty$, the asymptotically best lower and upper bounds for the density of sphere packings, due to Ball [Ba92] and to Kabatjanskiĭ and Levenšteĭn [KaL78], respectively, are

$$\zeta(d)\frac{d-1}{2^{d-1}} \leq \delta(B^d) \leq 2^{-(0.599+o(1))d}.$$

Problem 4   (*G. Fejes Tóth [FeT99], Pach*) *Do the limits*
$$\lim_{d\to\infty} \frac{\log \delta(B^d)}{d}, \quad \lim_{d\to\infty} \frac{\log \delta_L(B^d)}{d} \quad \text{exist? Are they equal?}$$

In the densest lattice packing of equal balls in $\mathbb{R}^3$, the centers of the balls lie at the vertices and at the centers of the faces of a cubic lattice. Somewhat surprisingly, the covering of space obtained by putting around these points equal balls of minimum radii so that together they cover $\mathbb{R}^3$ is not the most economical covering with unit balls. As was proved by Bambah [Ba54], in the thinnest lattice covering, the centers of the balls lie at the vertices and at the centers of the cubes of a cubic lattice. The dual of Kepler's conjecture is completely open:

Conjecture 5 *The (lower) density of any covering of $\mathbb{R}^3$ with unit balls is at least $5\sqrt{5}\pi/24$; i.e., we have $\theta(B^3) = \theta_L(B^3)$.*

The best known lower bounds on the covering densities of spheres were established by Coxeter, Few, and Rogers [CoFR59]. In particular, they proved

$$\theta(B^3) \geq \frac{3\sqrt{3}}{2}\left(3\arccos\frac{1}{3} - \pi\right) = 1.432\ldots.$$

As we have already mentioned in the introduction to this chapter, in his famous lecture at the International Congress in 1900, Hilbert drew the attention of the mathematical community to the problem of dense sphere packings. It seems that the following closely related question he asked has been (undeservedly) forgotten.

Problem 6 *(Hilbert [Hi900]) Let T^d denote the regular d-dimensional simplex of side 1. Determine or estimate $\delta(T^d)$, the maximum density of a packing of congruent copies of T^d, at least for $d = 3$.*

Correcting an error of Minkowski, Groemer [Gr62] and Hoylman [Ho70] showed that $\delta_L(T^3) = 18/49$. However, it is not known whether $\delta_T(T^3) = \delta_L(T^3)$. It follows from the results of Rogers and Shephard [RoS57] that the lattice packing density and the translative packing density of the regular d-dimensional simplex satisfy

$$\frac{2(d!)^2}{(2d)!} \leq \delta_L(T^d) \leq \delta_T(T^d) \leq \frac{2^d(d!)^2}{(2d)!} \qquad \text{for every } d.$$

The densest known packing of congruent copies of T^3 is the union of four lattice arrangements, and its density is $2/3 > 18/49$.

Problem 7 *(L. Moser [Mo66]) Estimate the maximum number $f(d)$ of congruent copies of a d-dimensional regular simplex T^d that can be packed into a cube of the same side length. Is it true that*

$$\lim_{d\to\infty} \frac{\log f(d)}{d} = \infty?$$

[Ba92]   K.M. BALL: A lower bound for the optimal density of lattice
           packings, *Internat. Math. Res. Notices* **92** (1992) 217–221.

[Ba54]   R.P. BAMBAH: On lattice coverings by spheres, *Proc. Nat.
           Inst. Sci. India* **20** (1954) 25–52.

[Be02]   K. BEZDEK: Improving Rogers' upper bound for the density
           of unit ball packings via estimating the surface area of Voronoi
           cells from below in Euclidean d-space for all $d \geq 8$, *Discrete
           Comput. Geom.* **28** (2002) 75–106.

[BeKM91]   A. BEZDEK, W. KUPERBERG, E. MAKAI JR.: Maximum
           density space packing with parallel strings of spheres, *Discrete
           Comput. Geom.* **6** (1991) 277–283.

[BlF64]   M.N. BLEICHER, L. FEJES TÓTH: Circle-packings and circle-
           coverings on a cylinder, *Michigan Math. J.* **11** (1964) 337–341.

[Bö01]   K. BÖRÖCZKY: Edge close ball packings, *Discrete Comput.
           Geom.* **26** (2001) 59–71.

[Bö86]   K. BÖRÖCZKY: Closest packing and loosest covering of the
           space with balls, *Studia Sci. Math. Hungar.* **21** (1986) 79–89.

[Bö75]   K. BÖRÖCZKY: Research Problem 12, *Period. Math. Hun-
           gar.* **6** (1975) p.109.

[BöS03]   K. BÖRÖCZKY, L. SZABÓ: Arrangements of thirteen points
           on a sphere, in: *Discrete Geometry — In honor of W. Ku-
           perbergs 65th birthday*, A. Bezdek, ed., Marcel Dekker, 2003,
           111–184.

[Co02]   H. COHN: New upper bounds on sphere packings II, *Geom.
           Topol.* **6** (2002) 329–353.

[CoE03]   H. COHN, N.D. ELKIES: New upper bounds on sphere pack-
           ings I, *Ann. Math. (2)* **157** (2003) 689–714.

[CoK05]   H. COHN, A. KUMAR: Optimality and uniqueness of the
           Leech lattice among lattices, manuscript,
           http://research.microsoft.com/~cohn/Leech

[CoS99]   J.H. CONWAY, N.J.A. SLOANE: *Sphere Packings, Lattices
           and Groups* (3rd edition), Springer-Verlag, 1999.

[CoS96]   J.H. CONWAY, N.J.A. SLOANE: The antipode construction
           for sphere packings, *Invent. Math.* **123** (1996) 309–313.

[CoS95]   J.H. CONWAY, N.J.A. SLOANE: What are all the best sphere
           packings in low dimensions? *Discrete Comput. Geom.* **13**
           (1995) 383–403.

[CoFR59] H.S.M. COXETER, L. FEW, C.A. ROGERS: Covering space with equal spheres, *Mathematika* **6** (1959) 147–157.

[FeT99] G. FEJES TÓTH: Recent progress on packing and covering, in: *Advances in Discrete and Computational Geometry*, *Contemporary Math.* **223** AMS 1999, 145–162.

[FeT76] L. FEJES TÓTH: Close packing and loose covering with balls, *Publ. Math. Debrecen* **23** (1976) 323–326.

[FeT69] L. FEJES TÓTH: Remarks on a theorem of R. M. Robinson, *Studia Sci. Math. Hungar.* **4** (1969) 441–445.

[Fe05] S.P. FERGUSON: Sphere packings, V, *Discrete Comput. Geom.*, to appear; see arXiv:math.MG/9811077

[FeH05] S.P. FERGUSON, T.C. HALES: A formulation of Kepler conjecture, *Discrete Comput. Geom.*, to appear; see arXiv:math.MG/9811072

[Ga831] C.F. GAUSS: Untersuchungen über die Eigenschaften der positiven ternären quadratischen Formen von Ludwig August Seber, *Göttingische gelehrte Anzeigen* 9. Juli 1831, see: Werke, Band 2, 2. Aufl. Göttingen 1876, 188–196; also *J. Reine Angew. Math.* **20** (1840) 312–320.

[Gr62] H. GROEMER: Über die dichteste gitterförmige Lagerung kongruenter Tetraeder, *Monatshefte Math.* **66** (1962) 12–15.

[Ha05a] T.C. HALES: Overview of the Kepler conjecture, *Discrete Comput. Geom.*, to appear; see arXiv:math.MG/9811071

[Ha05b] T.C. HALES: Sphere packings, III, *Discrete Comput. Geom.*, to appear; see arXiv:math.MG/9811075

[Ha05c] T.C. HALES: Sphere packings, IV, *Discrete Comput. Geom.*, to appear; see arXiv:math.MG/9811076

[Ha05d] T.C. HALES: The Kepler conjecture, *Discrete Comput. Geom.*, to appear; see arXiv:math.MG/9811078

[Ha00] T.C. HALES: Cannonballs and honeycombs, *Notices Amer. Math. Soc.* **47** (2000) 440–449.

[Ha97a] T.C. HALES: Sphere packings, I, *Discrete Comput. Geom.* **17** (1997) 1–51.

[Ha97b] T.C. HALES: Sphere packings, II, *Discrete Comput. Geom.* **18** (1997) 135–149.

[Ha94] T.C. HALES: The status of the Kepler conjecture, *Math. Intelligencer* **16** (1994) 47–58.

[HeM05] A. HEPPES, E. MAKAI JR.: String packing, manuscript.

[Hi900]   D. HILBERT: Mathematische Probleme. (Lecture delivered at the International Congress of Mathematicians, Paris 1900.) *Göttinger Nachrichten* (1900) 253–297, or *Gesammelte Abhandlungen* Band 3, 290–329. English translation in *Bull. Amer. Math. Soc.* (1902) 437–479. and *Bull. Amer. Math. Soc. (New Ser.)* **37** (2000) 407–436.

[Ho70]   D.J. HOYLMAN: The densest lattice packing of tetrahedra, *Bull. Amer. Math. Soc.* **76** (1970) 135–137.

[Hs01]   W.-Y. HSIANG: *Least Action Principle of Crystal Formation of Dense Packing Type and Kepler's Conjecture*, World Sci. Publishing 2001.

[Hs93a]   W.-Y. HSIANG: On the sphere packing problem and the proof of Kepler's conjecture, *Internat. J. Math.* **4** (1993) 739–831.

[Hs93b]   W.-Y. HSIANG: On the sphere packing problem and the proof of Kepler's conjecture, in: *Differential Geometry and Topology* (Alghero, 1992), World Sci. Publishing 1993, 117–127.

[KaL78]   G.A. KABATJANSKIĬ, V.I. LEVENŠTEĬN: Bounds for packings on a sphere and in space (in Russian), *Problemy Peredači Informacii* **14** (1978) 3–25. English translation: *Problems of Information Transmission* **14** (1978) 1–17.

[Ke611]   J. KEPLER: *Strena seu de nive sexangula*. Tampach, Frankfurt, 1611. English translation: *The Six-Cornered Snowflake*. Oxford, 1966.

[La02]   J.C. LAGARIAS: Bounds for the local density of sphere packings and the Kepler conjecture, *Discrete Comput. Geom.* **27** (2002) 165–193.

[Le02]   WING-LUNG LEE: *Thirteen spheres problem and Fejes Tóth conjecture* (Ph.D. Thesis.) Hong Kong University of Science and Technology, October 2002.

[Li86]   J.H. LINDSEY: Sphere packing in R^3, *Mathematika* **33** (1986) 417–421.

[Mo66]   L. MOSER: *Poorly Formulated Unsolved Problems in Combinatorial Geometry*. Mimeographed, 1966. Reprinted [Mo91].

[Mo91]   W.O.J. MOSER: Problems, problems, problems, *Discrete Appl. Math.* **31** (1991) 201–225.

[Mu93]   D.J. MUDER: A new bound on the local density of sphere packings, *Discrete Comput. Geom.* **10** (1993) 351–375.

[Mu88]   D.J. MUDER: Putting the best face on a Voronoi polyhedron, *Proc. London Math. Soc.* 3. Ser. **56** (1988) 329-348.

[**Ro58**]   C.A. ROGERS: The packing of equal spheres, *Proc. London Math. Soc.* 3. Ser. **8** (1958) 609–620.

[**RoS57**]   C.A. ROGERS, G.C. SHEPHARD: The difference body of a convex body, *Arch. Math.* **8** (1957) 220–223.

[**Va95**]   A. VARDY: A new sphere packing in 20 dimensions, *Invent. Math.* **121** (1995) 119–133.

1.9 The Densest Packings of Specific Convex Bodies

Despite many significant results and powerful techniques developed in the last fifty years, the problem of calculating the packing densities of even relatively simple three-dimensional convex bodies remains an extremely difficult task.

We start with a provocative problem.

Problem 1   *Give nontrivial examples of bounded convex bodies C in three-dimensional space that are not tiles (i.e., $\delta(C) \neq 1$) but whose packing densities $\delta(C)$ can be exactly determined.*

As far as we can tell, there is only one such example known. The Dirichlet–Voronoi cells of the face-centered cubic lattice (whose vertices are the gridpoints of a cubic integer lattice, together with the centers of the faces of the cubes) are congruent copies of a rhombic dodecahedron C that form a lattice tiling of $\mathbb{R}^3$. A. Bezdek [Be94] proved that slightly truncating C leaves the structure of the densest packing of congruent copies of the resulting body C' unchanged, and therefore the packing density of C' is $\mathrm{Vol}(C')/\mathrm{Vol}(C)$.

If we drop the condition that C must be bounded, then there is another example, due to A. Bezdek and W. Kuperberg [BeK90], for which $\delta(C)$ can be determined. Given a plane convex body K and $h > 0$, let $C(K, h)$ and $C(K, \infty)$ denote the right cylinders based on K with height h and with infinite height (in both directions), respectively. It was shown in [BeK90] that $\delta(C(B^2, \infty)) = \delta(B^2) = \pi/\sqrt{12}$; i.e., the density of a packing of congruent (right) circular cylinders of infinite length cannot exceed the packing density of the circle B^2 in the plane. Obviously, this density can be attained if the cylinders are in parallel position.

However, it is not clear whether this result can be generalized to infinite cylinders over an arbitrary convex base. A general difficulty is that we know very little about the structure of cylinder packings in space. For instance, it is not obvious at first glance whether there exists any packing in $\mathbb{R}^3$ with congruent copies of $C(B^2, \infty)$ that does not contain two parallel cylinders but has positive density. This fact has been established by K. Kuperberg [Ku90]. Her construction was refined in [GrP97] to obtain such a packing with density larger than 0.416.

She reported the following unsolved problem.

Problem 2   *(C.A. Rogers) Let $\mathcal{C}$ be a packing of congruent circular cylinders in $\mathbb{R}^3$ such that no two of them are parallel. Is it true that for any $\epsilon > 0$ and for any $r > 0$, there exists a* ball B of radius r such that the density of $\mathcal{C}$ restricted to B is at most ϵ?

Problem 3   (A. Bezdek and W. Kuperberg [BeK91]; Wilker [Wi87])
Is it true that for every plane convex body K:
(1)   there exists $h > 0$ such that $\delta(C(K, h)) = \delta(K)$?
(2) $\delta(C(K, \infty)) = \delta(K)$?

Of course, an affirmative answer to (1) also implies (2).

For circular cylinders, Wilker made the following stronger conjecture.

Conjecture 4   (Wilker [Wi87]) $\delta(C(B^2, h)) = \delta(B^2)$ for every $h > 0$.

As was shown by A. Bezdek and W. Kuperberg [BeK91], if K is a sufficiently "elongated" ellipse (the ratio of its axes is larger than $\sqrt{3} + 1$) and h is very small, then $\delta(C(K, h)) > \delta(K)$.

Using the same idea, they proved that $\delta(E) > \delta_L(E)$ for any sufficiently "elongated" ellipsoid E in $\mathbb{R}^d$, provided that $d \geq 3$. (This is in sharp contrast to the situation in the plane.) For $d = 3$, this construction was slightly modified by Wills [Wi91], who found a packing of congruent ellipsoids with density larger than $0.7585 > \pi/\sqrt{18} \approx 0.7404$. An even better packing of density roughly 0.7707 was constructed by Donev et al. [DoSC04]. These results suggest the following problems.

Problem 5   Let $d \geq 3$ be fixed. Does there exist a small positive ϵ with the property that $\delta(E) = \delta(B^d)$ for every d-dimensional ellipsoid E whose axes have lengths at least 1 and at most $1 + \epsilon$?

Problem 6   Let $d \geq 3$ be fixed. Does there exist a small positive ϵ with the property that $\delta(E) \leq 1 - \epsilon$ for every d-dimensional ellipsoid E?

One might expect that it is much easier to compute the lattice packing density $\delta_L(C)$ of a given convex body C than to determine the packing density $\delta(C)$. In fact, the densest lattice packings of spheres are known for every $d \leq 8$. The case $d = 2$ was settled by Lagrange [La773], the case $d = 3$ by Gauss [Ga831], $d = 4, 5$ by Korkin and Zolotareff [KoZ872], [KoZ873], [KoZ877], and the cases $d = 6, 7, 8$ by Blichfeldt [Bl29], [Bl34]. Minkowski [Mi904] determined the lattice packing density of the octahedron, while Groemer [Gr62] and Hoylman [Ho70] solved the problem for the tetrahedron and the cuboctahedron. Mahler [Ma47] showed that the lattice packing density of the three-dimensional circular cylinder satisfies

$$\delta_L(C(B^2, 1)) = \delta_L(B^2) = \pi/\sqrt{12}.$$

Chalk and Rogers [ChR48] and Yeh [Ye48] managed to generalize this result to all three-dimensional convex cylinders. They proved that

$$\delta_L(C(K,h)) = \delta_L(K) \quad \text{for any } h > 0$$

holds for every plane convex body K. As far as we know, apart from polytopes, this is the only large class of three-dimensional convex bodies whose lattice packing densities can be determined (or reduced to a planar problem). On the other hand, Betke and Henk [BeH00] designed an efficient algorithm for computing the density of a densest lattice packing of any three-dimensional convex polytope. As an application, they determined the densest lattice packings of all regular and Archimedean polytopes. Whitworth [Wh51] computed the density of the densest lattice packing of a double cone whose base is a circular disk.

The following conjecture is probably due to Mahler and Hlawka.

Conjecture 7   *Let $C^d(K,h)$ denote a d-dimensional cylinder of height $h > 0$ based on a $(d-1)$-dimensional convex body K. Then $\delta_L(C^d(K,h)) = \delta_L(K)$.*

For $d > 3$, the only nontrivial result in this direction was obtained by Woods [Wo58], who verified this conjecture when $K = B^3$.

One can ask the analogous question for coverings of the space with cylinders and ellipsoids. The difficulty is that except for B^2 and for sufficiently "fat" convex bodies (see section 1.4 and below), we do not have a full solution of the covering problem in the plane.

Problem 8   *Decide (at least for $d = 3$) whether $\theta(C^d(B^{d-1}, \infty)) = \theta_L(B^{d-1})$, where $C^d(B^{d-1}, \infty)$ denotes the two-way infinite cylinder based on the $(d-1)$-dimensional unit ball.*

As was pointed out by G. Fejes Tóth and W. Kuperberg [FeK95], in dimension $d \geq 3$ every strictly convex smooth body C has an affine image C' with covering density

$$\theta(C') < \theta_L(C').$$

Problem 9   *Is it true that $\theta(E) = \theta_L(E) = 2\pi/\sqrt{27}$ for every ellipse E in the plane?*

It is quite astonishing (and sad) that we are unable to answer this question. The only steps in this direction were taken by Heppes [He03] and G. Fejes Tóth [FeT05], who proved that the answer is in the affirmative if E is sufficiently "fat," i.e., its shorter axis is at least 0.741 times the longer one.

[BeH00]   U. BETKE, M. HENK: Densest lattice packings of 3-polytopes, *Comput. Geom.* **16** (2000) 157–186.

[FeT05]   G. Fejes Tóth: Covering with fat convex discs, *Discrete Comput. Geom.*, to appear.

[Be94]   A. Bezdek: A remark on the packing density in the 3-space, in: *Intuitive Geometry* (Szeged, 1991), K. Böröczky et al., eds., *Colloq. Math. Soc. János Bolyai* **63** (1994) 17–22.

[BeK91]   A. Bezdek, W. Kuperberg: Packing Euclidean space with congruent cylinders and with congruent ellipsoids, in: *Applied Geometry and Discrete Mathematics, DIMACS Ser. Discrete Math. Theoret. Comput. Sci.* **4**, Amer. Math. Soc. 1991, 71–80.

[BeK90]   A. Bezdek, W. Kuperberg: Maximum density space packing with congruent circular cylinders of infinite length, *Mathematika* **34** (1990) 74–80.

[Bl34]   H.F. Blichfeldt: The minimum values of positive quadratic forms in six, seven and eight variables, *Math. Zeitschrift* **39** (1934) 1–15.

[Bl29]   H.F. Blichfeldt: The minimum value of quadratic forms and the closest packing of spheres, *Math. Ann.* **101** (1929) 605–608.

[ChR48]   J.H.H. Chalk, C.A. Rogers: The critical determinant of a convex cylinder, *J. London Math. Soc.* **23** (1948) 178–187. Corrigendum, ibid. **24** (1949) p. 240.

[DoSC04]   A. Donev, F.H. Stillinger, P.M. Chaikin, S. Torquato: Unusually dense crystal packings of ellipsoids, *Phys. Rev. Lett.* **92**, 255506 (2004).

[FeK95]   G. Fejes Tóth, W. Kuperberg: Thin non-lattice covering with an affine image of a strictly convex body, *Mathematika* **42** (1995) 239–250.

[Ga831]   C.F. Gauss: Untersuchungen über die Eigenschaften der positiven ternären quadratischen Formen von Ludwig August Seber, *Göttingische gelehrte Anzeigen* 9. Juli 1831, see: *Werke*, Band 2, 2. Aufl. Göttingen 1876, 188–196; also *J. Reine Angew. Math.* **20** (1840) 312–320.

[GrP97]   C. Graf, P. Paukowitsch: Möglichst dichte Packungen aus kongruenten Drehzylindern mit paarweise windschiefen Achsen, *Elemente Math.* **52** (1997) 71–83.

[Gr62]   H. Groemer: Über die dichteste gitterförmige Lagerung kongruenter Tetraeder, *Monatshefte Math.* **66** (1962) 12–15.

[He03]   A. Heppes: Covering the plane with fat ellipses without non-crossing assumption, *Discrete Comput. Geom.* **29** (2003)

477–481.

[Ho70]   D.J. HOYLMAN: The densest lattice packing of tetrahedra, *Bull. Amer. Math. Soc.* **76** (1970) 135–137.

[KoZ877]   A.N. KORKIN, E. ZOLOTAREFF: Sur les formes quadratiques positive, *Math. Ann.* **11** (1877) 242–292.

[KoZ873]   A.N. KORKIN, E. ZOLOTAREFF: Sur les formes quadratiques, *Math. Ann.* **6** (1873) 366–389.

[KoZ872]   A.N. KORKIN, E. ZOLOTAREFF: Sur les formes quadratiques positives quaternaires, *Math. Ann.* **5** (1872) 581–583.

[Ku90]   K. KUPERBERG: A nonparallel cylinder packing with positive density, *Mathematika* **37** (1990) 324–331.

[La773]   J.L. LAGRANGE: Recherches d'arithmétique, *Nouveaux Mem. Acad. Roy. Sci. et Belles-Lettres de Berlin*, 1773, 265–312. Also in: *Oeuvres* **3**, 693–758.

[Ma47]   K. MAHLER: On lattice points in a cylinder, *Quart. J. Math.* **17** (1947) 16–18.

[Mi904]   H. MINKOWSKI: Dichteste gitterförmige Lagerung kongruenter Körper, *Nachr. Königl. Ges. Wiss. Göttingen, Math.-Phys. Kl.* (1904) 311–355. Also in: *Ges. Abh.* **2**, 3–42.

[Sch02]   A. SCHÜRMANN: Dense ellipsoid packings, *Discrete Math.* **247** (2002) 243–249.

[Wh51]   J.V. WHITWORTH: The critical lattices of the double cone, *Proc. London Math. Soc.* 2. Ser. **53** (1951) 422–443.

[Wi87]   J.B. WILKER: Problem 2, in: *Intuitive Geometry* (Siófok, 1985), K. Böröczky et al., eds., *Colloq. Math. Soc. János Bolyai* **48** (1987) p. 700.

[Wi91]   J.M. WILLS: An ellipsoid packing in E^3 of unexpected high density, *Mathematika* **38** (1991) 318–320.

[Wo58]   A.C. WOODS: The critical determinant of a spherical cylinder, *J. London Math. Soc* **33** (1958) 357–368.

[Ye48]   Y. YEH: Lattice points in a cylinder over a convex domain, *J. London Math. Soc.* **23** (1948) 188–195.

1.10 Linking Packing and Covering Densities

From the point of view of both economical packing and covering, the best convex bodies are the *tiles* (or *space-fillers*), whose congruent copies can fill the whole space without gaps and (full-dimensional) overlaps. But is it true that if a convex body does not permit a very dense packing, then it cannot permit a very thin covering either, and vice versa? It appears that until very recently this natural question had completely escaped the attention of researchers.

Conjecture 1 (*W. Kuperberg [Ku88]*) *Let $d \geq 2$ be fixed. Then for any $\epsilon > 0$ there exists a $\delta > 0$ with the property that for every d-dimensional convex body C*
  (1) $\delta(C) \leq 1 - \epsilon$ *implies* $\theta(C) \geq 1 + \delta$,
  (2) $\theta(C) \geq 1 + \epsilon$ *implies* $\delta(C) \leq 1 - \delta$.

In fact, Kuperberg has also suggested that we should attempt to characterize, at least for $d = 2$, the sets Ω_d (and Ω_d^*) consisting of points $(x, y) \in \mathbb{R}^2$ such that $x = \delta(C)$ and $y = \theta(C)$ for some (centrally symmetric) convex body $C \subseteq \mathbb{R}^d$.

Problem 2 (*G. Fejes Tóth–W. Kuperberg [FeTK93]*)
  (1) *Is it true that Ω_d and Ω_d^* are convex sets?*
  (2) *Is it true that Ω_d and Ω_d^* are closed sets?*
  (3) *Is it true that $\theta(C) = \frac{4}{3}\delta(C)$ is reached only for* $\theta(C) = \theta(B^2) = \frac{2\pi}{3\sqrt{3}}$, $\delta(C) = \delta(B^2) = \frac{\pi}{2\sqrt{3}}$?

The last question is motivated by the following result, which can be deduced from a nice observation of Lázár [Lá47] (see also L. Fejes Tóth [FeT72]). For any centrally symmetric convex body C in the plane,

$$\theta(C) \leq \frac{4}{3}\delta(C).$$

This inequality was generalized to all convex plane bodies by W. Kuperberg [Ku87]. A similar inequality for the 3-dimensional case was given by Smith [Sm00], who showed that for every centrally symmetric convex body in $\mathbb{R}^3$, $\theta_L(C) \leq 4\delta_L(C)$. It seems very likely that the coefficient 4 can be replaced by a smaller one.

A first step toward a solution of the above conjecture was made by Ismailescu [Is01], who showed that for every centrally symmetric plane convex body C,

$$1 - \delta(C) \leq \theta_L(C) - 1 \leq 1.25\sqrt{1 - \delta(C)}.$$

Moreover, here the order of magnitude of the dependence of the bound on $\delta(C)$ cannot be improved.

It is a remarkable property of the Euclidean plane that there is a densest packing $\mathcal{C}$ and a thinnest covering $\mathcal{C}'$ with circles such that the sets of centers of $\mathcal{C}$ and $\mathcal{C}'$ coincide. The question arises whether the same phenomenon can occur on the sphere or in spaces of higher dimension.

The circle packing problem on the sphere (the problem of Tammes, mentioned earlier in Section 1.6) can be reformulated as follows. Find a set P_n of n points on the sphere that maximizes the minimum distance determined by them. In the corresponding covering problem we are looking for a set Q_n of n points that minimizes the maximum distance of a point on the sphere to the nearest element of Q_n. If P_n and Q_n can be chosen to be the same set, then n is called a *favorable number*, and $P_n = Q_n$ is a *favorable point set*. G. Fejes Tóth and L. Fejes Tóth [FeTF80] have shown that there are only finitely many favorable numbers, and they proposed the following conjecture.

Conjecture 3   *(G. Fejes Tóth–L. Fejes Tóth [FeTF80]): The only values for which there exist favorable n-element point sets on the sphere are $n = 2, 3, 4, 5, 6,$ and 12.*

We cannot resist mentioning the following interpretation of the above problem, which was popularized by Meschkowski [Me60]. Assume that on a certain planet n hostile dictators want to build their castles as far from one another as possible. Thinking about the future, they decide that one day they might form an alliance in order to jointly control the planet. Can they find locations for their residences optimal in both senses? A closely related question was answered in [FeT71].

Let $\mathcal{C}$ be an arrangement of convex bodies in $\mathbb{R}^d$. The *gap size* (or *closeness*, in most places in the literature) of $\mathcal{C}$ is defined as the supremum of the radii of the balls disjoint from every member of $\mathcal{C}$. Similarly, one can define the *overlap size* (or *looseness*) of $\mathcal{C}$ as the supremum of the radii of all balls that belong to the intersection of two members of $\mathcal{C}$. These notions were introduced independently by Ryškov [Ry70] and by L. Fejes Tóth [FeT76], [FeT78].

Böröczky [Bö86] settled a conjecture of L. Fejes Tóth by showing that the minimum gap size of a packing of unit balls in $\mathbb{R}^3$ is $\sqrt{5/3} - 1 = 0.29099\ldots$, and in the uniquely determined extremal configuration the centers of the balls are placed at the vertices and at the centers of all elementary cubes of a cubic lattice of side length $4/\sqrt{3}$. Of course, this also gives a tight bound for the minimum overlap size of a covering with equal balls. It is easy to deduce from Böröczky's theorem that there is no favorable point set in $\mathbb{R}^3$.

Problem 4   *Let $d \geq 4$. Decide whether there exists a favorable point set in $\mathbb{R}^d$, i.e., whether there exist a densest packing with* *equal balls and a thinnest covering with equal balls whose sets of centers are the same.*

Using the techniques of Delone and Ryškov [DeR63] and Baranovskiĭ and Ryškov [BaR75], [BaR76], [BaR79], which were developed for the solution of the problem of thinnest lattice covering, Horváth [Ho80] determined the minimum gap size in a lattice packing of unit balls in $\mathbb{R}^4$ and $\mathbb{R}^5$.

Problem 5   *Let C be a centrally symmetric convex body in the plane. Is it true that the overlap size of any covering of the plane with translates of C is at least as large as the minimum overlap size of a lattice covering with C?*

The corresponding statement for *lattice packings* was established by L. Fejes Tóth [FeT78], who has also exhibited some examples of centrally symmetric convex bodies C in the plane with the property that the gap size of a packing with congruent copies (not necessarily translates) of C is minimized by a nonlattice packing. (See also [Be80].)

Linhart [Li78] studied some closely related questions that can be regarded as variants of the above problem for other metrics. Given a packing $\mathcal{C}$ of congruent copies of a convex body that contains the origin 0, let $\rho(\mathcal{C})$ denote the smallest number ρ such that by enlarging every member of $\mathcal{C}$ to $1 + \rho$ times its original size from its point corresponding to 0, we obtain a covering of the space. Furthermore, let $\rho_T(C)$ (and $\rho_L(C)$) denote the minimum of $\rho(\mathcal{C})$ over all translative packings (respectively lattice packings) of C. Clearly, for balls, $\rho_T(B^d) \leq \rho_L(B^d)$ is the the radius of the largest ball that can be placed in the complement of every packing of unit balls. Linhart proved that $\rho_T(C) \leq 1/2$ for every C in the plane, with equality if and only if C is a triangle.

For centrally symmetric C, we obviously have that $\rho_T(C) \leq 1$. Rogers [Ro50] and Butler [Bu72] proved respectively that $\rho_L(C) < 2$ and that $\rho_L(C) \leq 1 + o(1)$ as the dimension d tends to infinity. (See also [Ba90].) Zong [Zo02a], [Zo03] showed that $\rho_T(C) = \rho_L(C) \leq 0.2$ for every centrally symmetric C in the plane and that $\rho_L(C) \leq 3/4$ in three-dimensional space. It would be interesting to determine the *largest* values of $\rho_T(C)$ and $\rho_L(C)$. In the plane these two values coincide. The best known example there is the regular octagon, for which this number is $\frac{2-\sqrt{2}}{2+\sqrt{2}} \approx 0.17157$. As we have mentioned before, for unit balls B^d, it is known that $\rho_T(B^2) = \rho_L(B^2) = 2/\sqrt{3} - 1$ and $\rho_T(B^3) = \rho_L(B^3) = \sqrt{5/3} - 1$ [FeT76], [Bö86]. Adapting Rogers's bound, Henk [He95] proved the inequality $\rho_L(B^d) \leq \sqrt{21}/2 - 1$ for every d.

Problem 6   *Does there exist any centrally symmetric convex body C such that $\rho_L(C) \geq 1$?*

Many believe that the answer to the last question is yes. That is, there exists a higher-dimensional centrally symmetric convex C such that to each of its lattice packings one can add an extra translate of C that has no interior point in common with any other element. Of course, if this is true, then for this body the maximum density of a translative packing is not attained for a lattice packing. As pointed out by Zong [Zo02b], [Zo03], no such body exists in $\mathbb{R}^d$ for $d \leq 3$.

[Ba90]   W. Banaszczyk: On the lattice packing–covering ratio of finite-dimensional normed spaces, *Colloq. Math.* **59** (1990), no. 1, 31–33.

[BaR79]   E.P. Baranovskiĭ, S.S. Ryškov: Classical methods in the theory of lattice packings (in Russian), *Uspehi Mat. Nauk* **34** (1979) 3–63. English translation: *Russian Math. Surveys* **34** (1979) 1–68.

[BaR76]   E.P. Baranovskiĭ, S.S. Ryškov: C-types of n-dimensional lattices and five-dimensional primitive parallelohedra (with an application to coverings theory) (in Russian), *Dokl. Akad. Nauk SSSR Trudy Otdel. Leningrad Mat. Inst. Steklova* **137**, 131 pp. Izdat. Nauka, Moscow, 1976. English translation: *Proc. Steklov Inst. Math.* **137** (1978).

[BaR75]   E.P. Baranovskiĭ, S.S. Ryškov: Solution of the problem of the least dense lattice covering of five-dimensional space by equal spheres (in Russian), *Dokl. Akad. Nauk SSSR* **222** (1975) 39–42. English translation: *Soviet Math. Dokl.* **16** (1975) 586–590.

[Be80]   A. Bezdek: Remark on the closest packing of convex discs, *Studia Sci. Math. Hungar.* **15** (1980) 283–285.

[Bö86]   K. Böröczky: Closest packing and loosest covering of the space with balls, *Studia Sci. Math. Hungar.* **21** (1986) 79–89.

[Bu72]   G.J. Butler: Simultaneous packing and covering in Euclidean space, *Proc. London Math. Soc.* 3. Ser. **25** (1972) 721–735.

[DeR63]   B.N. Delone, S.S. Ryškov: Solution of a problem on the least dense lattice covering of a 4-dimensional space by equal spheres (in Russian), *Dokl. Akad. Nauk SSSR* **152** (1963) 523–524. English translation: *Soviet Math. Dokl.* **4** (1963) 1333–1334.

[FeTF80]   G. Fejes Tóth, L. Fejes Tóth: Dictators on a planet, *Studia Sci. Math. Hungar.* **15** (1980) 313–316.

[FeTK93] G. FEJES TÓTH, W. KUPERBERG: A survey of recent results in the theory of packings and coverings, in: *New Trends in Discrete and Computational Geometry*, J. Pach, ed., *Algorithms Comb. Ser.* **10** Springer-Verlag 1993, 251–279.

[FeT78] L. FEJES TÓTH: Remarks on the closest packing of convex discs, *Comment. Math. Helv.* **53** 536–541.

[FeT76] L. FEJES TÓTH: Close packing and loose covering with balls, *Publ. Math. Debrecen* **23** (1976) 323–326.

[FeT72] L. FEJES TÓTH: *Lagerungen in der Ebene, auf der Kugel und im Raum* (2. Aufl.), Springer-Verlag 1972.

[FeT71] L. FEJES TÓTH: Perfect distribution of points on a sphere, *Period. Math. Hungar.* **1** (1971) 25–33.

[He95] M. HENK: *Finite and Infinite Packings*, Habilitationsschrift, Universität Siegen, 1995.

[Ho89] J. HORVÁTH: *Several Problems of n-Dimensional Discrete Geometry* (in Russian), Doctoral dissertation, Hungar. Acad. Sci., Budapest, 1989.

[Ho80] J. HORVÁTH: Narrow lattice packing of unit balls in the space E^n (in Russian), in: *Geometry of Positive Quadratic Forms'*, *Trudy Mat. Inst. Steklova* **152** (1980) 216–231. English translation: *Proc. Steklov Math. Inst.* **152** (1980) 237–254.

[Is01] D. ISMAILESCU: Inequalities between the lattice packing and lattice covering densities of a plane centrally symmetric convex body, *Discrete Comput. Geom.* **25** (2001) 365–388.

[Ku88] W. KUPERBERG: Personal communication, 1988.

[Ku87] W. KUPERBERG: An inequality linking packing and covering densities of plane convex bodies, *Geometriae Dedicata* **23** (1987) 59–66.

[Lá47] D. LÁZÁR: Sur l'approximation des courbes convexes par des polygones, *Acta Univ. Szeged Sect. Sci. Math.* **11** (1947) 129–132.

[Li78] J. LINHART: Closest packings and closest coverings by translates of a convex disc, *Studia Math. Hungar.* **13** (1978) 157–162.

[Me60] H. MESCHKOWSKI: *Ungelöste und unlösbare Probleme der Geometrie*, Vieweg-Verlag, 1960.

[Ro50] C.A. ROGERS: A note on coverings and packings, *J. London Math. Soc.* **25** (1950) 327–331.

[**Ry70**]   S.S. RYŠKOV: The polyhedron $\mu(m)$ and certain extremal problems of the geometry of numbers (in Russian), *Dokl. Akad. Nauk SSSR* **194** (1970) 514–517.

[**Sm00**]   E.H. SMITH: A bound on the ratio between the packing and covering densities of a convex body, *Discrete Comput. Geom.* **23** (2000) 325–331.

[**Zo03**]   C. ZONG: Simultaneous packing and covering in three-dimensional Euclidean space, *J. London Math. Soc.* 2. Ser. **67** (2003) 29–40.

[**Zo02a**]   C. ZONG: Simultaneous packing and covering in the Euclidean plane, *Monatshefte Math.* **134** (2002) 247–255.

[**Zo02b**]   C. ZONG: From deep holes to free planes, *Bull. Amer. Math. Soc.* (New Ser.) **39** (2002) 533–555.

1.11 Sausage Problems and Catastrophes

Many years ago, in the problem column of *Periodica Mathematica Hungarica*, L. Fejes Tóth [FeT75] posed the following innocent-looking question. How should we arrange n nonoverlapping unit balls in $\mathbb{R}^d$ so as to minimize the volume of the convex hull of their union? The question did not find geometers completely unprepared. Finite packings and coverings had been extensively studied before, especially in the plane. In particular, L. Fejes Tóth [FeT49] had shown that if $\mathcal{C}_n$ is a packing of $n \geq 2$ unit circles, then the area of the convex hull of their union satisfies

$$\text{area}\left(\text{conv}\left(\bigcup \mathcal{C}_n\right)\right) > \sqrt{12}n.$$

Groemer [Gr60] and Oler [Ol61] independently improved this bound to

$$\text{area}\left(\text{conv}\left(\bigcup \mathcal{C}_n\right)\right) \geq \sqrt{12}n + \frac{2-\sqrt{3}}{2}\text{peri}\left(\text{conv}\left(\bigcup \mathcal{C}_n\right)\right) + \sqrt{3}(\pi-2) - \pi,$$

which is already tight. (Here peri stands for the perimeter.) For simpler proofs and some generalizations consult Folkman and Graham [FoG69], Graham, Witsenhausen, and Zassenhaus [GrWZ72], Molnár [Mo79], Schürmann [Schü00]. Using this result and some deeper observations, Wegner [We86] showed that for a very large infinite class of integers n (the smallest possible exception is 121), the best packings of n unit circles minimizing the area of their convex hulls are hexagonal pieces of the densest circle packing. Böröczky Jr. and Ruzsa [BöR05] have shown that the proportion of possible exceptional values of n is smaller than 5%. It is conjectured that Wegner's construction is optimal for every n.

In $\mathbb{R}^3$ (and $\mathbb{R}^4$) the situation is somewhat different. It seems that the optimal configurations up to 56 balls (respectively several thousand balls) are the so called *sausages*, i.e., strings of consecutively touching unit balls whose centers are collinear. Increasing the number of balls, Wills [Wi83b] discovered a strange and rapidly progressing phenomenon that is often referred to as the *sausage catastrophe*: the optimal configurations start to form dense clusters [GaW92], [Scho00], [GaZ92]. We have no chance of determining their exact shapes without finding a transparent solution of Kepler's conjecture. In crystallography, the asymptotic shape of the extreme configurations, as the number of balls tends to infinity, is called the *Wulff shape*. See Wulff [Wu901] and Wills [Wi83b], [Wi96], [Wi97], [Wi98], [Wi00]. Unfortunately, even for small values of n, we have to deal with optimization problems of more variables than we can handle. The first open questions are the following.

Problem 1   *(K. Böröczky Jr. [Bö93], Betke–Gritzmann [BeG84]) Does the volume of the convex hull of the union of five nonoverlapping unit balls in $\mathbb{R}^3$ (and $\mathbb{R}^4$) attain its minimum for a sausage?*

Despite the uncertainty around the fate of Kepler's conjecture, it may be possible to answer the following question.

Problem 2   *Is it true that*

$$\frac{n \, \mathrm{Vol}(B^3)}{\mathrm{Vol}\left(\mathrm{conv}\left(\bigcup \mathcal{C}_n\right)\right)} < \delta(B^3)$$

holds for any packing $\mathcal{C}_n$ of $n > 3$ unit balls in $\mathbb{R}^3$?

In order to understand the sudden phase transition called "sausage catastrophe," we have to establish some structural properties of the extremal configurations similar to the previously mentioned result of Wegner [We86] for the planar case. A first step may be to verify the following conjecture.

Conjecture 3   *Let $\mathcal{C}_n$ be a packing of n unit balls in $\mathbb{R}^3$ for which the volume of the set $U_n = \mathrm{conv}\left(\bigcup \mathcal{C}_n\right)$ is minimal. Then the inradius of U_n tends to infinity as $n \to \infty$.*

For dimensions $d \geq 5$, L. Fejes Tóth made his famous *Sausage conjecture*, stating that in $\mathbb{R}^d$ no catastrophe occurs: the sausage is always best.

Conjecture 4   *(L. Fejes Tóth [FeT75]) Let n be a fixed positive integer, $d \geq 5$. The volume of the convex hull of any system of n nonoverlapping unit balls in $\mathbb{R}^d$ is at least*

$$2(n - 1) \, \mathrm{Vol}(B^{d-1}) + \mathrm{Vol}(B^d).$$

Equality can hold only if the centers of the balls are equally spaced on a line at distance 2.

Improving some preliminary results in [BeGW82], [KlPW84], Betke and Gritzmann [BeG84] succeeded in proving the conjecture in the special case in which the dimension d' of the smallest flat (affine subspace) containing the centers of the balls is at most $\frac{7}{12}(d-1)$, or when $d' \leq \min(9, d-1)$. Shortly thereafter, Gritzmann [Gr86] showed that the sausage conjecture is true up to a constant factor. More precisely, he proved that the volume of the convex hull of any packing of n balls in $\mathbb{R}^d$ is at least

$$\frac{2n \, \mathrm{Vol}(B^{d-1}) + \mathrm{Vol}(B^d)}{2 + \sqrt{2} + 2/\sqrt{d-1}}.$$

The real breakthrough was achieved by Betke, Henk, and Wills [BeHW94], who proved the conjecture for $d \geq 13387$. Later, Betke and Henk [BeH98] extended the argument to all $d \geq 42$. Despite considerable efforts to settle this problem, a big "slice" of the sausage conjecture has remained unsolved: we do not know what happens if $5 \leq d \leq 41$.

Wills [Wi82] formulated his so-called Sausage-skin conjecture, which is intimately related to L. Fejes Tóth's problem.

Conjecture 5 *(Wills [Wi82]) Let n be a fixed positive integer and $d \geq 7$. The surface area of the convex hull of any packing of n unit balls in $\mathbb{R}^d$ whose centers lie in a hyperplane is at least $2(n-1)\operatorname{Surf}(B^{d-1}) + \operatorname{Surf}(B^d)$. Equality can hold only if the centers of the balls are equally spaced on a line at distance 2. ($\operatorname{Surf}(\cdot)$ denotes the surface area.)*

Notice that the assumption that the centers of the balls do not span the whole space is necessary, because the statement is false for large clusters.

The analogous problems for coverings are also interesting. Wills posed the following question [Wi83a]. What is the maximum volume of a d-dimensional convex body that can be covered by n unit balls? Although in the plane L. Fejes Tóth [FeT49] (see also [FeT87]) essentially answered this question in the asymptotic sense, some attractive problems remain unsolved. It is not hard to show (see, e.g., [Bö04]) that there exist absolute constants $c, c' > 0$ such that the inradius (circumradius) of every plane convex body K of maximum area that can be covered by n unit circles is at least $c\sqrt{n}$ (respectively at most $c'\sqrt{n}$). It is most likely that the shape of K must be close to a regular hexagon.

Problem 6 *Let $n = 6\binom{k+1}{2} + 1$ for some positive integer k, and consider an arrangement of n unit circles (disks) that cover a plane convex body of maximum area. Is it true that the convex hull of the centers of the circles is a regular hexagon of side length $\sqrt{3}k$?*

The answer is not known even for $k = 1$.

In higher dimensions, it seems that the following special sausage-like arrangement may be optimal. Let $\mathcal{C}_n^d$ denote a system of n unit balls in $\mathbb{R}^d$ whose centers lie equally spaced along a line so that the distance between two adjacent centers is $2/\sqrt{d}$.

Conjecture 7 *(Wills [Wi83a]) Let n be a fixed positive integer and $d \geq 5$. If C is a d-dimensional convex body covered by an arrangement $\mathcal{C}$ of n unit balls, then the volume of C cannot exceed the volume of the largest convex body covered by $\mathcal{C}_n^d$. Moreover, equality can hold only if $\mathcal{C} = \mathcal{C}_n^d$.*

One of the inherent difficulties is that in general, it is hard to determine the largest convex body covered by a given arrangement.

For a convex body $C \subseteq \mathbb{R}^d$ and $0 \leq j < d$, let us define the j-skeleton of C as the set of all points of C that do not lie in the interior of any $(j+1)$-dimensional convex subset of C. The set $C + rB^d$ of all points of $\mathbb{R}^d$ whose distance from C is at most r is called the *parallel body* of C with radius r. It is well known (Hadwiger [Ha57], McMullen [McM75]) that $\mathrm{Vol}(C + rB^d)$ is a polynomial of degree d in r:

$$\mathrm{Vol}(C + rB^d) = \sum_{j=0}^{d} (\mathrm{Vol}(B^{d-j}))(\mathrm{Vol}_j\, C) r^{d-j}.$$

The coefficients $\mathrm{Vol}_j\, C$ are called the *intrinsic volumes* of C. Obviously, $\mathrm{Vol}_0\, C = 1$, $\mathrm{Vol}_{d-1}\, C$ is half of the surface area of C, and $\mathrm{Vol}_d\, C = \mathrm{Vol}(C)$.

Now we are in a position to state the dual counterpart of the sausage-skin conjecture in a more general form.

Conjecture 8   *(G. Fejes Tóth–Gritzmann–Wills [FeT*84]) Let $n \geq 2$ and $1 \leq j < d$ be fixed integers. If C is a d-dimensional convex body of maximum intrinsic volume $\mathrm{Vol}_j\, C$ such that the j-dimensional skeleton of C can be covered by an arrangement $\mathcal{C}$ of n unit balls, then the centers of these balls lie equally spaced on a line segment of suitable length.*

For $d = 2$, $j = 1$, G. Fejes Tóth, Gritzmann, and Wills [FeT*84] settled this conjecture by proving that the maximum of the perimeters of all convex bodies in the plane whose boundary can be covered by n unit circles is

$$4\left(\sqrt{n^2 - 1} + \arcsin\frac{1}{n}\right).$$

In the uniquely determined optimal arrangement, the centers of the circles are equally spaced at the distance $2\sqrt{1 - \frac{1}{n^2}}$ on a line.

For more partial results supporting the above conjectures, consult G. Fejes Tóth, Gritzmann, and Wills [FeT*89] and the survey articles of Gritzmann and Wills [GrW86], [GrW93].

Instead of venturing further into higher dimensions, we close this section with three challenging problems that are open even in the plane. Bateman and Erdős [BaE51] raised the following question. How should we arrange n nonoverlapping unit balls in $\mathbb{R}^d$ so as to minimize the *diameter* of the convex hull of their union? In the plane, for $n \leq 9$ the problem was solved in [BaE51] and [BeF99]. On the other hand, as n tends to infinity, by density considerations, the convex hull of the unit circles in an optimal arrangement must not be far from a circular disk of area $\sqrt{12}n$. This means that the optimal arrangements must resemble a circular piece of the densest hexagonal packing of unit disks. However, as was proved by Schürmann [Schü02], if n is sufficiently large, none of the optimal arrangements can be completed to a perfect hexagonal packing. Nevertheless, Erdős (personal communication) conjectured that almost all circles in such an arrangement touch 6 others. We state the simplest unsolved question.

Problem 9   *(Erdős) Is it true that every packing C_n of n unit disks in the plane for which the diameter of conv $(\bigcup C_n)$ is as small as possible has three mutually touching members if n is sufficiently large?*

It may be interesting to study similar questions in other normed spaces, e.g., in a Minkowski plane whose unit disk is some centrally symmetric plane convex body. Brass and Swanepoel have made the following conjecture.

Problem 10   *(Brass, Swanepoel) Is it true that in every Minkowski plane the diameter of the convex hull of any packing of seven unit disks is at least six?*

Equivalently, is it true that for any set of seven points in a Minkowski plane, the largest distance divided by the shortest distance between them is at least two?

Many of the above questions make sense and have been studied for packings and coverings with not necessarily spherical objects.

Problem 11   *(K. Böröczky Jr. [Bö03]) Is it true that for any strictly convex body D in the plane, there exists a $c(D) > 0$ such that for all n the convex body K_n of minimum area that contains n nonoverlapping congruent copies of D has inradius$(K_n) \geq c(D)\sqrt{n}$.*

The same question can be asked about the inradius of a convex body of maximum area that can be covered by n congruent copies of D. Some partial results can be found in [Bö03], [Bö04]. In particular, the answer to the problem is known to be affirmative if D is centrally symmetric or if only translative packings are allowed.

[BaE51]   P. BATEMAN, P. ERDŐS: Geometrical extrema suggested by a lemma of Besicovitch, *Amer. Math. Monthly* **58** (1951) 306–314.

[BeGW82] U. BETKE, P. GRITZMANN, J.M. WILLS: Slices of L. Fejes Tóth's sausage conjecture, *Mathematika* **29** (1982) 194–201.

[BeG84]   U. BETKE, P. GRITZMANN: Über L. Fejes Tóth's Wurstvermutung in kleinen Dimensionen, *Acta Math. Hungar.* **43**

(1984) 299–307.

[BeH98]   U. BETKE, M. HENK: Finite packings of spheres, *Discrete Comput. Geom.* **19** (1998) 197–227.

[BeHW94]   U. BETKE, M. HENK, J.M. WILLS: Finite and infinite packings, *J. Reine Angew. Math.* **453** (1994) 165–191.

[BeF99]   A. BEZDEK, F. FODOR: Minimal diameter of certain sets in the plane, *J. Combinatorial Theory Ser. A* **85** (1999) 105–111.

[Bö04]   K. BÖRÖCZKY JR.:   *Finite Packing and Covering*, Cambridge University Press, 2004

[Bö03]   K. BÖRÖCZKY JR.:   Finite packing and covering by congruent convex domains, *Discrete Comput. Geom.* **30** (2003) 185–193.

[Bö93]   K. BÖRÖCZKY JR.:   About four-ball packings, *Mathematika* **40** (1993) 226–232.

[BöR05]   K. BÖRÖCZKY JR., I.Z. RUZSA: Note on an inequality of Wegner, manuscript.

[FeT87]   G. FEJES TÓTH: Finite coverings by translates of centrally symmetric convex domains, *Discrete Comput. Geom.* **2** (1987) 353–363.

[FeT*89]   G. FEJES TÓTH, P. GRITZMANN, J.M. WILLS: Finite sphere packing and covering, *Discrete Comput. Geom.* **4** (1989) 19–40.

[FeT*84]   G. FEJES TÓTH, P. GRITZMANN, J.M. WILLS: Sausage-skin problems for finite coverings, *Mathematika* **31** (1984) 117–136.

[FeT75]   L. FEJES TÓTH: Research problem 13, *Periodica Math. Hungar.* **6** (1975) 197–199.

[FeT49]   L. FEJES TÓTH: Über die dichteste Kreislagerung und dünnste Kreisüberdeckung, *Comment. Math. Helv.* **23** (1949) 342–349.

[FoG69]   J.H. FOLKMAN, R.L. GRAHAM: A packing inequality for compact convex subsets of the plane, *Canad. Math. Bull.* **12** (1969) 745–752.

[GaW92]   P.M. GANDINI, J.M. WILLS: On finite sphere-packings, *Math. Pannon.* **3** (1992) 19–29.

[GaZ92]   P.M. GANDINI, A. ZUCCO: On the sausage catastrophe in 4-space, *Mathematika* **39** (1992) 274–278.

[GrWZ72]   R.L. GRAHAM, H.S. WITSENHAUSEN, H. ZASSENHAUS: On tightest packings on the Minkowski plane, *Pacific J. Math.* **41** (1972) 699–715.

[Gr86]   P. Gritzmann: Finite packing of equal balls, *J. London Math. Soc.* 2. Series **33** (1986) 543–553.

[GrW93]   P. Gritzmann, J.M. Wills: Finite packing and covering, in: *Handbook of Convex Geometry*, P.M. Gruber et al., eds., Elsevier 1993, 861–897.

[GrW86]   P. Gritzmann, J.M. Wills: Finite packing and covering, *Studia Sci. Math. Hungar.* **21** (1986) 149–162.

[Gr60]   H. Groemer: Über die Einlagerung von Kreisen in einen konvexen Bereich, *Math. Zeitschrift* **73** (1960) 285–294.

[Ha57]   H. Hadwiger: *Vorlesungen über Inhalt, Oberfläche und Isoperimetrie*, Springer-Verlag 1970.

[KlPW84] P. Kleinschmidt, U. Pachner, J.M. Wills: On L. Fejes Tóth's sausage conjecture, *Israel J. Math.* **47** (1984) 216–226.

[McM75]   P. McMullen: Non-linear angle-sum relations for polyhedral cones and polytopes, *Proc. Cambridge Phil. Soc.* **78** (1975) 247–261.

[Mo79]   J. Molnár: On the packing of unit circles in a convex domain, *Ann. Univ. Sci. Budapest Eőtvős, Sect. Math.* **22** (1979) 113–123.

[Ol61]   N. Oler: An inequality in the geometry of numbers, *Acta Math.* **105** (1961) 19–48.

[Scho00] P. Scholl: *Finite Sphere Packings* (Diploma Thesis, in German), Universität Siegen, 2000.

[Schü02] A. Schürmann: On extremal finite packings, *Discrete Comput. Geom.* **28** (2002) 389–403.

[Schü00] A. Schürmann: On parametric density of finite circle packings, *Beiträge Algebra Geom.* **41** (2000) 329–334.

[We86]   G. Wegner: Über endliche Kreispackungen in der Ebene, *Studia. Sci. Math. Hungar.* **21** (1986) 1–28.

[Wi00]   J.M. Wills: The Wulff-shape of large periodic sphere packings, in: *Discrete Mathematical Chemistry*, Amer. Math. Soc., *DIMACS Ser. Discrete Math. Theoret. Comput. Sci.* **51** (2000) 367–375.

[Wi98]   J.M. Wills: Spheres and sausages, crystals and catastrophes – and a joint packing theory, *Math. Intelligencer* **20** (1998) 16–21.

[Wi97]   J.M. Wills: On large lattice packings of spheres, *Geometriae Dedicata* **65** (1997) 117–126.

[**Wi96**]   J.M. WILLS: Lattice packings of spheres and the Wulff-shape, *Mathematika* **43** (1996) 229–236.

[**Wi83a**]   J.M. WILLS: Research Problem 33, *Periodica Math. Hungar.* **14** (1983) 189–191.

[**Wi83b**]   J.M. WILLS: Research Problem 35, *Periodica Math. Hungar.* **14** (1983) 312–314.

[**Wi82**]   J.M. WILLS: Research Problem 30, *Periodica Math. Hungar.* **13** (1982) 75–76.

[**Wu901**]   G. WULFF: Zur Frage der Geschwindigkeit des Wachstums und der Auflösung der Krystallflächen, *Zeitschrift Krystallographie Mineral.* **34** (1901) 499.

2. Structural Packing and Covering Problems

2.1 Decomposition of Multiple Packings and Coverings

Let k be a natural number and $D \subseteq \mathbb{R}^d$ an arbitrary domain. An arrangement $\mathcal{C}$ of convex bodies in $\mathbb{R}^d$ is said to form a k-fold covering of D if every point of D belongs to at least k members of $\mathcal{C}$, and it is called a k-fold packing if every point belongs to the interior of at most k members of $\mathcal{C}$. Thus, the usual coverings and packings are the same as the 1-fold (or *simple*) coverings and packings, respectively. The survey paper of G. Fejes Tóth [FeT83] gives an almost complete account of the known results about the thinnest k-fold coverings and densest k-fold packings of $\mathbb{R}^d$ with congruent copies of a convex body $\mathcal{C}$. These questions are usually quite difficult, and they often require somewhat technical extensions of the methods developed for the case $k = 1$ (see, e.g., Few [Fe64], [Fe67], [Ga96], [Bl99].) G. Fejes Tóth [FeT76], [FeT77], [FeT79] generalized the concept of Voronoi–Dirichlet cell decomposition to show that the densities $\delta^k(B^2)$ and $\theta^k(B^2)$ of the densest k-fold packing and thinnest k-fold covering of the plane with unit circles (disks) satisfy

$$\delta^k(B^2) \leq \frac{\pi}{6} \cot \frac{\pi}{6k} < k - \frac{\pi^2}{108k},$$

$$\theta^k(B^2) \geq \frac{\pi}{3} \csc \frac{\pi}{3k} > k + \frac{\pi^2}{54k},$$

for every k. Improving some earlier results of Cohn [Co76] and Groemer [Gr86], Bolle [Bo84], [Bo89] and Huxley [Hu93] showed that for the corresponding densities $\delta_L^k(B^2)$ and $\theta_L^k(B^2)$, restricted to *lattice* packings and coverings, respectively, we have

$$k - ck^{\frac{23}{73+\varepsilon}} \leq \delta_L^k(B^2) \leq k - c'k^{\frac{1}{4}},$$

$$k + c'k^{\frac{1}{4}} \leq \theta_L^k(B^2) \leq k + ck^{\frac{23}{73+\varepsilon}},$$

for any $\varepsilon > 0$ and for suitable $c, c' > 0$ depending only on ε. Moreover, the order of magnitude of the error term of the upper (lower) estimate in the first (respectively second) inequality cannot be improved for an infinite sequence of integers. Many of the proofs of these bounds require deep analytical tools developed by Hardy, Littlewood, Bombieri, Iwaniec, and others.

A structural approach to density problems for multiple covering and packing is suggested by the following interesting observation of Bielecki

[Bi48] and Rado [Ra48]: Any k-fold covering (packing) of the line with intervals can be decomposed into k simple coverings (packings). Therefore, if we wish to determine the minimum (maximum) density $\theta_L^k(C)$ $(\delta_L^k(C))$ of a k-fold lattice covering (packing) of $\mathbb{R}^d$ with a convex body C, then we can attempt to decompose it into as many (as few) simple coverings (packings) as possible, and then to apply the known bounds for the densities of the simple arrangements. This approach was initiated by Pach [Pa80]. Using this method, G. Fejes Tóth [FeT84] showed that $\delta_L^3(C) = 3\delta_L(C)$ for every centrally symmetric plane convex body C, because any threefold lattice packing of the plane with C is the union of three simple lattice packings. Furthermore, every fourfold lattice packing of the plane with C splits into two twofold lattice packings. In view of the equation $\delta_L^2(C) = 2\delta_L(C)$, due to Dumir and Hans-Gill [DuH72b], this implies that $\delta_L^4(C) = 4\delta_L(C)$. When $C = B^2$ is the circle, these relations had been established earlier, using different methods, by Heppes [He59]. It is interesting to note that, although we have $\delta_L^2(B^2) = 2\delta_L(B^2)$, it is *not* true that every twofold lattice packing with unit circles can be decomposed into two simple packings. However, Schmitz [Sch91] proved that if a twofold circle packing can be obtained as the union of two lattice arrangements, then it can always be decomposed into *three* simple packings. Some special decomposition results can be found in [Sch92] and [Te94a]. (See Blundon [Bl63], [Bl64], Yakovlev [Ya83], Temesvári [Te94b] for precise values of $\delta_L^k(B^2)$ when $5 \leq k \leq 9$; and see [DuH72a] for a proof of the equation $\theta_L^2(C) = 2\theta_L(C)$ when C is centrally symmetric, and [Li83], [Ya84], [TeHY87] for algorithms to compute the maximum density of a k-fold lattice packing in the plane by equal circles.)

There are two difficulties with the above approach. First, no decomposition results are known for k-fold lattice arrangements when $k > 4$. In fact, we have $\delta_L^k(B^d) > k\delta_L(B^d)$, unless $d = 2$ and $k \leq 4$ [FeTF75], [Yan83]. Second, the problem becomes essentially different if we do not restrict ourselves to lattice arrangements. It is known [He55] that $\delta^k(B^2) > k\delta(B^2)$ for every $k > 1$; nevertheless, we cannot even solve the following simple problem.

Problem 1   *Determine $\delta^2(B^2)$ and $\theta^2(B^2)$, the densities of the densest double packing and the thinnest double covering of the plane with equal circles.*

The best known constructions are due to Heppes [He55] and Danzer [Da60]. Horváth and Temesvári [HoT83] proved that the configuration of Heppes depicted in the figure is optimal among double packings that can be obtained as the union of two lattice packings.

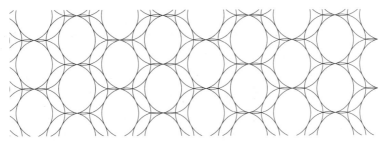

THE DENSEST KNOWN DOUBLE PACKING BY CIRCLES

Conjecture 2   *(Pach [Pa80]) For any centrally symmetric plane convex body C and any positive integer r, there exists an integer $k = k(C, r)$ such that every k-fold covering of the plane with translates of C can be decomposed into r coverings.*

Pach [Pa86] proved this conjecture in the special case of C a centrally symmetric polygon (e.g., a square). Unfortunately, his argument cannot be extended to the general case by a compactness argument, because the constant $k(C, r)$ yielded by the proof tends to infinity as the number of sides of C increases. Mani-Levitska and Pach [MaP86] showed that $k(B^2, 2) \leq 33$; i.e., every 33-fold covering of the plane with unit circles can be decomposed into two coverings.

An apparent difficulty with this problem is that due to the complicated combinatorial structure of the intersection patterns, it is very hard to establish nontrivial lower bounds for $k(C, r)$. As a matter of fact, lacking good constructions, we are unable to exclude the possibility that the answer to the following much stronger question is also yes.

Problem 3   *[Pa86] Is it true that for any plane convex body C and for any integer r, there exists an integer $k' = k'(C, r)$ such that every k'-fold covering of the plane with homothetic copies of C can be decomposed into r coverings?*

In particular, we do not know whether $k'(B^2, 2)$ exists, i.e., whether every sufficiently "thick" covering of the plane with not necessarily equal circles splits into two coverings.

It is not too difficult to show that there exists no k' such that any k'-fold covering of the plane with strips of unit width or with axis-parallel rectangles can be decomposed into two coverings. On the other hand, there exists such an integer for half-planes (or, in higher dimensions, half-spaces) rather than strips. (Recall that a strip is the set of all points lying between two parallel lines.)

Perhaps it will be not too difficult to settle the following related problem.

Problem 4   *(Pach) Does there exist an integer k such that any k-fold covering of the plane with axis-parallel squares can be decomposed into 2 coverings?*

The next question is "dual" to the previous one in the sense that the roles of points and rectangles are interchanged.

Problem 5   *(Pach) Does there exist an integer k with the following property: any finite set of points in the plane can be colored by two colors so that every axis-parallel rectangle that covers at least k points contains points of both colors?*

Surprisingly, Mani-Levitska and Pach [MaP86] proved that for every k there exists a k-fold covering of $\mathbb{R}^3$ with equal balls that cannot be split into two coverings, and a similar theorem is true in higher dimensions.

The corresponding questions for k-fold packings are much easier. For a convex body $C \subseteq \mathbb{R}^d$, let $\ell(C) = (R(C))^d \operatorname{Vol}(B^d)/\operatorname{Vol}(C)$, where $R(C)$ denotes the circumradius of C. Pach [Pa80] observed that if $\mathcal{C}$ is a k-fold packing of (not necessarily congruent) convex bodies such that $\ell(C) \leq L$ for every $C \in \mathcal{C}$, then $\mathcal{C}$ can be decomposed into $3^d Lk$ or fewer simple packings. Obviously, the condition on $\ell(C)$ cannot be dropped, because we can choose a twofold packing of n long and thin pairwise crossing rectangles in the plane, and their arrangement cannot be decomposed into fewer than n packings. An amusing fact, which is an easy corollary to the four color theorem, is that every twofold packing of the plane with pairwise noncrossing convex bodies can be split into four packings. (Two convex bodies, C_1 and C_2, are said to *cross* if both $C_1 \setminus C_2$ and $C_2 \setminus C_1$ are disconnected.)

[Bi48]   A. Bielecki: Problem 56, *Colloq. Math.* **1** (1948) 333–334.

[Bl99]   V.M. Blinovsky: Multiple packing of the Euclidean sphere, *IEEE Trans. Inform. Theory* **45** (1999) 1334–1337.

[Bl64]   W.J. Blundon: Some lower bounds for the density of multiple packing, *Canad. Math. Bull.* **7** (1964) 565–572.

[Bl63]   W.J. Blundon: Multiple packing of circles in the plane, *J. London Math. Soc.* **38** (1963) 176–182.

[Bo89]   U. Bolle: On the density of multiple packings and coverings of convex discs, *Studia Sci. Math. Hung.* **24** (1989) 119–126.

[Bo84]   U. Bolle: Über die Dichte mehrfacher gitterförmiger Kreisanordnungen in der Ebene, *Studia Sci. Math. Hung.* **19** (1984) 275–284.

[Co76]   M. Cohn: Multiple lattice covering of space, *Proc. London Math. Soc.* 3. Ser. **32** (1976) 117–132.

[Da60]   L. Danzer: Drei Beispiele zu Lagerungsproblemen, *Arch. Math.* **11** (1960) 159–165.

[DuH72b]   V.C. Dumir, R.J. Hans-Gill: Lattice double coverings in the plane, *Indian J. Pure Appl. Math.* **3** (1972) 466–480.

[DuH72a]   V.C. Dumir, R.J. Hans-Gill: Lattice double packings in the plane, *Indian J. Pure Appl. Math.* **3** (1972) 481–487.

[FeT84]   G. Fejes Tóth: Multiple lattice packings of symmetric convex domains in the plane, *J. London Math. Soc.* 2. Ser. **29** (1984) 556–561.

[FeT83]   G. Fejes Tóth: New results in the theory of packing and covering, in: *Convexity and its Applications*, P.M. Gruber et al., eds., Birkhäuser 1983, 318–359.

[FeT79]   G. Fejes Tóth: Multiple packing and covering of spheres, *Acta Math. Acad. Sci. Hungar.* **34** (1979) 165–176.

[FeT77]   G. Fejes Tóth: A problem connected with multiple circle-packings and circle-coverings, *Studia Sci. Math. Hungar.* **12** (1977) 447–456.

[FeT76]   G. Fejes Tóth: Multiple packing and covering of the plane with circles, *Acta Math. Acad. Sci. Hungar.* **27** (1976) 135–140.

[FeTF75]   G. Fejes Tóth, A. Florian: Mehrfache gitterförmige Kreis- und Kugelanordnungen, *Monatshefte Math.* **79** (1975) 13–20.

[Fe67]   L. Few: Multiple packing of spheres: a survey, in: *Proc. Colloq. on Convexity*, Københavens Univ. Math. Inst. 1965, 88–93.

[Fe64]   L. Few: Multiple packing of spheres, *J. London Math. Soc.* **39** (1964) 51–54.

[Ga96]   Sh.I. Galiev: Multiple packings and coverings of a sphere (Russian), *Diskret. Mat.* 1996, 148–160, translation in: *Discrete Math. Appl.* **6** (1996) 413–426.

[Gr86]   H. Groemer: Multiple packings and coverings, *Stud. Sci. Math. Hungar.* **21** (1986) 189–200.

[He59]   A. Heppes: Mehrfache gitterförmige Kreislagerungen in der Ebene, *Acta Math. Acad. Sci. Hungar.* **10** (1959) 141–148.

[He55]   A. Heppes: Über mehrfache Kreislagerungen, *Elemente Math.* **10** (1955) 125–127.

[HoT83]   J. Horváth, Á.H. Temesvári: Über Dichte und Enge von doppelgitterförmigen zweifachen Kreispackungen, *Studia Sci.*

Math. Hungar. **18** (1983) 253–268.

[**Hu93**]   M.N. HUXLEY: Exponential sums and lattice points, *Proc. London. Math. Soc. 3. Ser.* **66** (1993), 279–301.

[**Li83**]   J. LINHART: Ein Methode zur Berechnung der Dichte einer dichtesten gitterförmigen k-fachen Kreispackung, *Ber. Math. Inst. Univ. Salzburg* **1-2** (1983), 11–40.

[**MaP86**]   P. MANI-LEVITSKA, J. PACH: Decomposition problems for multiple coverings of unit balls, manuscript 1986.

[**Pa86**]   J. PACH: Covering the plane with convex polygons, *Discrete Comput. Geom.* **1** (1986) 73–81.

[**Pa80**]   J. PACH: Decomposition of multiple packing and covering, in: *2. Kolloq. über Diskrete Geom.*, Inst. Math. Univ. Salzburg 1980, 169–178.

[**Ra48**]   R. RADO: Covering theorems for ordered sets, *Proc. London Math. Soc. 2. Ser.* **50** (1948) 509–535.

[**Sch92**]   M. SCHMITZ: Die Zerlegung spezieller Einheitskreisüber-deckungen in drei Einheitskreispackungen, *Beiträge Algebra Geom.* **33** (1992) 17–37.

[**Sch91**]   M. SCHMITZ: Die Zerlegung von doppelgitterförmigen 2-fachen Einheitskreispackungen in drei Einheitskreispackungen, *Beiträge Algebra Geom.* **32** (1991) 71–86.

[**Te94a**]   Á.H. TEMESVÁRI: Über die dünnste doppelgitterförmige 2-fache Überdeckung mit einem zentralsymmetrischen konvexen Bereich, *Beiträge Algebra Geom.* **35** (1994) 45–54.

[**Te94b**]   Á.H. TEMESVÁRI: Die dichteste gitterförmige 9-fache Kreis-packung, *Rad Hrvatske Akad. Znan. Umjet.* **467** (1994) 95–110.

[**TeHY87**]   Á.H. TEMESVÁRI, J. HORVÁTH, N.N. YAKOVLEV: A method for finding the densest lattice k-fold packing of circles (in Russian), *Mat. Zametki* **41** (1987) 625–636.

[**Ya84**]   N.N. YAKOVLEV: A method of finding the densest lattice k-packing on a plane (in Russian), *Functional Analysis and its Applications in Mechanics and Probability Theory (Moscow, 1983)*, Moskov. Gos. Univ. 1984, 170–171.

[**Ya83**]   N.N. YAKOVLEV: The densest lattice 8-packing on a plane (in Russian), *Vestnik Moskov. Univ. Ser. I Mat. Mekh.* 1983/5, 8–16.

[**Yan80**]   L.J. YANG: Multiple lattice packings and coverings of spheres, *Monatshefte Math.* **89** (1980) 69–76.

2.2 Solid and Saturated Packings and Reduced Coverings

It is an important direction of research in geometry to investigate which results in Euclidean space (plane) can be extended to spaces of constant curvature (spheres and hyperbolic planes). Thus, it is natural to ask, for example, what is the maximum density of a packing of equal circles in the hyperbolic plane? However, Böröczky [Bö74], [Bö78] has shown that any attempt to give a reasonable *global* definition of density in the hyperbolic plane is destined to fail. He constructed a packing C of equal circles and a tiling $\mathcal{T}$ with congruent convex cells such that each cell $T \in \mathcal{T}$ contains exactly one member $C \in \mathcal{C}$. Now it would seem to be natural to define the density of $\mathcal{C}$ as $\frac{\text{area}(C)}{\text{area}(T)}$. But Böröczky came up with another tiling $\mathcal{T}'$, satisfying the same conditions with respect to $\mathcal{C}$, whose cells have an area different from area(T)! A similar paradoxical phenomenon was discovered by K. Bezdek [Be84]. (See [BoR04], [BoR03] for an alternative attempt to define the density of certain "nice" packings in the hyperbolic plane.)

To avoid these difficulties and to capture the densest packings by a *local* structural property, L. Fejes Tóth [FeT68] introduced the following definition: A packing (covering) of space with convex bodies is called *solid* if any other packing (covering) $\mathcal{C}'$ that can be obtained from $\mathcal{C}$ by rearranging a finite number of its members is congruent to $\mathcal{C}$. This notion clearly applies to the hyperbolic plane. Moreover, in Euclidean space any solid packing (covering) of congruent copies of a domain has maximum (minimum) density.

A tiling that consists of congruent copies of a regular m-gon meeting edge-to-edge is called *regular*, and it is denoted by the *Schläfli symbol* $\{m,n\}$, where n is the number of polygons meeting at a vertex. It is well known (see, e.g., L. Fejes Tóth [FeT64]) that for any pair of integers $m, n > 1$, there exists a

$$\text{regular tiling } \{m,n\} \left\{ \begin{array}{l} \text{on the sphere} \\ \text{in the Euclidean plane} \\ \text{in the hyperbolic plane} \end{array} \right\} \text{ if } \frac{1}{m} + \frac{1}{n} \left\{ \begin{array}{l} > 1/2, \\ = 1/2, \\ < 1/2 \end{array} \right. .$$

If a tiling $\mathcal{T}$ consists of congruent copies of at least two different kinds of regular polygons, and for any two vertices of $\mathcal{T}$ there is an isometry that carries one to the other and $\mathcal{T}$ into itself, then $\mathcal{T}$ is called *Archimedean*. A tiling is called *trihedral* if the number of faces (polygons) meeting at each vertex is 3. It follows from a theorem of Imre [Im64] that the inscribed (circumscribing) circles of the faces of any regular trihedral tiling $\{m,3\}$ form a solid packing (covering). As pointed out earlier, on the sphere and in the plane every solid packing (covering) has maximum (minimum) density, so that this result can be regarded as a generalization of Thue's (Kershner's) theorem determining the maximum (minimum) density of a packing (covering) of equal circles in the plane.

Conjecture 1   (L. Fejes Tóth [FeT68]) The inscribed (circumscribing) circles of the faces of an Archimedean tiling form a solid packing (covering) if and only if it is a trihedral tiling.

Many special cases of this conjecture have been confirmed by L. Fejes Tóth [FeT68], G. Fejes Tóth [FeT74], Heppes [He92], Heppes and Kertész [HeK97], Florian [Fl01], and Florian and Heppes [FlH03]. See [Fl00] for a recent survey. In the Euclidean plane, all cases of the conjecture have been verified for packings, so the next natural step is to settle the covering part of the conjecture (see [FlH00], [FlH03] for partial results).

In his Habilitationsschrift (Universität Göttingen 1963), Danzer [Da86] (see also [Bö83]) showed that the densest packing of 11 equal circles on the sphere is uniquely determined and can be obtained from the densest packing of 12 circles (i.e., the inscribed circles of the 12 faces of the tiling $\{5,3\}$) by simply removing one circle. This means that the packing formed by the incircles of $\{5,3\}$ is not only solid, but remains solid even after the removal of a circle. L. Fejes Tóth conjectured that the same is true for any regular tiling $\{m,3\}$ with $m > 5$. For $m \geq 8$ this was proved by A. Bezdek [Be79].

Conjecture 2   [FeT80] The packing $\mathcal{C}^-$ obtained from the densest lattice packing of unit circles in the plane by removing one of its elements is solid.

Bárány and Dolbilin [BáD88] showed that the conjecture is "locally" true: every packing that can be obtained from $\mathcal{C}^-$ by moving a finite number of its elements to a distance at most $1/40$ is congruent to $\mathcal{C}^-$. Heppes [He94] proved that the packing $\mathcal{H}$ of hexagons obtained from the regular tiling $\{6,3\}$ of the Euclidean plane by removing one of its elements is *translation solid*. That is, every packing resulting from it by a finite rearrangement through translations is congruent to $\mathcal{H}$.

Define the *degree of solidity* of a solid packing $\mathcal{C}$ as the largest integer k $(1 \leq k \leq |\mathcal{C}| - 1)$ with the property that any packing obtained from $\mathcal{C}$ by removing $k - 1$ elements and rearranging finitely many others is congruent to a subcollection of $\mathcal{C}$. Thus, the above conjecture states that the degree of solidity of the densest lattice packing of circles is at least two. (Obviously, it cannot be larger than two.) L. Fejes Tóth [FeT80] proved that the degree of solidity of the packing formed by the inscribed circles of the tiling $\{m,3\}$ tends to infinity as $m \to \infty$. He raised the following question.

Problem 3   (L. Fejes Tóth [FeT80]) Does there exist a packing of congruent convex bodies in the Euclidean plane whose degree of solidity is at least three?

G. Fejes Tóth, W. Kuperberg, and, independently, Pach noticed that there is another rather weak assumption that guarantees that a packing (covering) has maximum (minimum) density. A k-fold covering $\mathcal{C}$ of $\mathbb{R}^d$ with congruent copies of C is called *reduced* if no element can be removed without leaving some point covered only $k-1$ times. It is *i-reduced* if no i elements of it can be replaced by $i-1$ congruent copies of $\mathcal{C}$ without destroying its property of being a k-fold covering. Finally, $\mathcal{C}$ is *completely reduced* if it is *i-reduced* for every finite i.

Similarly, a k-fold packing $\mathcal{C}$ is *saturated* (*i-saturated*) if it cannot be augmented with an extra copy of C (none if its subcollections of size $i-1$ can be replaced by a larger number of copies of C) without creating a point belonging to the interior of more than k members. It is *completely saturated* if it is *i-saturated* for every finite i.

Obviously, every k-fold covering in which each point is covered a finite number of times contains a reduced subcollection. Similarly, every k-fold packing of congruent bodies is contained in a saturated one. It is easy to see that any completely reduced k-fold covering of $\mathbb{R}^d$ with congruent bodies has minimum density among all k-fold coverings, and any completely saturated k-fold packing of congruent bodies has maximum density among all k-fold packings. This suggests that a natural way of attacking density problems would be to study (completely) reduced coverings and saturated packings. G. Fejes Tóth, G. and W. Kuperberg [FeT*98] proved that every convex body in Euclidean or hyperbolic d-space admits a completely reduced covering and a completely saturated packing. It is not hard to extend their argument to show the existence of completely reduced *k-fold* coverings and completely saturated *k-fold* packings, for every natural number k. Using measure-theoretic techniques, Bowen [Bo03] generalized these results to packings and coverings with not necessarily convex bodies (to connected compact sets identical to the closure of their interiors).

It is not difficult to show that if a thinnest k-fold covering of $\mathbb{R}^d$ (a densest k-fold packing of $\mathbb{R}^d$) with congruent copies of C is periodic, then it is necessarily completely reduced (saturated). (Recall that a d-dimensional arrangement $\mathcal{C}$ is periodic if there exist d linearly independent translations of $\mathbb{R}^d$ that leave $\mathcal{C}$ invariant.) In the sequel, we consider only the case $k=1$, but all of our questions can be generalized to arbitrary k.

Let C be a convex body in $\mathbb{R}^d$. For a fixed $i>0$, let $\theta_i(C)$ denote the supremum of the densities of all i-reduced coverings of space with congruent copies of C. Analogously, let $\delta_i(C)$ stand for the infimum of the densities of all i-saturated packings of congruent copies of C. It is not hard to see that we have

$$\lim_{i\to\infty}\theta_i(C)=\theta(C),$$

$$\lim_{i\to\infty}\delta_i(C)=\delta(C),$$

and both sequences are monotonic. Moreover, it was shown in [FeT*98] that in both cases the rate of convergence is $O(1/i^{1/d})$ as i tends to infinity, but it is not clear whether this bound is tight.

For simplicity, let us consider the case in which $d = 2$ and $C = B^2$ is the unit disk. Clearly, we have $\theta_1(B^2) = \infty$, and there are also 2-reduced coverings by circles of arbitrarily high density, so $\theta_2(B^2) = \infty$. Thus we can ask for the smallest integer $i(B^2) = i$ such that $\theta_i(B^2) < \infty$. G. Fejes Tóth and G. and W. Kuperberg [FeT*98] proved that $i(B^d) \leq d + 1$ in all dimensions. (In fact, they established an upper bound in terms of a quantity closely related to the *Hadwiger number*.) In particular, we have $i(B^2) = 3$. It may be true that $i(B^d) = 3$ for every $d \geq 2$.

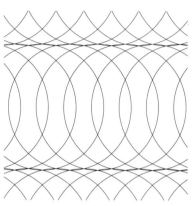

TWO-REDUCED COVERING
OF HIGH DENSITY

Problem 4   (*G. Fejes Tóth, G. and W. Kuperberg [FeT*98]*) *Given a sufficiently dense covering of* $\mathbb{R}^d$ *with unit balls, can one always make a new covering by replacing three balls by two?*

Similar questions can be asked for saturated packings. Since a packing of unit balls is one-saturated if and only if it becomes a covering by replacing each of its elements with an open concentric ball of radius two, the infimum of the densities of all one-saturated packings of unit balls satisfies $\delta_1(B^d) = \theta(B^d)/2^d$. In particular, we have $\delta_1(B^2) = (2\pi/\sqrt{27})/4 = \pi/\sqrt{108}$. These are the only known exact values of $\delta_i(B^d)$ or $\theta_i(B^d)$. The first interesting open question is the following.

Conjecture 5   [*FeT*98*]

$$\delta_2(B^2) = \frac{(3 - \sqrt{5})\pi}{\sqrt{27}}.$$

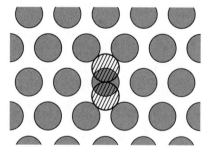

CONJECTURED TWO-SATURATED
PACKING OF MINIMUM DENSITY

Heppes [He01] proved this for lattice packings: there is no two-saturated lattice packing of circles whose density is smaller than $(3 - \sqrt{5})\pi/\sqrt{27}$.

Another variant of this question was proposed by L. Fejes Tóth and Heppes [FeTH80]: Determine the density $\delta_k^*(B^2)$ of the thinnest packing of unit circles in the plane with the property that any unit circle that does not belong to the packing intersects the interior of at least k of its elements. Clearly, we have $\delta_1^*(B^2) = \delta_1(B^2) = \pi/\sqrt{108}$. A. Bezdek [Be90] proved that $\delta_2^*(B^2) = \pi/\sqrt{27}$, and it was shown in [FeTH80] that $\delta_3^*(B^2) = \pi/\left(\sqrt{3}+2\right)$. There is no packing that has the required property for $k \geq 4$.

Problem 6   *[FeT*98] Determine $\theta_3(B^2)$, the supremum of the densities of all three-reduced coverings of the plane with unit circles.*

It is known that
$$\theta_3(B^2) \leq \frac{6\pi}{\sqrt{12} - \pi} \approx 58.45,$$
but this inequality is obviously far from sharp.

We close this section with a curious problem from [FeT*98]. Although the existence of completely saturated packings (completely reduced coverings) is known for *all* bodies, it is often quite difficult to decide whether a *particular* arrangement has this property. For circles in the plane, we know that the density of a packing (covering) is maximized (minimized) for the well-known hexagonal lattice arrangements. Therefore, these arrangements are obviously completely saturated (completely reduced).

Problem 7   *[FeT*98] Does there exist a completely saturated packing (completely reduced covering) of the plane with unit circles that is not a lattice packing (covering)?*

A possible candidate that may be completely saturated is shown below. Take the densest (hexagonal) lattice packing of unit circles, and slightly shift one-half of it along two adjacent rows of circles so that these rows remain in contact. The resulting packing is not a lattice packing. In fact, L. Fejes Tóth conjectured that the same packing is even solid.

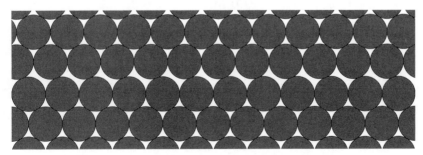

POSSIBLY A COMPLETELY SATURATED PACKING

[BáD88]   I. BÁRÁNY, N.P. DOLBILIN: A stability property of the densest circle packing, *Monatshefte Math.* **106** (1988) 107–114.

[Be90]   A. BEZDEK: Double-saturated packing of unit disks, *Period. Math. Hungar.* **21** (1990) 189–203.

[Be79]   A. BEZDEK: Solid packing of circles in the hyperbolic plane, *Studia Sci. Math. Hungar.* **14** (1979) 203–207.

[Be84]   K. BEZDEK: Ausfüllungen in der hyperbolischen Ebene durch endliche Anzahl kongruenter Kreise, *Ann. Univ. Sci. Budapest. Eötvös Sect. Math.* **27** (1984) 113–124.

[Bö83]   K. BÖRÖCZKY: The problem of Tammes for $n = 11$, *Studia Sci. Math. Hungar.* **18** (1983) 165–171.

[Bö78]   K. BÖRÖCZKY: Packing of spheres in spaces of constant curvature, *Acta Math. Acad. Sci. Hungar.* **32** (1978) 243–261.

[Bö74]   K. BÖRÖCZKY: Sphere packings in spaces of constant curvature I (in Hungarian), *Mat. Lapok* **25** (1974) 265–306.

[Bo03]   L. BOWEN: On the existence of completely saturated packings and completely reduced coverings, *Geometriae Dedicata* **98** (2003) 211–226.

[BoR04]   L. BOWEN, C. RADIN: Optimally dense packings of hyperbolic space, *Geometriae Dedicata* **104** (2004) 37–59.

[BoR03]   L. BOWEN, C. RADIN: Densest packing of equal spheres in hyperbolic space, *Discrete Comput. Geom.* **29** (2003) 23–39.

[Da86]   L. DANZER: Finite point-sets on S^2 with minimum distance as large as possible, *Discrete Math.* **60** (1986) 3–66.

[FeT74]   G. FEJES TÓTH: Solid sets of circles, *Studia Sci. Math. Hungar.* **9** (1974) 101–109.

[FeT*98]   G. FEJES TÓTH, G. KUPERBERG, W. KUPERBERG: Highly saturated packings and reduced coverings, *Monatshefte Math.* **125** (1998) 127–145.

[FeT80]   L. FEJES TÓTH: Solid packing of circles in the hyperbolic plane, *Studia Sci. Math. Hungar.* **15** (1980) 299–302.

[FeT68]   L. FEJES TÓTH: Solid circle-packings and circle-coverings, *Studia Sci. Math. Hungar.* **3** (1968) 401–409.

[FeT64]   L. FEJES TÓTH: *Regular Figures*, Pergamon Press 1964.

[FeTH80]   L. FEJES TÓTH, A. HEPPES: Multisaturated packings of circles, *Studia Sci. Math. Hungar.* **15** (1980) 303–307.

[Fl01]   A. FLORIAN: Packing of incongruent circles on the sphere, *Monatshefte Math.* **133** (2001) 111–129.

[Fl00]   A. Florian: Some recent results on packing and covering
            with incongruent circles, in: *3rd Internat. Conf. in Stochas-
            tic Geom., Convex Bodies and Empirical Measures, Part II*
            (Mazara del Vallo, 1999), P.M. Gruber, ed., *Rend. Circ. Mat.
            Palermo (2) Suppl.* **65** part 2 (2000) 93–104.

[FlH03]   A. Florian, A. Heppes: On the non-solidity of some pack-
            ings and coverings with circles, in: *Discrete Geometry: In
            Honor of W. Kuperberg's 60th Birthday*, A. Bezdek, ed., Mar-
            cel Dekker 2003, 279–290.

[FlH00]   A. Florian, A. Heppes: Solid coverings of the Euclidean
            plane with incongruent circles, *Discrete Comput. Geom.* **23**
            (2000), 225–245.

[He01]   A. Heppes: On the density of 2-saturated lattice packings of
            discs, *Monatshefte Math.* **134** (2001) 51–66.

[He94]   A. Heppes: On the solidity of the hexagonal tiling, in: *Intu-
            itive Geometry* (Szeged, 1991), K. Böröczky et al., eds., *Colloq.
            Math. Soc. János Bolyai* **63** (1994) 151–154.

[He92]   A. Heppes: Solid circle packings in the Euclidean plane,
            Discrete Comput. Geom. **7** (1992) 29–43.

[HeK97]   A. Heppes, G. Kertész: Packing circles of two different
            sizes on the sphere, in: *Intuitive Geometry* (Budapest, 1995),
            Bolyai Soc. Math. Studies **6** (1997) 357–365.

[Im64]   M. Imre: Kreislagerungen auf Flächen konstanter Krüm-
            mung, *Acta Math. Acad. Sci. Hungar.* **15** (1964) 115–121.

2.3 Stable Packings and Coverings

A packing $\mathcal{C}$ of convex bodies in $\mathbb{R}^d$ is said to be *k-stable* if any k-element subset of $\mathcal{C}$ is kept fixed by the rest. In other words, $\mathcal{C}$ is k-stable if for any $\mathcal{C}' = \{C_1, \ldots, C_k\} \subseteq \mathcal{C}$ there exists $\epsilon > 0$ with the property that we cannot move the members of $\mathcal{C}'$ less than ϵ without overlapping one another or some other member of $\mathcal{C}$. More precisely, if C_i' is a congruent copy of C_i whose (say) Hausdorff distance from C_i is less than ϵ ($i = 1, \ldots, k$) and $(\mathcal{C} \setminus \mathcal{C}') \cup \{C_1', C_2', \ldots, C_k'\}$ is a packing, then $C_i' = C_i$ for all i. If $\mathcal{C}$ is k-stable for every k, then it is said to be *finitely stable*. If, in addition, we can choose an $\epsilon > 0$ satisfying the above conditions for all k-member subfamilies (or all finite subfamilies) $\mathcal{C}' \subseteq \mathcal{C}$, then $\mathcal{C}$ is *uniformly k-stable* (*uniformly finitely stable*, respectively). A (uniformly) one-stable packing is called *(uniformly) stable*.

Comparing these definitions to the concept of *solidity* introduced in the previous section, it seems likely that every solid packing of congruent convex bodies is uniformly finitely stable. In particular, together with the results stated there, this would imply that the densest lattice packing of circles is uniformly finitely stable, which is known to be true. In fact, it was proved by Bárány and Dolbilin [BáD88] that it remains uniformly finitely stable even after the removal of a circle.

Place a ball of radius $1/\sqrt{2}$ around every integer point in $\mathbb{R}^d$ whose sum of coordinates is even. A. and K. Bezdek and Connelly [BeB*98] proved that for $d = 2$, the resulting packing is finitely stable, but not uniformly. If $d \geq 3$, however, we obtain uniformly finitely stable sphere packings. For $d \geq 4$, the density of these packings is not maximal. Therefore, they cannot be solid.

Problem 1   (Pach) *For $d = 2$ and 3, do there exist uniformly finitely stable packings with unit balls that are not solid?*

Obviously, any member of a stable packing of convex bodies in $\mathbb{R}^d$ has at least $d + 1$ neighbors, i.e., touches at least $d + 1$ other members. The following question arises: what is the minimum density of a stable packing of unit balls in $\mathbb{R}^d$? However, Böröczky [Bö64] (see also L. Fejes Tóth [FeT72]) has shown that the answer is 0 already for $d = 2$. Böröczky's example is not *uniformly* stable. In fact, it is easy to see that every uniformly stable packing with unit disks in the plane has positive density.

Problem 2   *For a fixed $\epsilon > 0$, find the infimum of the densities of all uniformly stable packings with unit disks in which no member can be moved less than ϵ without colliding with the others.*

Problem 3   *(Connelly [Co87]) Does there exist a finitely stable packing of equal circles in the plane with density 0?*

Solving a problem of L. Fejes Tóth [FeT69], Wegner [We71] constructed for every $n \geq 3$, $n \neq 11$, a finitely stable packing $\mathcal{C}_n$ of congruent convex bodies in the plane such that each of them has exactly n neighbors.

Problem 4   *(Wegner [We71]) Does there exist*
                     (1) a finitely stable packing, or
                     (2) a tiling
                     in the plane consisting of congruent copies of a convex body such that each of them touches exactly 11 others?

In higher than 3-dimensional spaces even the following simple question is open.

Problem 5   *Is the densest lattice packing of balls in $\mathbb{R}^d$ (uniformly) finitely stable for every $d \geq 4$?*

Stable packings of finitely many circles on a sphere or of balls in a bounded container were considered by Connelly [Co88] and L. Fejes Tóth [FeT85]. A packing of bodies in a container is *stable* if each of its members is kept fixed by its neighbors and by the container.

Problem 6   *Let $d(n)$ denote the minimum density of a stable packing of unit circles in a square of side length n. Determine*

$$\lim_{n \to \infty} nd(n) > 0.$$

It follows from Böröczky's construction [Bö64] that the limit in question is bounded.

It was shown by de Bruijn [dBr54] (and rediscovered many times since then) that there is no stable packing in the plane consisting of finitely many convex bodies. As a matter of fact, a much stronger statement is true: given any direction, any finite packing of plane convex bodies has a member that can be translated to infinity in this direction without disturbing the other members. As L. Fejes Tóth and Heppes [FeTH63] observed, a similar theorem holds for infinite packings of convex bodies in the plane: given any direction, $\mathcal{C}$ can be decomposed into two nonempty subfamilies such that one of them can be translated in this direction to an arbitrary distance without collision.

Surprisingly, the situation is radically different in $\mathbb{R}^3$. L. Fejes Tóth and Heppes [FeTH63] exhibited a stable packing of twelve congruent tetrahedra, while Danzer [Dan67] constructed a stable packing of twelve congruent centrally symmetric convex dodecahedra. They are probably even two-stable! (See Shephard [Sh70] for a similar construction.)

Conjecture 7   *(L. Fejes Tóth [FeT67]; Danzer [Dan67]) There exists a stable packing of congruent ellipsoids in $\mathbb{R}^3$ consisting of finitely many (probably at most 12) members.*

Conjecture 8   *(L. Fejes Tóth [FeT67]; Danzer [Dan67]) For every $d \geq 4$, there exists a stable packing of finitely many congruent centrally symmetric convex bodies in $\mathbb{R}^d$.*

Problem 9   *(Pach) Does there exist a stable packing in $\mathbb{R}^d$, $d \geq 3$, consisting of finitely many homothetic copies of a given convex body C?*

As L. Fejes Tóth and Heppes [FeTH63] have noted, no such packing exists if C is an ellipsoid. (See also Dawson [Daw84].)

A finite packing $\mathcal{C}$ of not necessarily convex bodies (say, Jordan domains) in $\mathbb{R}^d$ is said to be *rigid* if there exists a positive ϵ such that any packing obtained from $\mathcal{C}$ by moving its members by less than ϵ is congruent to $\mathcal{C}$. So a rigid packing cannot be taken apart even if we may simultaneously move all bodies, whereas a stable packing cannot be taken apart if we may move only one body. ("rigidity" is essentially the same as "finite stability" restricted to finite packings.) A body C is *star-shaped*, if it has an interior point x such that the closed segment $[x, c]$ is contained in C for every $c \in C$. The body C is called *directionally convex* if there is a direction such that any line parallel to this direction intersects C in a connected (possibly empty) set.

De Bruijn [dBr54] observed that no finite packing of star-shaped bodies is rigid. To see this, slightly "blow up" the whole arrangement from an arbitrary point of the plane and then contract every member back to its original size from its star-center. As it was repeatedly discovered, the resulting simultaneous translations always separate the members from one another.

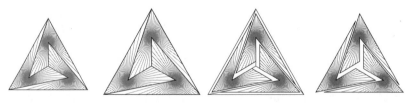

Separating star-shaped sets by simultaneous translations:
global blowing up, then contracting each set to its center

On the other hand, L. Fejes Tóth and Heppes [FeTH63] exhibited a collection of four directionally convex bodies in the plane that form a rigid packing. They are directionally convex in different directions. No packing of sets that are directionally convex in the same direction can be rigid.

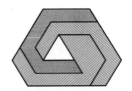

THREE RIGIDLY INTERLOCKED
DIRECTIONALLY CONVEX SETS

Similar questions can be asked for coverings. We call a covering $\mathcal{C}$ of $\mathbb{R}^d$ *k-stable* if every sufficiently small rearrangement of k of its members leaves a piece of $\mathbb{R}^d$ uncovered. If $\mathcal{C}$ is k-stable for every k, then it is said to be finitely stable. Uniformly finitely stable coverings can be defined analogously to the case of packings. One-stable coverings are simply called stable. It is not difficult to show that the thinnest lattice covering of the plane with circles is uniformly finitely stable.

[BáD88]   I. BÁRÁNY, N.P. DOLBILIN: A stability property of the densest circle packing, *Monatshefte Math.* **106** (1988) 107–114.

[BeB*98]   A. BEZDEK, K. BEZDEK, R. CONNELLY: Finite and uniform stability of sphere packings, *Discrete Comput. Geom.* **20** (1998) 111–130.

[Bö64]   K. BÖRÖCZKY: Über stabile Kreis- und Kugelsysteme, *Ann. Univ. Sci. Budapest. Eötvös, Sect. Math.* **7** (1964) 79–82.

[dBr54]   N.G. DE BRUIJN: Aufgaben 17 und 18, *Nieuw Archief voor Wiskunde* **2** (1954) p.67. Solution appeared in: *Wiskundige Opgaven met de Oplossingen* **20** (1955) 19–20.

[Co88]   R. CONNELLY: Rigid circle and sphere packings. I. Finite packings (dual French-English text), *Structural Topology* **14** (1988) 43–60.

[Co87]   R. CONNELLY: Uniformly stable circle packings, *Tagungsberichte Math. Forschungsinst. Oberwolfach,* Diskrete Geometrie 1987.

[Dan67]   L. DANZER: To Problem 8, in: *Proc. Colloq. Convexity* (Copenhagen, 1965), W. Fenchel, ed., Univ. Copenhagen 1967, 312–313.

[Daw85]   R. DAWSON: On the mobility of bodies in $\mathbb{R}^n$, *Math. Proc. Cambridge Philos. Soc.* **98** (1985) 403–412. Corrigenda: ibid, **99** (1986) 377–379.

[Daw84]   R. DAWSON: On removing a ball without disturbing the others, *Math. Mag.* **57** (1984) 27–30.

[FeT85]   L. FEJES TÓTH: Stable packing of circles on the sphere (dual French–English text), *Structural Topology* **11** (1985) 9–14.

[FeT72]   L. FEJES TÓTH: *Lagerungen in der Ebene, auf der Kugel und im Raum* (2. Auflage), Springer-Verlag 1972.

[FeT69]   L. FEJES TÓTH: Scheibenpackungen konstanter Nachbarnzahl, *Acta Math. Acad. Sci. Hungar.* **20** (1969) 375–381.

[FeT67]   L. FEJES TÓTH: Problem 8, in: *Proc. Colloq. Convexity* (Copenhagen, 1965), W. Fenchel, ed., Univ. Copenhagen 1967, p. 312.

[FeTH63] L. FEJES TÓTH, A. HEPPES: Über stabile Körpersysteme, *Compositio Math.* **15** (1963) 119–126.

[Sh70]   G.C. SHEPHARD: On a problem of Fejes Tóth, *Studia Sci. Math. Hungar.* **5** (1970) 471–473.

[We71]   G. WEGNER: Bewegungsstabile Packungen konstanter Nachbarnzahl, *Studia Sci. Math. Hungar.* **6** (1971) 431–438.

2.4 Kissing and Neighborly Convex Bodies

For any convex body $C \subseteq \mathbb{R}^d$, let $N(C)$ denote the maximum number of nonoverlapping congruent copies of C that can be arranged so that each of them is touching (kissing) C. This number is called the *Newton number* of C, referring to a wild dispute between Isaac Newton and David Gregory about the exact value of $N(B^3)$. Newton's claim that $N(B^3) = 12$ was established only 180 years later by Hoppe [Ho874], but even his argument is incomplete. The first correct proof was given by Schütte and van der Waerden [ScvW53]. Shortly thereafter, an elegant argument was found by Leech [Le56] (see [An04], [Bö03], [Bö04], [Mu05] for alternative proofs). The problem of estimating $N(B^d)$ in higher dimensions is closely related to the question of densest sphere packings (mentioned earlier), and the gap between the corresponding lower and upper bounds is enormous. However, we know the exact answer in $\mathbb{R}^4$, in $\mathbb{R}^8$ and, due to the existence of some extremely symmetric lattices discovered by Leech [Le64], in $\mathbb{R}^{24}$. According to Musin's recent result [Mu05], [Mu03], we have $N(B^4) = 24$, while Odlyzko and Sloane [OdS79] and Levenšteĭn [Le79] showed that $N(B^8) = 240$ and $N(B^{24}) = 196560$ (see also [ErZ92], [BaS81]). Many upper and lower estimates on $N(B^d)$, up to 128 dimensions, are given in [Bo94] and [EdRS98], respectively. Most of them are based on the "linear programming method" of Delsarte [De72] and Odlyzko–Sloane [OdS79], and on the existence of certain codes, respectively (see [CoS98], [ErZ01] for surveys).

Problem 1 *Are there infinitely many dimensions d for which there exists a lattice packing of unit balls in $\mathbb{R}^d$ whose every member is tangent to precisely $N(B^d)$ others?*

Watson [Wa71] proved that no such lattice packing exists for $d = 9$.

The investigation of Newton numbers of plane convex bodies other than the circle was initiated by L. Fejes Tóth [FeT67], who established the inequality

$$N(C) \le (4 + 2\pi)\frac{\mathrm{diam}(C)}{\mathrm{width}(C)} + 2 + \frac{\mathrm{width}(C)}{\mathrm{diam}(C)},$$

where $\mathrm{diam}(C)$ is the diameter of C, and $\mathrm{width}(C)$ is the smallest distance between two distinct parallel supporting lines of C. This bound is tight in many different cases.

In the case in which C is of *constant width*, i.e., the distance between any two distinct parallel supporting lines is the same, Schopp [Sch70] proved that $N(C) \le 7$, and this bound cannot be improved.

Some other general bounds for $N(C)$ in the plane were given by Hortobágyi [Ho72], [Ho75]. The Newton numbers of regular n-gons have been determined by Böröczky [Bö71] and Linhart [Li73], while for rectangles and isosceles triangles various estimates were given in [Li77], [We92], [KeM97] and [KeM*02]. Halberg, Levin, and Straus [HaLS59] proved that $N(C) \geq N(B^d)$ for any d-dimensional convex body C, i.e., in all dimensions the ball "likes to be crowded least."

A slightly different notion was introduced by Hadwiger [Ha57a]. For any convex body $C \subset \mathbb{R}^d$, define $H(C)$, the *Hadwiger number* of C, as the largest number of nonoverlapping *translates* of C that can be brought into contact with C. (See Zong [Zo98] for a survey.)

It is easy to see that it is sufficient to determine $H(C)$ for centrally symmetric convex bodies because $H(C) = H(C')$ for any C, where C' is the *Minkowski symmetrization* of C, i.e., $C' = \frac{1}{2}(C + (-C))$. Based on this observation, the upper bound

$$H(C) \leq 3^d - 1$$

follows almost immediately for every convex body C, since all translates touching a 0-symmetric convex body C are contained in $3C$ (see Hadwiger [Ha57a] and Grünbaum [Gr61]). This bound is tight for parallelepipeds. On the other hand, denoting by $H_L(C)$ the *lattice kissing number* of C, i.e., the largest number of nonoverlapping translates of C that can touch a given member of a lattice packing of C, by a theorem of Swinnerton-Dyer [SwD53] (see also [Zo95]), we have that

$$d(d+1) \leq H_L(C) \leq H(C),$$

for every convex body C. In fact, every member of any *densest* lattice packing of C is tangent to at least $d(d+1)$ other members. Minkowski proved that $H_L(C) \leq 2^{d+1} - 2$ for every strictly convex C, and it is not hard to construct d-dimensional examples of convex bodies, $d \geq 3$, for which $H_L(C) \neq H(C)$; see [Zo94], [Ta98a]. Moreover, Gruber [Gru86] showed that in any densest lattice packing of a "typical" convex body in $\mathbb{R}^d$, every member is adjacent to at most $2d^2$ others. Bourgain (see [FüL94]) and Talata [Ta98b] have established an exponential lower bound for min $H(C)$ over all d-dimensional convex bodies C. In particular, this implies that in sufficiently high-dimensional spaces, $H(C) \neq d(d+1)$. Asymptotically better (exponential) lower bounds are known for the Hadwiger numbers of d-dimensional simplices, balls, and so-called superballs [Ta00], [LaZ99]. Talata [Ta99a] determined the Hadwiger number of the tetrahedron $T^3 \subset \mathbb{R}^3$: we have $H(T^3) = H_L(T^3) = 18$.

Problem 2   *Determine the Newton number of the regular tetrahedron in $\mathbb{R}^3$.*

Problem 3   *Does the Newton number $N(C^3)$ of the three-dimensional cube C^3 coincide with its Hadwiger number $H(C^3) = 26$?*

Problem 4   *(Grünbaum [Gr61]) Characterize those convex bodies $C \subseteq \mathbb{R}^d$ for which $H(C) = d(d+1)$.*

In the plane it is known that $H(C) = 6$ for every convex C other than a parallelogram. In fact, $H(C) \geq 6$ holds for any *Jordan region* $C \subseteq \mathbb{R}^2$, i.e., for any region bounded by a simple Jordan curve (Halberg, Levin, Straus [HaLS59]). However, we do not have any upper bound for this class.

Problem 5   *(A. Bezdek, Pach)   Can the Hadwiger numbers of all Jordan regions in the plane be bounded from above by an absolute constant?*

In fact, this constant is probably eight, which is already reached by the parallelograms. We have no example that shows that Jordan regions behave differently from convex sets in this problem, but the argument in the proof for the upper bound breaks down completely. For convex sets C, any translate of C that touches C must be contained in the blown-up set $3C$, but that is not true if the set is not convex. A first step to a proof that works also for nonconvex sets might be the result of A. Bezdek and K. and W. Kuperberg [BeKK95], who showed that the maximum number of pairwise touching translates of a Jordan region C in the plane is four. So in this respect Jordan regions behave in the same way as convex sets.

Conjecture 6   *(A. Bezdek, K. and W. Kuperberg [BeKK95]) For any star-shaped set C in the plane, we have $H(C) \leq 8$.*

The upper bound $H(C) \leq 75$ for star-shaped sets was obtained by A. Bezdek [Be97]. On the other hand, K. and W. Kuperberg [KuK94] showed that the maximum number of nonoverlapping translates of a star-shaped region that have a point in common is 4. In three-dimensional space, it is easy to see that the Hadwiger number of star-shaped sets is unbounded. Moreover, K. and W. Kuperberg [KuK94] have given examples of arbitrary many nonoverlapping translates of a star-shaped three-dimensional set meeting in a point.

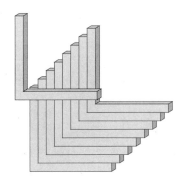

STAR-SHAPED SET TOUCHED
BY MANY TRANSLATES

Problem 7   *Does there exist an absolute constant $\varepsilon > 0$ such that the Hadwiger number of any strictly convex body C in $\mathbb{R}^d$ is at most $(3 - \varepsilon)^d$?*

Talata [Ta05] has constructed strictly convex bodies whose Hadwiger numbers are roughly $7^{d/2} \approx 2.645^d$. Moreover, for these bodies $H(C)/H_L(C)$ tends to infinity exponentially fast.

Problem 8   *(Zong [Zo99]) Does there exist an absolute constant $c > 0$ such that the lattice kissing number $H_L(C)$ of any convex body C in $\mathbb{R}^d$ is at least 3^{cd}?*

Zong conjectures that the answer to this question is in the negative, in which case it would be interesting to know whether the assertion is true at least for infinitely many d.

A related question was considered by Tietze [Ti905], [Ti59]. A family $\mathcal{C}$ of (nonoverlapping) d-dimensional convex bodies is called *neighborly* if the intersection of any two members of $\mathcal{C}$ is a $(d-1)$-dimensional set. It is easy to see that any neighborly family in the plane has at most four members. On the other hand, Tietze and (independently, but 40 years later) Besicovitch [Be47] constructed a neighborly family in $\mathbb{R}^3$ consisting of infinitely many convex polytopes.

However, under certain restrictions on the members, one can bound the size of a neighborly family. Perles [Pe84] proved than any neighborly family of convex polytopes in $\mathbb{R}^d$ has at most 2^m members, provided that each member has at most m facets, i.e., $(d-1)$-dimensional faces. In particular, any neighborly family of simplices in $\mathbb{R}^d$ consists of at most 2^{d+1} members. Zaks [Za81] and Dekster[De82] constructed such a family with 2^d members, and Zaks [Za91] showed that for $d = 3$ this is best possible.

Problem 9   *(Bagemihl [Ba56]) What is the maximum number of elements in a neighborly family of simplices in d-dimensional Euclidean space? Is it 2^d?*

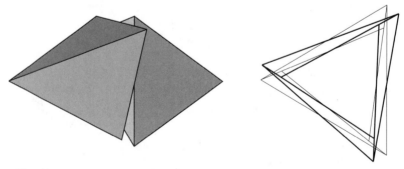

EIGHT NEIGHBORLY SIMPLICES: TWO SIMPLICES, EACH SUBDIVIDED
INTO FOUR SIMPLICES, TOUCHING ALONG A ROTATED FACE

Zaks called a family of convex polytopes in $\mathbb{R}^d$ *nearly neighborly* if for any pair of its members there exists a hyperplane that separates them and contains a facet of each of them. He proved that in three-dimensional space any nearly neighborly family of convex polytopes has at most 14 members.

Conjecture 10 *(Zaks [Za79], [Za87]) Every finite neighborly family of convex polytopes has a member with a facet such that no point of its relative interior belongs to the boundary of any other member of the family.*

Imposing the additional condition that all members of the family must be congruent, we obtain a different set of interesting problems.

Conjecture 11 *(Zaks [Za86], [Za87]; Linhart) There exists a constant c such that no neighborly family of congruent convex bodies in $\mathbb{R}^3$ has more than c members.*

However, as Zaks [Za86] has shown, for dimensions greater than three there is no such upper bound. The best lower bound in $\mathbb{R}^3$ is also due to Zaks [Za79], who constructed eight congruent neighborly polytopes.

It follows from an elegant result of Danzer and Grünbaum [DaG62] that no neighborly family of translates of a given convex polytope in $\mathbb{R}^d$ can have more than 2^d members. For $d = 2$ and 3, Grünbaum [Gr63] established the better bounds three and five, respectively, which cannot be improved.

Conjecture 12 *(Erdős–Füredi [ErF83]) There exists a positive constant δ such that for every d, any neighborly family of translates of a given d-dimensional polytope has at most $(2 - \delta)^d$ members.*

On the other hand, it can be easily deduced from the results of Erdős and Füredi [ErF83] that there exist neighborly families of translates of a d-dimensional polytope consisting of $(1 + \varepsilon)^d$ members for some $\varepsilon > 0$.

We get some interesting versions of the above questions by replacing the condition that any two members of the family meet along a $(d - 1)$-dimensional set with the weaker assumption that any two members touch each other. Of course, every such family of translates of C can have at most $H(C)$ (the Hadwiger number of C) members. For historical reasons, the maximum number of pairwise touching translates of a convex body is often called its *Petty number* (see [Pe71] and [Bl53] for a much earlier analysis of this concept). For a centrally symmetric convex body C this number is equal to the maximum cardinality of an *equilateral set* in the normed space whose unit ball is C; see Section 5.2. Danzer and Grünbaum [DaG63] proved that the Petty number of any d-dimensional convex body

C is at most 2^d, which is attained for parallelepipeds. For parallelepipeds, it is not even necessary that they are all of the same size. This suggests that the same upper bound might even hold for homothets, that is, scaled translates.

Conjecture 13 *(K. Bezdek, Pach, [BeC88]) Any family of pairwise touching homothetic copies of a centrally symmetric convex body C has at most 2^d members.*

It is easy to establish the weaker upper bound $3^d - 1$. For the case of balls $(C = B^d)$, the much better bound $d + 2$ holds. To see this, it is enough to observe that applying an inversion around the point of tangency of any two members of a family of n pairwise touching balls of arbitrary radii, we map them into an arrangement of $n - 2$ congruent balls simultaneously touching two parallel hyperplanes. Thus, the centers of these balls must form an at most $(d - 1)$-dimensional simplex, and we have $n - 2 \leq d$ [BeC88].

Problem 14   *(Littlewood [Li68]) What is the maximum number of congruent infinite circular cylinders that can be arranged in $\mathbb{R}^3$ so that every pair is touching? Is it 7?*

A. Bezdek [Be05] proved that this number cannot exceed 24. There are several possible types of arrangements of six mutually touching infinite circular cylinders, and the ones we know are "flexible" with one degree of freedom. It is a well-known geometric puzzle to find seven finite congruent mutually touching circular cylinders.

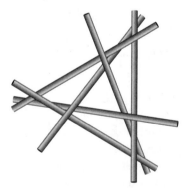

SIX PAIRWISE TOUCHING CYLINDERS

The d-dimensional *crosspolytope* is the convex hull of d pairwise orthogonal segments of equal length sharing the same midpoint.

Conjecture 15 *The maximum number of pairwise touching translates of the d-dimensional crosspolytope is $2d$, for every d.*

Alon and Pudlák [AlP03] established the upper bound $O(d \log d)$. The conjecture has been proved for $d \leq 3$ by Bandelt, Chepoi, and Laurent

[BaCL98] and for $d = 4$ by Koolen, Laurent, and Schrijver [KoLS00]. In the latter paper it was also shown that for every d, the maximum number of pairwise touching translates of the crosspolytope that all share a boundary point is $2d$.

Hortobágyi [Ho75] has suggested that the following "asymmetric" version of the Newton (Hadwiger) problem may also be of some interest. Given two d-dimensional solids C and D, determine the maximum number $N(C, D)$ (or $(H(C, D))$) of nonoverlapping congruent copies (or translates) of C that can be arranged so that each of them touches D. Obviously, $N(C, C) = N(C)$ is the Newton number ($H(C, C) = H(C)$ is the Hadwiger number) of C. A general upper estimate for $N(C, D)$ was given by Wegner [We92]. Talata [Ta99b] proved that $H(C, -C) \leq 3^d - 1$ for any convex $C \subset \mathbb{R}^d$. In [BoF93] and [BöL*00], some general inequalities were established for $N(\alpha C, C)$, where $0 < \alpha \leq 1$ and C is a d-dimensional convex body, centrally symmetric about the origin.

Let C_r denote an infinite circular cylinder of radius r in $\mathbb{R}^3$.

Conjecture 16 (*W. Kuperberg [Ku90]*) $N(C_1, B^3) = 6$.

Improving on earlier results of Heppes and Szabó [HeS91], Brass and Wenk [BrW00] proved that $N(C_1, B^3) \leq 7$.

Problem 17   (*W. Kuperberg [Ku90]*) *For any* $k \geq 3$, *determine the largest number* $r(k) = r$ *for which one can arrange* k *nonoverlapping congruent copies of* C_r *in* $\mathbb{R}^3$ *so as to touch the unit ball.*

The exact value of $r(k)$ is known only for $k = 3$ and 4. Conjecture 16 is equivalent to the inequality $r(7) < 1$. It would be interesting to describe at least the asymptotic behavior of $r(k)$ as k tends to infinity.

SIX UNIT CYLINDERS CAN TOUCH A UNIT BALL

Many of the packing problems described in this section have their natural dual counterparts for covering. In particular, Hadwiger [Ha57b] has also formulated the following question: Given a convex body C, what is the smallest number of its *translates* whose union covers a small neighborhood of C? This important and well-studied parameter, called the *Hadwiger cov-*

ering number of C, is closely related to Borsuk-type problems and questions about illumination. They will be discussed in detail in the next chapter. However, as far as we know, the dual counterpart of the Newton number has received much less attention. G. Fejes Tóth and G. and W. Kuperberg [FeT*98] define the *Newton covering number* $N^*(C)$ of C as the smallest number of *congruent copies* of C whose union covers a neighborhood of C. Obviously, this number never exceeds the Hadwiger covering number of C. They raise a number of interesting questions.

Problem 18   *[FeT*98] Among all convex bodies in $\mathbb{R}^d$, which one has the greatest Newton covering number, and what is this maximum?*

Problem 19   *[FeT*98] Let C^{d-1} denote the $(d-1)$-dimensional unit cube, and let $P^d(h)$ be the right pyramid of height h based on C^{d-1}.*

  (1)   *Determine the Newton covering number $N^*(C^d)$ of the cube.*

  (2)   *Is the equation $N^*(P^d(h)) = N^*(C^{d-1}) + 1$ always true?*

  (3)   *Does $N^*(P^d(h))$ really depend on h?*

[AlP03]   N. ALON, P. PUDLÁK: Equilateral sets in l_p^n, *Geom. Funct. Anal.* **13** (2003) 467–482.

[An04]   K.M. ANSTREICHER: The thirteen spheres: A new proof, *Discrete Comput. Geom.* **31** (2004) 613–625.

[Ba56]   F. BAGEMIHL: A conjecture concerning neighborly tetrahedra. *Amer. Math. Monthly* **63** (1956) 328–329.

[BaCL98]   H.-J. BANDELT, V. CHEPOI, M. LAURENT: Embedding into rectilinear spaces, *Discrete Comput. Geom.* **19** (1998) 595–604.

[BaS81]   E. BANNAI, N.J.A. SLOANE: Uniqueness of certain spherical codes, *Canad. J. Math* **33** (1981) 437–449.

[Be47]   A.S. BESICOVITCH: On Crum's problem, *J. London Math. Soc.* **22** (1947) 285–287.

[Be05]   A. BEZDEK: On the number of mutually touching cylinders, in: *Combinatorial and Computational Geometry*, J.E. Goodman et al., eds., Cambridge Univ. Press, *MSRI Publications* **52** (2005), to appear.

[Be97]   A. BEZDEK: On the Hadwiger number of a starlike disk,
           in: *Intuitive Geometry (Budapest 1995)*, Bolyai Soc. Math.
           Studies **6** (1997) 237–245.

[BeKK95]   A. BEZDEK, K. KUPERBERG, W. KUPERBERG: Mutually
           contiguous translates of a plane disk, *Duke Math. J.* **78** (1995)
           19–31.

[BeC88]   K. BEZDEK, R. CONNELLY: Intersection points, *Ann. Univ.*
           Sci. Budapest Eötvös, Sect. Math. **31** (1988) 115–127.

[Bl53]   L.M. BLUMENTHAL: *Theory and Applications of Distance*
           Geometry, Clarendon Press 1953.

[BoF93]   V. BOJU, L. FUNAR: Generalized Hadwiger numbers for
           symmetric ovals, *Proc. Amer. Math. Soc.* **119** (1993) 931–
           934.

[Bö03]   K. BÖRÖCZKY: The Newton–Gregory problem revisited, in:
           Discrete Geometry: In Honor of W. Kuperberg's 60th Birth-
           day, A. Bezdek, ed., Marcel Dekker 2003, 103–110.

[Bö71]   K. BÖRÖCZKY: Über die Newtonsche Zahl regulärer Vielecke,
           Period. Math. Hungar. **1** (1971) 113–119.

[Bö04]   K. BÖRÖCZKY JR.:   *Finite Packing and Covering*, Cambridge
           Univ. Press, 2004.

[BöL*00]   K. BÖRÖCZKY JR., D.G. LARMAN, S. SEZGIN, C. ZONG:
           On generalized kissing numbers and blocking numbers, in: *3rd*
           Internat. Conf. Stochastic Geometry, Convex Bodies and Em-
           pirical Measures, Part II (Mazara del Vallo 1999), P.M. Gru-
           ber, ed., *Rend. Circ. Mat. Palermo (2) Suppl.* **65** part 2,
           (2000) 39–57.

[Bo94]   P. BOYVALENKOV: Small improvements of the upper bounds
           of the kissing numbers in dimensions 19, 21 and 23, *Atti Sem.*
           Mat. Fis. Univ. Modena **42** (1994) 159–163.

[BrW00]   P. BRASS, C. WENK: On the number of cylinders touching
           a ball, *Geometriae Dedicata* **81** (2000) 281–284.

[CoS98]   J.H. CONWAY, N.J.A. SLOANE: *Sphere Packings, Lattices,*
           and Groups, Springer-Verlag 1998.

[DaG62]   L. DANZER, B. GRÜNBAUM: Über zwei Probleme bezüglich
           konvexer Körper von P. Erdős und von V.L. Klee. *Math. Z.*
           79 (1962) 95–99.

[De82]   B.V. DEKSTER: An example of 2^n pairwise neighboring n-
           simplices in $\mathbb{R}^n$, $n \geq 2$, Preprint Univ. Toronto 1982.

[De72]   P. DELSARTE: Bounds for unrestricted codes by linear pro-
           gramming, *Philips Res. Rep.* **27** (1972) 272–289.

[EdRS98] Y. EDEL, E.M. RAINS, N.J.A. SLOANE: On kissing numbers in dimensions 32 to 128, *Electron. J. Combin.* **5** (1988), #R22.

[ErF83] P. ERDŐS, Z. FÜREDI: The greatest angle among n points in the d-dimensional Euclidean space, *Ann. Discrete Math.* **17** (1983) 275–283.

[ErZ01] T. ERICSON, V. ZINOVIEV: *Codes on Euclidean Spheres*, North-Holland 2001.

[ErZ92] T. ERICSON, V. ZINOVIEV: On spherical codes generating the kissing number in dimensions 8 and 24, *Discrete Mathematics* **106/107** (1992) 199–207.

[FeT*98] G. FEJES TÓTH, G. KUPERBERG, W. KUPERBERG: Highly saturated packings and reduced coverings, *Monatshefte Math.* **125** (1998) 127–145.

[FeT67] L. FEJES TÓTH: On the number of equal discs that can touch another of the same kind, *Studia Sci. Math. Hungar.* **2** (1967) 363–367.

[FüL94] Z. FÜREDI, P.A. LOEB: On the best constant for the Besicovitch covering theorem, *Proc. Amer. Math. Soc.* **121** (1994) 1063–1073.

[Gru86] P.M. GRUBER: Typical convex bodies have surprisingly few neighbours in densest lattice packings, *Studia Sci. Math. Hungar.* **21** (1986) 163–173.

[Gr63] B. GRÜNBAUM: Strictly antipodal sets, *Israel J. Math* **1** (1963) 5–10.

[Gr61] B. GRÜNBAUM: On a conjecture of Hadwiger, *Pacific J. Math* **11** (1961) 215–219.

[Ha57a] H. HADWIGER: Über Treffanzahlen be translationsgleichen Eikörpern, *Arch. Math.* **8** (1957) 212–213.

[Ha57b] H. HADWIGER: Ungelöstes Problem Nr. 20, *Elem. Math.* **12** (1957) 121.

[HaLS59] C.J.A. HALBERG JR., E. LEVIN, E.G. STRAUS: On contiguous congruent sets in Euclidean space, *Proc. Amer. Math. Soc.* **10** (1959) 335–344.

[HeS91] A. HEPPES, L. SZABÓ: On the number of cylinders touching a ball, *Geometriae Dedicata* **40** (1991) 111–116.

[Ho874] R. HOPPE: Bemerkung der Redaktion, *Archiv Math. Physik (Grunert)* **56** (1874) 307–312.

[Ho75] I. HORTOBÁGYI: Über die Scheibenklassen bezügliche Newtonsche Zahl der konvexen Scheiben, *Ann. Univ. Sci. Bu-*

dapest Eőtvős, Sect. Math. **18** (1975) 123–127.

[Ho72] I. Hortobágyi: The Newton number of convex plane regions (in Hungarian), Mat. Lapok **23** (1972) 313–317.

[KeM97] A. Kemnitz, M. Möller: On the Newton number of rectangles, in: *Intuitive Geometry* (Budapest, 1995), I. Bárány et al., eds., *Bolyai Soc. Math. Studies* **6** (1997) 373–381.

[KeM*02] A. Kemnitz, M. Möller, D. Wojzischke: Bounds for the Newton number in the plane, *Results Math.* **41** (2002) 128–139.

[KoLS00] J. Koolen, M. Laurent, A. Schrijver: Equilateral dimension of the rectilinear space, *Designs, Codes Cryptography* **21** (2000) 149–164.

[Ku90] W. Kuperberg: Problem 3.3, *DIMACS report on Workshop on Polytopes and Convex Sets*, Rutgers Univ. 1990.

[KuK94] K. Kuperberg, W. Kuperberg: Translates of a starlike plane region with a common point, in: *Intuitive Geometry* (Szeged, 1991), K. Böröczky et al., eds., *Coll. Math. Soc. János Bolyai* **63**, North-Holland 1994, 205–216.

[LaZ99] D.G. Larman, C. Zong: On the kissing number of some special convex bodies, *Discrete Comput. Geom.* **21** (1999) 233–242.

[Le64] J. Leech: Some sphere packings in higher space, *Canad. J. Math.* **16** (1964) 657–682.

[Le56] J. Leech: The problem of thirteen spheres, *Math. Gazette* **40** (1956) 22–23.

[Le79] V.I. Levenšteĭn: On bounds for packings in *n*-dimensional Euclidean space, *Soviet Math. Dokl.* **20** (1979) 417–421.

[Li77] J. Linhart: Scheibenpackungen mit nach unten beschränkter Nachbarnzahl, *Studia Sci. Math. Hungar.* **12** (1977) 281–293.

[Li73] J. Linhart: Die Newtonsche Zahl von regelmässigen Fünfecken, *Period. Math. Hungar.* **4** (1973) 315–328.

[Li68] J.E. Littlewood: *Some Problems in Real and Complex Analysis*, D. C. Heath and Co. Raytheon Education Co., 1968

[Mu05] O.R. Musin: The kissing number in four dimensions, arXiv: math.MG/0309430, manuscript.

[Mu03] O.R. Musin: The problem of the twenty-five spheres, *Russ. Math. Surv.* **58** (2003), 794–795.

[OdS79]   A.M. Odlyzko, N.J.A. Sloane: New bounds on the unit spheres that can touch a unit sphere in n-dimensions, *J. Combinatorial Theory Ser. A* **26** (1979) 210–214.

[Pe84]   M.A. Perles: At most 2^{d+1} neighborly simplices in E^d, *Ann. Discrete Math.* **20** (1984) 253–254.

[Pe71]   C.M. Petty: Equilateral sets in Minkowski space, *Proc. Amer. Math. Soc.* **29** 369–374.

[Sch70]   J. Schopp: Über die Newtonsche Zahl einer Scheibe konstanter Breite, *Studia Sci. Math. Hungar.* **5** (1970) 475–478.

[ScvW53]   K. Schütte, B.L. van der Waerden: Das Problem der dreizehn Kugeln, *Math. Ann.* **125** (1953) 325–334.

[SwD53]   H.P.F. Swinnerton-Dyer: Extremal lattices of convex bodies, *Proc. Cambridge Phil. Soc.* **49** (1953) 161–162.

[Ta05]   I. Talata: On Hadwiger numbers of direct products of convex bodies, in: *Combinatorial and Computational Geometry*, J.E. Goodman et al., eds., Cambridge Univ. Press, *MSRI Publications* **52** (2005), to appear.

[Ta00]   I. Talata: A lower bound for the translative kissing numbers of simplices, *Combinatorica* **20** (2000) 281–293.

[Ta99a]   I. Talata: The translative kissing number of tetrahedra is 18, *Discrete Comput. Geom.* **22** (1999) 231–248.

[Ta99b]   I. Talata: On extensive subsets of convex bodies, *Period. Math. Hungar.* **38** (1999) 231–246.

[Ta98a]   I. Talata: On a lemma of Minkowski, *Period. Math. Hungar.* **36** (1998) 199–207.

[Ta98b]   I. Talata: Exponential lower bound for the translative kissing numbers of d-dimensional convex bodies, *Discrete Comput. Geom.* **19** (1998) 447–455.

[Ti59]   H. Tietze: *Gelöste und ungelöste mathematische Probleme aus alter und neuer Zeit*, 2. Aufl., Verlag C.H. Beck, 1959. English translation: *Famous Problems in Mathematics*, Graylock Press, 1965.

[Ti905]   H. Tietze: Über das Problem der Nachbargebiete im Raum, *Monatshefte Math. Phys.* **16** (1905) 211–216.

[Vo36]   H. Voderberg: Zur Zerlegung der Umgebung eines ebenen Bereiches in kongruente, *Jahresber. Deutsch. Math.-Verein.* **46** (1936) 229–231.

[Wa71]   G.L. Watson: The number of minimum points of positive quadratic forms, *Dissertationes Math. Rozprawy Mat.* **84**

(1971) 1–43.

[We92] G. Wegner: Relative Newton numbers, *Monatshefte Math.* **114** (1992) 149–160.

[Za91] J. Zaks: No nine neighborly tetrahedra exist, *Memoirs Amer. Math. Soc.* **447**, Amer. Math. Soc. 1991.

[Za87] J. Zaks: Neighborly families of congruent convex polytopes, *Amer. Math. Monthly* **94** (1987) 151–155.

[Za86] J. Zaks: Arbitrarily large neighborly families of symmetric convex polytopes, *Geometriae Dedicata* **20** (1986) 175–179.

[Za81] J. Zaks: Neighborly families of 2^d d-simplices in E^d, *Geometriae Dedicata* **11** (1981) 505–507.

[Za79] J. Zaks: Bounds on neighborly families of convex polytopes, *Geometriae Dedicata* **8** (1979) 279–296.

[Zo99] C. Zong: *Sphere Packings*, Springer-Verlag 1999.

[Zo98] C. Zong: The kissing numbers of convex bodies—a brief survey, *Bull. London Math. Soc.* **30** (1998) 1–10.

[Zo95] C. Zong: A few remarks on kissing numbers of convex body, *Anz. Österreich. Akad. Wiss. Math.-Naturw. Kl.* **132** (1995) 11–15.

[Zo94] C. Zong: An example concerning the translative kissing number of a convex body, *Discrete Comput. Geom.* **12** (1994) 183–188.

2.5 Thin Packings with Many Neighbors

Two members of a packing are said to be *neighbors* if they touch each other, i.e., if they have a boundary point in common. If every member of a packing has at least k neighbors, we call it a *k-neighbor* packing [FeT69].

Taking two adjacent rows of circles in the densest lattice packing, we obtain a four-neighbor packing of equal circles with density 0 in the plane. On the other hand, L. Fejes Tóth [FeT73] proved that the density of any five-neighbor packing of equal circles is at least $\sqrt{3}\pi/7$, and this bound can be attained. It was proved in [FeTF91] that if the infimum of the ratios of the radii of any two circles in a five-neighbor circle packing $\mathcal{C}$ is larger than

$$
h = \frac{1}{8}\left(\left(23 + \sqrt{\frac{7424}{27}}\right)^{1/3} + \left(23 - \sqrt{\frac{7424}{27}}\right)^{1/3} - 1\right) = 0.5329\ldots,
$$

then the lower density of $\mathcal{C}$ is positive.

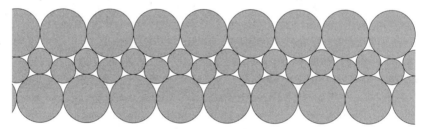

FIVE-NEIGHBOR PACKING OF DENSITY 0

Problem 1   (G. and L. Fejes Tóth [FeTF91]) Is it true that if the infimum of the ratios of the radii of any two circles in a five-neighbor circle packing $\mathcal{C}$ is larger than the above constant h, then the lower density of $\mathcal{C}$ is necessarily at least $\sqrt{3}\pi/7$?

In other words, is it true that the thinnest five-neighbor packing of circles with the property in Problem 1 consists of congruent circles?

Conjecture 2   (G. and L. Fejes Tóth [FeTF91]) If the infimum of the ratios of the radii of any two circles in a five-neighbor packing $\mathcal{C}$ of circles is larger than the above constant h, then $\mathcal{C}$ is necessarily connected.

G. and L. Fejes Tóth have verified that the last statement holds for circle packings in which the ratio of the radii of any two circles exceeds 0.7231. On the other hand, using circles of very different radii, it is easy to construct finite five-neighbor packings and hence also disconnected ones.

For each even $n \geq 12$, $n \neq 14$, Szabó [Sz94] described packings of n circles whose every member has precisely five neighbors, and showed that no such packing exists for any other value of n.

It readily follows from the fact that every planar graph with $n \geq 3$ vertices has at most $3n - 6$ edges that every 6-neighbor packing of circles (or of any other bodies in the plane) has infinitely many members. The only six-neighbor packing of *equal* circles is the densest lattice packing. Moreover, Bárány, Füredi, and Pach proved [BáFP84] that if $\mathcal{C}$ is a six-neighbor packing of (not necessarily equal) circles in the plane, then either all circles are congruent or $\mathcal{C}$ contains circles of arbitrarily small radii.

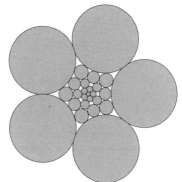

FIVE-NEIGHBOR PACKING
OF 26 DISCS

Many further results and references about six-neighbor circle packings can be found in Koebe [Ko36], Rodin, Sullivan [RoS87], Marden, Rodin [MaR89], Beardon, Stephenson [BeS90], and Sachs [Sa94].

It is very likely that similar phenomena occur for any convex body $C \subseteq \mathbb{R}^d$. In particular, there exists a smallest $k = k(C)$ with the property that the density of any k-neighbor packing with congruent copies of C is positive. Obviously, $k(C)$ cannot exceed $N(C)$, the Newton number of C, which is the maximum number of nonoverlapping congruent copies of C that can touch C. However, our knowledge about Newton numbers is very limited. As far as we know, the following simple question is open even in the plane.

Problem 3   (J. Pach) *For any convex body $C \subseteq \mathbb{R}^d$, let $N'(C)$ denote the largest integer k for which there exists a k-neighbor packing of congruent copies of C in $\mathbb{R}^d$. Is it true that the density of any $N'(C)$-neighbor packing of congruent copies of C is positive?*

Gács [Gá72] showed that if $N'(C) = N(C)$ for some plane convex body C, then this value is bounded from above by an absolute constant. Linhart [Li77] proved that the value of this constant is at most 21, and he has also constructed a tiling of the plane by congruent copies of an isosceles triangle T with base angles $\pi/6$ in which every member has precisely 21 neighbors. Wegner [We92] proved that for this triangle we have $N'(T) = N(T) = 21$. In three and higher dimensions it is not known whether there exist infinitely many integers that can occur as the common value of $N'(C) = N(C)$ for some convex body C.

The question becomes somewhat easier if we restrict our investigations to translative packings. It is not hard to see that any plane convex body C permits a six-neighbor lattice packing of density at least $1/2$ (see [FeT69]). Moreover, according to L. Fejes Tóth [FeT73], every six-neighbor packing of translates of any plane convex body C has density at least $1/2$. On the other hand, Makai [Ma87] proved a conjecture in [FeT73] by showing that for any plane convex body different from a parallelogram, the density of any five-neighbor packing with translates of C is at least $3/7$ (and at least $9/14$ for centrally symmetric C), with equality only if C is a triangle (affine image of a regular hexagon, respectively). Of course, C always permits a translative four-neighbor packing of density 0.

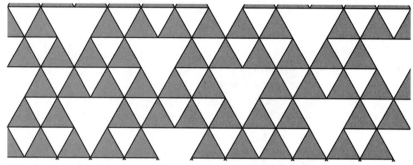

THINNEST FIVE-NEIGHBOR PACKING WITH CONGRUENT CONVEX BODIES

THINNEST FIVE-NEIGHBOR PACKING WITH CENTRALLY SYMMETRIC
CONGRUENT CONVEX BODIES

Solving a problem of L. Fejes Tóth and Sachs [FeTS76], G. Fejes Tóth [Fe81] and Sachs [Sa86] showed that any ten-neighbor packing of congruent balls in $\mathbb{R}^3$ has positive density (see also [BeB88]). This statement is a consequence of the fact that it is impossible to place ten points on the closed unit hemisphere in $\mathbb{R}^3$ so that the distance between any two of them is at least one. G. Fejes Tóth has also described a relatively thin 10-neighbor packing of density $\frac{5\pi}{18\sqrt{2}} = 0.617\ldots$. The arrangement consisting of two adjacent hexagonal layers in the densest lattice packing of balls is a

9-neighbor packing of density 0.

Problem 4   (L. Fejes Tóth, Sachs [FeTS76]) Determine the minimum
density of
(1) a ten-neighbor packing of congruent balls in $\mathbb{R}^3$,
(2) an eleven-neighbor packing of congruent balls in $\mathbb{R}^3$.

Conjecture 5   (L. Fejes Tóth [FeT89]) Any twelve-neighbor packing of
unit balls in $\mathbb{R}^3$ can be decomposed into parallel hexagonal
layers (as in the densest lattice packing of balls).

This conjecture would offer an unconventional alternative for attacking the sphere packing problem in $\mathbb{R}^3$. Note that, for example, in eight-
dimensional space the Newton number of the unit ball is 240, and according to a well-known result of Bannai and Sloane [BaS81], a 240-neighbor
packing of equal balls is essentially uniquely determined. The situation is
similar for $d = 2$ and 24.

Kertész [Ke94] proved that in every arrangement of nine points on
the closed unit hemisphere in $\mathbb{R}^3$ whose mutual distances are at least one,
six of the points must lie on the boundary. Thus, there exists no finite
nine-neighbor packing of unit balls.

Problem 6   [FeTH87] Does there exist a finite seven-neighbor (or even
eight-neighbor) packing of congruent balls in $\mathbb{R}^3$?

G. Fejes Tóth and Harborth [FeTH87] proved that the smallest packing in which every ball has precisely four (five) neighbors consists of six
(respectively twelve) balls. They also reported a construction by Wegner
of a six-neighbor packing consisting of 240 congruent balls, but this example, of course, is not known to be minimal. Harborth, Szabó, and Ujváry-
Menyhárt [HaSU02] studied the same problem for incongruent balls. They
constructed small (finite) examples for every $k \leq 10$ in which every ball has
precisely k neighbors. Using stereographic projection, the 120 spheres inscribed in the faces of a regular spherical polytope in $\mathbb{S}^3$ provide an example
in $\mathbb{R}^3$ where each ball has precisely twelve neighbors.

Problem 7   [HaSU02] Does there exist a finite packing of balls in $\mathbb{R}^3$
whose every member has precisely eleven neighbors?

Since the smallest ball in a finite packing has at most twelve neighbors,
there exists no finite 13-neighbor ball packing in $\mathbb{R}^3$.

For $d = 4$, Szabó [Sz91] showed that any 21-neighbor packing of equal
balls in $\mathbb{R}^4$ is of positive density, and there exists an 18-neighbor packing
with density 0. As in the three-dimensional case, the proof of the first
part of this result is based on the observation that if $\overline{g}(d)$ is the maximum
number of points that can be arranged on a closed unit hemisphere in $\mathbb{R}^d$ so

that their minimum (Euclidean) distance is 1, then any $(\bar{g}(d)+1)$-neighbor packing of B^d has positive density. Szabó showed that $\bar{g}(4) \leq 20$, but the best known lower bound is $\bar{g}(4) \geq 18$.

Thus, the following variant of the classical problem of Tammes, which seems to be of somewhat technical character, plays a crucial role here.

Problem 8   (L. Fejes Tóth, Sachs [FeT76]) Determine the maximum number $\bar{g}(d)$ (and $g(d)$) of points that can be placed on a closed (respectively open) unit hemisphere in $\mathbb{R}^d$ so that the Euclidean distance between any two of them is at least one. In particular, decide whether $\bar{g}(4) = 18$.

A possible "application" of bounds on $g(d)$ would follow from the well-known fact that any system of nonoverlapping unit balls in $\mathbb{R}^d$ can be colored by $g(d)$ colors so that no two balls of the same color touch each other. Kertész's result mentioned above states that $g(3) = 8$.

Problem 9   (Sachs [Sa68]) What is the smallest number of colors that is needed to color all elements of a packing of (congruent) balls in $\mathbb{R}^3$ so that no two balls that are tangent to each other receive the same color?

It follows, for instance, from the four color theorem for planar graphs that the answer to the corresponding question(s) in the plane is four.

Another interesting line of research initiated by L. Fejes Tóth was to maximize the *average* number of neighbors in a finite packing of equal balls. Clearly, in the plane the supremum of this quantity over all finite packings is six. However, the problem is already nontrivial for circle packings on the two-dimensional sphere. Here every circle can have at most five neighbors, but it is known [Ro69] that there are only five different finite packings in which every circle is tangent to precisely five others. K. Bezdek, Connelly, and Kertész [BeCK87] proved that there exist constants $\varepsilon, \rho > 0$ such that in any packing of equal circles of radius at most ρ on the two-dimensional unit sphere, the average number of neighbors of the elements does not exceed $5 - \varepsilon$. Based on a relatively "dense" construction, they posed the following question.

Problem 10   [BeCK87] Is it true that the lim sup of the average number of neighbors of a packing of circles of radius r on a two-dimensional unit sphere is 4.4, as r tends to zero?

G. Kuperberg and Schramm [KuS94] studied the same problem for finite packings of not necessarily equal balls in $\mathbb{R}^3$. In view of the Newton–Gregory controversy about the kissing number of the ball, it would be nice to answer the following question.

Problem 11 *Do there exist arbitrarily large finite packings of not necessarily congruent balls in $\mathbb{R}^3$, such that the supremum of the average number of neighbors of their elements is at least thirteen?*

The construction in [KuS94] shows that this supremum is at least $666/53 \approx 12.56$.

We know very little about densities of k-neighbor packings in higher dimensions. Alon [Al97] constructed finite packings of balls in $\mathbb{R}^d$, in which every ball has roughly $2^{\sqrt{d}}$ neighbors. Since we lack satisfactory constructions, we have only poor estimates for the minimum number $k(d)$ such that every $k(d)$-neighbor packing of equal balls in $\mathbb{R}^d$ has positive density. For translates of cubes instead of balls, L. Fejes Tóth and Sauer [FeTS77] established such a result (see also Chvátal [Ch75] for the planar case).

Plenty of problems remain unsolved even in the plane and in three-dimensional space. It is known that the only three-dimensional convex body that can be touched by 26 of its nonoverlapping translates is the parallelepiped (H. Groemer [Gr61]).

Problem 12 *(L. Fejes Tóth [FeT73]) Is it true that in three-dimensional space, parallelepipeds are the only convex bodies that*
  (1)   allow a 25-neighbor packing by translates?
  (2)   can be touched by 25 nonoverlapping translates?

Solving a problem of L. Fejes Tóth [FeT73], K. Bezdek and Brass [BeB03] proved that the number of nonoverlapping translates of a convex body $C \subset \mathbb{R}^d$ that can touch C and lie in a closed half-space bounded by a supporting hyperplane of C cannot exceed $2 \cdot 3^{d-1} - 1$. Equality holds here if and only if C is a parallelepiped. Since every packing of translates of C with density zero must contain such a configuration, it follows that any $2 \cdot 3^{d-1}$-neighbor translative packing by C has positive density. On the other hand, there exist $(2 \cdot 3^{d-1} - 1)$-neighbor packings with translates of a parallelepiped with density zero.

Problem 13 *(L. Fejes Tóth [FeT73]) What is the smallest number k, such that any k-neighbor packing of translates of a three-dimensional convex body with smooth boundary is of positive density?*

To see that $k \geq 11$, notice that from a tiling of space by truncated octahedra one can select a packing of density zero whose every member shares a two-dimensional boundary piece with ten others. One can preserve all face-to-face tangencies among these bodies while smoothening their boundaries [FeT73].

Problem 14   *[BeB03] What is the smallest number $K(d)$ with the property that for any convex body $C \subset \mathbb{R}^d$, the number of nonoverlapping translates that can touch C and lie in the open half-space bounded by a supporting hyperplane of C is at most $K(d)$?*

Clearly, $K(d) < 2 \cdot 3^{d-1} - 1$ for every d. Furthermore, every $(K(d)+1)$-neighbor packing of translates of a d-dimensional convex body is infinite.

Let $\ell(C)$ denote the largest integer ℓ for which there exists a finite ℓ-neighbor packing of translates of C. Talata [Ta02] proved that for $d = 3$, $\ell(C) = 13$ if C is a parallelepiped, and $\ell(C) = 10$ if C is any other convex cylinder. He also determined the density of the densest 17-neighbor packing of translates of a tetrahedron [Ta99]. It turns out that every such packing has at least one element that has more than seventeen neighbors. Talata [Ta99] conjectured that the maximum density of a translative packing of a tetrahedron is $18/49$, the same as the density of the densest lattice packing. In the densest lattice packing, every member has precisely fourteen neighbors. Paradoxically, it can be expected that if we require every member of a translative packing to have at least fifteen neighbors, then the maximum density *decreases*. In fact, Talata made the following attractive conjecture.

Conjecture 15 *[Ta99] The density of a 15-neighbor packing of translates of a tetrahedron cannot exceed $1/3$.*

[Al97]   N. ALON: Packings with large minimum kissing numbers, *Discrete Math.* **175** (1997) 249–251.

[BaS81]   E. BANNAI, N.J.A. SLOANE: Uniqueness of certain spherical codes, *Canad. J. Math.* **33** (1981) 437–449.

[BáFP84]   I. BÁRÁNY, Z. FÜREDI, J. PACH: Discrete convex functions and proof of the six circle conjecture of Fejes Tóth, *Canad. J. Math.* **36** (1984) 569–576.

[BeS90]   A.F. BEARDON, K. STEPHENSON: The uniformization theorem for circle packings, *Indiana Univ. Math. J.* **39** (1990) 1383–1425

[BeB88]   A. BEZDEK, K. BEZDEK: A note on the ten-neighbor packings of equal balls, *Beiträge Algebra Geom.* **27** (1988) 49–53.

[BeB03]   K. BEZDEK, P. BRASS: On k^+-neighbour packings and one-sided Hadwiger configurations, *Beiträge Algebra Geom.* **44** (2003) 493–498.

[BeCK87]   K. BEZDEK, R. CONNELLY, G. KERTÉSZ: On the average number of neighbors in a spherical packing of congruent circles, in: *Intuitive Geometry* (Siófok, 1985), K. Böröczky et al., eds., *Colloq. Math. Soc. János Bolyai* **48** (1987) 37–52.

[Ch75]   V. CHVÁTAL: On a conjecture of Fejes Tóth, *Period. Math. Hungar.* **6** (1975) 357–362.

[FeT81]   G. FEJES TÓTH: Ten-neighbor packing of equal balls, *Period. Math. Hungar.* **12** (1981) 125–127.

[FeTF91]   G. FEJES TÓTH, L. FEJES TÓTH: Remarks on 5-neighbor packings and coverings with circles, in: *Applied Geometry and Discrete Mathematics*, P. Gritzmann et al., eds., *DIMACS Ser. Discrete Math. Theoret. Comput. Sci.* **4**, Amer. Math. Soc. 1991, 275–288.

[FeTH87]   G. FEJES TÓTH, H. HARBORTH: Kugelpackungen mit vorgegebenen Nachbarnzahlen, *Studia Sci. Math. Hungar.* **22** (1987) 79–82.

[FeT89]   L. FEJES TÓTH: Research problem 44, *Period. Math. Hungar.* **20** (1989) 89–91.

[FeT73]   L. FEJES TÓTH: Five-neighbor packing of convex discs, *Period. Math. Hungar.* **4** (1973) 221–229.

[FeT69]   L. FEJES TÓTH: Scheibenpackungen konstanter Nachbarnzahl, *Acta Math. Acad. Sci. Hungar.* **20** (1969) 375–381.

[FeTS76]   L. FEJES TÓTH, H. SACHS: Research problem 17, *Period. Math. Hungar.* **7** (1976) 87–89.

[FeTS77]   L. FEJES TÓTH, N. SAUER: Thinnest packing of cubes with a given number of neighbours, *Canad. Math. Bull.* **20** (1977) 501–507.

[Gá72]   P. GÁCS: Packing of convex sets in the plane with a great number of neighbors, *Acta Math. Acad. Sci. Hungar.* **23** (1972) 383–388.

[Gr61]   H. GROEMER: Abschätzungen für die Anzahl der konvexen Körper, die einen konvexen Körper berühren, *Monatshefte Math.* **65** (1961) 74–81.

[HaSU02]   H. HARBORTH, L. SZABÓ, Z. UJVÁRY-MENYHÁRT: Regular sphere packings, *Arch. Math. (Basel)* **78** (2002) 81–89.

[Ke94]   G. KERTÉSZ: Nine points on the hemisphere, in: *Intuitive Geometry* (Szeged, 1991), K. Böröczky et al., eds., *Colloq. Math. Soc. János Bolyai* **63** (1994) 189–196.

[Ko36]   P. KOEBE: Kontaktprobleme der konformen Abbildung, *Ber. Sächs. Akad. Wiss. Leipzig, Math.-Phys. Kl.* **88** (1936) 141–164.

[KuS94]   G. KUPERBERG, O. SCHRAMM: Average kissing numbers
          for non-congruent sphere packings, *Math. Res. Lett.* **1** (1994)
          339–344.

[Li77]   J. LINHART: Scheibenpackungen mit nach unten beschränk-
          ter Nachbarnzahl, *Studia Sci. Math. Hungar.* **12** (1977) 281–
          293.

[Ma87]   E. MAKAI JR.:   Five-neighbor packing of convex plates, in:
          Intuitive Geometry (Siófok, 1985) K. Böröczky et al., eds.,
          Colloq. Math. Soc. János Bolyai **48** (1987) 373–381.

[MaR90]   A. MARDEN, B. RODIN: On Thurston's formulation and
          proof of Andreev's theorem, in: *Combinatorial Methods and
          Function Theory 1989*, S. Ruscheweyh et al., eds., Springer
          Lecture Notes Math. **1435** (1990) 103–115.

[Ro69]   R.M. ROBINSON: Finite sets of points on a sphere with each
          nearest to five others, *Math. Ann.* **179** (1969) 296–318.

[RoS87]   B. RODIN, D. SULLIVAN: The convergence of circle packings
          to the Riemann mapping, *J. Differential Geom.*, **26** (1987)
          349–360.

[Sa94]   H. SACHS: Coin graphs, polyhedra, and conformal mapping,
          Discrete Math. **134** (1994) 133–138.

[Sa86]   H. SACHS: No more than nine unit balls can touch a closed
          unit hemisphere, *Studia Sci. Math. Hungar.* **21** (1986) 203–
          206.

[Sa68]   H. SACHS: Problem 6, in: *Beiträge zur Graphentheorie vorge-
          tragen auf dem internationalen Kolloquium* (Manebach, DDR,
          Mai 1967), H. Sachs et al., eds., Teubner 1968, 225.

[Sz94]   L. SZABÓ: Regular circle packings, in: *Intuitive Geometry*
          (Szeged, 1991), K. Böröczky et al., eds., *Colloq. Math. Soc.
          János Bolyai* **63** (1994) 465–474.

[Sz91]   L. SZABÓ: 21-neighbor packing of equal balls in the 4-dimen-
          sional Euclidean space, *Geometriae Dedicata* **38** (1991) 193–
          197.

[Ta02]   I. TALATA: On minimum kissing numbers of finite translative
          packings of a convex body, *Beiträge Algebra Geom.* **43** (2002)
          501–511.

[Ta99]   I. TALATA: The translative kissing number of tetrahedra is
          18, *Discrete Comput. Geom.* **22** (1999) 231–248.

[We92]   G. WEGNER: Relative Newton numbers, *Monatshefte Math.*
          114 (1992) 149–160.

2.6 Permeability and Blocking Light Rays

A packing $\mathcal{C}$ of convex bodies lying between two parallel hyperplanes is called a *layer*. The distance d between the boundary hyperplanes is the *width* of the layer. Assume, for simplicity, that every bounded portion of a layer contains only finitely many members of $\mathcal{C}$. Following L. Fejes Tóth [FeT66], we define the *permeability* of the layer $\mathcal{C}$ as

$$p(\mathcal{C}) = \frac{d}{\inf \text{length}(\gamma)},$$

where the infimum is taken over all rectifiable curves γ connecting the boundary hyperplanes and avoiding the interior of every member of $\mathcal{C}$.

L. Fejes Tóth [FeT66] proved that in the plane the permeability of any layer of congruent circles is at least $\frac{\sqrt{27}}{2\pi}$ (see also [Fl78] for a more precise result, depending on the width). This bound cannot be improved, as shown by layers formed by many consecutive parallel strings of circles in the densest lattice packing of circles in the plane. L. Fejes Tóth also constructed layers of incongruent circles with arbitrarily large widths whose permeabilities are smaller than $0.8233 < \frac{\sqrt{27}}{2\pi} \approx 0.8269$. Danzer (unpublished) exhibited some better examples with permeabilities not exceeding 0.8224.

Problem 1   *(L. Fejes Tóth [FeT66]) Determine the infimum of the permeabilities of all layers formed by (not necessarily congruent) circles.*

L. Fejes Tóth [FeT68] (see also Bollobás [Bo68]) noticed that using incongruent *squares*, one cannot construct layers of substantially smaller permeabilities than using congruent ones.

Problem 2   *(L. Fejes Tóth [FeT68]) Characterize all convex bodies C in the plane that have the property that given any layer $\mathcal{C}$ of convex bodies similar (or homothetic) to C, then there is a layer of translates of C whose permeability is not larger than $p(\mathcal{C})$.*

It follows from the results of L. Fejes Tóth that the circle does not belong to this class, but every parallelogram does.

Given a convex body C, let $p_T(C) = \inf_{\mathcal{C}} p(\mathcal{C})$, where the infimum is taken over all layers $\mathcal{C}$ of *translates* of C. Furthermore, let $D(C)$ denote the *Minkowski symmetrization* of C, that is, let

$$D(C) = \frac{C + (-C)}{2}.$$

Answering a question of Gruber, Florian [Fl80] proved that $p_T(D(C)) = p_T(C)$ holds for every plane convex body C. Previously, this had been verified for bodies of constant width [Ho76] and for the regular triangle [Fl79].

For circles and squares, the above-mentioned results on permeabilities have been strengthened by G. Fejes Tóth [FeT78] and Pach [Pa77]. They proved that any pair of points at distance d from each other not belonging to the interior of any member of a packing C of congruent circles (respectively congruent squares), can be connected by a curve of length at most $\frac{2\pi}{\sqrt{27}} d + o(d)$ (resp., $\frac{3}{2} d + o(d)$) avoiding all "obstacles." Chan and Lam [ChaL93] and A. Bezdek [Be99] designed "on-line" algorithms to achieve these bounds, that is, to construct such connecting curves for an intelligent robot that can see and move in any direction but has no additional information about the location of the obstacles in advance. In fact, Bezdek's method also applies to packings of cubes in 3-dimensional space.

Problem 3   *Determine the infimum of the permeabilities of all layers of congruent balls or congruent cubes in 3-space.*

We do not have even a reasonable conjecture related to this problem. Apart from Bezdek's on-line algorithm, the only known method applicable to higher dimensional questions is the probabilistic approach of Pach [Pa77] (see also [FlG85]) for establishing some (rather weak) lower bounds.

A layer C of convex bodies is said to be a *k-fold cloud* (respectively *dark cloud*), if every straight line perpendicular to the boundary hyperplanes (respectively every straight line) intersects at least k members $C \in C$. Such an intersection point may belong to the boundary of C. It was shown by Heppes [He61] that the width of any k-fold cloud of unit circles is at least $(k-1)\sqrt{3} + 2$, with equality if C consists of k strings of circles from the densest lattice packing. For the three-dimensional case, L. Fejes Tóth [FeT59] proved that the smallest width of a 1-fold cloud of unit balls is $2 + \sqrt{2}$. Equality holds here only if C is the union of two square-lattice layers of balls such that each ball is tangent to precisely four balls in the other layer. Fitting together k such "simple" (1-fold) clouds, we obtain a k-fold cloud of width $(2k-1)\sqrt{2} + 2$. However, for $k > 1$, this bound is not tight. Heppes [He61] constructed a 2-fold cloud of width $2\sqrt{3} + 2$, which can be obtained as the union of three regular triangular lattice arrangements placed on top of one another, as shown in the figure.

Problem 4   *(L. Fejes Tóth) What is the smallest width of a dark cloud of unit spheres in three-dimensional Euclidean space?*

According to an observation of Heppes [He60], for any lattice packing of balls there is an infinite circular cylinder avoiding every member of the packing. Therefore, in contrast to the planar situation, by taking a few layers of balls in a lattice packing we never obtain a dark cloud. Actually, Heppes [He60] proved the stronger statement that one can also find *three* circular cylinders with the required property whose axes are noncoplanar.

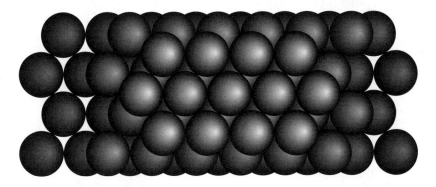

Twofold cloud of balls

Problem 5   (*L. Fejes Tóth*) *What is the smallest width of a twofold cloud of unit balls in three-dimensional Euclidean space?*

A local version of the above problem was suggested much earlier by H. Hornich (see [FeT64], p. 305 and [Zo99b]).

Problem 6   (*Hornich*) *What is the smallest number $h(d)$ of nonoverlapping unit balls that can "radially" hide a unit ball $B \subset \mathbb{R}^d$, in the sense that every ray emanating from the center of B meets the interior or the boundary of at least one of them?*

In the plane, this number is clearly 6. In three-dimensional space, Csóka [Cs77] and Danzer [Dan60] proved that $30 \leq h(3) \leq 42$.

Problem 7   (*L. Fejes Tóth [FeT59]*) *What is the smallest number $f(d)$ of nonoverlapping unit spheres that can hide a unit ball $B \subset \mathbb{R}^d$, in the sense that every ray emanating from the boundary of B meets the interior or the boundary of at least one of them?*

Obviously, we have that $f(d) \geq h(d)$ holds for every d and that $f(2) = h(2) = 6$. On the other hand, J. Schopp (cf. [BöS96]) showed that $f(3) \leq 326$. See [Zo97] for the best known general estimates.

Improving some earlier results by Bárány, Leader, and Zong (see Zong [Zo99a]), Talata [Ta00] and Böröczky Jr. and Tardos [BöT02] established various bounds of the form $2^{\Theta(d^2)}$ on both $f(d)$ and $h(d)$.

Similar questions can be asked about systems of congruent copies or translates of any convex body other than the sphere.

Conjecture 8   *(Böröczky Jr. [Bö04]) For any smooth strictly convex body*
C in the plane, the smallest set of translates that can hide
a copy of C has eight or nine members.

It is not hard to see that nine is an upper bound.

The *blocking number* $b(C)$ of a convex body C is defined as the small-est number of nonoverlapping translates that are in contact with C at its boundary and prevent any other translate from touching it [Zo95]. The number $b'(C)$ is defined similarly, except that now the translates of C are allowed to overlap. Clearly, we have $b'(C) \le b(C)$ for every C.

Problem 9   *(Dalla, Larman, Mani-Levitska, and Zong [DaL*00]) Is it*
true that $b'(C) = b(C)$ holds for every convex body C?

It was proved in [DaL*00] that the blocking number of the d-dimensional cube is 2^d $(d \ge 1)$ and that $b(B^3) = 6$ and $b(B^4) = 9$. We also have $b(B^2) = 4$, since the blocking number of every plane convex body was shown to be *four* by Zong [Zo93]. Apart from this, our knowledge about blocking num-bers is rather limited.

CENTRAL SQUARE BLOCKED
BY FOUR TRANSLATES

Another variant of the permeability problem was discussed in [FrPR84]. For a given layer $\mathcal{C}$ of congruent copies of a convex body C, let $\pi(\mathcal{C})$ denote the infimum of the total length of the segments $\ell \cap C$, $C \in \mathcal{C}$, over all lines ℓ, divided by the width of the layer. It was proved that in the plane $\sup_{\mathcal{C}} \pi(\mathcal{C}) = \delta(C)$ holds, where the supremum is taken over all layers of congruent copies of C, and $\delta(C)$ denotes the density of the densest packing with congruent copies of C.

One can try to construct a ball packing whose complement contains no long straight-line segments. Let $s(d)$ denote the infimum of the length of the longest "free" segment avoiding all members of $\mathcal{C}$ over all packings $\mathcal{C}$ of unit balls in $\mathbb{R}^d$. Using a probabilistic argument similar to the one applied by Frankl et al. [FrPR84], Böröczky Jr. and Tardos [BöT02] proved that $s(d) \le 2^{d(1+o(1))}$.

[Be99]   A. BEZDEK: An optimal route planning evading cubes in three space, *Beiträge Algebra Geom.* **40** (1999) 79–87.

[Bo68]   B. BOLLOBÁS: Remarks on a paper of L. Fejes Tóth, *Studia Sci. Math. Hungar.* **3** (1968) 373–379.

[Bö04]   K. BÖRÖCZKY JR.:   *Finite Packing and Covering*, Cambridge University Press, 2004.

[BöS96]   K. BÖRÖCZKY, V.P. SOLTAN: Translational and homothetic clouds for a convex body, *Studia Sci. Math. Hungar.* **32** (1996) 93–102.

[BöT02]   K. BÖRÖCZKY JR., G. TARDOS: The longest segment in the complement of a packing, *Mathematika* **49** (2002) 45–49.

[ChaL93]   K.-F. CHAN, T.W. LAM: An on-line algorithm for navigating in an unknown environment, *Internat. J. Comput. Geom. Appl.* **3** (1993) 227–244.

[Cs77]   G. CSÓKA: The number of congruent spheres that cover a given sphere of three-dimensional space is not less than 30 (in Russian), *Studia Sci. Math. Hungar.* **12** (1977) 323–334.

[DaL*00]   L. DALLA, D.G. LARMAN, P. MANI-LEVITSKA, C. ZONG: The blocking numbers of convex bodies, *Discrete Comput. Geom.* **24** (2000) 267–144.

[Dan60]   L. DANZER: Drei Beispiele zu Lagerungsproblemen, *Arch. Math.* **11** (1960) 159–165.

[FeT78]   G. FEJES TÓTH: Evading convex discs, *Studia Sci. Math. Hungar.* **13** (1978) 453–461.

[FeT68]   L. FEJES TÓTH: On the permeability of a layer of parallelograms, *Studia Sci. Math. Hungar.* **3** (1968) 195–200.

[FeT66]   L. FEJES TÓTH: On the permeability of a circle layer, *Studia Sci. Math. Hungar.* **1** (1966) 5–10.

[FeT64]   L. FEJES TÓTH: *Regular Figures*, Pergamon Press, Oxford, 1964.

[FeT59]   L. FEJES TÓTH: Verdeckung einer Kugel durch Kugeln, *Publ. Math. Debrecen* **6** (1959) 234–240.

[Fl80]   A. FLORIAN: Über die Durchlässigkeit einer Schicht konvexer Scheiben, *Studia Sci. Math. Hungar.* **15** (1980) 125–132.

[Fl79]   A. FLORIAN: Über die Durchlässigkeit gewisser Scheibenschichten, *Österreich. Akad. Wiss. Math.-Naturw. Kl. Sitzungsber. II* **188** (1979) 417–427.

[Fl78]   A. FLORIAN: On the permeability of layers of discs, *Studia Sci. Math. Hungar.* **13** (1978) 125–132.

[FlG85]   A. FLORIAN, H. GROEMER: Two remarks on the permeability of layers of convex bodies, *Studia Sci. Math. Hungar.* **20** (1985) 259–265.

[FrPR84]   P. FRANKL, J. PACH, V. RÖDL: How to build a barricade, *Monatshefte Math.* **98** (1984) 93–98.

[He61]   A. HEPPES: Über Kreis- und Kugelwolken, *Acta Math. Acad. Sci. Hungar.* **12** (1961) 209– 214.

[He60]   A. HEPPES: Ein Satz über gitterförmige Kugelpackungen, *Ann. Univ. Sci. Budapest. Eötvös Sect. Math.* **3–4** (1960) 89–90.

[Ho76]   I. HORTOBÁGYI: Über die Durchlässigkeit einer aus Scheiben konstanter Breite bestehenden Schicht, *Studia Sci. Math. Hungar.* **11** (1976) 383–387.

[Pa77]   J. PACH: On the permeability problem, *Studia Sci. Math. Hungar.* **12** (1977) 419–424.

[Ta00]   I. TALATA: On translational clouds for a convex body, *Geometriae Dedicata* **80** (2000) 319–329.

[Zo99a]   C. ZONG: *Sphere Packings*, Springer-Verlag, New York, 1999.

[Zo99b]   C. ZONG: A note on Hornich's problem, *Arch. Math.* **72** (1999) 127–131.

[Zo97]   C. ZONG: A problem of blocking light rays, *Geometriae Dedicata* **67** (1997) 117–128.

[Zo95]   C. ZONG: Some remarks concerning kissing numbers, blocking numbers and covering numbers, *Period. Math. Hungar.* **30** (1995) 233–238.

[Zo93]   C. ZONG: *Packing and Covering* (Ph.D. thesis), TU Wien, 1993.

3. Packing and Covering with Homothetic Copies

3.1 Potato Bag Problems

Given a convex body $C \subseteq \mathbb{R}^d$ and a positive real λ, any set of the form $\lambda C + x = \{\lambda c + x \mid c \in C\}$ for some $x \in \mathbb{R}^d$ is called a *homothetic copy* (or, briefly, a *homothet*) of C. The number $\lambda > 0$ is said to be the *coefficient* of the homothety. A system of convex bodies $C_1, C_2, \ldots$ *permits a covering* (respectively, *packing*) of C if one can choose suitable congruent copies C_i' of C_i ($i = 1, 2, \ldots$) such that $\cup_{i=1}^{\infty} C_i' \supseteq C$ (respectively, the sets C_i' have pairwise disjoint interiors and $\cup_{i=1}^{\infty} C_i' \subseteq C$). If instead of congruent copies, we allow only *translates* of C_i satisfying the above conditions, then we obtain a *translative covering* (respectively, *translative packing*).

The celebrated Auerbach–Banach–Mazur–Ulam theorem announced in the *Scottish Book* (Mauldin [Ma81], p. 74) states that for every positive integer d and for every $V > 0$, there exists a number $f_d(V)$ such that any system $\{C_1, C_2, \ldots\}$ of d-dimensional convex bodies of diameter at most 1 and with total volume at most V can be packed into a cube of side $f_d(V)$. The authors mention as a corollary that "one kilogram of potatoes can be packed into a finite sack." However, the determination of the best possible value of $f_d(V)$ seems to be an extremely difficult problem. (For results of this kind, see Groemer [Gr82], [Gr85], Makai–Pach [MaP83].)

The first published proof of this theorem is due to Kosiński [Ko57]; it is based on a result about *boxes* (i.e., rectangular parallelepipeds) whose faces are parallel to the coordinate planes (see also [La98]). The following improved version of this lemma was established by Moon and Moser [MoM67]: Any system of d-dimensional boxes $B_1, B_2, \ldots$ whose edge lengths are at most 1 and whose total volume is V can be packed in parallel positions into a box with sides $2, 2, \ldots, 2, 2(V+1)$. For $d = 2$, a sharper result was found by Meir and Moser [MeM68]. In particular, they proved that all squares of sides $1/i$ ($i = 2, 3, \ldots$) can be packed into a square of side $5/6$, and this bound cannot be improved.

The following three questions in the same vein are still unsettled.

Problem 1   (L. Moser) *Can one pack all $\frac{1}{i} \times \frac{1}{i+1}$ rectangles R_i ($i = 1, 2, 3, \ldots$) into a square of side 1?*

Problem 2   (Meir, L. Moser [MeM68]) *Can one pack all squares of sides $\frac{1}{i}$ ($i = 2, 3, \ldots$) into a rectangle of area $\frac{\pi^2}{6} - 1$?*

Problem 3   (Meir, L. Moser [MeM68]) *Can one pack all squares of sides $\frac{1}{2i+1}$ ($i = 1, 2, 3, \ldots$) into a rectangle of area $\frac{\pi^2}{8} - 1$?*

Clearly, we have

$$\frac{1}{1 \cdot 2} + \frac{1}{2 \cdot 3} + \frac{1}{3 \cdot 4} + \ldots = 1,$$

$$\frac{1}{2^2} + \frac{1}{3^2} + \frac{1}{4^2} + \ldots = \frac{\pi^2}{6} - 1,$$

$$\frac{1}{3^2} + \frac{1}{5^2} + \frac{1}{7^2} + \ldots = \frac{\pi^2}{8} - 1,$$

so that if the answer to any of the above questions is affirmative, then the corresponding claim cannot be improved.

Martin [Ma02] showed that, e.g., if we can prove that all rectangles R_i can be fit into a square of side $1 + \varepsilon$ for every $\varepsilon > 0$, then the answer to the first question is in the affirmative. The same holds for the other two problems: if all squares can be packed into a rectangle of area $A + \varepsilon$ for every $\varepsilon > 0$, then they can also be packed into a rectangle of area A. Note that this result does not follow from a straightforward compactness argument.

Concerning the first problem, Jennings [Je95], [Je94] proved that all rectangles R_i can be packed into a square of side $\frac{204}{203}$. Bálint [Bá98], [Bá92], [Bá90] showed that if $R_1, \ldots, R_{k-1}$ can be packed into a unit square for some k divisible by 4, then all R_i can be packed into a $1 \times \left(1 + \frac{6}{5k}\right)$ rectangle. Since he could verify the assumption up to $k = 500$, it follows that all R_i can be packed into a 1×1.0024 rectangle, and he also showed that they fit into a square of side length $\frac{501}{500}$. Concerning the third problem, Jennings [Je95] proved that all squares of sides $\frac{1}{2i+1}$ can be packed into a $\left(\frac{1}{3} + \frac{1}{5}\right) \times \left(\frac{1}{3} + \frac{1}{9}\right)$ rectangle, whose area, $\frac{32}{135}$, exceeds $\frac{\pi^2}{8} - 1$ only by less than $\frac{1}{299}$. Paulhus [Pa98] designed an algorithm that enabled him to get even closer to the (conjectured) optimal values in all three problems.

Answering a similar question of Coxeter [Co79], Boyd [Bo80] proved that the radius of the smallest disk into which all disks of radii $1/i$ ($i = 1, 2, \ldots$) can be packed is $3/2$. This result can also be deduced using rectangle packings [Bá92].

For packings and coverings with cubes, Meir and L. Moser [MeM68], generalizing some earlier results of Moon and L. Moser [MoM67], established the following two theorems, duals of each other, whose weaker versions have been rediscovered many times (see, e.g., Newman [Ne83], A. and K. Bezdek [BeB84], and Stong [St97]):

(1) Any system of d-dimensional cubes of sides $s_1 \geq s_2 \geq \ldots$ and with total volume $V = \sum_{i=1}^{\infty} s_i^d$ can be packed into a box of size $r_1 \times r_2 \times \cdots \times r_d$, provided that $s_1 \leq \min\{r_1, r_2, \ldots, r_d\}$ and

$$(r_1 - s_1)(r_2 - s_1) \cdots (r_d - s_1) \geq V - s_1^d.$$

We obtain as a corollary that any system of cubes with total volume 1 can be packed into a cube of volume 2^{d-1}.

(2) Any system of d-dimensional cubes of sides $s_1 \geq s_2 \geq \ldots$ and with total volume $V = \sum_{i=1}^{\infty} s_i^d$ can cover a box of size $r_1 \times r_2 \times \ldots \times r_d$, provided that

$$(r_1 + s_1)(r_2 + s_1) \ldots (r_d + s_1) \leq V + s_1^d.$$

We obtain as a corollary that any system of cubes with total volume 1 can be arranged so as to cover a cube of volume $\frac{1}{2^d-1}$.

Neither of the above results can be improved for finitely many equal cubes if they are required to be in parallel positions with the box. However, several exciting related questions remain open.

Problem 4   (L. Moser [Mo66], also Guy [Gu75]) Can one strengthen (1) and (2) by imposing further conditions on the k largest elements of the sequence s_i $(i = 1, 2, \ldots)$?

Problem 5   (L. Moser [Mo66], also Guy [Gu75]) Can any system of boxes with total volume 1 and maximum side length 1 be packed into a cube of volume 2^{d-1}?

For $d = 2$, Januszewski [Ja00] answered the last question in the affirmative: any sequence of rectangles of sides at most $\sqrt{2}$ whose total area is 1 can be packed into a square of area 2. He also showed [Ja02c] that any sequence of axis-parallel rectangles with sides no longer than 1 and with total area 1 permits a translative packing into a 1×2 rectangle in parallel position. It is easy to argue that there is no rectangle satisfying the same condition whose area is smaller than 2.

Problem 6   (Moon, L. Moser [MoM67]) Determine the smallest number A such that any system of squares of total area 1 can be packed parallelly into some rectangle of area A.

By (1), we have $A \leq 2$. Kleitman and Krieger [KlK70] showed that such a system can always be packed into a 1 by $\sqrt{3}$ rectangle. (The example of 3 squares of side lengths $1/\sqrt{3}$ shows that this is the smallest rectangle with the required property, one of whose sides is 1.) In a subsequent paper Kleitman and Krieger [KlK75] proved that any collection of squares with total area 1 can be packed into a $2/\sqrt{3} \times \sqrt{2}$ rectangle. This implies that $A \leq 2\sqrt{2}/\sqrt{3}$, which was improved by Novotný [No96] to 1.53. On the other hand, Moon and L. Moser [MoM67] noticed that $A > 1.2$ follows from considering two squares of sides x and y, where $x > y$, $x^2 + y^2 = 1$, and $x(x + y)$ is maximal. Considering four squares, one of side $1/\sqrt{2}$ and three of sides $1/\sqrt{6}$, Novotný [No95] showed that $A \geq \frac{2+\sqrt{3}}{3} > 1.244$. He

[No99] also proved that any system of at most five squares with total area 1 can be packed into a rectangle of area $\frac{2+\sqrt{3}}{3}$.

For any convex body C, let $v(C)$ denote the smallest number v such that any system of similar copies of C whose combined volume is 1 can be packed into a similar copy of C with volume v. Analogously, let $v_T(C)$ denote the smallest v such that any system of homothets of C with combined volume 1 permits a translative packing into a homothetic copy of C with volume v. Clearly, we have $v(C) \leq v_T(C)$, for every C.

Problem 7   (Soifer [So99a], Novotný [No01]) Is it true that $v(C) \leq 2$ for any plane convex body C?

For triangles, the same problem had been asked earlier by Richardson [Ri95]. He proved that any system of homothetic copies of a triangle Δ having combined volume 1 can be packed into a homothet of Δ whose area is 2, provided that we are allowed to use both translations and reflections. It is easy to see that, for example, for an equilateral triangle, $v_T(\Delta) \geq v(\Delta) = 2$ holds. Indeed, $v(\Delta) \geq 2$ follows from the fact that two equilateral triangles of unit sides cannot be packed into an equilateral triangle of side length $2 - \varepsilon$ for any $\varepsilon > 0$. In fact, Januszewski [Ja03b] proved that for all nonequilateral triangles we have $v(\Delta) < 2$.

Soifer originally thought that $v(C) = 2$ holds for every plane convex body C, but Novotný [No01] showed that for the $3^{1/4} \times 2^{1/4}$ rectangle R we have $v(R) = \sqrt{8/3}$. Later, Januszewski [Ja03a] found that for the right isosceles triangle $\Delta_{\pi/2}$ the inequality $v(\Delta_{\pi/2}) \leq \frac{2}{7-4\sqrt{2}} < \frac{3}{2}$ holds.

Conjecture 8   (Januszewski [Ja03a]) For the right isosceles triangle $\Delta_{\pi/2}$, we have
$$v(\Delta_{\pi/2}) = \frac{2}{7 - 4\sqrt{2}}.$$

Conjecture 9   (Januszewski [Ja03a]) For any triangle Δ, we have $v(\Delta) \geq v(\Delta_{\pi/2})$.

Conjecture 10 (Richardson [Ri95]) $v_T(\Delta) = 2$ for every triangle Δ.

Slightly improving Soifer's estimate [So99a], [So99b], Januszewski in [Ja02a] established the upper bound $v_T(\Delta) \leq 56/19$. Novotný [No01] believes that Richardson's conjecture is true in a more general form: $v_T(C) = 2$ for every plane convex body C.

Doyle, Lagarias, and Randall [DoLR92] investigated an interesting variant of the above problem. For any convex body C and integer $k > 1$, let $r(C, k)$ denote the smallest r such that one can pack k translates of C into rC.

Using the above notation, we obviously have that $kv_T(C) \geq r^d(C, k)$ for any d-dimensional convex body C. They found that for centrally symmetric plane convex bodies $r(C, 2) = 2$, $r(C, 6) = r(C, 7) = 3$, independently of the shape of C. It is easy to verify that $r(C, 2) = 2$ even if C is not centrally symmetric. For general convex bodies, it is known that $r(C, 3) \geq \sqrt{6}$ [Lá04]. Furthermore, we have $r(C, 5) \geq \frac{3+\sqrt{5}}{2}$, and $r(C, 6) \geq \frac{9+\sqrt{5}}{4}$, where equalities hold if C is an affinely regular pentagon [DoL95], [BöL05].

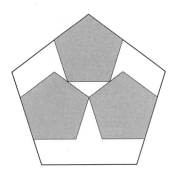

THREE HOMOTHETIC COPIES
IN A REGULAR PENTAGON

From the other direction, the only tight bound is $r(C, 5) \leq 3$ for any convex C [La94].

Conjecture 11 *(Lassak [La94])* $r(C, 3) \leq \sqrt{5}$ *for every plane convex body C, with equality when C is a regular pentagon. That is, one can always fit three nonoverlapping translates of C into $\sqrt{5}C$.*

Conjecture 12 *(Lángi and Lassak [LáL03]) For every centrally symmetric plane convex body, we have*

$$r(C, 3) \leq \frac{17}{5(1 + \sqrt{2})},$$

with equality for the regular octagon.

The proofs of (1) and (2) by Meir and L. Moser are algorithmic. Lassak and Zhang [LaZ91] suggested the *on-line* version of these questions. Imagine that the members of a sequence of convex bodies C_i are revealed one by one. We do not know what C_{i+1} is before we assign a position to C_i, which cannot be changed afterwards. In many cases, even with these restrictions one can construct (almost) optimal packings and coverings. For instance, Januszewski and Lassak [JaL97] obtained the following on-line version of (1) for cubes: any sequence of d-dimensional cubes of total volume at most $1/2^{d-1}$ permits an on-line packing into a unit cube. Of course, this result, just like (1), cannot be improved for cubes in parallel positions, as shown by the example of two congruent cubes of side length $1/2 + \varepsilon$, for any $\varepsilon > 0$. Other efficient on-line packing algorithms for boxes in parallel positions were given by Lassak [La97a], [La97b] and by Januszewski [Ja99].

Subsequently, Lassak [La02] showed that the d-dimensional unit cube can be on-line covered by any sequence of cubes of total volume at least $2^d +$

$\frac{5}{3}\left(1 + \frac{1}{2^d}\right)$, improving the results in [Ku94], [JaL94b], [JaL95], [JaL*96]. This comes surprisingly close to the sharp off-line bound $2^d - 1$ of Meir and L. Moser, which follows from (2).

Problem 13 *(Januszewski and Lassak [JaL94a]) Determine the smallest A such that any sequence of plane convex bodies of diameters at most one and total area A permits a (translative on-line) covering of the unit square.*

Januszewski [Ja96] proved that $A \leq 15$ (for translative on-line coverings). See [La97c] for a survey.

For problems and results concerning packings of boxes of integral side lengths, see de Bruijn [dBr69], Göbel [Gö79], Kotljar [Ko78], [Ko84], and Owings [Ow85].

[Bá98]   V. BÁLINT: Two packing problems, *Discrete Math.* **178** (1998) 233–236.

[Bá92]   V. BÁLINT: A packing problem and the geometrical series, in: *Proc. 4th Czechoslovakian Symp. on Combinatorics, Graphs and Complexity (Prachatice, 1990)*, J. Nešetřil et al., eds., *Ann. Discrete Math.* **51** (1992) 17–21.

[Bá90]   V. BÁLINT: A remark on a packing problem, *Práce Štúd. Vysokej Školy Doprav. Spojov Žiline Sér. Mat.-Fyz.* **8** (1990) 7–12.

[BeB84]   A. BEZDEK, K. BEZDEK: Eine hinreichende Bedingung für die Überdeckung des Einheitswürfels durch homothetische Exemplare im n-dimensionalen euklidischen Raum, *Beiträge Algebra Geom.* **17** (1984) 5–21.

[BöL05]   K. BÖRÖCZKY, Z. LÁNGI: On the relative distance of six points in a plane convex body, manuscript.

[Bo80]   D.V. BOYD: Solution to P-276, *Canad. Math. Bull.* **23** (1980) 251–252.

[dBr69]   N.G. DE BRUIJN: Filling boxes with bricks, *Amer. Math. Monthly* **76** (1969) 37–40.

[Co79]   H.S.M. COXETER: Problem P–276, *Canad. Math. Bull.* **22** (1979) 248.

[DoL95]   K. DOLIWKA, M. LASSAK: On relatively short and long sides of convex pentagons, *Geometriae Dedicata* **56** (1994) 221–224.

[DoLR92]   P.G. DOYLE, J.C. LAGARIAS, D. RANDALL: Self-packing of centrally symmetric convex bodies in R^2, *Discrete Comput. Geom.* **8** (1992) 171–189.

[**Gö79**] F. GÖBEL: Geometrical packing and covering problems, in: *Packing and Covering in Combinatorics*, A. Schrijver, ed., *Math. Centrum Tracts* **106** (1979) 179–199.

[**Gr85**] H. GROEMER: Coverings and packings by sequences of convex sets, in: *Discrete Geometry and Convexity*, J.E. Goodman et al., eds., *Ann. New York Acad. Sci.* **440** (1985) 262–278.

[**Gr82**] H. GROEMER: Covering and packing properties of bounded sequences of convex sets, *Mathematika* **29** (1982) 18–31.

[**Gu75**] R.K. GUY: Problems, in: *The Geometry of Linear and Metric Spaces*, L.M. Kelly, ed., Springer *Lecture Notes Math.* **490** (1975) 233–244.

[**Ja03a**] J. JANUSZEWSKI: Packing similar triangles into a triangle, *Period. Math. Hungar.* **46** (2003) 61–65.

[**Ja03b**] J. JANUSZEWSKI: Packing clones in triangles, *Geombinatorics* **13** 73–78.

[**Ja02a**] J. JANUSZEWSKI: A simple method of translative packing triangles in a triangle, *Geombinatorics* **12** (2002) 61–68.

[**Ja02b**] J. JANUSZEWSKI: Packing similar rectangles in a rectangle, *Geombinatorics* **12** (2002) 24–30.

[**Ja02c**] J. JANUSZEWSKI: Universal container for packing rectangles, *Colloq. Math.* **92** (2002) 155–160.

[**Ja00**] J. JANUSZEWSKI: Packing rectangles into the unit square, *Geometriae Dedicata* **81** (2000) 13–18.

[**Ja99**] J. JANUSZEWSKI: On-line packing regular boxes in the unit cube *Arch. Math. (Brno)* **35** (1999) 97–101.

[**Ja96**] J. JANUSZEWSKI: On-line covering of the unit square by a sequence of convex bodies, *Demonstratio Math.* **29** (1996) 155–158.

[**JaL97**] J. JANUSZEWSKI, M. LASSAK: On-line packing sequences of cubes in the unit cube, *Geometriae Dedicata* **67** (1997) 285–293.

[**JaL95**] J. JANUSZEWSKI, M. LASSAK: On-line covering the unit square by squares and the three-dimensional unit cube by cubes, *Demonstratio Math.* **28** (1995) 143–149.

[**JaL94a**] J. JANUSZEWSKI, M. LASSAK: On-line covering by boxes and by convex bodies, *Bull. Polish Acad. Sci. Math.* **42** (1994) 69–76.

[**JaL94b**] J. JANUSZEWSKI, M. LASSAK: On-line covering the unit cube by cubes, *Discrete Comput. Geom.* **12** (1994) 433–438.

[JaL*96]   J. Januszewski, M. Lassak, G. Rote, G. Woeginger: On-line q-adic covering by the method of the n-th segment and its application to on-line covering by cubes, *Beiträge Algebra Geom.* **37** (1996) 51–65.

[Je95]   D. Jennings: On packings of squares and rectangles, *Discrete Math.* **138** (1995) 293–300.

[Je94]   D. Jennings: On packing unequal rectangles in the unit square, *J. Combinatorial Theory Ser. A* **68** (1994) 465–469.

[KlK75]   D.J. Kleitman, M.M. Krieger: An optimal bound for two dimensional bin packing, in: *FOCS 1975* (Proc. 16th Annual Symposium on Foundations of Comp. Sci.), IEEE Computer Soc., 1975, 163–168

[KlK70]   D.J. Kleitman, M.M. Krieger: Packing squares in rectangles I, *Ann. New York Acad. Sci.* **175** (1970) 253–262.

[Ko57]   A. Kosiński: A proof of the Auerbach-Banach-Mazur-Ulam theorem on convex bodies, *Colloq. Math.* **4** (1957) 216–218.

[Ko84]   B.D. Kotljar: Packings of parallelotopes and certain other sets (in Russian), *Sibirsk. Mat. Zh.* **25** (1984) 222–225.

[Ko78]   B.D. Kotljar: Packings of rectangular parallelotopes (in Russian), *Kibernetika (Kiev)* **2** 1978 133–135. English translation in *Cybernetics* **14** (1978) 294–296.

[Ku94]   W. Kuperberg: On-line covering a cube by a sequence of cubes, *Discrete Comput. Geom.* **12** (1994) 83–90.

[La02]   M. Lassak: On-line algorithms for the q-adic covering of the unit interval and for covering a cube by cubes, *Beiträge Algebra Geom.* **43** (2002) 537–549.

[La98]   M. Lassak: The Auerbach-Banach-Mazur-Ulam problem on the packing of a potato sack (in Polish), *Wiadom. Mat.* **34** (1998) 49–59.

[La97a]   M. Lassak: On-line packing sequences of segments, cubes and boxes, *Beiträge Algebra Geom.* **38** (1997) 377–384.

[La97b]   M. Lassak: On-line potato-sack algorithm efficient for packing into small boxes, *Period. Math. Hung.* **34** (1997) 105–110.

[La97c]   M. Lassak: A survey of algorithms for on-line packing and covering by sequences of convex bodies, in: *Intuitive Geometry* (Budapest, 1995), I. Bárány et al., eds., *Bolyai Soc. Math. Stud.* **6** (1997) 129–157.

[La94]   M. Lassak: On five points in a plane convex body pairwise in at least unit relative distances, in: *Intuitive Geometry* (Szeged,

1991), K. Böröczky et al., eds., *Colloq. Math. Soc. János Bolyai* **63** (1994) 245–247.

[Lá04]   Z. LÁNGI: On seven points in the boundary of a plane convex body in large relative distances, *Beiträge Algebra Geom.* **45** (2004) 275–281.

[LáL03]   Z. LÁNGI, M. LASSAK: Relative distance and packing a body by homothetical copies, *Geombinatorics* **13** (2003) 29–40.

[LaZ91]   M. LASSAK, J. ZHANG: An on-line potato-sack theorem, *Discrete Comput. Geom.* **6** (1991) 1–7.

[MaP83]   E. MAKAI JR., J. PACH: Controlling function classes and covering Euclidean space, *Studia Sci. Math. Hungar.* **18** (1983) 435–459.

[Ma02]   G. MARTIN: Compactness theorems for geometric packings, *J. Combinatorial Theory Ser. A* **97** (2002) 225–238.

[Ma81]   R.D. MAULDIN: *The Scottish Book*, Birkhäuser 1981.

[MeM68]   A. MEIR, L. MOSER: On packing of squares and cubes, *J. Combinatorial Theory* **5** (1968) 126–134.

[MoM67]   J. MOON, L. MOSER: Some packing and covering theorems, *Colloq. Math.* **17** (1967) 103–110.

[Mo66]   L. MOSER: Poorly formulated unsolved problems of combinatorial geometry, mimeographed, 1966.

[Ne83]   D.J. NEWMAN: A covering problem, *SIAM Rev.* **25** (1983) 99–100.

[No01]   P. NOVOTNÝ: A note on packing clones, *Geombinatorics* **11** (2001) 29–30.

[No99]   P. NOVOTNÝ: On packing of four and five squares into a rectangle, *Note Matematica* **19** (1999) 199–206.

[No96]   P. NOVOTNÝ: On packing of squares into a rectangle, *Arch. Math. (Brno)* **32** (1996) 75–83.

[No95]   P. NOVOTNÝ: A note on a packing of squares, *Stud. Univ. Transp. Commun. Žilina Math.-Phys. Ser.* **10** (1995) 35–39.

[Ow85]   J. OWINGS: Tiling the unit square with squares and rectangles, *J. Combinatorial Theory Ser. A* **40** (1985) 156–160.

[Pa98]   M.M. PAULHUS: An algorithm for packing squares, *J. Combinatorial Theory Ser. A* **82** (1998) 147–157.

[Ri95]   T.J. RICHARDSON: Optimal packing of similar triangles, *J. Combinatorial Theory Ser. A* **69** (1995) 288–300.

[So99a]   A. SOIFER: Packing clones in convex figures, *Congressus Numerantium* **136** (1999) 65–72.

[**So99b**]   A. Soifer: Packing triangles in triangles, *Geombinatorics* **8** (1999) 110–115.

[**St97**]   R. Stong: Squares inside squares, *Geombinatorics* **7** (1997) 29–34.

3.2 Covering a Convex Body with Its Homothetic Copies

Let $f_d(C)$ denote the least positive number f with the property that any system $C_1, C_2, \ldots$ of homothetic copies of a convex body $C \subset \mathbb{R}^d$ whose total volume is at least f permits a translative covering of C. (Note that the existence of such a number follows from the "dual" Auerbach–Banach–Mazur–Ulam theorem; see Mauldin [Ma81].)

Meir and L. Moser [MeM68] showed that for the d-dimensional unit cube Q^d, we have $f_d(Q^d) = 2^d - 1$. The inequality $f_d(Q^d) \geq 2^d - 1$ follows from the fact that a cube cannot be covered by fewer than 2^d of its slightly smaller homothetic copies, because no pair of its vertices can be covered by the same copy. Since the problem remains invariant under any (volume-preserving) affine transformation of $\mathbb{R}^d$, we obtain that $f_d(P) = 2^d - 1$ for any parallelepiped P.

Conjecture 1 (*L. Fejes Tóth, 1984, personal communication*) *For any convex body C of area 1 in the plane, $f_2(C) \leq 3$.*

Evidently, $f_2(C) \geq 2$ for any plane convex body of unit area. On the other hand, reducing some earlier upper bounds in [BeB84], [BáB*93], [Ja01], Januszewski [Ja03] showed that $f_2(C) \leq 6.5$ for every such C. In $d \geq 3$ dimensions, he established the bound $f_d(C) \leq (d+1)^d - 1$.

Conjecture 2 (*V. Soltan, 1990, personal communication*) *Let C be a d-dimensional convex body that is covered by its homothetic copies $C_1, C_2, \ldots$ with positive coefficients $\lambda_1, \lambda_2, \ldots$, respectively, where each $\lambda_i < 1$. Then $\sum_{i=1}^{\infty} \lambda_i \geq d$.*

For $d = 2$, this has been proved by V. Soltan and Vásárhelyi [SoV93]. They also settled the special case in which one is allowed to use at most $d + 1$ homothets. (See also Fudali [Fu88].)

Given two convex bodies $C, D \subseteq \mathbb{R}^d$, let $f_d(C, D)$ denote the smallest positive number f with the property that any system $D_1, D_2, \ldots$ of homothetic copies of D with total volume at least f permits a translative covering of C. Obviously, $f_d(C, C) = f_d(C)$.

Januszewski [Ja02a] proved Böröczky's conjecture [MoP93] that if S and S' are unit squares such that one side of S' is parallel to a diagonal of S, then $f_2(S, S') = \frac{5}{2}$. He [Ja02b] also confirmed Bognár's conjecture [MoP93] that any system of squares of total area at least 2 permits a (not necessarily translative) covering of the unit square. This result is best possible, because the unit square cannot be covered by two squares of side $1 - \varepsilon$ ($\varepsilon > 0$).

Given a triangle $T \subseteq \mathbb{R}^2$ and a real α ($0 \leq \alpha < 2\pi$), let T^α denote the triangle obtained from T by a clockwise rotation with angle α. Füredi [Fü03] and Vásárhelyi [Vá84] proved that $f_2(T, T^0) = f_2(T) = 2$

and $f_2(T, T^\pi) = 4$, respectively, for any triangle T of unit area. Januszewski [Ja98] strengthened the latter statement by showing that if each member of a sequence is a homothet of T or T^π, and the total area of these triangles is at least 4, then they permit a translative covering of T.

Conjecture 3 *(A. and K. Bezdek [BeB84]) Let T be an equilateral triangle of unit area. Then $f_2(T, T^\alpha) \leq 4$ holds for any $0 \leq \alpha < 2\pi$.*

Vásárhelyi [Vá93] confirmed this conjecture for $\alpha \leq \pi/12$ and for $\alpha = \pi/6$. On the other hand, she showed that for any nonequilateral triangle T of area 1, there exists $0 < \alpha < \pi$ with $f_2(T, T^\alpha) > 4$.

Given a convex body C in the plane and an integer $k \geq 2$, let $\lambda_k(C)$ denote the smallest $\lambda > 0$ such that C can be covered by k homothetic copies of itself with coefficient λ. Let

$$\ell_k = \inf_C \lambda_k(C), \qquad L_k = \sup_C \lambda_k(C),$$

where the infimum and the supremum are taken over all plane convex bodies C. The investigation of these functions was initiated by Levi [Le54].

Obviously, we have $\ell_2 = L_2 = L_3 = 1$ and $\ell_k \geq 1/\sqrt{k}$, with equality if k is a perfect square. Belousov [Be77] and Krotoszyński [Kr87] proved that $\ell_3 = 2/3$ and $\ell_4 = \ell_5 = 1/2$, respectively. Lassak [La86] has shown that $L_4 = 1/\sqrt{2}$, and Levi [Le54] proved $L_7 = L_8 = 1/2$. These are the only exact values of ℓ_k and L_k that are known to us. Some estimates for $k \leq 9$ are given in Lassak [La87]. Lassak's conjecture that $\ell_6 = 1/2$ was disproved by Krotoszyński [Kr94], who established the inequality $\ell_6 \leq \frac{7+2\sqrt{6}}{25} < 0.476$. On the other hand, Lassak [La87] showed that for any centrally symmetric plane convex body C, we have the stronger inequality $\lambda_7(C) \geq 1/2$.

Problem 4   *What is the correct value of ℓ_6?*

Lassak and Vásárhelyi [LaV93] studied the same problem for coverings with homothets of $-C$, the rotation of C by π. Let

$$L_k^- = \sup_C \lambda_k^-(C),$$

where $\lambda_k^-(C)$ denotes the smallest $\lambda > 0$ such that C can be covered by k translates of $-\lambda C$. It is well known [Ne39] that $L_1^- = 2$. Januszewski, Lassak, and Vásárhelyi [JaL01], [LaV93] proved that $L_2^- = 4/3$ and $L_3^- = 1$. In all of these cases, equality is attained for triangles.

Conjecture 5 *([LaV93]) $L_4^- = 4/5$.*

This statement, if true, is sharp for triangles.

Very little is known about similar problems in higher dimensions. It is known that every three-dimensional convex body C can be covered by 16 smaller positive homothetic copies [Pa99], but here the homothety ratio depends on C. Under the requirement that the homothety ratio be universal, Lassak [La98] proved that such a covering is always possible using 24 smaller copies. Should this condition really make a difference?

Problem 6   *Let H_d denote the smallest number h for which every d-dimensional convex body can be covered by h smaller positive homothetic copies of itself. Let $\overline{H}_d$ be the smallest h for which there exists a positive $\lambda_d < 1$ such that every d-dimensional convex body can be covered by at most h of its homothetic copies with homothety ratio at most λ_d. Is it true that $H_d = \overline{H}_d$ for every d?*

Januszewski and Lassak [JaL01] proved that for every $k + l > d^d$, any d-dimensional convex body C can be covered by k translates of λC and l translates of $-\lambda C$, where $\lambda = 1 - \frac{1}{(d+1)d^d}$. In particular, we have $\overline{H}_d \leq d^d + 1$.

[BáB*93]   V. BÁLINT, A. BÁLINTOVÁ, M. BRANICKÁ, P. GREŠÁK, I. HRINKO, P. NOVOTNÝ, M. STACHO: Translative covering by homothetic copies, *Geometriae Dedicata* **46** (1993) 173–180.

[Be77]   J.F. BELOUSOV: Theorems on the covering of plane figures (in Russian), *Ukrain. Geom. Sb.* **20** (1977) 10–17.

[BeB84]   A. BEZDEK, K. BEZDEK: Eine hinreichende Bedingung für die Überdeckung des Einheitswürfels durch homothetische Exemplare im n-dimensionalen euklidischen Raum, *Beiträge Algebra Geom.* **17** (1984) 5-21.

[Fu88]   S. FUDALI: Two remarks about homothetic coverings, *Demonstratio Math.* **21** (1988) 387–392.

[Fü03]   Z. FÜREDI: Covering a triangle with homothetic copies, in: *Discrete Geometry—in Honor of W. Kuperberg's 65th Birthday*, A. Bezdek, ed., Marcel Dekker, 2003, 435–445.

[Ja03]   J. JANUSZEWSKI: Translative covering of a convex body with its positive homothetic copies, *Studia Sci. Math. Hungar.* **40** (2003) 341–348.

[Ja02a]   J. JANUSZEWSKI: Covering by sequences of squares, *Studia Sci. Math. Hungar.* **39** (2002) 179–188.

[Ja02b]   J. JANUSZEWSKI: Covering the unit square by squares, *Beiträge Algebra Geom.* **43** (2002) 411–422.

[Ja01]   J. JANUSZEWSKI: Translative covering by sequences of homothetic copies, *Acta Math. Hungar.* **91** (2001) 337–342.

[Ja98]   J. JANUSZEWSKI: Covering a triangle with sequences of its homothetic copies, *Period. Math. Hungar.* **36** (1998) 183–189.

[JaL01]   J. JANUSZEWSKI, M. LASSAK: Covering a convex body by its negative homothetic copies, *Pacific J. Math.* **197** (2001) 43–51.

[Kr94]   S. KROTOSZYŃSKI: Covering a quadrilateral by six homothetical copies, *Demonstratio Math.* **27** (1994) 487–492.

[Kr87]   S. KROTOSZYŃSKI: Covering a plane convex body with five smaller homothetical copies, *Beiträge Algebra Geom.* **25** (1987) 171–176.

[La98]   M. LASSAK: Covering a three-dimensional convex body by smaller homothetic copies, *Beiträge Algebra Geom.* **39** (1998) 259–262.

[La87]   M. LASSAK: Covering plane convex bodies with smaller homothetical copies, in: *Intuitive Geometry* (Siófok, 1985), K. Böröczky et al., eds., *Colloquia Math. Soc. János Bolyai* **48** (1987) 331–337.

[La86]   M. LASSAK: Covering a plane convex body by four homothetical copies with the smallest positive ratio, *Geometriae Dedicata* **21** (1986) 157–167.

[LaV93]   M. LASSAK, É. VÁSÁRHELYI: Covering a plane convex body with negative homothetical copies, *Studia Sci. Math. Hungar.* **28** (1993) 375–378.

[Le54]   F.W. LEVI: Ein geometrisches Überdeckungsproblem, *Arch. Math.* **5** (1954) 476–478.

[Ma81]   R.D. MAULDIN: *The Scottish Book*, Birkhäuser Verlag, 1981.

[MeM68]   A. MEIR, L. MOSER: On packing of squares and cubes, *J. Comb. Theory* **5** (1968) 126–134.

[MoP93]   W.O.J. MOSER, J. PACH: *Research Problems in Discrete Geometry, Part I: Packing and Covering*, DIMACS Tech. Report, 1993.

[Ne39]   B.H. NEUMANN: On some affine invariants of closed convex regions, *J. London Math. Soc.* **14** (1939) 262–272.

[Pa99]   I. PAPADOPERAKIS: An estimate for the problem of illumination of the boundary of a convex body in E^3, *Geometriae*

Dedicata **75** (1999), 275–285.

[SoV93]   V.P. SOLTAN, É. VÁSÁRHELYI: Covering a convex body by smaller homothetic copies, *Geometriae Dedicata* **45** (1993) 101–113.

[Vá93]   É. VÁSÁRHELYI: Covering of a triangle by homothetic triangles, *Studia Sci. Math. Hungar.* **28** (1993) 163–172.

[Vá84]   É. VÁSÁRHELYI: Über eine Überdeckung mit homothetischen Dreiecken, *Beiträge Algebra Geom.* **17** (1984) 61–70.

3.3 Levi-Hadwiger Covering Problem and Illumination

Given a d-dimensional convex body C, a set of the form $\lambda C + x = \{\lambda c + x \mid c \in C\}$, where $0 < \lambda < 1$ and $x \in \mathbb{R}^d$, is called a *smaller homothetic copy* of C. For a personal account of the interesting history of the following famous conjecture, see [BoG95].

Conjecture 1 *(Levi [Le55], Hadwiger [Ha57], Gohberg-Markus [GoM60]) Any convex body $C \subseteq \mathbb{R}^d$ can be covered by at most 2^d of its smaller homothetic copies.*

Let $H(C)$ denote the least integer H such that C can be covered by H smaller homothetic copies of itself. If C is a parallelepiped, then it is easy to see that $H(C) = 2^d$, and it is also conjectured that in all other cases $H(C) < 2^d$. For $d = 2$ this was shown by Levi [Le55].

In spite of the fact that this problem has attracted a lot of attention, surprisingly little is known about it even in three-dimensional space. Lassak [La84] has confirmed the conjecture for centrally symmetric convex bodies $C \subseteq \mathbb{R}^3$. For centrally symmetric polytopes in $\mathbb{R}^3$, another proof was found independently by P.S. Soltan and V.P. Soltan [SoS86]. K. Bezdek [Be93a] generalized this result to three-dimensional convex polytopes with any (affine) symmetry. For convex bodies $C \subset \mathbb{R}^3$ of constant width, Lassak [La97] proved that $H(C) \leq 6$. Another interesting three-dimensional special case was settled by Boltyanski [Bo01], who also claimed that for $d = 3$ he had verified the Levi–Hadwiger conjecture in its full generality [Bo00]. However, the best published general upper bound in $\mathbb{R}^3$ is that of Papadoperakis [Pa99], who proved that $H(C) \leq 16$.

Levi [Le55] had already observed that if C has a smooth boundary (i.e., no two supporting hyperplanes pass through the same boundary point of C), then $H(C) = d + 1$. Dekster [De94] proved that the same assertion is true under the weaker assumption that C has a smooth "belt." Schramm [Sch88] has confirmed the Levi–Hadwiger conjecture for convex bodies of constant width when d is sufficiently large, while Boltyanski and P.S. Soltan [BoS90], [BoS92] proved it for *zonoids* (limits of zonotopes, i.e., sets of the form $S_1 + \cdots + S_k$, where each $S_i \subset \mathbb{R}^d$ is a line segment). They extended Martini's result [Mar87], which showed that for zonotopes C other than parallelepipeds the stronger bound $H(C) \leq 3 \cdot 2^{d-2}$ holds. Boltyanski [Bo95] further generalized this statement to every "belt body" that is not a parallelepiped; the cases in which equality holds were characterized in [BoM01]. K. Bezdek and Bisztriczky [BeB97] proved the Levi–Hadwiger conjecture for duals of cyclic polytopes, while Talata [Ta99] found that for these bodies $H(C)$ can be bounded by a polynomial of d.

The currently best known general upper bound,

$$H(C) \leq 4^d (5d \ln d),$$

is a simple consequence of Rogers's results (see [BoMS97], [Bö04]). If C is centrally symmetric, the much better estimate

$$H(C) \leq 2^d (d \ln d + d \ln \ln d + 5d)$$

was obtained by Rogers (see Grünbaum [DaGK63]) and, in a slightly more general form, by Rogers and Zong [RoZ97].

A natural approach to disprove the Levi–Hadwiger conjecture would be the following. Try to construct a d-dimensional convex polytope P with the property that any two vertices x and y are *antipodal*; i.e., P lies in a "slab" between two parallel supporting hyperplanes of P passing through x and y. Obviously, for such a polytope $H(P)$ is at least as large as the number of its vertices. Hence, if such a P can have more than 2^d vertices, then the conjecture is false. However, as Danzer and Grünbaum [DaG62] proved by an elegant argument, any polytope with the above property has at most 2^d vertices.

Conjecture 2 *(Makai [Mak90], Makai, Martini [MaM91]) Let $0 \leq i \leq d$, and let $\mathcal{F}$ be a family of disjoint i-dimensional convex bodies with the property that for any $F_1 \neq F_2 \in \mathcal{F}$ there exist parallel hyperplanes H_1 and H_2, containing F_1 and F_2, respectively, such that all members of $\mathcal{F}$ lie in the slab between H_1 and H_2. Then $|\mathcal{F}| \leq 2^{d-i}$.*

For $i = d - 1$, the statement is evident, while for $i = d - 2$ it was verified by Talata (unpublished).

Boltyanski [Bo60] has suggested a completely different approach to the Levi–Hadwiger problem. Let $C \subseteq R^d$ be a convex body whose boundary is denoted by Bd C, and let s be a point (the source) in the exterior of C. We say that a point $b \in$ Bd C is *illuminated* by s if the ray (half-line) from s through b passes through some interior point of C, but there are no points of C on the segment $[s, b]$. Let $H'(C)$ denote the minimum number of sources outside C needed to illuminate every boundary point of C. Boltyanski has made the simple but rather surprising discovery that $H'(C) = H(C)$ for all convex bodies C. This result has been refined and extended to unbounded convex sets by P.S. Soltan [So63] and Boltyanski–P.S. Soltan [BoS78]. See [MaS99] for a survey of illumination problems.

In view of the equivalence between the covering and the illumination problems, Grünbaum [Gr67] suggested that we could obtain an upper bound on $H(C)$ by showing that every large set of sources illuminating

Bd C can be reduced. However, this is not the case. By properly placing n light sources near the vertices of a convex polytope C in $\mathbb{R}^d$, we obtain a system that cannot be reduced [Be93a].

Conjecture 3 *[Be97] With the possible exception of its vertices, the boundary of every d-dimensional convex polytope requires at most 2^{d-1} light sources to be illuminated.*

More generally, it can be conjectured that for every i, with the possible exception of its i-dimensional faces, every d-dimensional convex polytope requires at most 2^{d-1-i} light sources to be illuminated.

Instead of point sources, one can try to illuminate the boundary of a convex body C by the smallest number $H_i(C)$ of i-dimensional affine subspaces (flats) disjoint from C. We say that a boundary point of C is *illuminated by an affine subspace F* if it is illuminated by at least one point of F. K. Bezdek [Be92], [Be93b] proved that

$$H_i(C) \geq \left\lceil \frac{d+1}{i+1} \right\rceil,$$

for every d-dimensional convex body C and for every $0 \leq i < d$, and that equality holds for smooth C.

Conjecture 4 *(K. Bezdek–Kiss–Mollard [BeKM93]) Let $d \geq 2$. For every d-dimensional convex body C, we have $H_1(C) \leq \lceil \frac{2^d}{d+1} \rceil$.*

In [BeKM93], it was proved that the conjecture holds for zonoids if $d + 1$ is a power of 2, and that equality holds for parallelepipeds. In fact, K. Bezdek made a stronger conjecture.

Conjecture 5 *(K. Bezdek [Be94]) For any $d \geq 3$, $0 \leq i < d$, and for any d-dimensional convex body C, we have $H_i(C) \leq H_i(P)$, where P denotes a d-dimensional parallelepiped.*

In [Be94], this conjecture is verified for $i = d - 2$; that is, it is shown that $H_{d-2}(C) = 2$ for every C.

P.S. Soltan [So62], [So72] considered the analogous problems for *inner* illumination. Now the sources are placed on the boundary of C, and we say that $s \in \mathrm{Bd}\,C$ illuminates $b \in \mathrm{Bd}\,C$ if $s \neq b$ and the open segment (s, b) is completely contained in the interior of C. P.S. Soltan [So62] proved that any convex body $C \subset \mathbb{R}^d$ requires at most $d + 1$ light sources to be illuminated from within and that this is optimal if and only if C is a simplex. Grünbaum [Gr64] conjectured that if $S \subseteq \mathrm{Bd}\,C$ is a set of sources illuminating the boundary of C from inside such that no proper subset of S illuminates $\mathrm{Bd}\,C$, then $|S| \leq 2^d$. For $d = 3$, this was confirmed by V.P. Soltan [So95], while for larger values of d the conjecture turned out to be false [BoMS99].

There are also some interesting questions for covering convex sets with their *larger* homothetic copies.

Conjecture 6   *(V.P. Soltan [So88]) Every d-dimensional closed convex set C with nonempty interior (other than a bounded or unbounded cone) can be covered by at most d of its larger homothets whose homothety centers are outside of C.*

The statement is known to be true in the plane and in three-dimensional space [So88].

[Be97]   K. Bezdek: Light-sources that illuminate the boundary points all but the vertices of a convex polyhedron, *Period. Math. Hungar.* **34** (1997) 17–21.

[Be94]   K. Bezdek: On affine subspaces that illuminate a convex set, *Beiträge Algebra Geom.* **35** (1994) 131–139.

[Be93a]   K. Bezdek: Hadwiger-Levi's covering problem revisited, in: *New Trends in Discrete and Computational Geometry,* J. Pach, ed., *Algorithms Combin. Series* **10**, Springer-Verlag, 1993, 199–233.

[Be93b]   K. Bezdek: A note on the illumination of convex bodies, *Geometriae Dedicata* **45** (1993) 89–91.

[Be92]   K. Bezdek: On the illumination of smooth convex bodies, *Arch. Math. (Basel)* **58** (1992) 611–614.

[BeB97]   K. Bezdek, T. Bisztriczky: A proof of Hadwiger's covering conjecture for dual cyclic polytopes, *Geometriae Dedicata* **68** (1997) 29–41.

[BeKM93] K. Bezdek, Gy. Kiss, M. Mollard: An illumination problem for zonoids, *Israel J. Math.* **81** (1993) 265–272.

[Bo01]   V.G. Boltyanski: Solution of the illumination problem for bodies with md$M = 2$, *Discrete Comput. Geom.* **26** (2001) 527–541.

[Bo00]   V.G. Boltyanski: Solution of the illumination problem for three-dimensional convex bodies (in Russian), *Dokl. Akad. Nauk* **375** (2000) 298–301.

[Bo95]   V.G. Boltyanski: Solution of the illumination problem for belt-bodies (in Russian), *Mat. Zametki* **58** (1995) 505–511; English translation *Math. Notes* **58** (1995) 1029–1032.

[Bo60]   V.G. Boltyanski: The problem of illumination of the boundary of a convex body (in Russian), *Izv. Mold. Fil. Akad. Nauk SSSR* No. 10 (1960) 79–86.

[BoG95]   V.G. Boltyanski, I. Gohberg: Stories about covering and illuminating of convex bodies, Nieuw Arch. Wisk. (4) **13** (1995) 1–26.

[BoM01]   V.G. Boltyanski, H. Martini: Covering belt bodies by smaller homothetical copies, Beiträge Algebra Geom. **42** (2001) 313–324.

[BoMS99] V.G. Boltyanski, H. Martini, V.P. Soltan: On Grünbaum's conjecture about inner illumination of convex bodies, Discrete Comput. Geom. **22** (1999) 403–410.

[BoMS97] V.G. Boltyanski, H. Martini, V.P. Soltan: Excursions into Combinatorial Geometry, Springer-Verlag, 1997.

[BoS92]   V.G. Boltyanski, P.S. Soltan: A solution of Hadwiger's problem for zonoids, Combinatorica **12** (1992) 381–388.

[BoS90]   V.G. Boltyanski, P.S. Soltan: Solution of the Hadwiger problem for a class of convex bodies (in Russian) Dokl. Akad. Nauk SSSR **313** (1990) 528–532; English translation in Soviet Math. Dokl. **42** (1991) 18–22.

[BoS78]   V.G. Boltyanski, P.S. Soltan: Combinatorial geometry and convexity classes (in Russian), Uspehi Mat. Nauk **33** (1978) 3–42.

[Bö04]   K. Böröczky Jr.:   Finite Packing and Covering, Cambridge University Press, 2004

[DaG62]   L. Danzer, B. Grünbaum: Über zwei Probleme bezüglich konvexer Körper von P. Erdős und V. Klee, Math. Z. **79** (1962) 95–99.

[DaGK63] L. Danzer, B. Grünbaum, V. Klee: Helly's theorem and its relatives, in: Convexity, V. Klee, ed., AMS Proc. Symp. Pure Math. **7** (1963) 101–180.

[De94]   B.V. Dekster: Every convex n-dimensional body with a smooth belt can be illuminated by $n + 1$ directions, J. Geom. **49** (1994) 90–95.

[GoM60]   I. Gohberg, A.S. Markus: A certain problem about the covering of convex sets with homothetic ones (in Russian), Izv. Mold. Fil. Akad. Nauk SSSR **10**(76) (1960) 87–90.

[Gr67]   B. Grünbaum: Convex Polytopes, Wiley 1967, second edition: Springer-Verlag 2003.

[Gr64]   B. Grünbaum: Fixing systems and inner illumination, Acta Math. Acad. Sci. Hungar. **15** (1964) 161–163.

[Ha57]   H. Hadwiger: Ungelöste Probleme Nr. 20, Elemente Math. **12** (1957) 121.

[**La97**]   M. LASSAK: Illumination of three-dimensional convex bodies of constant width, in: *Proceedings of the 4-th International Congress of Geometry (Thessaloniki, 1996)*, Giachoudis-Giapoulis (1997) 246–250.

[**La84**]   M. LASSAK: Solution of Hadwiger's covering problem for centrally symmetric convex bodies in E^3, *J. London Math. Soc. (2)* **30** (1984) 501–511.

[**Le55**]   F.W. LEVI: Überdeckung eines Eibereiches durch Parallelverschiebungen seines offenen Kerns, *Arch. Math.* **6** (1955) 369–370.

[**Mak90**]   E. MAKAI JR.:   Problem posed at the Disc. Geom. Workshop, Budapest, 1990.

[**MaM91**]   E. MAKAI JR., H. MARTINI: On the number of antipodal pairs of points in finite subsets of $\mathbb{R}^d$, *DIMACS Ser. Discr. Math. Theor. Comp. Sci.* **4** (1991) 457–470.

[**Mar87**]   H. MARTINI: Some results and problems around zonotopes, in: *Intuitive Geometry (Siófok, 1985)*, K. Böröczky et al., eds., *Coll. Math. Soc. J. Bolyai* **48** (1987) 383–418.

[**MaS99**]   H. MARTINI, V.P. SOLTAN: Combinatorial problems on the illumination of convex bodies *Aequationes Math.* **57** (1999) 121–152.

[**Pa99**]   I. PAPADOPERAKIS: An estimate for the problem of illumination of the boundary of a convex body in E^3, *Geometriae Dedicata* **75** (1999) 275–285.

[**RoZ97**]   C.A. ROGERS, C. ZONG: Covering convex bodies by translates of convex bodies, *Mathematika* **44** (1997) 215–218.

[**Sch88**]   O. SCHRAMM: Illuminating sets of constant width, *Mathematika* **35** (1988) 180–189.

[**So72**]   P.S. SOLTAN: Covering of convex bodies by large homothetic ones (in Russian), *Mat. Zametki* **12** (1972) 85–90.

[**So63**]   P.S. SOLTAN: Towards the problem of covering and illumination of convex sets (in Russian), *Izv. Akad. Nauk Mold. SSR* **1** (1963) 49–57.

[**So62**]   P.S. SOLTAN: Illumination of the boundary of a convex body from within (in Russian), *Mat. Sbornik* **57** (1962) 443–448.

[**SoS86**]   P.S. SOLTAN, V.P. SOLTAN: On the x-raying of convex bodies, *Soviet Math. Dokl.* **33** (1986) 42–44.

[**So95**]   V.P. SOLTAN: On Grünbaum's problem about inner illumination of convex bodies, *Acta Math. Hungar.* **69** (1995) 15–25.

[**So88**]   V.P. SOLTAN: The covering of a three-dimensional unbounded convex body by large homothetic bodies (in Russian), *Mat. Issled.* **103**, *Voprosy Diskret. Geom.* (1988) 164–173.

[**Ta99**]   I. TALATA: Solution of Hadwiger-Levi's covering problem for duals of cyclic $3k$-polytopes, *Geometriae Dedicata* **74** (1999) 61–71.

3.4 Covering a Ball by Slabs

A *slab S* is defined as the set of all points lying between two parallel hyperplanes in $\mathbb{R}^d$. The distance between these hyperplanes is called the *width* of S. Thus, for $d = 2$ a slab is just a strip.

Tarski [Ta32] noticed that if we cover the unit circle by a system of strips, then their total width is at least 2, with equality only when the strips are parallel. His famous question, whether this result can be generalized to higher dimensions and to convex bodies other than the ball, is usually referred to as *Tarski's plank problem*. It was settled by Bang [Ba50], [Ba51], [Ba54], whose proof has been simplified and generalized by Fenchel [Fe51], Ohmann [Oh57], and Bognár [Bo61].

Conjecture 1   *(Bang [Ba51]) For any collection of slabs S_i ($i = 1, 2, 3, \ldots$) covering a convex body $C \subseteq \mathbb{R}^d$, we have*

$$\sum_{i=1}^{\infty} \frac{w_i}{w_i(C)} \geq 1,$$

where w_i is the width of S_i and $w_i(C)$ denotes the distance between the two supporting hyperplanes of C parallel to S_i.

The most general results of this type are due to Ball, who proved the conjecture for centrally symmetric bodies [Ba91], and established a similar result in complex Hilbert space [Ba01]. Hunter [Hu93] studied some extremal configurations in the plane for coverings with at most three strips (slabs).

The following equivalent formulation of Bang's conjecture was suggested by Davenport and Alexander [Al68]. Given a convex body C and $n - 1$ hyperplanes in Euclidean d-space, one can always find a homothet of C, at least $1/n$ times as large as C, whose interior does not meet any of the hyperplanes. A. and K. Bezdek [BeB95] proved a weaker result, suggested by Conway. Starting with a convex "potato" C of width 1, make $n-1$ plane cuts, cutting only one piece at a time. Then at least one of the resulting pieces ("chips") must contain a ball of diameter $1/n$, and hence its width is at least $1/n$. A. and K. Bezdek [BeB96] generalized this statement to the case in which the thickness of a chip is measured by the size of the largest homothetic copy of a fixed convex body that it contains.

Problem 2   *Let C be a convex body of unit width. Is it true that, for any decomposition of C into n convex pieces, at least one of the pieces must contain a full-dimensional ball of diameter $1/n$?*

In the special case when C a circular disk in the plane, an affirmative answer was given by A. Bezdek [Be05] and, independently, by Kertész (unpublished).

The following attractive stability question was raised by A. Bezdek.

Conjecture 3   (A. Bezdek [Be03]) Let C_ε denote a unit circle from which a concentric circle of radius $\varepsilon > 0$ has been removed. If ε is sufficiently small, then the total width of any system of strips that cover C_ε is at least 2. In other words, if a system of strips permits a covering of the annulus C_ε, then it can be rearranged to cover the whole unit circle.

In fact, A. Bezdek conjectures that the same statement is true for any annulus obtained from a convex body C by removing a homothet εC from its interior. He managed to verify this statement when C is a square and $\varepsilon \geq 1 - 1/\sqrt{2} \approx 0.29$.

Conjecture 4   (A. Bezdek [Be03]) Let R be a region obtained from the unit square by removing an axis-parallel square of side $1/2$ from its interior. Then the total width of any system of strips that cover R is at least 1.

L. Fejes Tóth [FeT73] asked whether any analogue of Tarski's result holds in spherical geometry. The set of points within spherical distance $w/2$ of a given great circle is called a *zone of width* w.

Conjecture 5   (L. Fejes Tóth [FeT73]) The total width of any collection of zones covering the surface of the three-dimensional unit ball is at least π. Equality may hold when the mid-great-circles of the zones pass through the same point.

As far as we know, the only results in this direction are due to Rosta [Ro72] and Linhart [Li74], who deal respectively with the special cases of three and four zones.

It is possible that the above conjecture can be further generalized. A closed subset C of the open hemisphere is called *convex* if for any pair of points of C, the shortest arc connecting them on the sphere (on the surface of B^3) belongs to C.

Problem 6   (L. Fejes Tóth [FeT73]) Is it true that the total width of any collection of zones covering a convex subset C of the sphere is at least as large as the smallest width of a single zone covering C?

Any two great circles divide the surface of the 3-dimensional ball into four regions, called *digons*. Another spherical analogue of the plank problem is the following.

Problem 7   (L. Fejes Tóth [FeT72]) Is it true that the total area of
any collection of digons covering a convex subset C of the
sphere is at least as large as the smallest area of a single
digon covering C?

Tarski's observation at the beginning of this section was generalized
by Páles [Pá92], as follows. Let $C_1 \subseteq C_2$ be concentric circles, and assume
that from any boundary point of C_2 the smaller circle can be seen at an
angle α. Then the total angle of any system of angular regions that cover
C_1 and whose apices lie on the boundary of C_2 is at least α.

All of the above questions can be extended to multiple coverings.

Problem 8   (Falconer [Fa80]) Let d and $k \geq 2$ be fixed. Determine
the largest number $F_d(k)$ with the property that for any
collection of slabs $S_i \subseteq \mathbb{R}^d$ $(i = 1, 2, 3, \ldots)$ such that every
point of the unit ball is covered by at least k of them,

$$\sum_{i=1}^{\infty} w_i \geq F_d(k),$$

where w_i denotes the width of S_i.

Notice that already for $k = 2$, the value of $F_d(k)$ is strictly smaller than k
times the width of the unit ball (which is 2).

Let C be a not necessarily full-dimensional convex set in the unit cube
$[0, 1]^d$ that meets all facets of the cube. For $1 \leq i \leq d$, a *generalized slab
of type* i is a set of the form $\{(x_1, \ldots, x_d) \in [0, 1]^d \mid x_i \in M\}$, for some
measurable subset $M \subseteq [0, 1]$. The *width* of such a generalized slab is the
measure of M. Aharoni, Holzman, Krivelevich, and Meshulam [AhH*02]
studied the following equivalent form of Bang's conjecture. Suppose that
a system of generalized slabs $S_1, \ldots, S_d$ covers C, where S_i is of type i for
all $1 \leq i \leq d$. Then the sum of their widths is at least 1.

Aharoni et al. have asked what happens if we further relax the notion
of a slab. A generalized slab of type i was associated with a $(0, 1)$-function
defined on the interval $[0, 1]$, whose value is 1 at every point of a measurable
set M. Replace each S_i by a nonnegative measurable function f_i. We
say that these functions form a *fractional cover* of the set $C \subseteq [0, 1]^d$ if
$\sum_{i=1}^{d} f_i(x_i) \geq 1$ holds for every $(x_1, \ldots, x_d) \in C$. The width of f_i is its
integral over the interval $[0, 1]$.

Conjecture 9   (Aharoni, Holzman, Krivelevich, Meshulam [AhH*02]) Let
C be a convex set in the unit cube $[0, 1]^d$ $(d \geq 2)$ that
meets all facets of the cube, and let $f_1, \ldots, f_d$ be nonnegative measurable functions that form a fractional cover of
C. Then the total width of these functions must be at
least $2/d$.

This conjecture is weaker than Bang's conjecture at the beginning of this section. Moreover, if true, the conjecture is sharp, as shown by the example where C is the $(d-1)$-dimensional simplex

$$\left\{(x_1,\ldots,x_d) \in [0,1]^d \;\middle|\; \sum_{i=1}^{d} x_i = 1\right\},$$

and $f_i(x_i) = 2/d - x_i$ if $0 \le x_i \le 2/d$ and $f_i(x_i) = 0$ otherwise. It was proved by Aharoni et al. that the statement holds if we replace the bound $2/d$ by $1/d$. Bang verified his conjecture for systems consisting of only two strips, which implies that the last conjecture holds for $d = 2$.

It is natural to ask under what circumstances one can formulate reverse statements that guarantee that certain systems of slabs can cover a convex set C. A sequence $C_1, C_2, \ldots$ of convex bodies in $\mathbb{R}^d$ is said to permit a *covering* (or a *translative covering*) of a set C if there are congruent copies (or translates) C_i' of C_i such that $\cup C_i' \supseteq C$.

Conjecture 10 *(Groemer [Gr81], Makai–Pach [MaP83]) For any $d \ge 2$, there exists a constant $c(d)$ such that any sequence of slabs in $\mathbb{R}^d$ with total width at least $c(d)$ permits a translative covering of the unit ball.*

In the plane, this statement was first proved by Makai–Pach [MaP83] and, independently, by Erdős–Straus (unpublished). The best known bound, $c(2) \le 6$, was established by Groemer [Gr84]. For higher dimensions we do not even know whether all sequences of slabs with infinite total width permit a translative covering of B^d. On the other hand, it is very easy to see that any such sequence permits a translative covering of the whole space with the exception of a set of Lebesgue measure 0. The truth of the above conjecture would be a first step toward proving that any sequence of convex bodies of diameter at most 1 and with infinite total volume permits a translative covering of $\mathbb{R}^d$. More generally, we can state the following conjecture:

Conjecture 11 *(Makai, Pach [MaP83]) For any $d \le d'$, there exists a constant $\gamma = \gamma(d, d')$ with the following property. A sequence of convex sets $C_1, C_2, \ldots \subseteq \mathbb{R}^{d'}$ permits a translative covering of a d-dimensional convex body $D \subseteq \mathbb{R}^{d'}$ if there are suitable translates $D_1, D_2, \ldots$ of D such that*

$$\sum_{i=1}^{\infty} \mathrm{Vol}_d(C_i \cap D_i) \ge \gamma.$$

We say that a sequence of slabs $S_1, S_2, \ldots$ permits a *nondissective* covering of a convex body $C \subseteq \mathbb{R}^d$ if there are suitable translates S_i' of S_i $(i = 1, 2, \ldots)$ such that $C \setminus (\cup_{i=1}^{j} S_i')$ is a connected set or empty for every j. Note that we are not allowed to change the ordering of the sequence. Let w_i denote the width of S_i. Groemer [Gr81] showed that if $\sum_{i=1}^{\infty} w_i^{(d+1)/2}$ is larger than a constant depending only on d, then $S_1, S_2, \ldots$ permits a nondissective covering of B^d. Moreover, the exponent $(d+1)/2$ cannot be replaced by any smaller number.

Some necessary and sufficient conditions for a sequence of convex bodies to permit a not necessarily translative covering of a given infinite convex cone or a convex body can be found in Makai and Pach [MaP83] and in Chakerian and Groemer [ChG78]. A comprehensive survey of related results is given in Groemer [Gr85] and in Chakerian and Groemer [ChG78].

[AhH*02] R. Aharoni, R. Holzman, M. Krivelevich, R. Meshulam: Fractional planks, *Discrete Comput. Geom.* **27** (2002) 585–602.

[Al68] R. Alexander: A problem about lines and ovals, *Amer. Math. Monthly* **75** (1968) 482–487.

[Ba01] K.M. Ball: The complex plank problem, *Bull. London Math. Soc.* **33** (2001) 433–442.

[Ba91] K.M. Ball: The plank problem for symmetric bodies, *Invent. Math.* **104** (1991) 535–543.

[Ba54] Th. Bang: Some remarks on the union of convex bodies, *Tolfte Skandinaviska Matematikerkongressen, Lund* (1954) 5–11.

[Ba51] Th. Bang: A solution of the plank problem, *Proc. Amer. Math. Soc.* **2** (1951) 990–993.

[Ba50] Th. Bang: On coverings by parallel strips, *Mat. Tidsskr. B* (1950) 49–53.

[Be05] A. Bezdek: On an area inequality related to Tarski's plank problem, manuscript, 2003.

[Be03] A. Bezdek: Covering an annulus by strips, *Discrete Comput. Geom.* **30** (2003) 177–180.

[BeB96] A. Bezdek, K. Bezdek: Conway's fried potato problem revisited, *Arch. Math. (Basel)* **66** (1996) 522–528.

[BeB95] A. Bezdek, K. Bezdek: A solution of Conway's fried potato problem, *Bull. London Math. Soc.* **27** (1995) 492–496.

[**Bo61**]   M. BOGNÁR: On W. Fenchel's solution of the plank problem, *Acta Math. Acad. Sci. Hungar.* **12** (1961) 269–270.

[**ChG78**]   G.D. CHAKERIAN, H. GROEMER: On coverings of Euclidean space by convex sets, *Pacific J. Math.* **75** (1978) 77–86.

[**Fa90**]   K.J. FALCONER: Function space topologies defined by sectional integrals and applications to an extremal problem, *Math. Proc. Cambridge Philos. Soc.* **87** (1990) 81–96.

[**FeT73**]   L. FEJES TÓTH: Exploring a planet, *Amer. Math. Monthly* **80** (1973) 1043–1044.

[**FeT72**]   L. FEJES TÓTH: *Lagerungen in der Ebene, auf der Kugel und im Raum* (2. Auflage), Springer-Verlag, 1972.

[**Fe51**]   W. FENCHEL: On Th. Bang's solution of the plank problem, *Mat. Tidsskr. B* 1951, 49–51.

[**Gr85**]   H. GROEMER: Coverings and packings by sequences of convex sets, in: *Discrete Geometry and Convexity*, J.E. Goodman et al., eds., *Annals New York Acad. Sci.* **440** (1985) 262–278.

[**Gr84**]   H. GROEMER: Some remarks on translative coverings of convex domains by strips, *Canad. Math. Bull.* **27** (1984) 233–237.

[**Gr81**]   H. GROEMER: On coverings of convex sets by translates of slabs, *Proc. Amer. Math. Soc.* **82** (1981) 261–266.

[**Hu93**]   H.F. HUNTER: Some special cases of Bang's inequality, *Proc. Amer. Math. Soc.* **117** (1993) 819–821.

[**Li74**]   J. LINHART: Eine extremale Verteilung von Grosskreisen, *Elemente Math.* **129** (1974) 57–59.

[**MaP83**]   E. MAKAI JR., J. PACH: Controlling function classes and covering Euclidean space, *Studia Sci. Math. Hungar.* **18** (1983) 435–459.

[**Oh57**]   D. OHMANN: Kurzer Beweis einer Abschätzung für die Breite bei Überdeckung durch konvexe Körper, *Arch. Math.* **8** (1957) 150–152.

[**Pá92**]   ZS. PÁLES: On a generalization of the plank problem, in: *General Inequalities, 6* (Oberwolfach, 1990), *Internat. Ser. Numer. Math.* **103**, Birkhäuser 1992, 473–476.

[**Ro72**]   V. ROSTA: An extremal distribution of three great circles (in Hungarian), *Mat. Lapok* **23** (1972) 161–162.

[**Ta32**]   A. TARSKI: Further remarks about the degree of equivalence of polygons (in Polish), *Parametr* **2** (1932) 310–314.

3.5 Point Trapping and Impassable Lattice Arrangements

Let $\mathcal{C} = \{C + \lambda \mid \lambda \in \Lambda\}$ be a d-dimensional lattice arrangement of a convex body $C \subseteq \mathbb{R}^d$, where Λ is a lattice and P denotes (one of) its fundamental parallelepiped(s). It is easy to see that the density of this arrangement in $\mathbb{R}^d$ is

$$d(\mathcal{C}, \mathbb{R}^d) = \frac{\mathrm{Vol}(C)}{\mathrm{Vol}(P)}.$$

Given any point set S in $\mathbb{R}^d$, one can define the *upper* and *lower density* of S in $\mathbb{R}^d$ by

$$\overline{d}^*(S, \mathbb{R}^d) = \limsup_{r \to \infty} \frac{\mathrm{Vol}(S \cap B(r))}{\mathrm{Vol}(B(r))}, \quad \underline{d}^*(S, \mathbb{R}^d) = \liminf_{r \to \infty} \frac{\mathrm{Vol}(S \cap B(r))}{\mathrm{Vol}(B(r))},$$

respectively, where $B(r)$ denotes the ball of radius r around the origin. If these two numbers coincide, then their common value $d^*(S, \mathbb{R}^d)$ is called the *density* of the point set S in $\mathbb{R}^d$.

Let $\cup\mathcal{C} = \cup_{\lambda \in \Lambda}(C + \lambda)$. Obviously,

$$d^*(\cup\mathcal{C}, \mathbb{R}^d) \leq d(\mathcal{C}, \mathbb{R}^d),$$

and equality holds if and only if $\mathcal{C}$ is a packing.

The arrangement $\mathcal{C}$ is said to be *point trapping* if every component of $\mathbb{R}^d \setminus (\cup\mathcal{C})$ is bounded, i.e., if no point in the complement of $\cup\mathcal{C}$ can be continuously moved to infinity while avoiding all members of $\mathcal{C}$.

For instance, if $C = \{(x_1, \ldots, x_d) \mid 0 \leq x_i \leq 1 \text{ for all } i\}$ is the unit cube and Λ consists of all integer points $(m_1, \ldots, m_d) \in \mathbb{Z}^d$ with $\sum_{i=1}^{d} m_i$ even, then the lattice arrangement (packing) $\mathcal{C} = \{C + \lambda \mid \lambda \in \Lambda\}$ is clearly point trapping.

Conjecture 1 *(Böröczky, Bárány, Makai, Pach [BöB*86])*
If $\mathcal{C} = \{C + \lambda \mid \lambda \in \Lambda\}$ is a point trapping lattice arrangement of a d-dimensional convex body C, then

$$d^*(\cup\mathcal{C}, \mathbb{R}^d) \geq \frac{1}{2}$$

and the minimum can be obtained only for the checkerboard-like arrangement of cubes (described above) or for its linear transforms ($d \geq 3$).

The weaker conjecture that $d(\mathcal{C}, \mathbb{R}^d) \geq 1/2$ under the same assumption was made by L. Fejes Tóth [FeT73], [FeT75]. Groemer [Gr66], [Gr68], [FeT73] settled the case $d = 2$. For $d \geq 3$, L. Fejes Tóth's conjecture was verified by Böröczky, Bárány, Makai, and Pach [BöB*86]. In fact, they proved the somewhat stronger result

$$d(\mathcal{C}, \mathbb{R}^d) + d^*(\cup \mathcal{C}, \mathbb{R}^d) \geq 1,$$

and pointed out that the result of Groemer and L. Fejes Tóth immediately implies that in the plane $d^*(\cup \mathcal{C}, \mathbb{R}^2) \geq 1/2$.

L. Fejes Tóth raised the following related question.

Problem 2   *Let C be a fixed convex body in $\mathbb{R}^d$, $d \geq 2$. Determine the smallest density of a point trapping lattice arrangement of C, at least in the special case of $C = B^d$ (the unit ball).*

In the plane it is an easy exercise to prove that the thinnest point trapping lattice arrangement of circles can be obtained by drawing unit circles around all points with even integral coordinates. The density of this arrangement is $\pi/4$. For the three-dimensional case, Bleicher [Bl75] showed that

$$d(\mathcal{C}, \mathbb{R}^3) \geq \frac{128\pi}{3\sqrt{7142 + 1802\sqrt{17}}}$$

for any point trapping lattice arrangement of unit balls, and this bound can be attained.

An equivalent formulation of the above problem is the following. Determine the thinnest lattice Λ (i.e., a lattice whose fundamental parallelepiped has as large a volume as possible) with the property that a given convex body C cannot move more than a finite distance without hitting a point of Λ, provided that C is not allowed to rotate. If rotation is permitted and Λ still satisfies this property, then Λ is called C-*trapping*.

Problem 3   *(Bleicher [Bl75]) For a given convex body $C \subseteq \mathbb{R}^d$ determine the thinnest C-trapping lattice(s).*

Problem 4   *(Bleicher [Bl75]) Determine the thinnest lattice(s) that can trap every convex body of constant width one.*

Recall that the *width* of a convex body C in a given direction v is the distance between the two parallel hyperplanes orthogonal to v and supporting C. A convex body has *constant width* 1 if its width in every direction is 1. If $C = B^d$, then Problem 3 is equivalent to Problem 2. Otherwise, the problem is completely open even in the plane. Obviously, any lattice that traps the incircle of a plane convex body C is also C-trapping, but in general this is not the most economical way of trapping C. A related question of G. Fejes Tóth was solved by G. Blind [Bli69] (see also [He67], [Bli76], [Bli77], for generalizations).

L. Fejes Tóth [FeT84] also suggested the investigation of Problem 2 restricted to so-called *compact* packings. A packing $\mathcal{C}$ is called compact if the neighbors of each $C_0 \in \mathcal{C}$ completely "enclose" C_0, that is, no point of C_0 can be continuously moved to infinity while avoiding all members of $\mathcal{C}$ tangent to C_0. L. Fejes Tóth proved that the lower density of any point trapping lattice packing of a centrally symmetric convex body C in the plane is at least $\frac{3}{4}$, with equality only for affinely regular hexagons (see also [BeC91]). Dropping the condition of central symmetry, the corresponding minimum becomes $\frac{1}{2}$ [BeBB86]. Both results hold in a more general setting: for any (not necessarily lattice) packing $\mathcal{C}$ of *homothetic* copies of C such that the ratio between the volumes of any two members is bounded from above by a constant.

K. Bezdek [Be87] showed that the density of every point trapping lattice packing of a d-dimensional centrally symmetric convex body C is at least $\frac{1}{1+2^{-\frac{1}{d-1}}}$.

The concept of trapping a point or a convex body can be modified and generalized in many different ways, but this often leads to hopelessly difficult problems. We mention only one generalization. Let us call an arrangement $\mathcal{C}$ of convex bodies in $\mathbb{R}^d$ *k-impassable* if every k-dimensional flat meets at least one member of $\mathcal{C}$.

Problem 5   *(G. Fejes Tóth [FeT76]) What is the minimum density of a k-impassable lattice arrangement of $\mathbb{R}^d$ with unit balls $(0 \leq k \leq d-1)$?*

For $k = 0$, this problem is identical to the problem of the thinnest lattice covering with balls, which is solved for $d \leq 5$ (Kershner [Ke39], Bambah [Ba54], Delone and Ryškov [DeR63], Baranovskiĭ and Ryškov [BaR76]). On the other hand, Makai [Ma78] pointed out that for $k = d - 1$ the question is equivalent to the problem of the densest lattice packing of balls; hence it is solved for $k = d - 1 \leq 7$ (see also [Ma74], [KaL88]). Thus, the first interesting case is $d = 3$, $k = 1$, in which Bambah and Woods [BaW94] proved that the density of any 1-impassable lattice arrangement of balls in $\mathbb{R}^3$ is at least $3\pi\sqrt{2}/4$. This bound is attained for the arrangement obtained from the densest lattice packing by a suitable enlargement of its balls. The case $d = 4$, $k = 1$ has been settled by Makai (personal communication), but little is known, e.g., for $d = 4$, $k = 2$.

Heppes [He60] showed that a lattice packing $\mathcal{C}$ of balls in $\mathbb{R}^3$ cannot be 1-impassable, and Hortobágyi [Ho71] proved that $\mathbb{R}^3 \setminus (\cup \mathcal{C})$ always contains three two-way infinite circular cylinders of radius $\frac{3\sqrt{2}}{4} - 1 = 0.0606\ldots$ whose axes are linearly independent. This bound cannot be improved.

Horváth and Ryškov [Ho70], [HoR75], [Ho97] generalized this result to higher dimensions. In particular, they have completely solved the analogous problem in four dimensions.

Conjecture 6   *(Makai [Ma78]) Let C be a $(d-1)$-impassable lattice packing of $\mathbb{R}^d$ with a convex body C. Then:*

     (1) $d(\mathcal{C}, \mathbb{R}^d) \geq \frac{d+1}{d!2^d}$, *with equality only if C is a simplex;*

     (2)   *if C is centrally symmetric, then $d(\mathcal{C}, \mathbb{R}^d) \geq \frac{1}{d!}$, with equality only for crosspolytopes.*

For $d = 2$, this conjecture was confirmed by L. Fejes Tóth and Makai [FeTM74] and by Makai [Ma78]. In the special case of $\cup\mathcal{C}$ a connected set, (2) is settled by a result of Groemer [Gr66], [Gr68] and L. Fejes Tóth [FeT73].

In sufficiently large dimensions d, no lattice packing of balls is 2-impassable. More precisely, it was shown in [HeZZ02] that for $d \geq 5$, in every d-dimensional lattice packing $\mathcal{C}$ of balls there is a so-called *free plane* of dimension two, that is, an affine subspace of dimension two that does not meet the interior of any member of $\mathcal{C}$. They have also shown that in 8 dimensions one can always find a three-dimensional free plane, and according to Henk [He05], in d dimensions there always exists a free plane of dimension at least $\frac{d}{\log_2 d}$. Thus, denoting by $\varphi(C)$ the largest dimension of a free plane that can be found in any lattice packing of C, we have that

$$\varphi(B^d) \geq \frac{d}{\log_2 d},$$

for every d. On the other hand, improving an earlier result of Hausel [Ha92], Henk, Ziegler, and Zong [HeZZ02] showed that there exists an absolute constant $\varepsilon > 0$ such that

$$\varphi(B^d) \leq (1 - \varepsilon)d,$$

for every d.

Problem 7   *Does there exist a positive constant ε such that in any lattice packing of B^d there is a free plane whose dimension is at least εd? In notation, is it true that $\varphi(B^d) \geq \varepsilon d$, for every d?*

It seems likely that the answer to the following question is yes, so that the ball B^d is extremal among all centrally symmetric convex bodies in d dimensions.

Problem 8   *(Zong [Zo02]) Is it true that $\varphi(C) \geq \varphi(B^d)$, for every d-dimensional convex body C?*

A closely related question, known as the "lonely runner problem," was raised by Wills [Wi68]. It plays a role in Diophantine approximation. Suppose that $d + 1$ athletes run around a race track of unit perimeter with different constant velocities, starting from the same point. Is it true that for each of them there will come a moment when his distance from all other competitors will be at least $\frac{1}{d+1}$? Since the $d + 1$ runners cut the total length 1 in $d + 1$ intervals, at each moment the average length of the interval between two runners is $\frac{1}{d+1}$. So the question is whether for each runner at some moment the interval in front as well as the interval behind will be simultaneously of at least average length.

Conjecture 9   *(Wills [Wi68], Cusick [Cu73]) For any d positive numbers, $x_1, \ldots, x_d$, there is a real t such that none of the numbers tx_i is closer to an integer than $\frac{1}{d+1}$.*

This conjecture has been verified for $d \le 4$ by Cusick and Pomerance [CuP84] and for $d = 5$ by Bohman, Holzman, and Kleitman [BoHK01]. Bienia, Goddyn et al. [BiG*98] found an elegant proof for the case $d = 4$. Extending the methods in [BiG*98], Chen and Cusick [ChC99] proved the weaker bound $\frac{1}{2d-3}$, instead of $\frac{1}{d+1}$, provided that $d \ge 4$ and $2d - 3$ is a prime. (The $\frac{1}{2d}$ bound is trivial.)

Why is the lonely runner problem related to the main topic of this section? It can be regarded as a special case of the "view-obstruction problem" of Cusick [Cu73]: For any d-dimensional convex body C, find the largest number $\alpha = \alpha(C)$ for which there is a half-line starting from the origin and lying in the positive cone

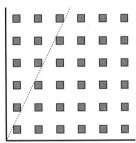

$$\{(x_1, \ldots, x_d) \mid x_i > 0, \ldots, x_d > 0\}$$

that does not meet the interior of any member of the lattice arrangement

$$\alpha C + \ \mathbb{Z}^d + \left(\tfrac{1}{2}, \ldots, \tfrac{1}{2}\right).$$

UNOBSTRUCTED PATH
CORRESPONDING TO ONE OF
THREE RUNNERS

The above conjecture can now be reformulated as follows: For the d-dimensional cube $C^d = [-1, 1]^d$ centered at the origin, we have

$$\alpha(C^d) = \frac{1}{2} - \frac{1}{d+1}.$$

Conjecture 10   *(Cusick [Cu73]) For $d \ge 2$, let $P(d)$ denote the rectangular box $\{(x_1, \ldots, x_d) \mid 0 \le x_i \le i(i = 1, \ldots, d)\}$. Then $\alpha(B^d)$ is equal to the Euclidean distance between the half-line $(tx_1, \ldots, tx_d)$, $t \ge 0$, and the closest point of $\mathbb{Z}^d + \left(\tfrac{1}{2}, \ldots, \tfrac{1}{2}\right)$ that belongs to $P(d)$.*

This conjecture has been proved for $d \leq$ 5 in [Cu73], [DuH86], [DuHW96]. Dumir, Hans-Gill, and Wilker [Du*93a] gave a complete analysis of planar view-obstruction problems for centrally symmetric sets C centered at the origin (see also [ChM02]). They generalized the problem to obstruction of subspaces of other dimensions [Du*93a] and to unit balls in other ℓ_p-norms [Du*93b], [Du*93c].

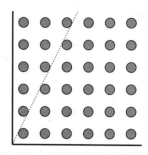

UNOBSTRUCTED RAY IN
MAXIMAL DISK ARRANGEMENT

It was suggested by L. Fejes Tóth [FeT74] that analogous questions can be asked for nonlattice arrangements. In particular, what are the thinnest k-impassable arrangements of unit balls in $\mathbb{R}^d$ ($d \geq 2$, $1 \leq k \leq d-1$)? It turns out that our usual notion of density is not appropriate for handling this problem. Erdős and Pach [ErP80] established the following result. Let $r_1, r_2, \ldots$ be an increasing sequence of positive numbers tending to infinity. Then there exists a k-impassable arrangement $\{B_1^d, B_2^d, \ldots\}$ of unit balls in $\mathbb{R}^d$ such that the center of B_i is at distance r_i from the origin if and only if $\sum_{i=1}^{\infty} r_i^{k-d} = \infty$. Some generalizations can be found in Makai and Pach [MaP83].

We close this section with a kind of converse of the last question. First we have to introduce a property that is, roughly speaking, the opposite of $(d-1)$-separability. Following Erdős (see A.W. Goodman and R.E. Goodman [GoG45]), we call an arrangement $\mathcal{C}$ of convex bodies in $\mathbb{R}^d$ *separable* if there exists a hyperplane avoiding the interior of every element of $\mathcal{C}$. Such a hyperplane is often called a *free* hyperplane. If any two elements of $\mathcal{C}$ can be separated by a free hyperplane, then $\mathcal{C}$ is called *totally separable*. G. Fejes Tóth and L. Fejes Tóth [FeTF73] proved that the maximum density of a totally separable packing with congruent copies of a centrally symmetric convex body C in the plane is $\frac{\text{area}(C)}{\text{area}(Q_C)}$, where Q_C denotes a quadrilateral of minimum area circumscribed around C. In particular, for $C = B^2$ this value is $\pi/4$. (See also A. Bezdek [Be83].)

Conjecture 11 *(G. Fejes Tóth, L. Fejes Tóth [FeTF73]) The maximum density of a totally separable packing of the plane with (not necessarily equal) circles is $\frac{11\pi}{24\sqrt{3}} = 0.831\ldots$.*
This bound is attained for the system of incircles of the cells in a tiling determined by three parallel families of lines that dissect the plane into regular hexagons and triangles with unit sides.

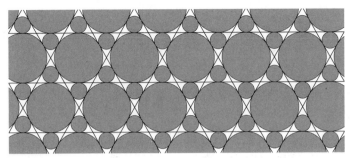

THE CONJECTURED DENSEST TOTALLY SEPARABLE CIRCLE PACKING

The only result in this direction is due to G. Fejes Tóth [FeT87], who showed that the density of a totally separable circle packing is at most 0.98. Notice that any plane section of a totally separable packing of balls in $\mathbb{R}^d$, $d \geq 3$, is a totally separable circle packing, and hence this upper bound remains valid in higher dimensions. Kertész [Ke88] proved that the maximum density of a totally separable packing of equal balls in $\mathbb{R}^3$ is $\pi/6$, and this can be attained if the centers of the balls form a cube-lattice.

The question about dense totally separable circle packings can also be asked in a slightly different way. Suppose the plane is cut by lines into cells such that each cell has diameter at most one, and each cell contains one inscribed circle. How should the lines and circles be chosen as to maximize the density of the resulting circle packing? This suggests a dual covering problem.

Conjecture 12 *(G. Fejes Tóth [FeT87]) If the plane is cut by lines into cells, each cell of diameter at most one, and each cell is covered by a distinct circle, then the density of the resulting covering of the plane is at least $\frac{5\pi}{6\sqrt{3}} = 1.511\ldots$.*

This bound is attained for the system of circumcircles of the cells in a tiling determined by three parallel families of lines that dissect the plane into regular hexagons and triangles of the same side length. G. Fejes Tóth [FeT87] proved that the above minimum is at least 1.01.

[Ba54]   R.P. BAMBAH: On the lattice covering by spheres, *Proc. Nat. Inst. Sci. India* **20** (1954) 25–52.

[BaW94]   R.P. BAMBAH, A.C. WOODS: On a problem of G. Fejes Tóth, *Proc. Indian Acad. Sci. Math. Sci.* **104** (1994) 137–156.

[BaR76]   E.P. BARANOVSKIĬ, S.S. RYŠKOV: *C*-types of *n*-dimensional lattices and five-dimensional primitive parallelohedra (with applications to the theory of coverings), *Dokl. Akad. Nauk SSSR*

Trudy Otdel. Leningrad Mat. Inst. Steklova **137** (1976) Izdat. Nauka, Moscow; English translation by R.M. Erdahl in: *Proc. Steklov Inst.* **137**, Amer. Math. Soc. 1978.

[Be83] A. Bezdek: Locally separable circle packings, *Studia Sci. Math. Hungar.* **18** (1983) 371–375.

[Be87] K. Bezdek: Compact packings in the Euclidean space, *Beiträge Algebra Geom.* **25** (1987) 79–84.

[BeBB86] A. Bezdek, K. Bezdek, K. Böröczky: *On compact packings, Studia Sci. Math. Hungar.* **21** (1986) 343–346.

[BeC91]] K. Bezdek, R. Connelly: Lower bounds for packing densities, *Acta Math. Hungar.* **57** (1991) 291–311.

[BiG*98] W. Bienia, L. Goddyn, P. Gvozdjak, A. Sebő, M. Tarsi: Flows, view-obstructions, and the lonely runner, *J. Combin. Theory Ser. B* **72** (1998) 1–9.

[Bl75] M.N. Bleicher: The thinnest three dimensional point lattice trapping a sphere, *Studia Sci. Math. Hungar.* **10** (1975) 157–170.

[Bli77] G. Blind: r-zugängliche Unterdeckungen der Ebene durch kongruente Kreise, II, *J. Reine Angew. Math.* **289** (1977) 1–29.

[Bli76] G. Blind: r-zugängliche Unterdeckungen der Ebene durch kongruente Kreise, I, *J. Reine Angew. Math.* **288** (1976) 1–23.

[Bli69] G. Blind: Über Unterdeckungen der Ebene durch Kreise, *J. Reine Angew. Math.* **236** (1969) 145–173.

[BoHK01] T. Bohman, R. Holzman, D.J. Kleitman: Six lonely runners, *Electron. J. Combin.* **8** (2001), Research Paper 3, 49 pp.

[BöB*86] K. Böröczky, I. Bárány, E. Makai Jr., J. Pach: Maximal volume enclosed by plates and proof of the chessboard conjecture, *Discrete Math.* **60** (1986) 101–120.

[ChC99] Y.-G. Chen, T.W. Cusick: The view-obstruction problem for n-dimensional cubes, *J. Number Theory* **74** (1999) 126–133.

[ChM02] Y.-G. Chen, A. Mukhopadhyay: The view-obstruction problem for polygons, *Publ. Math. Debrecen* **60** (2002) 101–105.

[Cu73] T.W. Cusick: View-obstruction problems, *Aequationes Math.* **9** (1973) 165–170.

[CuP84]   T.W. CUSICK, C. POMERANCE: View-obstruction problems, III, *J. Number Theory* **19** (1984) 131–139.

[DeR63]   B.N. DELONE, S.S. RYŠKOV: Solution of the problem on the least dense lattice covering of a 4-dimensional space by equal spheres (in Russian), *Dokl. Akad. Nauk SSSR* **152** (1963) 523–524.

[DuH86]   V.C. DUMIR, R.J. HANS-GILL: View-obstruction problems for 3-dimensional spheres, *Monatshefte Math.* **101** (1986) 279–289.

[DuHW96] V.C. DUMIR, R.J. HANS-GILL, J.B. WILKER: The view-obstruction problem for spheres in $\mathbb{R}^5$, *Monatshefte Math.* **122** (1996) 21–34.

[Du*93a] V.C. DUMIR, R.J. HANS-GILL, J.B. WILKER: Contributions to a general theory of view-obstruction problems, *Canad. J. Math.* **45** (1993) 517–536.

[Du*93b] V.C. DUMIR, R.J. HANS-GILL, J.B. WILKER: The view-obstruction constants for ℓ_p-balls in $\mathbb{R}^3$, *Expositiones Math.* **11** (1993) 407–417.

[Du*93c] V.C. DUMIR, R.J. HANS-GILL, J.B. WILKER: The ℓ_p family of view-obstruction problems, *Nieuw Arch. Wisk. (4)* **11** (1993) 225–239.

[ErP80]   P. ERDŐS, J. PACH: On a problem of L. Fejes Tóth, *Discrete Math.* **30** (1980) 103–109.

[FeT87]   G. FEJES TÓTH: Totally separable packing and covering with circles, *Studia Sci. Math. Hungar.* **22** (1987) 65–73.

[FeT76]   G. FEJES TÓTH: Research Problem 18, *Period. Math. Hungar.* **7** (1976) 89–90.

[FeTF73] G. FEJES TÓTH, L. FEJES TÓTH: On totally separable domains, *Acta Math. Acad. Sci. Hungar.* **24** (1973) 229–232.

[FeT84]   L. FEJES TÓTH: Compact packing of circles, *Studia Sci. Math. Hungar.* **19** (1984) 103–107.

[FeT75]   L. FEJES TÓTH: Research problem 14, *Period. Math. Hungar.* **6** (1975) 277–278.

[FeT74]   L. FEJES TÓTH: Remarks on the dual to Tarski's plank problem (in Hungarian), *Matematikai Lapok* **25** (1974) 13–20.

[FeT73]   L. FEJES TÓTH: On the density of a connected lattice of convex bodies, *Acta Math. Acad. Sci. Hungar.* **24** (1973) 373–376.

[FeTM74] L. Fejes Tóth, E. Makai Jr.:   On the thinnest non-separable lattice of convex plates, *Studia Sci. Math. Hungar.* **9** (1974) 191–193.

[GoG45] A.W. Goodman, R.E. Goodman: A circle covering theorem, *Amer. Math. Monthly* **52** (1945) 494–498.

[Gr68] H. Groemer: Einige Bemerkungen über zusammenhängende Lagerungen, *Monatshefte Math.* **72** (1968) 212–216.

[Gr66] H. Groemer: Zusammenhängende Lagerungen konvexer Körper, *Math. Zeitschr.* **94** (1966) 66–78.

[Ha92] T. Hausel: Transillumination of lattice packing of balls, *Studia Sci. Math. Hungar.* **27** (1992) 241–242.

[He05] M. Henk: Free planes in lattice sphere packings, *Adv. Geometry*, to appear.

[HeZZ02] M. Henk, G.M. Ziegler, C. Zong: On free planes in lattice ball packings, *Bull. London Math. Soc.* **34** (2002) 284–290.

[He67] A. Heppes: On the densest packing of circles not blocking each other, *Studia Sci. Math. Hungar.* **2** (1967) 257–263.

[He60] A. Heppes: Ein Satz über gitterförmige Kugelpackungen, *Ann. Univ. Sci. Budapest. Eötvös Sect. Math.* **3–4** (1960) 89–90.

[Ho71] I. Hortobágyi: Durchleuchtung gitterförmiger Kugelpackungen mit Lichtbündeln, *Studia Sci. Math. Hungar.* **56** (1971) 147–150.

[Ho97] J. Horváth: Eine Bemerkung zur Durchleuchtung von gitterförmigen Kugelpackungen, in: *Proceedings of the 4th Internat. Congress of Geom.* (Thessaloniki, 1996), Giachoudis-Giapoulis, Thessaloniki, 1997.

[Ho70] J. Horváth: Über die Durchsichtigkeit gitterförmiger Kugelpackungen, *Studia Sci. Math. Hungar.* **5** (1970) 421–426.

[HoR75] J. Horváth, S.S. Ryškov: Estimation of the radius of a cylinder that can be imbedded in every lattice packing of n-dimensional unit balls, *Math. Notes* **17** (1975) 72–75.

[KaL88] R. Kannan, L. Lovász: Covering minima and lattice point free convex bodies, *Ann. Math. (2)* **128** (1988) 577–602.

[Ke39] R.B. Kershner: The number of circles covering a set, *Amer. J. Math.* **61** (1939) 665–671.

[Ke88] G. Kertész: On totally separable packing of equal balls, *Acta Math. Hungar.* **51** (1988) 363–364.

[**Ma74**]   K. MAHLER: Polar analogues of two theorems of Minkowski, *Bull. Austr. Math. Soc.* **11** (1974) 121–129.

[**Ma78**]   E. MAKAI JR.:   On the thinnest non-separable lattice of convex bodies, *Studia Sci. Math. Hungar.* **13** (1978) 19–27.

[**MaP83**]   E. MAKAI JR., J. PACH: Controlling function classes and covering Euclidean space, *Studia Sci. Math. Hungar.* **18** (1983) 435–459.

[**Wi68**]   J.M. WILLS: Zur simultanen homogenen diophantischen Approximation, I, *Monatshefte Math.* **72** (1968) 254–263.

[**Zo02**]   C. ZONG: From deep holes to free planes, *Bull. Amer. Math. Soc.* **39** (2002) 533–555.

4. Tiling Problems

4.1 Tiling the Plane with Congruent Regions

The systematic study of the theory of tilings started in the nineteenth century, largely motivated by many physical experiments and theories about the structure of solid matter. Modern group theory, whose foundation was laid at about the same time by Lagrange, Abel, Galois, and Cauchy, appeared to be just the right tool for these investigations.

In 1849 Bravais, using symmetries in three-dimensional space, classified the structure of crystals. Felix Klein's famous Erlangen Program, 1872, and Hilbert's lecture at the International Congress of Mathematicians, 1900, were further important stimuli to the development of the theory of tilings.

Recall that a family $\mathcal{T}$ of d-dimensional bodies is called a *tiling* if it is a packing and a covering at the same time; i.e., the members of $\mathcal{T}$ fill $\mathbb{R}^d$ with no gaps or (d-dimensional) overlaps. The members of $\mathcal{T}$ are called *tiles*. An isometry (or congruent transformation) of $\mathbb{R}^d$ is said to be a *symmetry* of $\mathcal{T}$ if it takes every tile of $\mathcal{T}$ onto another tile. The set of all symmetries of $\mathcal{T}$ forms a group, the *symmetry group* $S(\mathcal{T})$ of $\mathcal{T}$. If $S(\mathcal{T})$ acts transitively on the tiles (i.e., for every $T_1, T_2 \in \mathcal{T}$ there is a symmetry of $\mathcal{T}$ taking T_1 onto T_2), then $\mathcal{T}$ is called an *isohedral tiling*.

In the context of his 18th problem, Hilbert asked whether there exists a body P such that space can be tiled with congruent copies of P but no such tiling is isohedral. A body with this property is called *anisohedral*. In fact, encouraged by some spectacular achievements by crystallographers, Hilbert conjectured that no such body exists. Moreover, he posed his question in a higher-dimensional setting because he was convinced that the planar version can be trivially settled by providing a complete characterization of all tilings of $\mathbb{R}^2$ with congruent tiles. One of his assistants, Reinhardt [Re29], even announced such a result.

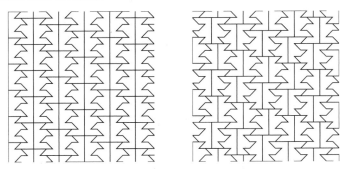

Two tilings with Heesch's anisohedral polygon

However, both Hilbert and Reinhardt were wrong. Reinhardt [Re28] found an anisohedral polytope in $\mathbb{R}^3$. The first planar example was discovered by Heesch [He35], while the first anisohedral convex *polygon* is among the pentagonal tiles constructed by Kershner [Ke68].

It is easy to check that if the plane can be tiled by congruent copies of a convex polygon with k sides, then $k \leq 6$. Many authors claimed to have a complete list of these polygons, but then someone (often an amateur) would discover a new type. (For an enjoyable account of the history of this problem, see Schattschneider [Sch81].)

Problem 1   *(Wang [Wa61]) Give a complete enumeration of all (convex) polygons whose congruent copies can tile the plane, or design an (efficient) algorithm to decide whether a given polygon belongs to this class.*

Hao Wang was a philosopher who was interested in tilings as a tool for studying decidability problems in propositional calculus. For an exact definition of the efficiency of an algorithm and for some negative results of this type, see Büchi [Bü62], Moore–Robson [MoR01]. Berger [Be66], who was actually Wang's student, and later Robinson [Ro71], proved that the problem whether it is possible to tile the plane with congruent copies taken from a finite *set* of polygons is undecidable. However, it is not known whether the problem is decidable for sets of polygons with a fixed number of sides and, in particular, for tilings with congruent copies of a single polygon.

If we restrict our attention to *polyominoes*, i.e., to unions of finitely many unit cells in the integer lattice whose boundary is a simple closed curve, then the situation becomes somewhat simpler. Wijshoff and van Leeuwen [WiL84] (see also [BeN91]) proved that a given polyomino admits a translative tiling if and only if it admits a lattice tiling. Consequently, there is a natural polynomial-time algorithm for deciding whether the plane can be tiled with translates of a given polyomino P of n sides. Keating and Vince [KeV99] designed a polynomial-time algorithm for deciding whether P admits a (not necessarily translative) isohedral tiling.

RECTANGLE TILED BY 76
CONGRUENT HEPTOMINOS

Problem 2   *Is it decidable whether a given polyomino admits a periodic tiling of the plane, that is, a tiling whose symmetry group contains two independent translations?*

For many more tiling problems with polyominoes, see [Go96].

Concerning Hilbert's question, Hirschhorn and Hunt [HiH85] determined all equilateral pentagons (convex or nonconvex) that tile the plane. Reinhardt [Re18] (also Heesch, Kienzle [HeK63], Bollobás [Bo63]) claimed that no convex hexagon is anisohedral.

Problem 3   *(Grünbaum, Shephard [GrS87], p. 497) Does there exist a nonconvex anisohedral hexagon?*

A group G of isometries of $\mathbb{R}^d$ is called *discrete* if every point x has a neighborhood containing no image of x under a transformation belonging to G other than itself. Bieberbach [Bi10], [Bi11], [Bi12] and Frobenius [Fr11] settled Hilbert's 18th problem by showing that (up to affine transformations) there are only finitely many discrete groups that can occur as the symmetry group of an isohedral tiling of $\mathbb{R}^d$ with congruent copies of a bounded set. Moreover, each of these groups contains d linearly independent translations; that is, all isohedral tilings are *periodic*.

As mentioned before, Hilbert's assumption that all tilings of the plane with congruent tiles can be easily characterized proved to be far too optimistic. We are unable to answer even the following simple question.

Problem 4   *(Grünbaum, Shephard [GrS87], p. 497) Is it true that every tiling $\mathcal{T}$ of the plane with congruent tiles can be approximated arbitrarily closely by a tiling $\mathcal{T}'$ with congruent polygons?*

More precisely, is it true that given $\mathcal{T}$ and $\epsilon > 0$, there exists $\mathcal{T}'$ such that for any $T \in \mathcal{T}$ one can find $T' \in \mathcal{T}'$ whose distance (say, Hausdorff) from T is less than ϵ? Our conventional methods of approximation seem to break down in this case.

The main difficulty is that there may exist uncountably many different tilings with congruent copies of the same set T. (Two tilings of the plane are considered different if there is no isometry of the plane taking one into the other.)

Problem 5   *(Grünbaum, Shephard [GrS77]) For every positive integer k, is it possible to find a body C_k in the plane such that there are exactly k different tilings with congruent copies of C_k?*

For all $k \leq 10$, Fontaine and Martin [FoM83], [FoM84] constructed polyominoes C_k with the desired property. Harborth [Ha77a], [Ha77b] and Schmitt [Sch86], [Sch87], [Sch88a], [Sch88b] described several constructions of pairs of bodies such that the number of different tilings of $\mathbb{R}^2$ using only tiles congruent to them is a given number. In particular, Schmitt [Sch87] found a pair of bodies admitting a countable infinity of tilings.

Conjecture 6 *(Grünbaum, Shephard [GrS87], p. 47) There is no C for which the number of different tilings of $\mathbb{R}^2$ with congruent copies of C is countably infinite.*

Dolbilin [Do95] proved that if for a given *set* of bodies there are only countably many different tilings with congruent copies of its elements that satisfy some "local matching rules" (i.e., the immediate neighborhoods of the copies fall into a bounded number of congruence classes), then one of these tilings must be periodic.

Let $\mathcal{C}_0 = \{C\} \subset \mathcal{C}_1 \subset \mathcal{C}_2 \subset \ldots \subset \mathcal{C}_k$ be a nested sequence of finite packings with congruent copies of a body C with the property that the closure of that part of the plane not covered by $\mathcal{C}_i$ is disjoint from all members of $\mathcal{C}_{i-1}$ $(i = 1, 2, \ldots, k)$. Then we say that C is *surrounded* by $\mathcal{C}_k$ k times. It is not difficult to find a body C that can be surrounded once by a packing with its congruent copies and yet the plane cannot be tiled with copies of C. On the other hand, it can be proved by standard compactness arguments that, if for every k there is a packing $\mathcal{C}_k$ surrounding C k times, then the plane can be tiled with congruent copies of C.

Problem 7   *(Heesch [He68], p. 23) Does there exist a positive integer k with the property that if a body C can be surrounded by a packing of its congruent copies k times, then the plane can be tiled with congruent copies of C?*

It is known that if such a number exists, it must be at least six. Ammann, Marshall, and Mann [Ma04] constructed examples that can be surrounded by their congruent copies three, four, and five times, and yet they do not tile the plane. It is easy to see that if it is undecidable whether a polygon of a given number of sides admits a tiling of the plane, the answer to the last question is in the negative [Ma04].

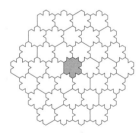

THE NOTCHED HEXAGON OF
AMMANN CAN BE SURROUNDED
AT MOST THREE TIMES

Note that the last question, as well as the previous three problems, is open even for convex polygons. We know almost nothing about their higher-dimensional versions. The known examples that can be surrounded several times without allowing a tiling of the plane are all nonconvex; by introducing a combinatorially unbalanced set of "notches" on the boundary of a polygon that otherwise would tile the plane, one can limit the size of tiling patches possible with that notched polygon.

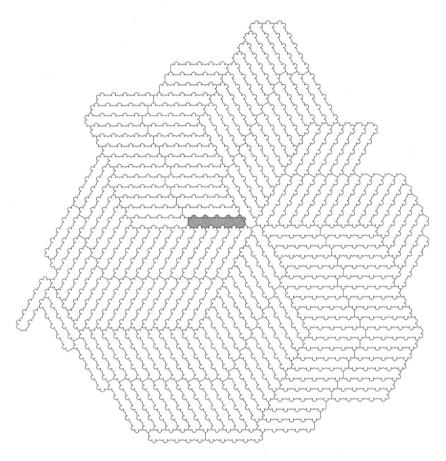

MANN'S NOTCHED UNION OF FIVE HEXAGONS
CAN BE SURROUNDED AT MOST FIVE TIMES

Voderberg [Vo36] discovered a tile with the remarkable property that two copies can completely surround two further copies. This solved a problem of Reinhardt [Re34]. Obviously, no convex tile can have this property, but in general, we do not know of any upper bound on the number of tiles that can be completely surrounded by two tiles.

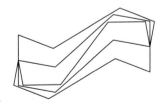

VODERBERG'S TILE:
MIDDLE TILES COMPLETELY
SURROUNDED BY OUTER TILES

Voderberg also discovered spiral-like tilings [Vo37], that can be done with the same tile, but many simpler constructions exist [GrS87].

Problem 8   (Voderberg [Vo37]; Grünbaum, Shephard [GrS87], p. 129)
              Does there exist for any $k \geq 3$ a tiling of the plane with
              congruent copies of a body C_k satisfying the following con-
              dition: there are two nonoverlapping copies C'_k and C''_k of
              C_k in the tiling such that the union of the bounded com-
              ponents of $\mathbb{R}^2 \setminus (C'_k \cup C''_k)$ can be tiled with exactly k con-
              gruent copies of C_k?

SIMPLE EXAMPLE OF VODERBERG-TYPE SPIRAL TILING

There are many other topological "degeneracies" that may theoret-
ically occur even in "well-behaved" tilings. One should attempt to rule
them out, and if there is a counterexample, this might lead to the discovery
of completely new types of tilings.

Conjecture 9   (Grünbaum, Shephard [GrS87], p. 26) Let $\mathcal{T}$ be a tiling
              of the plane with congruent copies of a (closed) body C.
              The neighborhood $N(T)$ of any $T \in \mathcal{T}$ is defined as the
              union of all members $T' \in \mathcal{T}$ with $T' \cap T \neq \emptyset$.
              (1) $N(T)$ is a simply connected set for every $T \in \mathcal{T}$
                     (i.e., it contains no holes)
              (2)   There exist no three members $T_1, T_2, T_3 \in \mathcal{T}$ such
                     that $N(T_1) = N(T_2) = N(T_3)$.

It is easy to construct a tiling of $\mathbb{R}^2$ with congruent triangles that has two members with identical neighborhoods.

[Be66]   R. BERGER: The undecidability of the domino problem, *Memoirs Amer. Math. Soc.* **66** (1966) 1–72.

[BeN91]   D. BEAUQUIER, M. NIVAT: On translating one polyomino to tile the plane, *Discrete Comput. Geom.* **6** (1991) 575–592.

[Bi12]   L. BIEBERBACH: Über die Bewegungsgruppen der euklidischen Räume. (Zweite Abhandlung.) Die Gruppen mit einem endlichen Fundamentalbereich., *Math. Ann.* **72** (1912) 400–412.

[Bi11]   L. BIEBERBACH: Über die Bewegungsgruppen der Euklidischen Räume (Erste Abhandlung), *Math. Ann.* **70** (1911) 297–336.

[Bi10]   L. BIEBERBACH: Über die Bewegungsgruppen des n-dimensionalen Euclidischen Raumes mit einem endlichen Fundamentalbereich, *Nachr. Gesell. Wiss. Göttingen, Math.-Phys. Klasse* 1910, 75–84.

[Bo63]   B. BOLLOBÁS: Filling the plane with congruent convex hexagons without overlapping, *Ann. Univ. Sci. Budapest. Eőtvős, Sect. Math.* **6** (1963) 117–123.

[Bü61]   R.J. BÜCHI: Turing-machines and the Entscheidungsproblem, *Math. Ann.* **148** (1962) 201–213.

[Do95]   N.P. DOLBILIN: The countability of a tiling family and the periodicity of a tiling, *Discrete Comput. Geom.* **13** (1995) 405–414.

[FoM84]   A. FONTAINE, G.E. MARTIN: Polymorphic polyominoes, *Math. Mag.* **57** (1984) 275–283.

[FoM83]   A. FONTAINE, G.E. MARTIN: Polymorphic prototiles, *J. Combinatorial Theory A* **34** (1983) 119–121.

[Fr11]   F.G. FROBENIUS: Über die unzerlegbaren diskreten Bewegungsgruppen, *Sitzungsber. Preuss. Akad. Wiss. Berlin Phys.-Math. Kl.* (1911) 654–655; Ges. Abh. Vol. 3: 507–518, Springer-Verlag, Berlin 1968.

[Go96]   S.W. GOLOMB: Tiling rectangles with polyominoes, *Math. Intelligencer* **18** (1996), no. 2, 38–47.

[GrS87]   B. GRÜNBAUM, G.C. SHEPHARD: *Tilings and Patterns*, Freeman, San Francisco 1987.

[GrS77]   B. GRÜNBAUM, G.C. SHEPHARD: Patch-determined tilings,
            Math. Gazette **61** (1977) 31–38.

[Ha77a]   H. HARBORTH: Prescribed numbers of tiles and tilings, Math.
            Gazette **61** (1977) 296–299.

[Ha77b]   H. HARBORTH: Nichtperiodische Parkettierungen der Ebene,
            Math. Naturwiss. Unterricht **30** (1977) 453–456.

[He68]   H. HEESCH: Reguläres Parkettierungsproblem, Westdeut-
            scher Verlag, 1968.

[He35]   H. HEESCH: Aufbau der Ebene aus kongruenten Bereichen,
            Nachr. Ges. Wiss. Göttingen, Neue Ser. **1** (1935) 115–117.

[HeK63]   H. HEESCH, O. KIENZLE: Flächenschluss. System der For-
            men lückenlos aneinanderschliessender Flachteile,   Springer-
            Verlag, 1963.

[HiH85]   M.D. HIRSCHHORN, D.C. HUNT: Equilateral convex pen-
            tagons which tile the plane, J. Combinatorial Theory Ser. A
            39 (1985) 1–18.

[Ke68]   R.B. KERSHNER: On paving the plane, Amer. Math. Monthly
            75 (1968) 839–844.

[KeV99]   K. KEATING, A. VINCE: Isohedral polyomino tiling of the
            plane, Discrete Comput. Geom. **21** (1999) 615–630.

[Ma04]   C. MANN: Heesch's tiling problem, Amer. Math. Monthly
            111 (2004) 509–517.

[MoR01]   C. MOORE, J.M. ROBSON: Hard tiling problems with simple
            tiles, Discrete Comput. Geom. **26** (2001) 573–590.

[Re34]   K. REINHARDT: Aufgabe 170, Jahresbericht Deutsch. Math.
            Verein. **44** (1934) 41.

[Re29]   K. REINHARDT: Über die Zerlegung der Euklidischen Ebene
            in kongruente Bereiche, Jahresbericht Deutsch. Math.-Verein.
            38 (1929) 12.

[Re28]   K. REINHARDT: Zur Zerlegung der euklidischen Räume in
            kongruente Polytope, Sitzungsber. Preuss. Akad. Wiss. Berlin
            (1928) 150–155.

[Re18]   K. REINHARDT: Über die Zerlegung der Ebene in Polygone
            (Inaugural-Dissertation, Univ. Frankfurt a. M.), R. Noske,
            Borna-Leipzig. 1918

[Ro71]   R.M. ROBINSON: Undecidability and nonperiodicity of tilings
            in the plane, Invent. Math. **12** (1971) 177–209.

[**Sch81**]   D. Schattschneider: In praise of amateurs, in: *The Math-ematical Gardner*, A. Klarner, ed., Wadsworth Int., Belmont (1981) 140–166.

[**Sch88a**]   P. Schmitt: Polymorphic sets of prototiles, *Sitzungsber. Abt II. Österr. Akad. Wiss. Math.-Naturwiss. Kl.* **197 (4-7)** (1988) 305–318.

[**Sch88b**]   P. Schmitt: Polymorphic prototiles for multiple tiling, *Anz. Österr. Akad. Wiss. Math.-Naturwiss. Kl.* **5** (1988) 79–84.

[**Sch87**]   P. Schmitt: σ-morphic sets of prototiles, *Discrete Comput. Geom.* **2** (1987) 271–295.

[**Sch86**]   P. Schmitt: Pairs of tiles admitting finitely or countably infinitely many tilings, *Geometriae Dedicata* **20** (1986) 133–142.

[**Vo37**]   H. Voderberg: Zur Zerlegung der Ebene in kongruente Bereiche in Form einer Spirale, *Jahresbericht Deutsch. Math. Verein.* **47** (1937) 159–160.

[**Vo36**]   H. Voderberg: Zur Zerlegung der Umgebung eines ebe-nen Bereiches in kongruente, *Jahresbericht Deutsch. Math. Verein.* **46** (1936) 229–231.

[**Wa61**]   H. Wang: Proving theorems by pattern recognition II, *Bell System Tech. J.* **40** (1961) 1–42.

[**WiL84**]   H.A.G. Wijshoff, J. van Leeuwen: Arbitrary versus pe-riodic storage schemes and tessellations of the plane using one type of polyomino, *Inform. and Control* **62** (1984) 1–25.

4.2 Aperiodic Tilings and Tilings with Fivefold Symmetry

As we have seen in the previous section, Hilbert's 18th problem led to the discovery of a convex pentagon P whose congruent copies can tile the plane, but no such tiling T is *isohedral*; i.e., one can always find two members of T such that no isometry of $\mathbb{R}^2$ takes one onto the other and every member of T onto another one (Kershner 1968). However, all known polygons P with the above property (convex or nonconvex) admit a periodic tiling of the plane.

The *symmetry group* $S(T)$ of a tiling T in $\mathbb{R}^d$ consists of all isometries that map T onto itself. A tiling is *periodic* if $S(T)$ contains d linearly independent translations. A tiling T is *aperiodic* if $S(T)$ does not contain *any* translations. (Warning: "aperiodic" is not the same as "not periodic"!)

Our last hope to obtain a decent characterization of all convex polygons whose congruent copies can tile the plane is to give a negative answer to part (1) of the following question.

Problem 1   *Does there exist*
(1) a convex polygon,
(2) any two-dimensional body
with the property that the plane can be tiled with its congruent copies, yet all such tilings are aperiodic?

The above question is perhaps the most outstanding unsolved problem in this area. One of the main difficulties is that it is hard to construct "nontrivial" aperiodic tilings in the plane (*per definitionem*). (See [Se99], [ScS97] for surveys.) As Wang [Wa63] pointed out, if there exists no algorithm for deciding whether there is a vertex-to-vertex tiling with congruent copies of a given (convex) polygon, then the answer to question (2) of Problem 1 is yes.

An ingenious method of constructing interesting tilings, many of which are aperiodic, was discovered in connection with a geometric puzzle by Langford [La40] and Sibson [Si40], and analyzed by Golomb [Go64], Valette and Zamfirescu [VaZ74], Giles [Gi79a], [Gi79b], [Gi79c], and Vince [Vi95]. Take a body C that can be dissected into k congruent parts each of which is similar to C. Put together k congruent copies of C "replicating" the dissection into smaller parts, and repeat this operation infinitely many times to obtain a tiling of $\mathbb{R}^2$ with congruent copies of C. A body C satisfying the above condition is called a *k-rep tile*.

Problem 2   *(Grünbaum, Shephard [GrS87], p. 525) Does there exist a two-dimensional body that is both a 2-rep tile and a 3-rep tile?*

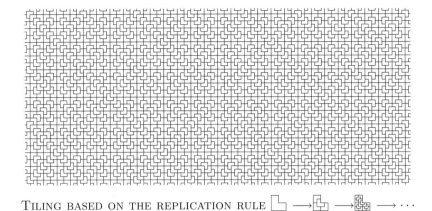

Tiling based on the replication rule ⌐ ⟶ ⌐ ⟶ ⌐ ⟶ · · ·

However, most known tilings of the plane with congruent tiles (including the ones that can be obtained by using rep tiles) have a relatively simple structure. In particular, they can be decomposed into a finite number of subfamilies such that every subfamily consists of translates. However, Versmissen [Ve91] constructed, for every n, a tiling of the plane with congruent copies of a right triangle with at least n pairwise nonparallel members. Even more surprisingly, Radin [Ra94] managed to use an idea of Conway to give a finite set S of polygons with the property that the plane can be tiled with their congruent copies, but in any such tiling each polygon occurs in *infinitely* many different orientations. In fact, the elements of the set S can be chosen to be similar triangles [Sa98].

Problem 3   *(Grünbaum, Shephard [GrS87], p. 37) Two members of a tiling are said to be equivalent if they are translates of each other. Let $e(\mathcal{T})$ denote the number of equivalence classes of $\mathcal{T}$ determined by this relation.*

(1)   *Is $e(\mathcal{T})$ finite for every tiling $\mathcal{T}$ of the plane with congruent polygons?*

(2)   *Is it true that for any polygon P there exists a constant $c = c(P)$ such that $e(\mathcal{T}) \leq c$ for every tiling of the plane with congruent copies of P?*

(3)   *What are the answers to the above questions if the polygons are assumed to be convex?*

A family $\{C_1, C_2, \ldots\}$ of d-dimensional bodies is called a *set of prototiles* if there exists a tiling of $\mathbb{R}^d$ in which every member is congruent to a C_i. One of the most remarkable discoveries in the theory of tilings is that there exist finite sets of prototiles that admit only aperiodic tilings. The first such set of prototiles (so-called *Wang tiles*) was found by Berger [Be66]. In these examples the prototiles were "marked" and had to obey certain "matching rules" (adjacency conditions). It is not difficult to see that by replacing the marked edges with complicated polygonal paths, one can get rid of the matching rules. However, the resulting prototiles have a large number of sides, and this procedure may also destroy some nice properties of the tiles (such as convexity). Smaller examples have been found by Robinson [Ro71] and Penrose [Pe74], [Pe78]. It is not hard to see that the constructions of Radin [Ra94] and Sadun [Sa98] mentioned above also have this property. An interesting dual characterization of the Penrose tilings, as projections of the square faces of some five-dimensional cubical surfaces, was found by de Bruijn [dBr81], [dBr86]. R. Ammann modified one of Penrose's tilings to produce a set of *three* convex polygons that admit only aperiodic tilings (Grünbaum, Shephard [GrS87], p. 548).

Conjecture 4   *(Grünbaum, Shephard [GrS87] p. 581) There exists no set of two convex prototiles in the plane that admit only aperiodic tilings.*

Note that it is easy to show that if a finite set of convex prototiles (polygons) does not admit a periodic tiling of $\mathbb{R}^2$, then all edge-to-edge tilings with their congruent copies must be aperiodic, i.e., they cannot have even one translational symmetry. (A tiling with polygons is *edge-to-edge* if no vertex of a polygon lies in the interior of an edge belonging to another polygon.)

Conjecture 5   *(Grünbaum, Shephard [GrS87] p. 151) If a finite set of prototiles admits a tiling of the plane whose group of symmetries contains a translation, then it also admits a periodic tiling.*

Many fascinating illustrations of aperiodic tilings are displayed in Gardner [Ga77], Grünbaum and Shephard [GrS87], and Schattschneider [Sc78]. Looking at the figures showing various tilings derived from Penrose's constructions, it is not hard to recognize their striking pentagonal symmetries.

One of the basic results in early geometric crystallography (Barlow [Ba894]; Coxeter, Moser [CoM80]; L. Fejes Tóth [FeT64]), the "crystallographic restriction," implies that no tiling of the plane has a five-fold rotational symmetry, provided that its symmetry group contains at least one translation. (A tiling, or in general a figure, is said to have *k-fold rotational symmetry* if it can be mapped onto itself by a rotation about some point of the plane through the angle $2\pi/k$.) However, this result does not exclude the existence of a tiling in which each individual tile has fivefold rotational symmetry. Danzer, Grünbaum, and Shephard [DaGS82] attributed the following problem to Kepler [Ke619].

Conjecture 6   *(Danzer, Grünbaum, Shephard [DaGS82]; Kepler [Ke619]) There does not exist a tiling of the plane whose every* *member (tile) is a body of diameter at most 1 with fivefold rotational symmetry.*

It is easy to construct a packing in $\mathbb{R}^2$ with regular pentagons of side at most 1 and only a set of Lebesgue measure 0 left uncovered.

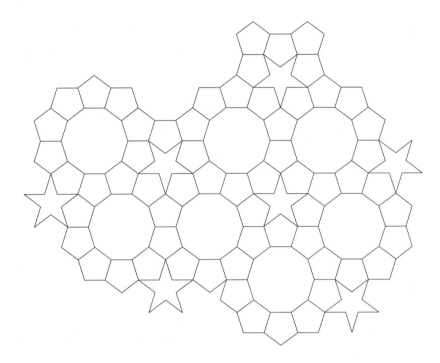

A PATCH TILED BY TILES WITH FIVEFOLD ROTATIONAL SYMMETRY

| [Ba894] | W. BARLOW: Über die geometrischen Eigenschaften starrer Strukturen und ihre Anwendung auf Kristalle, *Zeitschrift f. Kristallographie* **23** (1894) 1–63. |

[Be66] R. BERGER: The undecidability of the domino problem, *Memoirs Amer. Math. Soc.* No. **66** (1966) (72 pp).

[dBr86] N.G. DE BRUIJN: Quasicrystals and their Fourier transforms, *Nederl. Akad. Wetensch. Indag. Math.* **48** (1986) 123–152.

[dBr81] N.G. DE BRUIJN: Algebraic theory of Penrose's non-periodic tilings I, II, *Nederl. Akad. Wetensch. Indag. Math.* **43** (1981) 39–66.

[CoM80] H.S.M. COXETER, W.O.J. MOSER: *Generators and Relations for Discrete Groups* (4th edition), Springer-Verlag, 1980.

[DaGS82] L. Danzer, B. Grünbaum, G.C. Shephard: Can all tiles have five-fold rotational symmetry?, *Amer. Math. Monthly* **89** (1982) 568–570.

[FeT64] L. Fejes Tóth: *Regular Figures*, Pergamon, 1964.

[Ga77] M. Gardner: Extraordinary nonperiodic tiling that enriches the theory of tiles, *Scientific American* January 1977, 110–121.

[Gi79a] J. Giles Jr.: Infinite-level replicating dissections of plane figures, *J. Combinatorial Theory A* **26** (1979) 319–327.

[Gi79b] J. Giles Jr.: Constructions of replicating superfigures, *J. Combinatorial Theory A* **26** (1979) 328–334.

[Gi79c] J. Giles Jr.: Superfigures replicating with polar symmetry, *J. Combinatorial Theory A* **26** (1979) 335–337.

[Go64] S.W. Golomb: Replicating figures in the plane, *Math. Gazette* **48** (1964) 403–412.

[GrS87] B. Grünbaum, G.C. Shephard: *Tilings and Patterns*, Freeman, San Francisco 1987.

[Ke619] J. Kepler: *Harmonice Mundi*, Lincii. Annotated German translation: M. Caspar, *Welt Harmonik*, Oldenbourg, München 1939. Annotated English translation of Book II appears in J.V. Field: Kepler's star polyhedra, *Vistas in Astronomy* **23** (1979) 109–141.

[Ke68] R.B. Kershner: On paving the plane, *Amer. Math. Monthly* **75** (1968) 839–844.

[La40] C.D. Langford: Uses of a geometric puzzle, *Math. Gazette* **24** (1940) 209–211.

[Pe78] R. Penrose: Pentaplexity, *Eureka* **39** (1978) 16–32.

[Pe74] R. Penrose: The role of aesthetics in pure and applied mathematical research, *Bull. Inst. Math. Appl.* **10** (1974) 266–271.

[Ra94] C. Radin: The pinwheel tilings of the plane, *Ann. of Math.* (2) **139** (1994) 661–702.

[Ro71] R.M. Robinson: Undecidability tiling problems in the hyperbolic plane, *Inventiones Math.* **44** (1971) 259–264.

[Sa98] L. Sadun: Some generalizations of the pinwheel tiling, *Discrete Comput. Geom.* **20** (1998) 79–110.

[Sc78] D. Schattschneider: Tiling the plane with congruent pentagons, *Math. Mag.* **51** (1978) 29–44.

[ScS97]   D. SCHATTSCHNEIDER, M. SENECHAL: Tilings, in: *Handbook of Discrete and Computational Geometry*, J.E. Goodman, J. O'Rourke, eds., CRC Press, Boca Raton, 1997, 43–62.

[Se99]   M. SENECHAL: Periodic and aperiodic tilings of E^n, in: *Advances in Discrete and Computational Geometry* (South Hadley, MA, 1996), *Contemp. Math.* **223**, Amer. Math. Soc., Providence, 1999, 293–312.

[Si40]   R. SIBSON: Comments on Note 1464, *Math. Gazette* **24** (1940) 343.

[VaZ74]   G. VALETTE, T. ZAMFIRESCU: Les partages d'un polygone convexe en 4 polygones semblables au premier, *J. Combinatorial Theory B* **16** (1974) 1–16.

[Ve91]   K. VERSMISSEN: Monohedral periodic tilings of the plane with any number of aspects, *Discrete Math.* **93** (1991), 101–108.

[Vi95]   A. VINCE: Rep-tiling Euclidean space, *Aequationes Math.* **50** (1995) 191–213.

[Wa63]   H. WANG: Dominoes and the AEA case of the decision problem, in: *Proc. Sympos. Math. Theory of Automata* (New York, 1962), J. Fox, ed., Polytechnic Press, Brooklyn, 1963, 23–55.

4.3 Tiling Space with Polytopes

Many of the questions asked in a planar setting in the last two sections can be extended to higher dimensions. In view of the difficulties already present in the two-dimensional case, it seems to be a formidable task to characterize, for example, all convex polytopes whose congruent copies can tile $\mathbb{R}^3$. Here we would like to list some more modest questions whose solution does not seem to be hopeless.

Problem 1 *(Goldberg [Go74], Grünbaum, Shephard [GrS80]) Characterize all tetrahedra T with the property that $\mathbb{R}^3$ can be tiled with congruent copies of T.*

A tiling $\mathcal{T}$ of $\mathbb{R}^d$ with convex polytopes is called *face-to-face* if the intersection of any two members of $\mathcal{T}$ is a face of both of them. The set of all faces of all members of a face-to-face tiling $\mathcal{T}$, ordered by inclusion and completed with $\mathbb{R}^d$ (which is considered a $(d+1)$-dimensional face), forms the *face-lattice* of $\mathcal{T}$ and is denoted by $L(\mathcal{T})$. Obviously, $L(\mathcal{T}) \supseteq L(T)$ for every $T \in \mathcal{T}$.

A special class of tetrahedra admitting face-to-face tilings of $\mathbb{R}^3$ was characterized by Sommerville [So23a], [So23b], but it seems that his proof was incomplete (see Senechal [Se81] for historical details and further references). Many tilings of higher-dimensional spaces by congruent simplices have been constructed by Danzer [Da68], Baumgartner [Ba68], [Ba71], and Debrunner [Deb85].

For any polytope P we define $L(P)$, the *face-lattice* of P, as the set of all faces of P ordered by inclusion. (The empty set $\emptyset$ and P itself are considered -1- and d-dimensional faces, respectively.) If two polytopes have the same face-lattice, then they are called *combinatorially equivalent*.

Fedorov [Fe885], [Fe889] determined the five possible combinatorial types of three-dimensional polytopes admitting lattice tilings of $\mathbb{R}^3$. Minkowski [Mi897] proved that any d-dimensional convex polytope that admits a lattice tiling of $\mathbb{R}^d$ has at most $2^{d+1} - 2$ facets (i.e., $(d-1)$-dimensional faces). As a matter of fact, for a long time this was believed to be an upper bound on the number of facets of all d-dimensional polytopes that can tile $\mathbb{R}^d$. However, Föppl [Fö14], Nowacki [No35], Löckenhoff and Hellner [LöH41], Smith [Sm65], Štogrin [Št73], Koch and Fischer [KoF72], and Fischer [Fi79] have exhibited a number of examples in $\mathbb{R}^3$ with more facets. The present record is held by Engel [En81], [En86], who constructed several three-dimensional convex polytopes with 38 facets whose congruent copies can tile $\mathbb{R}^3$. All of these polytopes also admit *isohedral tilings*, i.e., tilings $\mathcal{T}$ with the property that for any two members $T_1, T_2 \in \mathcal{T}$ there exists an isometry that maps T_1 onto T_2 and $\mathcal{T}$ onto itself.

Problem 2   (*Brunner, Laves [BrL78]*) *Is there a finite upper bound for the number of faces of a three-dimensional convex polytope such that* $\mathbb{R}^3$ *can be tiled by its congruent copies?*

Problem 3   (*Engel [En81]*) *Does there exist a three-dimensional convex polytope with more than 38 facets that admits an isohedral tiling of* $\mathbb{R}^3$?

The only result in this direction is due to Delone [Del61], who generalized Minkowski's above-mentioned theorem by showing that if $\mathcal{T}$ is an isohedral tiling of $\mathbb{R}^d$ with congruent copies of a convex polytope P, then P has at most $2^d(1+t) - 2$ facets, where t denotes the minimum number of classes in a partition $\mathcal{T} = \mathcal{T}_1 \cup \mathcal{T}_2 \cup \ldots \cup \mathcal{T}_t$ with the property that any two elements belonging to the same class $\mathcal{T}_i$ are translates of each other. For $d = 3$, this implies that no polytope admitting an isohedral tiling of space can have more than $2^3(1 + 48) - 2 = 390$ facets.

A natural way to construct isohedral tilings in $\mathbb{R}^d$ is the following. Fix a discrete set of points $X \subseteq \mathbb{R}^d$ with the property that for any $x_1, x_2 \in X$ there exists an isometry of $\mathbb{R}^d$ taking x_1 to x_2 and X onto itself. (X is *discrete* if it has no point of accumulation.) Let $D(x)$ denote the Dirichlet–Voronoi region of x, i.e.,

$$D(x) = \{y \in \mathbb{R}^d \mid |y - x| \leq |y - x'| \text{ for all } x' \in X\}.$$

Obviously, the sets $D(x)$, $x \in X$, form an isohedral tiling of $\mathbb{R}^d$.

Problem 4   (*Voronoi [Vo08], [Vo09]*) *Is it true that for any convex polytope P that admits a tiling of $\mathbb{R}^d$ by translates, there is a d-dimensional lattice $\Lambda = \Lambda(P)$ whose Dirichlet–Voronoi region can be obtained from P by a suitable affine transformation?*

The answer is known to be yes for $d \leq 4$, and also if P admits a face-to-face isohedral tiling such that each vertex of P belongs to exactly $d+1$ tiles (Voronoi [Vo08], [Vo09]; Delone [Del29]; Žitomirskii [Ži29]). The statement is also true if P is a *zonotope*, i.e., the Minkowski sum of finitely many segments (see Shephard [Sh74]; McMullen [McM75]; Erdahl [Er99]).

Two face-to-face tilings $\mathcal{T}$ and $\mathcal{T}^\star$ with d-dimensional convex polytopes are said to be *dual* to each other if there exists a one-to-one mapping between their face-lattices $\phi : L(\mathcal{T}) \to L(\mathcal{T}^\star)$ that reverses inclusion, i.e.,

$$T_1 \subseteq T_2 \Leftrightarrow \phi(T_1) \supseteq \phi(T_2).$$

This definition can be easily extended to arbitrary (not necessarily face-to-face) tilings, by introducing new vertices and decomposing the faces into smaller ones, if necessary.

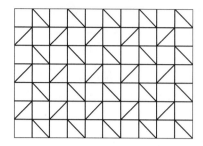 

A PAIR OF DUAL TILINGS

One of the obstacles in the way of further development in the theory of tilings is that there is no satisfactory metrical description even for tilings of relatively simple structure. (A positive answer to the last question would be an important step in this direction.) In lack of such a description, even the planar version of the following simple problem is unsolved.

Problem 5   *(Grünbaum, Shephard [GrS87] p. 174) Is it true that for any locally finite tiling $\mathcal{T}$ of $\mathbb{R}^d$ with convex polytopes there exists a dual tiling $\mathcal{T}^\star$ with convex polytopes?*

Of course, for $d = 2$ the answer is yes if $\mathcal{T}^\star$ is not required to be convex, but even this weaker statement is not clear in higher dimensions.

The following problem has attracted much attention recently.

Problem 6   *Does there exist a polytope $P \subseteq \mathbb{R}^3$ with the property that $\mathbb{R}^3$ can be tiled with congruent copies of P and yet none of these tilings is periodic?*

The discovery of a set of two "rhombohedra" (by R. Ammann) and a set of four tetrahedra with some matching rules (Danzer [Da89]) that admit no periodic tilings of $\mathbb{R}^3$ captured the interest of many physicists. This is due to the fact that both of these sets admit tilings with global icosahedral symmetry, which may provide an adequate mathematical model for the description of quasicrystals. The first quasicrystals were detected twenty years ago in certain aluminum–manganese alloys that solidify with a high degree of icosahedral symmetry (Schechtman, Blech, Gratias, Cahn [ScB*84]). Conway discovered a space filling convex polytope with nine faces that does not admit periodic tilings of $\mathbb{R}^3$ if no mirror image copies are allowed [ScS97].

Schulte [Sch03], [Sch84a], [Sch84b] suggested a combinatorial approach to the problem of aperiodicity, in which congruence of the tiles is replaced by combinatorial equivalence. We are interested in *locally finite face-to-face* tilings of $\mathbb{R}^d$ whose each cell is a convex polytope combinatorially equivalent to a polytope belonging to a given finite set $\mathcal{P}$ of so-called *combinatorial prototiles*. The set $\mathcal{P}$ is called *combinatorially aperiodic* if it admits such tilings but none of them has a combinatorial automorphism of infinite order. Schulte [Sch03] proved that there is no combinatorially aperiodic set of prototiles in the plane. We do not know whether such sets exist for any $d \geq 3$.

Conjecture 7   (Schulte [Sch03]) There are combinatorially aperiodic sets of prototiles in dimensions $d \geq 4$.

In fact, there might even exist a combinatorially aperiodic set of prototiles $\mathcal{P}$ such that every tiling admitted by $\mathcal{P}$ has a trivial combinatorial automorphism group. It is also possible that there is a one element set $\mathcal{P}$, that is, a single *monotile* with the above property.

It is known that for every $d \geq 3$ there exist convex d-dimensional polytopes that do not admit locally finite face-to-face tilings of $\mathbb{R}^d$ [Sch84a]. However, every *simplicial polytope* (all of whose faces are simplices) admits such a tiling.

Problem 8   (Schulte [Sch03]) Does every simplicial three-dimensional convex polytope admit a locally finite face-to-face tiling of $\mathbb{R}^3$ that has a combinatorial automorphism of infinite order?

In 1907, Minkowski conjectured that in any lattice tiling of $\mathbb{R}^d$ with congruent cubes, there are two cubes that share a facet ($(d-1)$-dimensional face). He verified the cases $d = 2$ and 3, but the statement in full generality was proved only much later by Hajós [Ha42]. Keller [Ke30] conjectured that the result remains true for any tiling of space with congruent cubes, which was proved by Perron [Pe40] for $d \leq 6$. However, in spaces of dimension 10 and higher, Keller's conjecture was disproved by Lagarias and Shor [LaS92], [LaS94], who used some methods developed by Corrádi and Szabó [CoS90], [Sz93], [Sz94]. In eight and nine dimensions the problem was resolved by Mackey [Ma02], so the only open case is $d = 7$.

Problem 9   (Keller) Is it true that in any tiling of seven-dimensional space with congruent cubes, there are two cubes that share a facet?

[**Ba71**]   L. Baumgartner: Zerlegung des n-dimensionalen Raumes in kongruente Simplexe, *Math. Nach.* **48** (1971) 213–224.

[**Ba68**]   L. Baumgartner: Zerlegung des vierdimensionalen Raumes in kongruente Fünfzelle, *Math.-Phys. Semesterberichte* **15** (1968) 76–86.

[**BrL78**]   G.O. Brunner, F. Laves: How many faces has the largest space-filling polyhedron?, *Zeitschr. Kristal.* **147** (1978) 39–43.

[**CoS90**]   K. Corrádi, S. Szabó: A combinatorial approach for Keller's conjecture, *Period. Math. Hungar.* **21** (1990) 95–100.

[Da89]   L. DANZER: Three-dimensional analogs of the planar Penrose tilings and quasicrystals, *Discrete Math.* **76** (1989) 1–7.

[Da68]   L. DANZER: Zerlegbarkeit endlichdimensionaler Räume in kongruente Simplices, *Math.-Phys. Semesterberichte* **15** (1968) p. 87

[Deb86]   H.E. DEBRUNNER: Tiling three-space with handlebodies, *Studia Sci. Math. Hungar.* **21** (1986) 201–202.

[Deb85]   H.E. DEBRUNNER: Tiling Euclidean d-space with congruent simplexes, in: *Discrete Geometry and Convexity*, J. Goodman et al., eds., *Annals New York Acad. Sci.* **440** (1985) 230–261.

[Del61]   B.N. DELONE: A proof of the fundamental theorem of the theory of stereohedra (in Russian), *Dokl. Akad. Nauk SSSR* **138** (1961) 1270–1272. English translation in *Soviet Math. Dokl.* **2** (1961) 812–815.

[Del29]   B.N. DELONE: Sur la partition regulière de l'espace a 4-dimensions I, II, *Izv. Akad. Nauk SSSR, Otdel. Fiz.-Mat. Nauk*, Ser. VII (1929) 79–110, 147–164.

[En86]   P. ENGEL: *Geometric Crystallography—An Axiomatic Introduction to Crystallography*, D. Reidel, 1986.

[En81]   P. ENGEL: Über Wirkungsbereichsteilungen von kubischer Symmetrie, *Zeitschr. Kristal.* **154** (1981) 199–215.

[Er99]   R.M. ERDAHL: Zonotopes, dicings, and Voronoi's conjecture on parallelohedra, *European J. Combin.* **20** (1999) 527–549.

[Fe889]   E.S. FEDOROV: Reguläre Plan- und Raumteilung. *Abh. Königl. Bayer Akad. Wiss., II. Kl.* **20** (1889) 465–588.

[Fe885]   E.S. FEDOROV: *Foundations of the Theory of Figures*, Acad. Sci. Peterburg 1885. Republished with comments, Akad. Nauk SSSR, 1953.

[Fi79]   W. FISCHER: Homogene Raumteilungen in konvexe Polyeder, in: Collected abstracts. Symp. Math. Crystall. Reideralp, Wallis, 1979.

[Fö14]   L. FÖPPL: Der Fundamentalbereich des Diamantgitters, *Phys. Zeitschrift* **15** (1914) 191–193.

[Go74]   M. GOLDBERG: Three infinite families of tetrahedral space-fillers, *J. Combin. Theory A* **16** (1974) 348–354.

[GrS87]   B. GRÜNBAUM, G.C. SHEPHARD: *Tilings and Patterns*, W.H. Freeman 1987.

[GrS80]   B. GRÜNBAUM, G.C. SHEPHARD: Tilings with congruent tiles, *Bull. Amer. Math. Soc. (N.S.)* **3** (1980) 951–973.

[Ha42]   G. HAJÓS: Über einfache und mehrfache Bedeckung des
            n-dimensionalen Raumes mit einen Würfelgitter, *Math. Z.* **47**
            (1942) 427–467.

[Ke30]   O.H. KELLER: Über die lückenlose Erfüllung des Raumes
            mit Würfeln, *J. Reine Angew. Math.* **163** (1930) 231–248.

[KoF72]   E. KOCH, W. FISCHER: Wirkungsbereichstypen einer ver-
            zerrten Diamantkonfiguration mit Kugelpackungscharakter,
            Zeitschr. Kristal. **135** (1972) 73–92.

[LaS94]   J.C. LAGARIAS, P.W. SHOR: Cube-tilings of $\mathbb{R}^n$ and non-
            linear codes, *Discrete Comput. Geom.* **11** (1994), 359–391.

[LaS92]   J.C. LAGARIAS, P.W. SHOR: Keller's cube-tiling conjecture
            is false in high dimensions, *Bull. Amer. Math. Soc. New Ser.*
            27 (1992) 279–283.

[LöH41]   H.D. LÖCKENHOFF, E. HELLNER: Die Wirkungsbereiche
            der invarianten kubischen Gitterkomplexe, *Neues Jahrbuch f.*
            Mineralogie – Monatshefte (1941) 155–174.

[Ma02]   J. MACKEY: A cube tiling of dimension eight with no face-
            sharing, *Discrete Comput. Geom.* **28** (2002) 275–279.

[McM75]   P. MCMULLEN: Space tiling zonotopes, *Mathematika* **22**
            (1975) 202–211.

[Mi897]   H. MINKOWSKI: Allgemeine Lehrsätze über die konvexen
            Polyeder, *Nachr. Ges. Wiss. Göttingen, Math.-Phys. Kl.* **2**
            (1897) 189–219. Gesammelte Abh. Bd. 2, 103–121, Teubner
            1911. Reprint: Chelsea 1967.

[No35]   W. NOWACKI: *Homogene Raumteilung und Kristallstruktur,*
            Dissertation, ETH Zürich, 1935.

[Pe40]   O. PERRON: Über lückenlose Ausfüllung des n-dimensionalen
            Raumes durch kongruente Würfel I, II, *Math. Z.* **46** (1940) 1–
            26, 161–180.

[ScS97]   D. SCHATTSCHNEIDER, M. SENECHAL: Tilings, in: *Hand-*
            book of Discrete and Computational Geometry, J.E. Good-
            man, J. O'Rourke, eds., CRC Press, Boca Raton, 1997, 43–62.

[ScB*84]   D. SCHECHTMAN, I. BLECH, D. GRATIAS, J.W. CAHN:
            Metallic phase with long-range orientational order and no trans-
            lational symmetry, *Phys. Review Letters* **53** (1984) 1951–1953.

[Sch03]   E. SCHULTE: Combinatorial aperiodicity of polyhedral pro-
            totiles, in: *Discrete Geometry—in Honor of W. Kuperberg's*
            65th Birthday, A. Bezdek, ed., Marcel Dekker, 2003, 397–406

[Sch84a]   E. SCHULTE: Nontiles and nonfacets for Euclidean space,
            spherical complexes and convex polytopes, *J. Reine Ange-*
            wandte Mathematik **352** (1984) 161–183.

[Sch84b]   E. SCHULTE: Tiling three-space by combinatorially equiva-
            lent convex polytopes, *Proc. London Math. Soc.* 3. Ser. **49**
            (1984) 128–140.

[Se81]   M. SENECHAL: Which tetrahedra fill space? *Math. Mag.* **54**
            (1981) 227–243.

[Sh74]   G.C. SHEPHARD: Space-filling zonotopes, *Mathematika* **21**
            (1974) 261–269.

[Sm65]   F.W. SMITH: The structure of aggregates — a class of 20-
            faced space filling polyhedra, *Canad. J. Physics* **43** (1965)
            2052–2055.

[So23a]   D.M.Y. SOMMERVILLE: Space-filling tetrahedra in Euclidean
            space, *Proc. Edinburgh Math. Soc.* **41** (1923) 49–57.

[So23b]   D.M.Y. SOMMERVILLE: Division of space by congruent tri-
            angles and tetrahedra, *Proc. Royal Soc. Edinburgh* **43** (1923)
            85–116.

[Št73]   M.I. ŠTOGRIN: Regular Dirichlet–Voronoi partitions for the
            second triclinic group (in Russian), *Trudy Mat. Inst. Steklov*
            123 (1973) English translation in: *Proc. Steklov Institute*
            Math. **123** (1973) Amer. Math. Soc. 1975.

[Št68]   M.I. ŠTOGRIN: Regular Dirichlet partitions for the second
            triclinic group (in Russian), *Dokl. Akad. Nauk SSSR* **183**
            (1968) 793–796.

[Sz94]   S. SZABÓ: *Algebra and Tiling — Homomorphisms in the*
            Service of Geometry, Mathematical Association of America,
            1994.

[Sz93]   S. SZABÓ: Cube tilings as contributions of algebra to geo-
            metry, *Beiträge Algebra Geom.* **34** (1993) 63–75.

[Vo09]   G.F. VORONOI: Nouvelles applications des paramètres conti-
            nus à la théorie des formes quadratiques. Domaines de formes
            quadratiques correspondant aux différent types des parallélo-
            èdres primitifs, *J. Reine Angew. Math.* **136** (1909) 67–181.

[Vo08]   G.F. VORONOI: Nouvelles applications des paramètres conti-
            nus à la théorie des formes quadratiques. Deuxième mémoire.
            Récherche sur les paralléloèdres primitifs, *J. Reine Angew.*
            Math. **134** (1908) 198–287.

[Ži29]   O.K. ŽITOMIRSKII: Verschärfung eines Satzes von Woronoi,
            Ž. Leningrad Fiz.-Mat. Obsč. **2** (1929) 131–151.

5. Distance Problems

5.1 The Maximum Number of Unit Distances in the Plane

The following problem of Erdős [Er46] is possibly the best known (and simplest to explain) problem in combinatorial geometry. How often can the same distance occur among n points in the plane? For $n \geq 4$, it is not possible that all $\binom{n}{2}$ point pairs determine the same distance, so it is a natural and nontrivial problem to find the maximum number of point pairs that determine the same distance δ. By scaling, we can assume without loss of generality that $\delta = 1$. Although this problem has received considerable attention, we are still far from the solution. Let $u(n)$ denote the maximum number of occurrences of the same (unit) distance among n points in the plane. We assume here, as always in combinatorial geometry, that the points are distinct. Allowing points to coincide would destroy the geometric aspect of the problem and essentially reduce it to a graph-theoretic problem. Given any point set S, we can define its so-called *unit distance graph*, connecting two elements of S by an edge if and only if their distance is one.

Problem 1 (*Erdős [Er46]*) *Give good asymptotic bounds for $u(n)$, the maximum number of occurrences of the same distance among n points in the plane.*

For small numbers n, one can determine the exact values of $u(n)$ by analyzing all possible point configurations, as was done by Schade in [Sc93]:

n	1	2	3	4	5	6	7	8	9	10	11	12	13	14
$u(n)$	0	1	3	5	7	9	12	14	18	20	23	27	30	33

For $n \leq 14$, Schade determined the extremal sets up to isomorphism of the unit distance graphs. Some of these examples are not rigid.

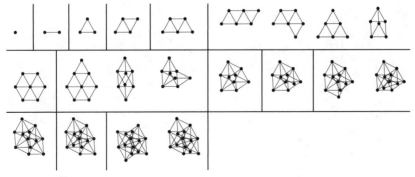

SETS WITH MAXIMUM NUMBER OF UNIT DISTANCES

Unfortunately, looking at small examples does not seem to give us much insight into the asymptotic behavior of $u(n)$. Erdős [Er46] proved that, for large n, taking a $\sqrt{n} \times \sqrt{n}$ piece ("section") of a properly scaled integer lattice gives the lower bound $u(n) > ne^{\frac{c \log n}{\log \log n}}$. This follows from the fact that the number of points at distance $\sqrt{r}$ from a given point of the integer lattice is proportional to the number of distinct ways in which r can be written as the sum of two integer squares. This number-theoretic function was investigated long ago, e.g., in connection with the four-squares theorem [PaA95].

The number of such distinct decompositions of r depends on the prime factors of r that are congruent to 1 (mod 4). Letting r_k denote the product of the first k primes in this residue class, we find that from each point of the integer lattice there are at least 2^{k+2} other points at distance $\sqrt{r_k}$. If we consider a $\lceil 2\sqrt{r_k} \rceil \times \lceil 2\sqrt{r_k} \rceil$ square section S of the integer lattice, then for each point p of S, at least a quarter of the circle of radius $\sqrt{r_k}$ around p belongs to S. Thus, S is a set of roughly $4r_k$ points whose elements all have at least 2^k and at most 2^{k+2} "neighbors" at distance $\sqrt{r_k}$. Using some standard estimates for the product of the first k primes from the congruence class 1 mod 4, we obtain the lower bound

$$u(n) \geq \Omega\left(ne^{\frac{c \log n}{\log \log n}}\right).$$

This lower bound has not been improved since 1946 and is conjectured to be asymptotically tight. Using a properly scaled section of the triangular lattice, we obtain asymptotically the same bound. We will refer to these constructions as "lattice section" constructions.

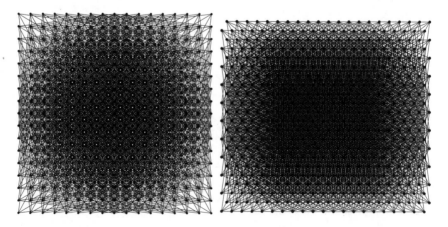

UNIT DISTANCES IN SQUARE AND TRIANGULAR LATTICE SECTIONS

Erdős's [Er46] initial upper bound, $u(n)$ $\leq O\left(n^{\frac{3}{2}}\right)$, is a simple consequence of the fact that any two (unit) circles intersect at most twice. It follows that the unit distance graph assigned to any set of points in the plane contains no $K_{2,3}$ (complete bipartite graph with two and three vertices in its classes). According to the Kő-

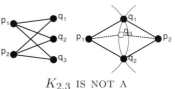

$K_{2,3}$ IS NOT A
UNIT DISTANCE GRAPH

vári–Sós–Turán theorem (see, e.g., [PaA95]), every $K_{2,3}$-free graph on n vertices has at most $O\left(n^{\frac{3}{2}}\right)$ edges. Further forbidden subgraphs were found by C. Purdy and G. Purdy [PuP88] and by Chilakamarri and Mahoney [ChM95]. In principle, one can describe the class of unit distance graphs by listing all forbidden subgraphs, but none of the known forbidden subgraphs leads to a better upper bound. It seems that purely graph-based methods cannot lead to any substantial improvement.

Erdős's upper bound on $u(n)$ was improved in several steps ([JóS75], [BeS84]) to $O\left(n^{\frac{4}{3}}\right)$. Today we have at least five different proofs for this estimate. The first proof was found by Spencer, Szemerédi, and Trotter [SpST84]. The second proof, due to Clarkson et al. [ClE*90], starts with a proper subdivision of the plane into smaller cells. In each part, one can apply a weaker Kővári–Sós–Turán-type (also called Canham-type) bound. This so-called cutting technique has become very successful in combinatorial and computational geometry. A similar proof is based on ε-nets and VC-dimension (see, e.g., [PaA95]). A fourth proof by Pach and Tardos [PaT05] uses a forbidden subgraph argument in ordered graphs. The most elegant proof, due to Székely [Sz97], is based on a general lower bound for the crossing number of graphs. It is a common feature of all these arguments that they explore no special properties of the Euclidean distance apart from the "combinatorial" structure of the cell decompositions of the plane induced by two intersecting unit circles. The constructions of Brass [Br98c] and Valtr [Va05] show that in this generality, for arbitrary metrics, none of the proofs can be improved (see Section 5.2). If any further improvement is possible, it must use some special property of the Euclidean plane.

Summarizing, we currently know that $u(n) \geq \Omega\left(ne^{\frac{c\log n}{\log\log n}}\right)$, and $u(n)$ $\leq O\left(n^{\frac{4}{3}}\right)$. It was conjectured by Erdős that $u(n) \leq O\left(n^{1+\varepsilon}\right)$ for all $\varepsilon > 0$, or even $u(n) \leq O\left(ne^{\frac{c\log n}{\log\log n}}\right)$, but we are still far from having a solution of this problem.

Another possible way to attack this problem is to establish some structural properties of the extremal sets. Brass [Br98a] showed that for $n > n_0(k)$, the extremal sets S of n points that determine the maximum number of unit distances parallel to an element of a set V of k prescribed directions (over all S and V) have a nice lattice section structure. Unfortunately, the threshold $n_0(k)$ is so big that one cannot answer the original question by taking the maximum over all k. It is known that all extremal sets for Problem 1 have small diameter [Br98d], unlike the lattice sections that occur in [Br98a] as extremal sets when the number of directions is fixed. Therefore, it is not quite clear whether one should expect any lattice-related structure from the extremal sets in the general case.

It would be interesting to find other good constructions. By some number-theoretic "coincidence," lattice section constructions determine a large number of unit distances. This might be misleading: it is possible that there are much better point sets with a completely different structure. It should be noted that, for increasing values of n, the lattice section constructions described above do not form a nested family in the usual sense: rather than taking larger and larger sections of the same object, we took the intersection of denser and denser lattices with the same square. The few known exact extremal sets found by Schade [Sc93] have a different structure: they are all integer combinations of four unit vectors [Br96]. Thus, it would be interesting to explore whether there is an efficient construction using projections of increasing sections of higher-dimensional lattices $\left\{ \sum_{i=1}^{k} a_i v_i \mid a_i \in \mathbb{Z} \right\}$ for some proper choice of the generating vectors $(v_i)_{i=1}^{k}$. It is not even clear that one can achieve by such a construction that the degrees of the vertices in the corresponding unit distance graphs are infinite.

Problem 2   (Brass) *Does there exist a finite set of rationally independent vectors $(v_i)_{i=1}^{k}$ in $\mathbb{R}^2$ such that their set of integer
combinations contains infinitely many points on the unit
circle?*

Projections of increasing sections of a fixed higher-dimensional lattice would provide a quite different construction of point sets with many unit distances.

Another well-known construction is the *iterated Minkowski sum construction*. If we have two sets S_1, S_2 with n_1, n_2 points and u_1, u_2 unit distances, respectively, then for almost all rotated images S_2' of S_2, the Minkowski sum $S_1 + S_2'$ consists of $n_1 n_2$ points that determine $n_1 u_2 + n_2 u_1$ unit distances (for some rotation angles, several points of the Minkowski sum may coincide). Taking the sum of many different rotated copies of an initial set, we obtain a set of n points with $\Theta(n \log n)$ unit distances. This is less than the number of unit distances in the lattice section construction, but the method can be applied in many other metric spaces. Using unit equilateral triangles as starting configurations, the resulting sets consist of 3^k points and determine $k3^k$ unit distances. The unique extremal set with nine points and 18 unit distances from Schade's [Sc93] list can be obtained as the sum of two unit equilateral triangles.

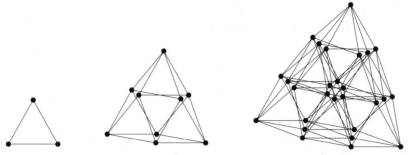

MINKOWSKI SUM OF UNIT EQUILATERAL TRIANGLES

An interesting variant of the problem on the maximum number of unit distances is the *favorite distances* problem. Let $\deg_S(p, r)$ denote the number of points in S at distance r from p. Using this notation, we have

$$u(n) = \max_{|S|=n} \frac{1}{2} \sum_{p \in S} \deg_S(p, 1).$$

Now, if we do not take the same (unit) distance for every point, but allow each point to have its own "favorite" distance, we obtain the function

$$u^{\text{favorite}}(n) = \max_{|S|=n} \sum_{p \in S} \max_r \deg_S(p, r).$$

This problem was studied by Avis, Erdős, and Pach [AvEP88], [ErP90]. In the plane, n points can determine $\Omega\left(n^{\frac{4}{3}}\right)$ favorite distances, which may be asymptotically optimal: choose $\frac{n}{2}$ points and $\frac{n}{2}$ lines with $cn^{\frac{4}{3}}$ incidences, and perform an inversion on a circle. This maps the lines to circles with distinct centers. Taking these $\frac{n}{2}$ centers together with the images of the $\frac{n}{2}$ original points, we obtain a set of n points showing that $u^{\text{favorite}}(n) \geq cn^{\frac{4}{3}}$. The order of magnitude of this bound is probably best possible. The currently best known upper bound of $O\left(n^{\frac{15}{11}+\varepsilon}\right)$ is due to Aronov and Sharir [ArS02].

Conjecture 3 $u^{\text{favorite}}(n) = O\left(n^{\frac{4}{3}}\right)$.

Connecting by a directed edge each point p to every point whose distance from p is the favorite distance assigned to p, we obtain a *favorite distance digraph*. Erdős suggested that instead of maximizing the number of edges in such a graph, we may try to maximize the minimum outdegree of its vertices. What is the largest k such that there is a set of n points in the plane with every element equidistant from at least k others? This problem is quite different from the previous one. In the construction described above, each of the $\frac{n}{2}$ points corresponding to centers of the circles is at the same distance from many other points. However, it is possible that there are no two points at the same distance from any of the remaining $\frac{n}{2}$ points. The best known lower bound is given by the lattice section construction, in which the outdegree of every point is $e^{c\frac{\log n}{\log\log n}}$. On the other hand, the upper bound of Aronov and Sharir [ArS02] on $u^{\text{favorite}}(n)$ implies that there is no point set with every element equidistant from $\Omega\left(n^{\frac{4}{11}+\varepsilon}\right)$ other elements. For some small values of n the problem has been solved by Erdős and Fishburn [ErF97].

Problem 4   *[ErF97] What is the smallest number $k = k(n)$ such that every set of n points in the plane has an element from which there are no k elements at the same distance?*

A similar looking question, for which one cannot obtain a lower bound by a lattice section construction, was raised by Erdős. A graph is k-regular if its every vertex has degree k.

Problem 5   *What is the smallest number n_k for which there exists a k-regular unit distance graph with n_k vertices?*

This question was studied by Harborth [Ha85], [Ha87]. For $k \le 5$, he determined the smallest point sets whose unit distance graphs are k-regular,

k	1	2	3	4	5
n_k	2	3	6	9	18

The extremal configurations turned out to be Minkowski sums of unit equilateral triangles if k is even, and sums of equilateral triangles with an additional unit distance pair if k is odd. The existence of regular induced subgraphs does not seem to follow from purely graph-theoretic considerations. The best construction we have is the iterated Minkowski sum of triangles: 3^k points such that the corresponding unit distance graph is $2k$-regular.

[ArS02]   B. Aronov, M. Sharir: Cutting circles into pseudo-segments and improved bounds for incidences, *Discrete Comput. Geom.* **28** (2002) 475–490.

[AvEP88]   D. Avis, P. Erdős, J. Pach: Repeated distances in space, *Graphs Combinatorics* **4** (1988) 207–217.

[Be83]   J. Beck: On the lattice property of the plane and some problems of Dirac, Motzkin and Erdős in combinatorial geometry, *Combinatorica* **3** (1983) 281–297.

[BeS84]   J. BECK, J. SPENCER: Unit distances, *J. Combinatorial Theory Ser. A* **37** (1984) 231–238.

[Br98a]   P. BRASS: On point sets with many unit distances in few directions, *Discrete Comput. Geom.* **19** (1998) 355–366.

[Br98b]   P. BRASS: Häufige Abstände in endlichen Punktmengen, *Jahrbuch Überblicke Mathematik* 1998, 66–75.

[Br98c]   P. BRASS: On lattice polyhedra and pseudocircle arrangements, in: *Charlemagne and His Heritage — 1200 Years of Civilization and Science in Europe. Vol 2: Mathematical Arts*, P.L. Butzer et al., eds., Brepols Verlag, 1998, 297–302.

[Br98d]   P. BRASS: On the diameter of sets with maximum number of unit distances, *Geombinatorics* **8** (1998) 149–153.

[Br96]   P. BRASS: Erdős distance problems in normed spaces, *Comput. Geom. Theory Appl.* **6** (1996) 195–214.

[Ch93]   K.B. CHILAKAMARRI: The unit-distance graph problem: a brief survey and some new results, *Bulletin of the ICA* **8** (1993) 39–60.

[ChM95]   K.B. CHILAKAMARRI, C.R. MAHONEY: Maximal and minimal forbidden unit-distance graphs in the plane, *Bulletin of the ICA* **13** (1995) 35–43.

[ClE*90]   K.L. CLARKSON, H. EDELSBRUNNER, L.J. GUIBAS, M. SHARIR, E. WELZL: Combinatorial complexity bounds for arrangements of curves and spheres, *Discrete Comput. Geom.* **5** (1990) 99–160.

[Er46]   P. ERDŐS: On sets of distances of n points, *Amer. Math. Monthly* **53** (1946) 248–250.

[ErF97]   P. ERDŐS, P. FISHBURN: Minimum planar sets with maximum equidistance counts, *Comput. Geom. Theory Appl.* **7** (1997) 207–218.

[ErP90]   P. ERDŐS, J. PACH: Variations on the theme of repeated distances, *Combinatorica* **10** (1990) 261–269.

[Ha87]   H. HARBORTH: Regular point sets with unit distances, in: *Intuitive Geometry* (Siófok, 1985), K. Böröczky et al., eds., *Colloquia Math. Soc. János Bolyai* **48** (1987) 239–253.

[Ha85]   H. HARBORTH: Äquidistante, reguläre Punktmengen, in: *2. Kolloquium Geometrie und Kombinatorik 1983*, TH Karl-Marx-Stadt, 1985, 81–86.

[JóS75]   S. JÓZSA, E. SZEMERÉDI: The number of unit distances on the plane, in: *Infinite and Finite Sets, Vol. 2*, A. Hajnal et al., eds., *Colloquia Math. Soc. János Bolyai* **10** (1975), North-Holland, 939–950.

[**PaA95**] J. Pach, P.K. Agarwal: *Combinatorial Geometry*, Wiley, New York 1995.

[**PaT05**] J. Pach, G. Tardos: Forbidden paths and cycles in ordered graphs and matrices, manuscript.

[**PuP88**] C. Purdy, G. Purdy: Minimal forbidden distance one graphs, *Congressus Numerantium* **66** (1988) 165–171.

[**Sc93**] C. Schade: *Exakte Maximalzahlen gleicher Abstände*, Diploma thesis directed by H. Harborth, Techn. Univ. Braunschweig 1993.

[**SpST84**] J. Spencer, E. Szemerédi, W.T. Trotter: Unit distances in the Euclidean plane, in: *Graph Theory and Combinatorics*, B. Bollobás, ed., Academic Press, London, 1984, 293–303.

[**Sz97**] L.A. Székely: Crossing numbers and hard Erdős problems in discrete geometry, *Comb. Probab. Comput.* **6** (1997) 353–358.

[**Va05**] P. Valtr: Strictly convex norms allowing many unit distances and related touching questions, manuscript 2005.

5.2 The Number of Equal Distances in Other Spaces

The main question of the previous section makes sense in any metric space: what is the maximum number of occurrences of a given distance Δ among n distinct points of a given space? If the space is homogeneous with respect to scaling, this number does not depend on the distance, so we can again restrict ourselves to counting unit distances ($\Delta = 1$).

The most natural candidates are higher-dimensional Euclidean spaces, especially the three-dimensional space. Finite-dimensional normed spaces as well as spherical spaces have also been studied. In the latter case, i.e., for points on the unit sphere, we cannot assume without loss of generality that $\Delta = 1$.

Let $u_d(n)$ denote the maximum number of unit distances among n points in d-dimensional Euclidean space.

Problem 1   [Er60] Give good asymptotic bounds for $u_3(n)$, the maximum number of occurrences of the same distance among n points in three-dimensional space.

In three-dimensional space, again large sections of properly scaled integer lattices provide the asymptotically best known constructions. Clearly, the number of distinct distances in the lattice section $\{1, \ldots, n^{\frac{1}{3}}\}^3$ is $O(n^{\frac{2}{3}})$, so it follows by the pigeonhole principle that $u_3(n) \geq \Omega(n^{\frac{4}{3}})$. Using better estimates on the number of decompositions of a positive integer k as the sum of three squares, Erdős proved that the same construction actually gives $u_3(n) \geq \Omega(n^{\frac{4}{3}} \log \log n)$. The first nontrivial upper bound, $u_3(n) \leq O(n^{\frac{5}{3}})$, was also obtained by Erdős [Er60]. It is a consequence of the fact that the complete bipartite graph $K_{3,3}$ with three vertices in each of its classes is not realizable with unit distances in three-dimensional space, because three unit spheres have at most two common intersection points. Then one can apply the Kővári–Sós–Turán theorem (see [PaA95]), which implies that any graph on n vertices not containing $K_{3,3}$ as a subgraph has at most $O(n^{\frac{5}{3}})$ edges. This was improved in several steps ([Be83], [Chu89]) to the present record $u_3(n) \leq O(n^{\frac{3}{2}} \beta(n))$, held by Clarkson, Edelsbrunner, Guibas, Sharir, and Welzl [ClE*90]. Here $\beta(n)$ is an extremely slowly growing function related to the inverse Ackermann function ($\beta(n) = 2^{O((\alpha(n))^2)}$, where α is the inverse Ackermann function). The $\beta(n)$-factor is an annoying artifact of the proof. It shows up in many arguments based on subdividing an arrangement into cells of constant complexity, i.e., into cells whose total number of faces of all dimensions is bounded from above by a constant. Although this is a very small factor, removing it from the upper bound would be of some interest.

Somewhere between the two- and three-dimensional cases is the problem of bounding the number of occurrences of a distance Δ on a fixed sphere. This problem also serves as an important building block for the corresponding questions in higher dimensions. Equivalently, we can count the number of occurrences of a fixed angular distance among a set of n directions in three-dimensional space. It turns out that the maximum of this quantity strongly depends on the ratio of the distance Δ to the radius of the sphere, or, equivalently, on the angle whose occurrences we are counting [ErHP89]. The maximum number of orthogonal pairs (unit distances among points on a sphere of radius $\frac{1}{\sqrt{2}}$) is known to be $\Theta(n^{\frac{4}{3}})$. An orthogonal pair (p, q) can be interpreted as an incidence between p and the great circle around q, so by projection this problem is equivalent to bounding the number of point–line incidences in the plane.

For any other angle $\gamma \neq \frac{\pi}{2}$ (that corresponds to unit distances on a sphere whose radius is different from $\frac{1}{\sqrt{2}}$), we are far from having good estimates for the maximum number of occurrences of γ. Erdős, Hickerson, and Pach [ErHP89] established the slightly superlinear lower bound of $\Omega(n \log^* n)$ for this quantity, where $\log^* n = k$ if

$$
\left. 2^{2^{2^{\cdot^{\cdot^{\cdot^2}}}}} \right\} k \text{ times} \;\; \leq n < \;\; \left. 2^{2^{2^{\cdot^{\cdot^{\cdot^2}}}}} \right\} k+1 \text{ times}.
$$

This lower bound has been improved by Swanepoel and Valtr [SwV04] to $\Omega(n\sqrt{\log n})$. The only known upper bound is still $O(n^{\frac{4}{3}})$, the same as for orthogonal pairs. One of the difficulties is that most of the small examples of unit distance graphs in the plane are also realizable on spheres of any radius; therefore, it is hard to explore any special property of the sphere. Almost the same can be said about the lower-bound constructions: they depend only on the existence of sufficiently many pairwise commuting isometries; hence they work in a very large class of metric spaces.

Problem 2   *[ErHP89] Give good asymptotic bounds for the maximum number of occurrences of a fixed angle $\gamma \neq \frac{\pi}{2}$ among n directions in three-dimensional space.*

We believe that the $\Omega(n\sqrt{\log n})$ lower bound can be much improved. It might be interesting to characterize the class of graphs that can be realized by all sufficiently small angular distances.

Problem 3   *(Brass) Give good asymptotic bounds for the maximum number of edges in a graph of n vertices that can be realized as a unit distance graph on every sphere of sufficiently large radius.*

In higher dimensions ($d \geq 4$), the problem becomes much simpler. The order of magnitude of $u_d(n)$ was determined by Erdős [Er60]. The following asymptotically optimal construction was found by Lenz.* For $d \geq 4$, choose a system of $\left\lfloor \frac{d}{2} \right\rfloor$ mutually orthogonal circles in d-dimensional space with a common center and with radius $\frac{1}{\sqrt{2}}$. Clearly, any two points on distinct circles are at distance one. For odd d, one of the circles can be replaced by a three-dimensional sphere. This construction gives $\frac{1}{2}\left(1 - \frac{1}{\left\lfloor \frac{d}{2} \right\rfloor}\right) n^2 + o(n^2)$ unit distances, and the coefficient of the quadratic term is best possible. Some further unit distances (giving lower-order terms) can be obtained by optimizing the choice of points on the circles and on the sphere. For the upper bound, Erdős [Er60] used the following forbidden subgraph argument. It is not hard to verify that a complete $\left(\left\lfloor \frac{d}{2} \right\rfloor + 1\right)$-partite graph $K_{3,3,\dots,3}$ with three elements in each of its classes is not realizable as a unit distance graph in d-dimensional space. Therefore, one can apply the Erdős–Stone theorem from extremal graph theory [PaA95] to conclude that

$$u_d(n) \leq \frac{1}{2}\left(1 - \frac{1}{\left\lfloor \frac{d}{2} \right\rfloor}\right) n^2 + o(n^2).$$

For even dimensions d, this upper bound can be slightly tightened. In this case, the complete $(\frac{d}{2} + 1)$-partite graph $K_{3,\dots,3,1}$ is not realizable as a unit distance graph in d-dimensional space, and a corresponding graph-theoretic result of Erdős and Simonovits [Si68] implies that $u_d(n) \leq \mathrm{ex}(n, K_{\frac{d}{2}+1}) + n$, whenever n is sufficiently large. Here $\mathrm{ex}(n, K_{\frac{d}{2}+1})$ denotes the so-called Turán function, i.e., the maximum number of edges in a $K_{\frac{d}{2}+1}$-free graph on n vertices. (This number coincides with the maximum number of edges in the Lenz construction, not counting any unit distances that may occur among points on the same circle.) Thus, for even dimensions d, we obtain the almost tight bounds $\mathrm{ex}(n, K_{\frac{d}{2}+1}) + n - \frac{d}{2} \leq u_d(n) \leq \mathrm{ex}(n, K_{\frac{d}{2}+1}) + n$, if n is sufficiently large [Er67]. The odd-dimensional case remained open until Erdős and Pach [ErP90] proved that the error term was $\Theta(n^{\frac{4}{3}})$. Summarizing: for $d \geq 4$, we have

$$u_d(n) = \frac{1}{2}\left(1 - \frac{1}{\left\lfloor \frac{d}{2} \right\rfloor}\right) n^2 + \begin{cases} n - O(1) & \text{for } d \geq 4 \text{ even,} \\ \Theta(n^{\frac{4}{3}}) & \text{for } d \geq 5 \text{ odd.} \end{cases}$$

For any fixed dimension $d \geq 4$, one can probably determine even the exact value of $u_d(n)$. This has been worked out for $d = 4$ by Brass [Br97] and

---

* The construction was communicated by H. Lenz to Erdős. There is no paper by Lenz describing this construction.

van Wamelen [vW99], who proved by induction that for $n \geq 10$,

$$u_4(n) = \left\lfloor \frac{1}{4}n^2 \right\rfloor + \begin{cases} n & \text{if } n \text{ is divisible by 8 or 10,} \\ n-1 & \text{otherwise.} \end{cases}$$

The same method can probably be used in any other even-dimensional case $d \geq 4$, but it may require a lot of work to get the induction started.

Problem 1 can be raised in any (finite-dimensional) normed space. The two-dimensional case of so-called *Minkowski planes* was discussed by Brass [Br96a]. He noticed that all known proofs of the $O(n^{\frac{4}{3}})$ upper bound valid in the Euclidean case directly generalize to every strictly convex normed plane, i.e., to any normed plane whose unit circle is strictly convex. In this generality the bound is best possible: Brass [Br98] and Valtr [Va05] exhibited strictly convex normed planes that allow at least $cn^{\frac{4}{3}}$ unit distances among n points. The unit circle of any normed plane P that is not strictly convex contains a line segment. Depending on the maximum length of such a line segment, the exact maximum number of unit distances among n points of P is either $\lfloor \frac{1}{4}n^2 \rfloor$ or $\lfloor \frac{1}{4}(n^2 + n) \rfloor$ for $n \geq 10$.

The unit circles of the strictly convex norms constructed by Brass [Br98] are strictly convex curves interpolating lattice polygons with $n^{\frac{1}{3}}$ vertices contained in an $n^{\frac{1}{2}} \times n^{\frac{1}{2}}$ section of a square lattice. As n tends to infinity these curves converge to circles, so the corresponding normed planes get arbitrarily close to the Euclidean plane. For each n, Brass's construction uses a different norm to obtain $\Omega(n^{\frac{4}{3}})$ unit distances among n points. Valtr [Va05] found a single norm for which n points can determine this many unit distances. In this norm, the unit circle consists of two parabolic arcs $\{(x_1, x_2) \mid x_1 \in [-1, 1], x_2 = 1 - x_1^2 \text{ or } x_2 = -1 + x_1^2\}$. The number of unit distances determined by the points $\left(\frac{i}{k}, \frac{j}{k^2} \right)$, where $|i| < k$, $|j| < k^2$ and $k = n^{\frac{1}{3}}$, is $\Omega(n^{\frac{4}{3}})$.

Almost nothing is known about higher-dimensional normed spaces other than the Euclidean and spherical ones. The maximum number of unit distances among n points in any normed space grows quadratically in n if the norm is not strictly convex. However, as shown by the Euclidean spaces of dimension $d \geq 4$, it may also grow quadratically for strictly convex norms. In four-dimensional space, we do not know of any norm for which this function is subquadratic. On the other hand, we are not aware of any general lower bound better than $\Omega(n \log n)$, which follows from the iterated Minkowski sum construction described in the previous section. A particularly simple special case of this construction that works in all normed spaces (despite the fact that in general, we cannot simply "rotate" a point set without distorting unit distances) is the following: for any k distinct unit vectors, $e_1, \ldots, e_k$, their 2^k subset sums induce at least $k2^{k-1}$ unit distances. This point set can be interpreted as the Minkowski sum $\{0, e_1\} + \cdots + \{0, e_k\}$.

Problem 4   *(Brass) Find a general construction of n points with strict-ly more than $\Omega(n \log n)$ unit distances that can be carried out in every normed space of a given dimension $d \geq 3$.*

Problem 5   *(Brass) For $d \geq 4$, does there exist a norm in d-dimensional space with respect to which the maximum number of oc-currences of the unit distance among n points is subquad-ratic?*

Conjecture 6   *(Brass) For any d and for any sufficiently large $n \geq n_0(d)$, the d-dimensional normed space in which the largest num-ber of unit distances among n points attains its maximum is L^d_∞.*

Any application of forbidden subgraph arguments to obtain an upper bound on the number of unit distances determined by a point set in a higher-dimensional normed space is hindered by our lack of understanding of the structure of possible realizations of complete bipartite graphs as unit distance graphs. Assuming strict convexity does not help here: even in three-dimensional space, with respect to some strictly convex norms, the complete bipartite graph $K_{\frac{n}{2},\frac{n}{2}}$ is realizable as a unit distance graph. For if f and g are two strictly convex functions on $[-1, 1]$, one can define a strictly convex surface patch by $(x, y, f(x) + g(y))$ that can be embedded in the unit sphere of a strictly convex norm. With respect to such a norm, the points $(x, 0, f(x))$ and $(0, -y, -g(y))$ all have the same distance, since their difference $(x, y, f(x) + g(y))$ belongs to the unit sphere.

Petty's following problem illustrates how little we know about unit distances in normed spaces: Does every d-dimensional normed space admit an equilateral simplex? Equivalently, does every d-dimensional convex body have $d + 1$ pairwise touching translates?

At first glance, it may appear that the answer must be yes. Petty [Pe71] proved this for $d \leq 3$. However, for $d \geq 4$ he found a "near-counter-example" consisting of four translates of a double cone over a $(d - 1)$-dimensional ball. Take two such double cones that touch each other apex-to-apex. It is possible to add two further translates that touch both double cones and each other. It is not hard to see that no three such translates can be added, in any dimension.

Petty's example shows that in some normed spaces of arbitrarily high dimensions there are maximal equilateral simplices consisting of only four points.

Thus, not every equilateral simplex can be completed to a full-dimensional equilateral simplex. However, such an extension is always possible if the norm does not differ much from the Euclidean norm [Br99], [De00]: a celebrated theorem of Dvoretzky implies that for every k there exists a $d = d(k)$ such that any d-dimensional normed space admits an equilateral k-dimensional simplex. This is a very weak result; it proves the existence of an equilateral $c(\log d)^{\frac{1}{3}}$-simplex, instead of an equilateral d-simplex.

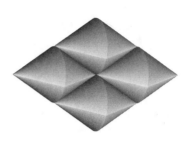

PETTY'S
NEAR-COUNTEREXAMPLE

Petty's conjecture is open for every $d \geq 4$ (see [Sw04] for a survey).

Conjecture 7 *[Pe71] Every d-dimensional normed space has $d+1$ points with all pairwise distances one.*

Most of the variants of the unit distance problem discussed in the last section for the Euclidean plane make perfect sense in any other normed space. They usually become simpler in higher dimensions, especially when there exists a quadratic lower bound construction, as in the case of the three-dimensional "favorite distances" problem. Again, let $\deg_S(p, r)$ denote the number of points in S at distance r from p, and let

$$u_d^{\text{favorite}}(n) = \max_{|S|=n} \sum_{p \in S} \max_r \deg_S(p, r),$$

where S is a set of points in d-dimensional Euclidean space.

We conjectured in the previous section that $u_2^{\text{favorite}}(n) = O(n^{\frac{4}{3}})$, but $u_3^{\text{favorite}}(n)$ is already at least $\frac{n^2}{4}$, as shown by the following example. Select $\frac{n}{2}$ points on a circle C and $\frac{n}{2}$ points on a line perpendicular to C and passing through its center. For higher dimensions $d \geq 4$, one can give a Lenz-type construction (see Section 5.1).

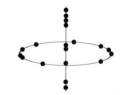

n POINTS WITH $\frac{n^2}{4}$
FURTHEST-NEIGHBOR PAIRS

The problem of estimating $u_d^{\text{favorite}}(n)$ was studied by Avis, Erdős, and Pach [AvEP88], [ErP90], who determined the coefficients of the quadratic terms. The known results are

$$\Omega(n^{\frac{4}{3}}) \leq u_2^{\text{favorite}}(n) \leq O(n^{\frac{15}{11}+\varepsilon}),$$

$$\tfrac{1}{4}n^2 + \tfrac{3}{2}n \leq u_3^{\text{favorite}}(n) \leq \tfrac{1}{4}n^2 + O(n^{2-\varepsilon}),$$

$$\left(1 - \frac{1}{\lfloor \frac{d}{2} \rfloor}\right)n^2 \leq u_d^{\text{favorite}}(n) \leq \left(1 - \frac{1}{\lfloor \frac{d}{2} \rfloor}\right)n^2 + o(n^2) \quad \text{for } d \geq 4.$$

One of the difficulties is that for solving the favorite distances problem one needs Turán-type results for directed graphs, and in this respect the theory is less developed than for graphs (see [BrS02]).

The problem about the maximal minimum outdegree of the favorite distance digraph is even more interesting in three-dimensional space than in the plane. In the above example, which consists of $\frac{n}{2}$ points on a circle and $\frac{n}{2}$ points on its axis, half of the points of the corresponding favorite distance digraph have outdegree $\frac{n}{2}$, while for all points on the circle there are at most only four other points at the same distance; i.e., their outdegrees do not exceed four. Therefore, this example, which asymptotically maximizes the average outdegree, is very bad from the point of view of maximizing the minimum outdegree. The currently best known lower bound for the maximal minimum outdegree, $\Omega(n^{\frac{1}{3}} \log \log n)$, follows from the lattice section construction. Erdős, Harcos, and Pach [ErHP99] obtained an upper bound of $O(n^{\frac{3}{5}} \beta(n))$, where $\beta(n)$ again denotes the very slowly growing function that has already occurred in the corresponding bound for $u_3(n)$.

Problem 8   *[ErHP99] Establish good bounds on the largest number $\Delta_3(n)$ such that there exist n points in three-dimensional space, each of which is at the same distance from at least $\Delta_3(n)$ others.*

In dimensions $d \geq 4$, the Lenz construction is again asymptotically extremal: there are at most $\left(1 - \frac{1}{\left\lfloor \frac{d}{2} \right\rfloor}\right) n + o(n)$ points equidistant from each of its elements.

A related extremal problem discussed by Brass [Br00] is to maximize the higher moments of the degree sum. Let

$$u_d^\alpha(n) = \max_{|S|=n} \frac{1}{2} \sum_{p \in S} \left(\deg_S(p, 1) \right)^\alpha.$$

Curiously, there is a simple solution for $d = 2, 3$ and $\alpha \geq d$. In this case, $u_d^\alpha(n) = \Theta(n^\alpha)$ because it is enough to maximize the number of subgraphs isomorphic to $K_{1,d}$. For $d \geq 4$, the Lenz construction shows that $u_d^\alpha(n) = \Theta(n^{\alpha+1})$, for every α. Thus, the only interesting instances are $d = 2$, $\alpha \in [1, 2)$ and $d = 3$, $\alpha \in [1, 3)$. In these cases, one can obtain some nontrivial upper bounds using Hölder's inequality, but they are certainly not best possible.

[AvEP88]   D. AVIS, P. ERDŐS, J. PACH: Repeated distances in space, *Graphs Combinatorics* **4** (1988) 207–217.

[Be83]   J. BECK: On the lattice property of the plane and some problems of Dirac, Motzkin and Erdős in combinatorial geometry, *Combinatorica* **3** (1983) 281–297.

[Br00]   P. BRASS: Exact point pattern matching and the number of congruent triangles in a three-dimensional pointset, *ESA 2000*, Springer LNCS **1879**, 112–119.

[Br99]   P. BRASS: On equilateral simplices in normed spaces, *Beiträge Algebra Geometrie* **40** (1999) 303–307.

[Br98]   P. BRASS: On lattice polyhedra and pseudocircle arrangements, in: *Charlemagne and his Heritage — 1200 Years of Civilization and Science in Europe. Vol 2: Mathematical Arts.*, P.L. Butzer et al., eds., Brepols Verlag 1998, 297–302.

[Br97]   P. BRASS: On the maximum number of unit distances among n points in dimension four, in: *Intuitive Geometry*, I. Bárány et al., eds., *Bolyai Soc. Mathematical Studies* **6** (1997) 277–290; note the correction of one case by K. Swanepoel in the review MR 98j:52030.

[Br96a]   P. BRASS: Erdős distance problems in normed spaces, *Comput. Geom. Theory Appl.* **6** (1996) 195–214.

[Br96b]   P. BRASS: *Extremale Konstruktionen in der kombinatorischen Geometrie*, Habilitationsschrift, Universität Greifswald 1996.

[BrS02]   W.G. BROWN, M. SIMONOVITS: Extremal multigraph and digraph problems, in: *Paul Erdős and his Mathematics, Vol. 2*, G. Halász et al., eds., *Bolyai Soc. Mathematical Studies* **11** (2002) 157–203.

[Chi93]   K.B. CHILAKAMARRI: The unit-distance graph problem: a brief survey and some new results, *Bulletin of the ICA* **8** (1993) 39–60.

[Chi91]   K.B. CHILAKAMARRI: Unit distance graphs in Minkowski metric spaces, *Geometriae Dedicata* **37** (1991) 345–356.

[Chu89]   F.R.K. CHUNG: Sphere-and-point incidence relations in high dimensions with applications to unit distances and furthest-neighbour pairs, *Discrete Comput. Geom.* **4** (1989) 183–190.

[ClE*90]   K.L. CLARKSON, H. EDELSBRUNNER, L.J. GUIBAS, M. SHARIR, E. WELZL: Combinatorial complexity bounds for arrangements of curves and spheres, *Discrete Comput. Geom.* **5** (1990) 99–160.

[De00]   B.V. DEKSTER: Simplexes with prescribed edge lengths in Minkowski and Banach spaces, *Acta Mathematica Hungarica*

86 (2000) 343–358.

[Er67]   P. Erdős: On some applications of graph theory to geometry, *Canad. J. Math.* **19** (1967) 968–971.

[Er60]   P. Erdős: On sets of distances of n points in Euclidean space, *Magyar Tudományos Akadémia Matematikai Kutató Intézet Közleményei* **5** (1960) 165–169.

[Er46]   P. Erdős: On sets of distances of n points, *Amer. Math. Monthly* **53** (1946), 248–250.

[ErHP99]   P. Erdős, G. Harcos, J. Pach: Popular distances in 3-space, *Discrete Math.* **200** (1999) 95–99.

[ErHP89]   P. Erdős, D. Hickerson, J. Pach: A problem of Leo Moser about repeated distances on the sphere, *Amer. Math. Monthly* **96** (1989) 569–575.

[ErP90]   P. Erdős, J. Pach: Variations on the theme of repeated distances, *Combinatorica* **10** (1990) 261–269.

[PaA95]   J. Pach, P.K. Agarwal: *Combinatorial Geometry*, Wiley, New York 1995.

[Pe71]   C.M. Petty: Equilateral sets in Minkowski spaces, *Proc. Amer. Math. Soc.* **29** (1971) 369-374.

[Si68]   M. Simonovits: A method for solving extremal problems in graph theory, stability problems, in: *Theory of Graphs* (Proc. Colloq., Tihany, 1966), Academic Press 1968, 279–319.

[Sw04]   K.J. Swanepoel: Equilateral sets in finite-dimensional normed spaces, arXiv: [PDF] math.MG/0406264.

[SwV04]   K.J. Swanepoel, P. Valtr: The unit distance problem on spheres, in: *Towards a Theory of Geometric Graphs*, J. Pach, ed., *Contemporary Mathematics* **342**, AMS 2004, 273–279.

[Va05]   P. Valtr: Strictly convex norms allowing many unit distances and related touching questions, manuscript 2005.

[vW99]   P. van Wamelen: The maximum number of unit distances among n points in dimension four, *Beiträge Algebra Geometrie* **40** (1999) 475–477.

5.3 The Minimum Number of Distinct Distances in the Plane

In his classical paper [Er46] initiating the investigation of the number of occurrences of the unit distance among n points, Erdős also raised another equally natural question: at least how many distinct distances must occur among n points in the plane? The maximum number $u(n)$ of unit distances and the minimum number $v(n)$ of distinct distances among n points in the plane are related via the pigeonhole principle, which implies that $u(n)v(n) \geq \binom{n}{2}$. Not surprisingly, the conjectured asymptotically extremal point sets are also the same: for instance, a $\sqrt{n} \times \sqrt{n}$ "section" (piece) of the integer lattice. This construction gives the upper bound $v(n) \leq O\left(\dfrac{n}{\sqrt{\log n}}\right)$, which is asymptotically equal to the number of integers smaller than $2n$ that are representable as the sum of two squares. As a first step toward showing the tightness of this estimate, Erdős [Er75] made the weaker conjecture that $v(n) \geq \Omega(n^{1-\varepsilon})$ for any positive ε, but this also seems to be difficult to prove. The easy lower bound in [Er46], $v(n) \geq \Omega(\sqrt{n})$, was improved in several steps [Mo52], [Be83], [Ch84], [ChST92], [Sz97]. The best currently known estimate follows by combining results of Solymosi–C. Tóth [SoT01] and Katz–Tardos [Ta03], [KaT04]: $v(n) \geq c_\varepsilon n^{\frac{48-14e}{55-16e}-\varepsilon} \geq \Omega(n^{0.8641})$ for every positive ε, where e stands for the basis of natural logarithm.

Conjecture 1 *(Erdős) The minimum number of distinct distances determined by n points in the Euclidean plane is $\Theta\left(\dfrac{n}{\sqrt{\log n}}\right)$.*

The first few exact values of the function $v(n)$ were determined in [ErF96]:

n	1	2	3	4	5	6	7	8	9	10	11	12	13
$v(n)$	0	1	1	2	2	3	3	4	4	5	5	5	6

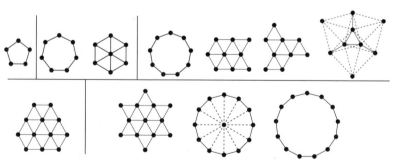

SETS WITH k DISTINCT DISTANCES, $2 \leq k \leq 6$,
AND MAXIMUM NUMBER OF POINTS

Motivated by the known extremal sets, Erdős and Fishburn [ErF96] made the strong conjecture that among the extremal sets for $v(n)$ there is always a lattice section example.

Problem 2   *[ErF96] Does there exist for every sufficiently large n, an n-element subset of a triangular lattice that determines only $v(n)$ distinct distances?*

Erdős [Er88] also made a number of weaker conjectures on the structure of the extremal sets. For instance, he conjectured that all extremal sets with n elements can be covered by $O(\sqrt{n})$ lines, or at least there is a line containing $\Omega(\sqrt{n})$ points. He also conjectured that no (large) extremal set can contain the vertex set of a regular pentagon.

Problem 3   *[Er88] Which sets can never occur as subsets of a set with n points and $v(n)$ distinct distances, for n sufficiently large?*

Since the function v grows sublinearly, it is convenient to consider the inverse question: what is the maximum cardinality of a *k-distance set*, i.e., a set of points that determines at most k distinct distances? In higher dimensions, for small values of k, the problem is related to the existence of certain codes. In the plane, k-distance sets have been classified in a number of papers [Pi93], [HaP96], [ErF96], [ErF97]. A 2-distance set in the plane has at most five points [Ke47].

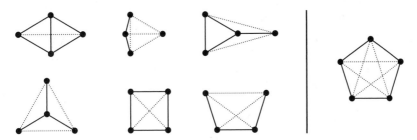

TWO-DISTANCE SETS WITH FOUR AND FIVE POINTS

Akutsu, Tamaki, and Tokuyama [AkTT98] raised an interesting problem that can be regarded as a common generalization of the questions about $u(n)$ and $v(n)$. They defined the function $f^{\text{AkTT}}(n,k)$ as the maximum combined multiplicity of the k most frequent distances among n points in the plane. Clearly, we have $f^{\text{AkTT}}(n,1) = u(n)$ and $f^{\text{AkTT}}(n,v(n)) = \binom{n}{2}$. Akutsu et al. showed that $f^{\text{AkTT}}(n,k) \leq O\left(\min(n^{\frac{10}{7}}k^{\frac{5}{7}}, n^2)\right)$, which implies the previously known inequality $v(n) \geq \Omega(n^{\frac{4}{5}})$. This was improved by Solymosi, Tardos, and Tóth [SoTT02] to

$$f^{\text{AkTT}}(n,k) \leq O(n^{\frac{12}{7}}k^{\frac{1}{3}}).$$

A related conjecture of Erdős [Er90] on the sum of the squared multiplicities of the distances determined by an n-element point set states that $\sum_{r>0}\left(\sum_{p\in S}\deg_S(p,r)\right)^2 \leq O(n^3(\log n)^\alpha)$ for some $\alpha > 0$. For this function, Akutsu et al. [AkTT98] established the upper bound of $O(n^{3.2})$, improving an earlier result of Thiele [Th95]. If no three points are collinear, Thiele found the better bound $O(n^3)$. This cannot be improved, as shown by the regular n-gons [Th95].

Problem 4   *[Er90] Is it true that for any set S of n points in the plane, the sum of the squared multiplicities of the distances determined by S is $O(n^3(\log n)^\alpha)$ for some $\alpha > 0$?*

The problem on the number of distinct distances has some stronger local variants. For any $p \in S$, let $\mathrm{dd}_S(p)$ denote the number of distinct distances from p to all other points of S. Obviously, we have

$$v(n) \geq \min_{|S|=n} \max_{p\in S} \mathrm{dd}_S(p).$$

Note that $v(n)$ can be much larger than the right-hand side if the sets of distances measured from different points are far from equal. Indeed, the two values are not necessarily equal: the eight-point set depicted in the figure (a square with an equilateral triangle on each side) determines four distinct distances, but there are only three different distances from each point [Fi02], [XuD02].

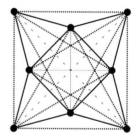

4 DISTINCT DISTANCES, BUT
ONLY 3 FROM EACH POINT

Nevertheless, Erdős [Er75] conjectured that the asymptotic behavior of the two functions is the same, so that every set S of n points has an element $p \in S$ with $\mathrm{dd}_S(p) \geq \frac{cn}{\sqrt{\log n}}$, and in a stronger version, we have $\sum_{p\in S}\mathrm{dd}_S(p) \geq \frac{cn^2}{\sqrt{\log n}}$. Most papers establishing lower bounds on $v(n)$ are really concerned with the local version of the problem. In particular, Katz and G. Tardos [KaT04] proved that one can always find a point $p \in S$ with $\mathrm{dd}_S(p) \geq \Omega(n^{0.8641})$.

The problem of distinct distances has an interesting bipartite version: determine the minimum number of distinct distances between an m-element point set A and an n-element point set B in the plane [El95]. That is, we count only those distances that are determined by two points belonging to different sets. Clearly, for any fixed $m \geq 2$, there are at least $\sqrt{n/2}$

distinct distances realized between A and B. Indeed, given $a_1, a_2 \in A$, assign to each $b \in B$ the ordered pair $(|b - a_1|, |b - a_2|)$. Denoting the maximum number of distinct distances from a point of A by Δ, the number of ordered pairs assigned to the elements of B is at most Δ^2, and at most two points of B are labeled by the same pair. Thus, we have $2\Delta^2 \geq n$, which implies that $\Delta \geq \sqrt{n/2}$. On the other hand, Elekes [El95] constructed for any fixed $m \geq 2$, an m-element set A and an n-element set B with only $O(\sqrt{n})$ distinct distances between them, as n tends to infinity. Hence, the bipartite version of the distinct distances problem behaves quite differently from its classical version, because the number of distinct distances between two elements of $A \cup B$ is much larger than the number of distinct distances between A and B.

However, in Elekes's constructions, A is collinear and B is in a very special position. Actually, the arrangement of the points in B and the $O(m\sqrt{n})$ circles around the elements of A can be interpreted as a "hyperbolic image" of the usual arrangement of n lattice points and m straight lines with many incidences, embedded in the Poincaré model of the plane. It follows from the results of Elekes and E. Szabó (see [El02]) that it is not a coincidence that in this construction A and B have some special properties. If the points of A are not all collinear, the number of distinct distances between A and any set B of n points is $\Omega(n^{0.51})$. All of these asymptotic bounds apply to the case in which $|A| = m \geq 2$ is fixed and n tends to infinity. Erdős [Er88] raised the same problem for $m = n$: determine the asymptotic behavior of the number of distinct distances between two sets of equal cardinality n. Does this function grow more slowly than $v(n)$?

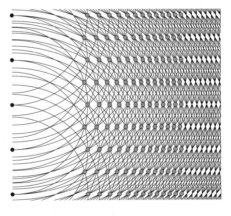

SET PAIR WITH FEW
DISTINCT BIPARTITE DISTANCES

Finally, there are many interesting questions related to the distribution of distances in small subsets of the underlying set. What is the minimum number of distinct distances determined by an n-element set if we impose various restrictions on the distances induced by its small subsets? Let $\varphi(n, k, l)$ denote the minimum number of distinct distances in a set of n points in the plane with the property that any k points determine at least l distinct distances [Er95]. Erdős conjectured that $\varphi(n, 3, 3)$ grows superlinearly and $\varphi(n, 4, 5)$ grows quadratically. He also asked whether $\varphi(n, 5, 9)$

grows quadratically in n [Er86]. It is not easy to understand the effect of such conditions. For instance, the integer lattice contains no equilateral triangle, which implies that $\varphi(n, 3, 2) \leq \frac{cn}{\sqrt{\log n}}$. Hence, $\varphi(n, 3, 2)$ probably does not differ much from $v(n)$, whereas the best known construction of Erdős et al. [ErF*93] for $\varphi(n, 3, 3)$ proves that $\varphi(n, 3, 3) \leq O(ne^{O(\sqrt{\log n})})$.

Conjecture 5   *[Er95] The minimum number of distinct distances determined by any set of n points in the plane that does not contain an isosceles triangle grows superlinearly.*

Conjecture 6   *[Er95] The minimum number of distinct distances determined by a set of n points in the plane whose four-element subsets all determine at least five distinct distances is $\Omega(n^2)$.*

Erdős [Er86] also conjectured that a set of n points that contains neither three elements that induce an equilateral triangle nor four elements that induce a square determine strictly more than $\frac{cn}{\sqrt{\log n}}$ distinct distances for every constant c. This suggests the following question: Are there lattices that allow lattice section constructions with the same asymptotic number, $O\left(\frac{n}{\sqrt{\log n}}\right)$, of distinct distances as in the integer lattice or in the triangular lattice, but that do not contain a square lattice or a triangular lattice as a sublattice?

There are several other problems regarding distances induced by subsets of the underlying set.

Problem 7   *[AvEP91] What is the largest number $f(n)$ such that any set of n points in the plane has an $f(n)$-element subset in which all distances are distinct?*

Clearly, the number of distinct distances determined by an $f(n)$-element subset with the required property is $\binom{f(n)}{2}$. Therefore, we have that $f(n) = O\left(\frac{n^{\frac{1}{2}}}{(\log n)^{\frac{1}{4}}}\right)$, and we cannot hope for any better lower bound on $f(n)$ than $c\sqrt{v(n)}$. But even for lattice sections, the situation is not quite clear. We do not know, even asymptotically, the maximum cardinality of a subset of the $n^{\frac{1}{2}} \times n^{\frac{1}{2}}$ integer lattice in which all distances are distinct [ErG70]. In the case of the lattice, the best known lower bound on this function is $\Omega(n^{\frac{1}{3}})$ [Th95], [LeT95] (see Section 10.3). The best known general lower bound is $f(n) = \Omega(n^{\frac{1}{4}})$. For the same function in d-dimensional space, we have the lower bound $\Omega(n^{\frac{1}{3d-2}})$ [Th95], [LeT95]. We also have a threshold: improving some earlier bounds of Avis et al. [AvEP91], Thiele and Lefmann [Th95], [LeT95] showed that for $k \leq cn^{\frac{3}{16}}$, almost all k-

element subsets of any n-element set in the plane have the property that all distances determined by them are distinct.

One can also investigate the analogous question in which instead of the restriction that all distances determined by the selected subset be distinct, we require only that no three of its elements induce an isosceles triangle.

Problem 8   *What is the largest number $\tilde{f} = \tilde{f}(n)$ such that any set of n points in the plane has an $\tilde{f}$-element subset with no three points that induce an isosceles triangle?*

Obviously, we have $\tilde{f}(n) \geq f(n)$ for every n.

Fishburn [Fi98] called a set k-*isosceles* if all of its k-point subsets induce at least one isosceles triangle. The structure of 4-isosceles sets in the plane was studied in [Fi98], [BáK01], [Ko01], [XuD02], [XuD04]. The function $\tilde{f}$ can be regarded as the inverse of the maximum size of a k-isosceles set in the plane. It follows from the lower bound on $f(n)$ mentioned above that any k-isosceles set has at most $O(k^4)$ points. Using the bound of Pach and Tardos [PaT03] on the maximum number of isosceles triangles determined by a point set, this estimate can be further improved. Applying a randomized selection process combined with the so-called deletion method, one can conclude that $\tilde{f}(n) \geq \Omega(n^{0.432})$. Thus, every k-isosceles set has at most $O(k^{2.315})$ elements. On the other hand, the regular n-gon shows that $\tilde{f}(n) \leq O(n^{0.5})$.

[AkTT98]   T. AKUTSU, H. TAMAKI, T. TOKUYAMA: Distribution of distances and triangles in a point set and algorithms for computing the largest common point sets, *Discrete Comput. Geom.* **20** (1998) 307–331.

[AvEP91]   D. AVIS, P. ERDŐS, J. PACH: Distinct distances determined by subsets of a point set in space, *Comput. Geom. Theory Appl.* **1** (1991) 1–11.

[BáK01]   V. BÁLINT, Z. KOJDJAKOVÁ: Answer to one of Fishburn's questions, *Arch. Math. (Brno)* **37** (2001) 289–290.

[Be83]   J. BECK: On the lattice property of the plane and some problems of Dirac, Motzkin and Erdős in combinatorial geometry, *Combinatorica* **3** (1983) 281–297.

[Ch84]   F.R.K. CHUNG: The number of different distances determined by n points in the plane, *J. Combinatorial Theory Ser. A* **36** (1984) 342–354.

[ChST92]   F.R.K. CHUNG, E. SZEMERÉDI, W.T. TROTTER: The number of different distances determined by a set of points in the Euclidean plane, *Discrete Comput. Geom.* **7** (1992) 1–11.

[ClE*90] K.L. CLARKSON, H. EDELSBRUNNER, L.J. GUIBAS, M. SHA-
RIR, E. WELZL: Combinatorial complexity bounds for arrangements of curves and spheres, *Discrete Comput. Geom.* **5**
(1990) 99–160.

[El02] G. ELEKES: Sums versus products in number theory, algebra
and Erdős geometry, in: *Paul Erdős and his Mathematics,*
Vol. 2, G. Halász et al., eds., *Bolyai Society Math. Studies* **11**
(2002) 241–290.

[El95] G. ELEKES: Circle grids and bipartite graphs of distances,
Combinatorica **15** (1995) 167–174.

[Er95] P. ERDŐS: Some of my recent problems in combinatorial
number theory, geometry and combinatorics, in: *Graph The-*
ory, Combinatorics, Algorithms and Applications. Vol. 1.,
Y. Alavi et al., eds., Wiley 1995, 335–349.

[Er90] P. ERDŐS: Some of my favorite unsolved problems, in: *A*
Tribute to Paul Erdős, A. Baker et al., eds., Cambridge Univ.
Press 1990, 467–478.

[Er88] P. ERDŐS: Some old and new problems in combinatorial geometry, in: *Applications of Discrete Mathematics,* R.D. Ringeisen et al., eds., SIAM 1988, 32–37.

[Er86] P. ERDŐS: On some metric and combinatorial geometric
problems, *Discrete Math.* **60** (1986) 147–153.

[Er85] P. ERDŐS: Problems and results in combinatorial geometry,
in: *Discrete Geometry and Convexity,* J.E. Goodman et al.,
eds., *Annals New York Acad. Sci.* **440** (1985) 1–11.

[Er75] P. ERDŐS: On some problems of elementary and combinatorial geometry, *Ann. Mat. Pura Appl. Ser. IV* **103** (1975)
99–108.

[Er46] P. ERDŐS: On sets of distances of n points, *Amer. Math.*
Monthly **53** (1946) 248–250.

[ErF97] P. ERDŐS, P. FISHBURN: Distinct distances in finite planar
sets, *Discrete Math.* **175** (1997) 97–132.

[ErF96] P. ERDŐS, P. FISHBURN: Maximum planar sets that determine k distances, *Discrete Math.* **160** (1996) 115–125.

[ErF*93] P. ERDŐS, Z. FÜREDI, J. PACH, I.Z. RUZSA: The grid
revisited, *Discrete Math.* **111** (1993) 189–196.

[ErG70] P. ERDŐS, R.K. GUY: Distinct distances between lattice
points, *Elemente Math.* **25** (1970) 121–123.

[Fi02] P. FISHBURN: A remarkable eight-point planar configuration,
Discrete Math. **252** (2002) 103–122.

[Fi98]   P. FISHBURN: Isosceles planar subsets, *Discrete Comput. Geom.* **19** (1998) 391–398.

[HaM94]   H. HARBORTH, M. MÖLLER: Ebene geradlinige Darstellungen der platonischen Graphen mit wenigen verschiedenen Kantenlängen, *Abh. Braunschweig. Wiss. Ges.* **45** (1994) 7–20.

[HaP96]   H. HARBORTH, L. PIEPMEYER: Three distinct distances in the plane, *Geometriae Dedicata* **61** (1996) 315–327.

[HaP93]   H. HARBORTH, L. PIEPMEYER: Two-distance sets and the golden ratio, in: *Applications of Fibonacci numbers, Vol. 5*, G.E. Bergum et al., eds., Kluwer Academic Publishers 1993, 279–288.

[KaT04]   N.H. KATZ, G. TARDOS: A new entropy inequality for the Erdős distance problem, in: *Towards a Theory of Geometric Graphs*, J. Pach, ed., *Contemporary Mathematics* **342**, AMS 2004, 119–126.

[Ke47]   L.M. KELLY: Isosceles n-points, *Amer. Math. Monthly* **54** (1947) 227–229.

[Ko01]   Z. KOJDJAKOVÁ: There are 7-point 4-isosceles planar sets with no 4 points in a circle, *Studies Univ. Žilina, Math. Series* **14** (2001) 11–12.

[LeT95]   H. LEFMANN, T. THIELE: Point sets with distinct distances, *Combinatorica* **15** (1995) 379–408.

[Mo52]   L. MOSER: On different distances determined by n points, *Amer. Math. Monthly* **59** (1952) 85–91.

[PaT03]   J. PACH, G. TARDOS: Isosceles triangles determined by a planar point set, *Graphs Combinatorics* **18** (2002) 769–779.

[Pi93]   L. PIEPMEYER: *Punktmengen mit minimaler Anzahl verschiedener Abstände*, Dissertation, TU Braunschweig 1993.

[SoT01]   J. SOLYMOSI, C.D. TÓTH: Distinct distances in the plane, *Discrete Comput. Geom.* **25** (2001) 629–634.

[SoTT02]   J. SOLYMOSI, G. TARDOS, C.D. TÓTH: The k most frequent distances in the plane, *Discrete Comput. Geom.* **28** (2002) 639–648.

[Sz97]   L.A. SZÉKELY: Crossing numbers and hard Erdős problems in discrete geometry, *Comb. Probab. Comput.* **6** (1997) 353–358.

[Ta03]   G. TARDOS: On distinct sums and distinct distances, *Advances in Mathematics* **180** (2003) 275–289.

[**Th95**]   T. THIELE: *Geometric selection problems and hypergraphs*, Dissertation, Freie Universität Berlin 1995.

[**XuD04**]   C. XU, R. DING: More about 4-isosceles planar sets, *Discrete Comput. Geom.* **31** (2004) 655–663.

[**XuD02**]   C. XU, R. DING: About 4-isosceles planar sets, *Discrete Comput. Geom.* **27** (2002) 287–290.

5.4 The Number of Distinct Distances in Other Spaces

Erdős's question about the minimum number of distinct distances among n points has been studied not only in the plane, but in many other metric spaces, including higher-dimensional Euclidean and spherical spaces and normed spaces.

Let $v_d(n)$ denote the minimum number of distinct distances determined by n points in d-dimensional Euclidean space. Erdős [Er46] proved that $\Omega(n^{\frac{1}{d}}) \leq v_d(n) \leq O(n^{\frac{2}{d}})$, where the upper bound follows by counting the number of distinct distances in an $n^{\frac{1}{d}} \times \ldots \times n^{\frac{1}{d}}$ "section" of the integer lattice.

Conjecture 1   *[Er46] For $d \geq 3$, the minimum number of distinct distances determined by n points in d-dimensional Euclidean space satisfies $v_d(n) = \Theta\left(n^{\frac{2}{d}}\right)$.*

In three-dimensional space, the lower bound $v_3(n) \geq \Omega(n^{\frac{1}{2}}(\beta(n))^{-1})$ follows by the pigeonhole principle from the bound of Clarkson, Edelsbrunner et al. [ClE*90] on the maximum number of unit distances determined by a point set. This was first improved by Aronov et al. [ArP*04] to $v_3(n) \geq \Omega(n^{0.5461})$, and then by Solymosi and Vu [SoV05] to $v_3(n) \geq \Omega(n^{0.5643})$. Thus, the currently best bounds are $\Omega(n^{0.5643}) \leq v_3(n) \leq O(n^{\frac{2}{3}})$. For higher dimensions, Solymosi and Vu [SoV05] proved the lower bound $v_d(n) \geq \Omega\left(n^{\frac{2}{d}(1-\frac{1}{d+2})}\right)$, which almost matches the $O(n^{\frac{2}{d}})$ upper bound. Slightly better bounds were proved in the special case of "dense" sets, i.e., for n-element sets in which the ratio of the largest distance to the smallest distance between the elements remains under a fixed constant times $n^{\frac{1}{d}}$ [SoV04].

The same questions have been studied for spheres. For the sphere in three-dimensional Euclidean space, the $\Omega(n^{\frac{4}{5}})$ lower bound of Székely [Sz97], as well as the best known bound of $\Omega(n^{0.8641})$ due to Katz and G. Tardos can be directly copied from the planar case. A sequence of n equidistant points on a circle define only $O(n)$ distinct distances. For higher-dimensional spheres, Erdős et al. [ErF*93] constructed upper bounds of $O(\frac{n}{\log \log n})$ for the sphere in four-dimensional Euclidean space, and $O(n^{\frac{2}{d-2}})$ for dimensions $d \geq 5$. For high dimensions, this upper bound is again almost matched by the lower bound of Solymosi and Vu [SoV05].

Since the functions v_d grow sublinearly, the inverse functions seem to be more manageable. That is, we want to estimate the largest number of points in d-dimensional space that determine at most k distinct distances. Such a set is called a *k-distance set*. In higher dimensions, for small k, these questions are related to the existence of certain codes. There has been much interest in two-distance sets in d-dimensional space, which were first studied by Kelly [Ke47], and later by many authors [Cr62], [LaRS77], [Bl84], [HaP93], [Pi93]. For results on k-distance sets for larger values of k, consult [BaB81], [Ba82], [BaBS83].

Problem 2   *What is the maximum cardinality of a k-distance set in d-dimensional Euclidean space?*

Only few exact answers to this question are known. In the plane, the maximum cardinality of k-distance sets is known for $k \leq 6$, as described in the previous section. In three-dimensional space, the maximum cardinality of a two-distance set is six [Cr62], which can be reached in several ways, e.g., by pyramids over a regular pentagon, or by an octahedron. The regular pentagon is the unique two-distance set of maximum cardinality in the plane [Ke47]. So the next open problems are to determine the maximum cardinality of a three-distance set in three-dimensional space and of a two-distance set in four-dimensional space. The best known general upper bound is that a k-distance set in d-dimensional space has at most $\binom{k+d}{k}$ points [BaBS83], which agrees with Erdős's [Er46] lower bound on v_d, and is almost certainly not optimal for large values of k. However, when k is small and d is large, the above upper bound of Bannai, Bannai, and Stanton is fairly good. For $k = 1$ it gives $d + 1$, which is the correct value, as shown by an equilateral simplex, and for $k = 2$ it becomes $\binom{d+2}{2}$, while the set of midpoints of the edges of an equilateral simplex give a lower bound of $\binom{d+1}{2}$. Neither of these bounds is tight in the plane. A useful tool for studying the realizability of two-distance sets is Menger's criterion for the embeddability of metric spaces in Euclidean d-dimensional space, using the Cayley–Menger-determinant. With this tool, Harborth and Piepmeyer [HaP93] studied the maximum size of two-distance sets depending on the distance ratio.

A special case of the two-distance set problem is to determine the maximum number of *equiangular* lines, i.e., the maximum number of lines through a point in d-dimensional space such that any two of them span the same angle. Representing each line by a unit vector, we obtain a set of points on the unit sphere with the property that the (angular) distance between any two of them is either α or $\pi - \alpha$, for a suitable $0 < \alpha < \pi$. For infinitely many values of d, de Caen [dCa00] established the lower bound $\frac{2}{9}(d + 1)^2$ on the maximum number of equiangular lines in d-dimensional space. The best known upper bound, $\frac{1}{2}(d+1)d$, due to Lemmens and Seidel [LeS73], is only slightly better than the general upper bound $\binom{d+2}{2}$, valid for two-distance sets. We also know several exact values of this function and many lower bound constructions for special values of d [Ha48], [vLS66], [LeS73], [Ho78], [Kö89].

d	2	3	4	5	6	7–13	14	15	16–20	21	22	23
max # equiang. lines	3	6	6	10	16	28	?	36	?	126	176	276

Problem 3   *What is the maximum cardinality of a set of equiangular lines in d-dimensional space?*

A local variant of the two-distance set problem was investigated by Croft [Cr62]: What is the largest number of points in d-dimensional space such that any three of them determine at most two distinct distances?

If the number of points and the dimension are allowed to grow simultaneously, we obtain several interesting problems. The d-dimensional cube is a set of 2^d points that determine d distinct distances. Erdős [Er85] asked whether every set of 2^d points in d-dimensional space determines $\Omega(d)$ distinct distances. This question was answered by Swanepoel, who observed that the lower bound $0.29d$ easily follows from the result of Bannai, Bannai, and Stanton [BaBS83] on the cardinality of k-distance sets. The value of the best multiplicative constant remains to be found.

The questions above can be asked for any normed space (or indeed, for any metric space). Swanepoel [Sw99] formulated the following attractive conjecture: The cardinality of every k-distance set in any d-dimensional normed space is at most $(k + 1)^d$. This number is attained by the set $\{0, \ldots, k\}^d$ in the L_∞ metric.

Conjecture 4   *[Sw99] Every k-distance set in any d-dimensional normed space has cardinality at most $(k + 1)^d$, with equality only if the unit ball is an affine cube.*

This conjecture was proved by Swanepoel [Sw99] in the planar case. It is also known to be true for $k = 1$: according to an old result of Petty [Pe71], the maximum cardinality of an equilateral set in a d-dimensional normed space is 2^d. The only published upper bound for higher dimensions is $\min\left(2^{kd}, (k + 1)^{\frac{1}{2}(11^d - 9^d)}\right)$ [Sw99]. Unfortunately, at the "low end" we have no nontrivial construction of point sets with few distinct distances. Clearly, a system of n equidistant collinear points in any normed space determines at least $n - 1$ distinct distances, and we are not aware of any better general construction.

Conjecture 5   *(Brass) In any normed space of dimension $d \geq 2$, the minimum number of distinct distances determined by n points is $o(n)$.*

Problem 6   *(Brass) Does there exist a strictly convex normed plane in which the minimum number of distinct distances is $O(n^{1-\varepsilon})$ for some positive ε?*

[ArP*04] B. ARONOV, J. PACH, M. SHARIR, G. TARDOS: Distinct distances in three and higher dimensions, *Comb. Probab. Comput.* **13** (2004) 283–293.

[Ba82]   E. BANNAI: On s-distance subsets in real hyperbolic space, *Hokkaido Math. J.* **11** (1982) 201–204.

[BaB81]   E. BANNAI, E. BANNAI: An upper bound for the cardinality of an s-distance subset in real Euclidean space, *Combinatorica* **1** (1981) 99–102.

[BaBS83] E. BANNAI, E. BANNAI, D. STANTON: An upper bound for the cardinality of an s-distance subset in real euclidean space II, *Combinatorica* **3** (1983) 147–152.

[Bl84]   A. BLOKHUIS: A new upper bound for the cardinality of 2-distance sets in Euclidean space, in: *Convexity and Graph Theory, Ann. Discrete Math.* **20** (1984) 65–66.

[dCa00]   D. DE CAEN: Large equiangular sets of lines in Euclidean space, *Electron. J. Comb.* **7** (2000) Research paper R55.

[ClE*90] K.L. CLARKSON, H. EDELSBRUNNER, L.J. GUIBAS, M. SHARIR, E. WELZL: Combinatorial complexity bounds for arrangements of curves and spheres, *Discrete Comput. Geom.* **5** (1990) 99–160.

[Cr62]   H.T. CROFT: 9-point and 7-point configurations in 3-space, *Proc. London Math. Soc.* **12** (1962) 400–424, Corrigendum. ibid. 13, 384 (1963).

[Er85]   P. ERDŐS: Problems and results in combinatorial geometry, in: *Discrete Geometry and Convexity*, J.E. Goodman et al., eds., *Annals New York Acad. Sci.* **440** (1985) 1–11.

[Er46]   P. ERDŐS: On sets of distances of n points, *Amer. Math. Monthly* **53** (1946) 248–250.

[ErF*93] P. ERDŐS, Z. FÜREDI, J. PACH, I.Z. RUZSA: The grid revisited, *Discrete Mathematics* **111** (1993) 189–196.

[Ha48]   J. HAANTJES: Equilateral point sets in elliptic two- and three-dimensional spaces, *Nieuw Arch. Wisk.* **22** (1948) 355–362.

[HaP93]   H. HARBORTH, L. PIEPMEYER: Two-distance sets and the golden ratio, in: *Applications of Fibonacci Numbers, Vol. 5*, G.E. Bergum et al., eds., Kluwer Academic Publishers 1993, 279–288.

[Ho78] S.G. HOGGAR: Equiangular lines, in: *Combinatorics, Proc. Fifth Hungarian Colloq., Vol. 1* (Keszthely 1976), A. Hajnal et al., eds., *Colloq. Math. Soc. János Bolyai* **18** (1978) 599–606.

[Ke47] L.M. KELLY: Isosceles n-points, *Amer. Math. Monthly* **54** (1947) 227–229.

[Kö89] H. KÖNIG: Gleichwinklige Geraden und Punktverteilungen auf der Kugel, *Math. Naturwiss. Unterricht* **42** (1989) 6–10.

[LaRS77] D.G. LARMAN, C.A. ROGERS, J.J. SEIDEL: On two-distance sets in Euclidean space, *Bull. London Math. Soc.* **9** (1977) 261–267.

[LeS73] P.W.H. LEMMENS, J.J. SEIDEL: Equiangular lines, *J. Algebra* **24** (1973) 494–512.

[vLS66] J.H. VAN LINT, J.J. SEIDEL: Equilateral point sets in elliptic geometry, *Indagationes Math.* **28** (1966) 335–348.

[Pe71] C.M. PETTY: Equilateral sets in Minkowski spaces, *Proc. Amer. Math. Soc.* **29** (1971), 369-374.

[Pi93] L. PIEPMEYER: *Punktmengen mit minimaler Anzahl verschiedener Abstände*, Dissertation, TU Braunschweig, 1993.

[SoV05] J. SOLYMOSI, V. VU: Near-optimal bounds for the number of distinct distances in high dimensions, manuscript.

[SoV04] J. SOLYMOSI, V. VU: Distinct distances in high-dimensional homogeneous sets, in: *Towards a Theory of Geometric Graphs*, J. Pach, ed., *Contemporary Mathematics* **342**, AMS 2004, 259–268.

[Sz97] L.A. SZÉKELY: Crossing numbers and hard Erdős problems in discrete geometry, *Comb. Probab. Comput.* **6** (1997) 353–358.

[Sw99] K.J. SWANEPOEL: Cardinalities of k-distance sets in Minkowski spaces, *Discrete Math.* **197-198** (1999) 759–767.

5.5 Repeated Distances in Point Sets in General Position

In most problems discussed in previous subsections in this chapter, the extremal configurations had very special structural properties. For instance, they had many collinear points (in lattice section constructions; see Section 5.1) or many cocircular points (in the Lenz construction, described in Section 5.2), so one can ask what happens if one excludes these constructions by requiring the point sets to be in "general position"?

There are several possible definitions of general position that might be of interest, and each of them could be combined with each question discussed in the previous sections. Given a condition γ, let $u^\gamma(n)$ and $v^\gamma(n)$ denote the maximum number of occurrences of the unit distance and the minimum number of distinct distances determined by n points satisfying condition γ. As we have seen in Sections 5.1 and 5.3, if we make no special assumption on our point sets, the conjectured extremal configurations for u and v in the plane are lattice sections with many collinear points. Thus, the first natural problem is to study these questions under the condition that no three elements of the point set lie on a line. The iterated Minkowski sum construction (Section 5.1) shows that constant times $n \log n$ unit distances are possible, i.e., we have $u^{\text{no-3-coll}}(n) \geq \Omega(n \log n)$. Erdős conjectured that the order of magnitude of this function is higher: $\limsup_{n\to\infty} \frac{u^{\text{no-3-coll}}(n)}{n \log n} = \infty$. We do not know any upper bound better than the general estimate $u^{\text{no-3-coll}}(n) \leq u(n) \leq n^{\frac{4}{3}}$ mentioned in Section 5.1. For the number of distinct distances, the regular polygons show that $v^{\text{no-3-coll}}(n) \leq \lfloor \frac{n}{2} \rfloor$ holds. Erdős conjectured that $v^{\text{no-3-coll}}(n) \geq \frac{1}{2}n - o(n)$. The best known lower bound, $v^{\text{no-3-coll}}(n) \geq \frac{1}{3}n$, follows by double-counting isosceles triangles. The simple proof due to Szemerédi was included in a paper of Erdős [Er75].

Conjecture 1   *[Er75] Any set of n points in the plane, no three of which are on a line, determines at least $\frac{1}{2}n - o(n)$ distinct distances.*

Another natural definition for a planar point set to be in "general position" is that no four of its points are on a circle. Then, of course, we have $u^{\text{no-4-circ}}(n) \leq \frac{3}{2}n$ and $v^{\text{no-4-circ}}(n) \geq \frac{n-1}{3}$. The former bound is almost tight; a point set obtained from a big portion of the vertex set of the regular hexagonal tiling by a slight deformation shows that $u^{\text{no-4-circ}}(n) \geq \frac{3}{2}n - O(\sqrt{n})$. However, the latter inequality is almost certainly not sharp. We do not know any argument showing that there exists $\varepsilon > 0$ such that $v^{\text{no-4-circ}}(n) \geq \left(\frac{1}{3} + \varepsilon\right)n$ holds. Equidistant points on a line show that $v^{\text{no-4-circ}}(n) \leq n - 1$.

Problem 2   *(Erdős) Find the best constant c such that n points in the plane, no four on a circle, determine at least $(c + o(1))n$ distinct distances.*

Combining both assumptions, that is, considering only sets with no three points on a line and no four on a circle, the number of unit distances stays the same as under the latter restriction. However, the construction of sets with few distinct distances satisfying both conditions becomes quite complicated. Here the best known construction [ErF*93] gives $v^{\text{strong-gen-pos}}(n) \leq O\left(n e^{O(\sqrt{\log n})}\right)$. From the other direction we have only the trivial lower bound $v^{\text{strong-gen-pos}}(n) \geq \frac{n-1}{3}$. In fact, the construction of Erdős et al. [ErF*93] has a stronger property: it gives n points with only $O\left(n e^{O(\sqrt{\log n})}\right)$ distinct difference vectors. The problem of minimizing the number of difference vectors is not too interesting for "unrestricted" point sets: any set of n points determines at least $n-1$ distinct difference vectors, and the same vector can occur at most $n - 1$ times. Both of these bounds are attained for n equidistant points along a line.

We can also define "general-position" point sets by requiring that no four elements induce a parallelogram or, equivalently, that no two difference vectors be the same. In particular, under this condition, a unit distance graph cannot contain C_4 as a subgraph. This excludes the iterated Minkowski sum and several other vector sum constructions. A planar version of the construction of Erdős, Hickerson, and Pach [ErHP89] shows that $u^{\text{no-parallel}}(n) \geq \Omega(n \log^* n)$. This was improved by Swanepoel and Valtr [SwV04] to $u^{\text{no-parallel}}(n) \geq \Omega(n\sqrt{\log n})$. Under this "no parallelogram" restriction, we do not know any set with a subquadratic number of distinct distances.

Problem 3   *[ErHP89] Is it true that the number of distinct distances determined by n points in the plane, no four of which induce a parallelogram, satisfies the relation*
$$v^{\text{no-parallel}}(n) \geq \Omega(n^2)?$$

A completely different kind of restriction on the structure of the point sets was suggested by the following observation of Purdy: Given two lines in the plane that are orthogonal or parallel to each other, it is possible to select n points on each of them so that the number of distinct distances between points on different lines is $O(n)$. Purdy conjectured that if the two lines are neither orthogonal nor parallel, the number of distinct distances between two n-element sets on them is strictly superlinear. This was proved by Elekes and Rónyai [ElR00]. Elekes [El99] strengthened this result by obtaining an $\Omega(n^{\frac{5}{4}})$ lower bound, which represents a considerable separation between the cases when the angle between the two lines is $0, \pi/2$, or anything else. However, in the latter case the true number of distinct distances between points on different lines may well be almost quadratic. The best known upper bound, $\dfrac{cn^2}{\sqrt{\log n}}$, is realized by two sets of points on lines that intersect at an angle $\frac{2}{3}\pi$.

Conjecture 4 *(Elekes [El99]) The number of distinct distances determined by two collinear n-element point sets in the plane whose lines are neither parallel nor orthogonal is at least $\Omega(n^{2-\varepsilon})$, for every $\varepsilon > 0$.*

Since many conjecturally optimal constructions on distance problems are grid-like, it may be worth restricting our attention to point sets that can be obtained by taking the Cartesian product of one-dimensional sets. Under this assumption, of course, we cannot expect any better lower bound on the number of distinct distances determined by n points than the general upper bound of $\dfrac{n}{\sqrt{\log n}}$, which is conjectured to be tight. This bound is realized by the lattice section construction, which gives a Cartesian product. For Cartesian products, however, it might be easier to establish the optimality of this construction than in the general case.

Conjecture 5 *(Pach) For any set A of n real numbers, the Cartesian product $A \times A$ determines at least $\Omega\left(\dfrac{n^2}{\sqrt{\log n}}\right)$ distinct distances.*

[El99]   G. ELEKES: A note on the number of distinct distances, *Period. Math. Hungar.* **38** (1999) 173–177.

[ElR00]   G. ELEKES, L. RÓNYAI: A combinatorial problem on polynomials and rational functions, *J. Combinatorial Theory Ser. A* **89** (2000) 1–20.

[Er75]   P. ERDŐS: On some problems of elementary and combinatorial geometry, *Ann. Mat. Pura Appl. Ser. IV* **103** (1975) 99–108.

[ErF*93]   P. ERDŐS, Z. FÜREDI, J. PACH, I.Z. RUZSA: The grid revisited, *Discrete Mathematics* **111** (1993) 189–196.

[ErHP89]   P. ERDŐS, D. HICKERSON, J. PACH: A problem of Leo Moser about repeated distances on the sphere, *Amer. Math. Monthly* **96** (1989) 569–575.

[SwV04]   K.J. SWANEPOEL, P. VALTR: The unit distance problem on spheres, in: *Towards a Theory of Geometric Graphs*, J. Pach, ed., *Contemporary Mathematics* **342**, AMS 2004 273–279.

5.6 Repeated Distances in Point Sets in Convex Position

In this section, we discuss the unit distance problem and the problem of distinct distances under the restriction that the underlying point set is in *convex position*; that is, its elements form the vertex set of a convex polygon (or polytope, in higher dimensions). This assumption completely changes the problems.

The oldest result here is due to Leo Moser [Mo52], who gave a lower bound on $v^{\text{conv}}(n)$, the minimum number of distinct distances among n points in convex position in the plane. The problem was completely solved by Altman [Al63], [Al72], who showed that $v^{\text{conv}}(n) = \lfloor \frac{n}{2} \rfloor$, with the only extremal sets being regular n-gons, and, for n even, regular $(n+1)$-gons with one point removed.

The problem of determining $u^{\text{conv}}(n)$, the maximum number of unit distances among n points in convex position in the plane, turns out to be much more difficult. Here the best known construction gives $u^{\text{conv}}(n) \geq 2n - 7$ [EdH91] (improving [ErM59]), essentially by winding a path of unit distances around a Reuleaux triangle of unit diameter.

<small>CONVEX POLYGON WITH $2n - 7$ UNIT DISTANCES</small>

The best known upper bound is $u^{\text{conv}}(n) \leq O(n \log n)$. It was originally obtained by Füredi [Fü90], who reduced the question to an extremal problem for matrices. Later, a very simple proof by induction was found by Brass and Pach [BrP01]. Another proof of the same bound follows from the Turán-type extremal theory for *convex geometric graphs*, as developed in [BrKV03].

Conjecture 1   *Among the vertices of any convex n-gon, the same distance cannot occur more than $2n$ times.*

If we do not insist on "strictly" convex position, and require only that every element of S belong to the boundary of the convex hull of S, then we say that S is in *weakly convex position*. A strip of the triangular lattice satisfies this weaker assumption, and it realizes $2n - 3$ unit distances. The upper bound remains asymptotically the same as for point sets in (strictly) convex position.

Surprisingly, the same problem for centrally symmetric convex polygons is much simpler. It was essentially solved by Ábrego and Fernández-

Merchant [ÁbF02], who established a $2n - O(1)$ upper bound and a $2n - O(\sqrt{n})$ lower bound.

Problem 2   *Do there exist centrally symmetric convex n-gons with at least $2n - O(1)$ unit distances between their vertices?*

Many variants and small cases of the above questions have been considered by Erdős and Fishburn [ErF94], [ErF95a], [ErF95b], [ErF96]. The most famous among them is probably the following local version of the problem of unit distances: Does there exist a number k with the property that any set S of n points in convex position in the plane has an element such that any circle around it passes through at most k other points?

Erdős originally made this conjecture for $k = 2$, but soon Danzer found infinitely many counterexamples. The conjecture is still open for $k = 3$. The construction of Danzer was published in a survey by Erdős [Er87]. It uses $l \geq 3$ concentric Reuleaux-type $(2m + 1)$-gons inscribed into one another so that the vertices of each of them lie on the sides of the previous one, and the last $(2m + 1)$-gon is rotated in such a way that its points, together with the elements of the first one, generate some additional equal distances.

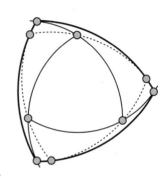

CONVEX POLYGON IN WHICH
EACH VERTEX IS EQUIDISTANT
TO THREE OTHERS

In this construction, each vertex has three other vertices equidistant from it, but these distances are not the same for vertices on different Reuleaux polygons. Fishburn and Reeds [FiR92] found a convex polygon with the stronger property that its unit distance graph is 3-regular.

Conjecture 3   *(Erdős [Er87]) Every convex n-gon in the plane has a vertex from which there are no four other vertices at the same distance.*

The problem concerning the minimum number of distinct distances also has a local variant. The following strengthening of the statement $v^{\mathrm{conv}}(n) = \lfloor \frac{n}{2} \rfloor$ was conjectured by Erdős.

Conjecture 4   *(Erdős [Er75]) Every set S of n points in convex position in the plane has an element $p \in S$ such that $\mathrm{dd}_S(p)$, the number of distinct distances from p, is at least $\lfloor \frac{n}{2} \rfloor$.*

Even stronger variants were proposed by Erdős and Fishburn, such as $\sum_{p \in S} \mathrm{dd}_S(p) \geq \binom{n}{2}$. They received some attention [Fi95], [Fi97], [Fi00].

As for Erdős's original conjecture, the best known lower bound is due to Dumitrescu [Du04], who proved that there is a point $p \in S$ with

$$\mathrm{dd}_S(p) \geq \left\lceil \frac{13n - 6}{36} \right\rceil.$$

The higher-dimensional analogues of the convex unit distance problem appear to be less interesting. Convexity stops posing a restriction in dimensions $d \geq 4$, because the Lenz sets are in convex position. It seems to represent only a minor restriction in three-dimensional space, because n points on the sphere admit $\Omega(n^{\frac{4}{3}})$ unit distances, which is only slightly less than the $\Omega(n^{\frac{4}{3}} \log \log n)$ realized by the lattice section construction. However, the higher-dimensional versions of the problem of distinct distances appear to be quite interesting. The best known constructions, due to Erdős et al. [ErF*93], are point sets on the sphere. They show that the number of distinct distances among n points on the sphere can be smaller than $O\left(\frac{n}{\log \log n}\right)$ in four-dimensional space and $O\left(n^{\frac{2}{d-2}}\right)$ in d dimensions, $d > 4$. We are not aware of any lower bounds on the number of distinct distances determined by the vertex set of a convex polytope that make use of the convex position.

Problem 5   *Give good bounds for the number of distinct distances determined by n points in convex position in d-dimensional Euclidean space, for $d \geq 3$.*

[ÁbF02]   B.M. ÁBREGO, S. FERNÁNDEZ-MERCHANT: The unit distance problem for centrally symmetric convex polygons, *Discrete Comput. Geom.* **28** (2002) 467–473.

[Al72]   E. ALTMAN: Some theorems on convex polygons, *Canad. Math. Bull.* **15** (1972) 329–340.

[Al63]   E. ALTMAN: On a problem of P. Erdős, *Amer. Math. Monthly* **70** (1963) 148–157.

[BrKV03] P. BRASS, G. KÁROLYI, P. VALTR: A Turán-type extremal theory for convex geometric graphs, in: *Discrete and Computational Geometry — The Goodman-Pollack Festschrift*, B. Aronov et al., eds., Springer, 2003, 275–300.

[BrP01]   P. BRASS, J. PACH: The maximum number of times the same distance can occur among the vertices of a convex n-gon is $O(n \log n)$, *J. Combinatorial Theory Ser. A* **94** (2001) 178–179.

[Du04]   A. DUMITRESCU: On distinct distances from a vertex in
           a convex polygon, in: SCG 2004 (Proc. Twentieth Annual
           Symp. Comput. Geom.) ACM-Press 2004, 57–59.

[EdH91]   H. EDELSBRUNNER, P. HAJNAL: A lower bound on the num-
           ber of unit distances between the points of a convex polygon,
           J. Combinatorial Theory Ser. A 56 (1991) 312–316.

[Er87]   P. ERDŐS: Some combinatorial and metric problems in geo-
           metry, in: Intuitive Geometry, (Siófok, 1985), K. Böröczky et
           al., eds., Colloq. Math. Soc. János Bolyai 48 (1987) 167–177.

[Er75]   P. ERDŐS: On some problems of elementary and combina-
           torial geometry, Ann. Mat. Pura Appl. Ser. IV 103 (1975)
           99–108.

[ErF96]   P. ERDŐS, P. FISHBURN: Convex nonagons with five inter-
           vertex distances, Geometriae Dedicata 60 (1996) 317–332.

[ErF95a]   P. ERDŐS, P. FISHBURN: Multiplicities of interpoint dis-
           tances in finite planar sets, Discrete Appl. Math. 60 (1995)
           141–147.

[ErF95b]   P. ERDŐS, P. FISHBURN: Intervertex distances in convex
           polygons, Discrete Appl. Math. 60 (1995) 149–158.

[ErF94]   P. ERDŐS, P. FISHBURN: A postscript on distances in convex
           n-gons, Discrete Comput. Geom. 11 (1994) 111–117.

[ErF*93]   P. ERDŐS, Z. FÜREDI, J. PACH, I.Z. RUZSA: The grid
           revisited, Discrete Math. 111 (1993) 189–196.

[ErHP89]   P. ERDŐS, D. HICKERSON, J. PACH: A problem of Leo
           Moser about repeated distances on the sphere, Amer. Math.
           Monthly 96 (1989) 569–575.

[ErM59]   P. ERDŐS, L. MOSER: Problem 11, Canad. Math. Bull. 2
           (1959) p. 43.

[ErP76]   P. ERDŐS, G. PURDY: Some extremal problems in geometry
           IV, Congressus Numerantium 17 (Proc. 7th South-Eastern
           Conf. Combinatorics, Graph Theory, and Computing, 1976)
           307–322.

[Fi00]   P. FISHBURN: On an Erdős problem for distinct distances in
           convex polygons, Geombinatorics 10 (2000) 17–23.

[Fi97]   P. FISHBURN: Distances in convex polygons, in: The Math-
           ematics of Paul Erdős II, R.L. Graham et al., eds., Springer
           1997, 284–293.

[Fi95]   P. FISHBURN: Convex polygons with few intervertex dis-
           tances, Comput. Geom. Theory Appl. 5 (1995) 65–93.

[**FiR92**]   P. Fishburn, J.A. Reeds: Unit distances between vertices of a convex polygon, *Comput. Geom. Theory Appl.* **2** (1992) 81–91.

[**Fü90**]   Z. Füredi: The maximum number of unit distances in a convex n-gon, *J. Combinatorial Theory Ser. A* **55** (1990) 316–320.

[**Mo52**]   L. Moser: On the different distances determined by n points, *Amer. Math. Monthly* **59** (1952) 85–91.

5.7 Frequent Small Distances and Touching Pairs

An old variant of the question of unit distances is the following. Consider the smallest distance among the elements of a point set S, and count how many unordered pairs in S determine this distance. What is the maximum, $s(n)$, of this quantity over all n-element point sets in the plane? The question was asked in two equivalent forms: by Erdős [Er75], as a problem on the number of occurrences of the minimal distance, and, independently, by Reutter [Re72], who was interested in finding the maximum number of touching pairs among n nonoverlapping congruent disks. This question was completely solved by Harborth [Ha74], and the answer is $s(n) = \lfloor 3n - \sqrt{12n - 3} \rfloor$.

All extremal configurations can be obtained as "sections" (i.e., connected pieces) of the triangular lattice. The almost perfect hexagonal sections obtained by following a spiral path outward are always extremal. If n is of the form $3k^2 + 3k + 1$ for some positive integer k, then there exists a section that spans a regular hexagon, and this configuration is the unique extremal set of this size. Otherwise, one can construct further extremal sets from the "spiral" example by shifting the boundary points around as long as the perimeter does not increase [Ku94].

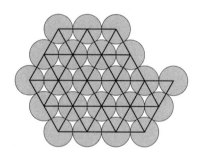

n UNIT DISKS WITH
$$\lfloor 3n - \sqrt{12n - 3} \rfloor$$
TOUCHING PAIRS

Most of the related problems can be interpreted in two ways: for smallest distances as well as for touching pairs in a packing of congruent disks (see Section 2.4). This immediately shows that the *smallest distance graph* on S whose edge set consists of all point pairs in S at minimum distance is planar and that the maximum degree of its vertices is at most six. Furthermore, each point of degree six and its neighbors form the vertex set and the center of a regular hexagon.

Several variations of this problem have been studied. It is simple to show that the maximum number of occurrences of the smallest distance among n points in (strictly) convex position is $n + 2$ for $n \geq 6$.

Brass [Br92c] answered a question of Erdős and Pach [ErP90] by showing that the maximum number of smallest distances among n points, k of which are vertices of the convex hull, is at most $3n - 2k + 4$. This bound is attained for infinitely many pairs n, k, but only when k is close to n.

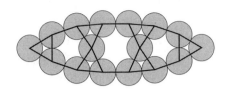

n POINTS, k OF WHICH ARE
VERTICES OF THE CONVEX HULL,
WITH $3n - 2k + 4$ SMALLEST DISTANCES

If no three points are on a line, then no vertices of the smallest distance graph have degree six, so the maximum degree is at most five. The smallest-distance graph is planar, all triangular faces are equilateral, and all quadrilateral faces are rhombi. Three consecutive (equilateral) triangles around a point would determine three collinear points. Thus, if all points are of degree five, the numbers of sides of the faces around each vertex are 3, 3, 4, 3, 4 (and the graph is isomorphic to an Archimedean tiling). However, any sufficiently large patch of this structure also contains three collinear points, so it must have points with degree smaller than five. Therefore, we have $s^{\text{no-3-coll}}(n) \le (2 + \frac{1}{2} - \varepsilon)n$ for some $\varepsilon > 0$. This was improved by Tóth [Tó97] to $s^{\text{no-3-coll}}(n) \le (2 + \frac{3}{7})n$. He also found a construction showing $s^{\text{no-3-coll}}(n) \ge (2 + \frac{5}{16})n - O(\sqrt{n})$.

Problem 1   *(Brass) Determine the largest constant c such that there exist n-element point sets in the plane with no three collinear points in which the smallest distance occurs at least $(c + o(1))n$ times.*

The question concerning the maximum number of occurrences of the smallest distance makes sense in any other metric space. It was completely answered by Brass [Br96] for all normed planes. If the unit disk of the norm is a parallelogram (isometric to L_∞), the maximum number of smallest distances among n points is $\lfloor 4n - \sqrt{28n - 12} \rfloor$, the extremal sets being octagonal sections of the integer lattice. In all other cases, the maximum number is $\lfloor 3n - \sqrt{12n - 3} \rfloor$, as in the Euclidean case. These problems are related to edge-isoperimetric inequalities in Cayley graphs on lattices; similar inequalities occur in [HaH76], [Br96], and [Br98]. In higher dimensions, almost nothing is known even in the Euclidean case. The translative kissing (Hadwiger) number provides a trivial upper bound for the maximum degree in the smallest-distance graph (see Section 2.4), and we obtain a little improvement on the corresponding upper bound on the number of edges using the isoperimetric inequality. In particular, in the three-dimensional Euclidean space we have $s_3(n) = 6n - \Theta(n^{\frac{2}{3}})$, but no conjecture similar to Harborth's formula in the plane is known for the exact value of $s_3(n)$.

Problem 2   *(Harborth) Determine the maximum number of smallest distances among n points in three-dimensional Euclidean space, and describe the extremal configurations.*

For higher-dimensional normed spaces, or equivalently for touching pairs in higher-dimensional packings of translates of a convex body, some initial results were obtained by K. Bezdek [Be02]. The most promising cases are the higher-dimensional L_∞-metrics, since in these spaces one can normalize all point configurations by transforming them ("pushing them together") to nearby lattice subsets.

Problem 3   *(Brass) Determine the maximum number of smallest distances among n points in L_∞^d, and describe the extremal configurations.*

The analogous problems were also studied on the sphere, where the situation is similar to the planar case with the assumption that no three points are collinear. The trivial upper bound $s^{\text{sphere}}(n) \le \frac{5}{2}n$ follows from the fact that the maximum degree of the vertices of the smallest-distance graph is at most five. This estimate was improved to

$$s^{\text{sphere}}(n) \le \left(\frac{5}{2} - \frac{1}{1682}\right) n$$

by K. Bezdek et al. [BeCK87]. They also found a construction that gives the lower bound $s^{\text{sphere}}(n) \ge \left(2 + \frac{1}{5}\right) n$.

Problem 4   *[BeCK87] Determine the smallest c such that n points on the sphere determine at most $(c + o(1))n$ smallest distances.*

Caterpillars that are realizable as smallest-distance graphs in the Euclidean plane were studied by Harary et al. [HaRM96]. The general problem of characterizing graphs realizable as smallest-distance graphs seems hopelessly difficult. A notable exception is the case of the L_∞ norm: every smallest-distance graph with respect to this norm in the plane is a subgraph of the L_∞-smallest-distance graph of the integer lattice, as can be seen by "pushing together" the corresponding translative packing of squares [Br96].

The bipartite version of the smallest-distance problem is still wide open. Given a set of m red and n blue points, consider the smallest distance between two points of different colors. What is the maximum number of occurrences of this distance between bichromatic pairs over all m-element and n-element point sets? In the plane this number must clearly be at most linear in $m + n$, because the corresponding bichromatic smallest-distance edges, when drawn by line segments, do not cross. However, in three-dimensional space we have only a $O(n^{\frac{2}{3}}m^{\frac{2}{3}} + n + m)$ upper bound and a linear lower bound [EdS91].

Conjecture 5 *Between two sets of m red and n blue points in three-dimensional Euclidean space there are only $O(m + n)$ bichromatic smallest-distance pairs.*

Given a set of n points, connect every element p by a directed edge to each of its nearest neighbors, that is, to all points whose distance from p is minimum. This defines the nearest-neighbor graph, really a digraph, of the set. Find a good upper bound on the number of edges of this digraph. Here the packing argument can be applied to bound only the indegree: the number of points q for which a given p is a nearest neighbor is at most the kissing number of the ball. So in the planar case there are fewer than $6n$ nearest-neighbor pairs, and there are $6n - O(\sqrt{n})$ such pairs in any not too elongated piece of the triangular lattice. We do not have an exact formula for this number because the proof of Harborth's theorem [Ha74] mentioned at the beginning of this section fails to generalize to this situation.

A class of graphs similar to smallest-distance graphs in the plane is the set of crossing-free unit-distance graphs. They can also be interpreted as configurations that can be realized by matchsticks [Ha94]. By planarity, any such graph with n vertices has at most $3n - 6$ edges. It seems very likely that the maximum number of edges in a crossing-free unit-distance graph is again $\lfloor 3n - \sqrt{12n - 3} \rfloor$, but the proof of Harborth [Ha74] for smallest distances does not seem to generalize to this case.

Another natural generalization of smallest-distance graphs is the following. Connect two points of an n-element set S in the plane by an edge if and only if their distance is at most some fixed multiple $1 + \varepsilon$ of the smallest distance. We can think of such a graph as the structure of tangencies among n slightly elastic nonoverlapping disks. One can choose this multiple as large as 1.15 before the maximum degree can increase to seven [HaKS01]. Hence, one would expect that, at least for some very small $\varepsilon > 0$, the maximum number of edges of a graph on n vertices constructed in this way is still $\lfloor 3n - \sqrt{12n - 3} \rfloor$, but Harborth's proof [Ha74] again breaks down.

Conjecture 6 *(Brass) There is an $\varepsilon > 0$ such that any set of n points in the plane, with smallest distance one, contains at most $\lfloor 3n - \sqrt{12n - 3} \rfloor$ distances smaller than $1 + \varepsilon$.*

Vesztergombi [Ve87] investigated the number of second-smallest and jth-smallest distances. Her conjecture on the number of second-smallest distances was proved by Brass [Br92a]: $s_{2nd}(n) = \frac{24}{7}n - O(\sqrt{n})$. The extremal sets are essentially sections of a packing of regular 12-gons, with a distance ratio $\frac{d_1}{d_2} = 2 \sin \frac{\pi}{12}$. For all other ratios of smallest to second-smallest distances, only the trivial $3n - O(\sqrt{n})$ lower bound can be realized.

The strong dependence of the answer to the last question on the distance ratio $\frac{d_1}{d_2}$ led Csizmadia [Cs99] to study the numbers of smallest and second-smallest distances for a fixed distance ratio. He obtained bounds on the total number of occurrences of these two distances depending on n and on the distance ratio. The maximum is $6n - O(\sqrt{n})$, and this bound can be attained only for $\frac{d_1}{d_2} = \frac{1}{\sqrt{3}}$. Otherwise, the maximum is $\frac{33}{7}n - O(\sqrt{n})$, which is asymptotically tight only when $\frac{d_1}{d_2} = 2\sin\frac{\pi}{12}$. For all other distance ratios, he obtained better bounds, but they are probably not sharp.

Brass [Br92b] obtained some further bounds for the number of jth-smallest distances, and proposed some conjectures for the cases $j = 3$ and 4. A useful method for the solution of problems of this kind is to select a maximal planar subgraph of the graph of relevant distances and then to bound the number of remaining edges [Br92a]. The effectiveness of this approach can be explained by the fact that in the graph of small distances there are relatively few crossings between the edges.

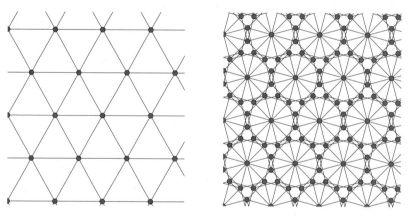

SETS WITH MAXIMUM NUMBERS OF SMALLEST
AND SECOND-SMALLEST DISTANCES

SETS WITH MANY 3RD-SMALLEST, 4TH-SMALLEST DISTANCES

Another metrically defined graph representing proximity of points is the "sphere-of-influence graph," which is defined for a finite point set S by connecting two points $x_1, x_2 \in S$ with an edge if and only if

$$d(x_1, x_2) \leq \min_{\substack{x \in S \\ x \neq x_1}} d(x_1, x) + \min_{\substack{x \in S \\ x \neq x_2}} d(x_2, x).$$

This graph, originally defined by Toussaint, has received considerable attention (see [MiQ94] for a survey). The number of edges of a sphere-of-influence graph on n vertices is $O(n)$ in any Euclidean space of fixed dimension and, indeed, in any finite-dimensional normed space. However, the correct multiplicative constant is not known, even in the planar case.

Conjecture 7   *The number of edges in the sphere-of-influence graph of a set of n points in the plane is less than $9n$.*

The best known upper bound is $15n$ [So99]. Several variants of this question were discussed by Guibas et al. [GuPS94].

A popular problem of Erdős, reported by Pollack [Po85], is about the minimum independence number of smallest-distance graphs: What is the largest number of points one can select from any set S of n points in the plane such that the smallest distance among the selected points is strictly larger than the smallest distance among the elements of the original point set? Taking n equilateral triangles far away from one another, we obtain the upper bound $\lfloor \frac{1}{3}n \rfloor$. This was improved to $\lfloor \frac{6}{19}n \rfloor$ by Chung and Graham and by Pach, and then to $\lfloor \frac{5}{16}n \rfloor$ by Pach and Tóth [PaT96], provided that n is sufficiently large. On the other hand, Pollack [Po85] noticed that recursively selecting vertices from the convex hull of the point set and removing their neighbors, we obtain a subset $S' \subseteq S$ of at least $\lceil \frac{1}{4}n \rceil$ elements, in which the smallest distance is strictly larger than in S. Csizmadia [Cs98] and Swanepoel [Sw02] improved this bound to $\lceil \frac{9}{35}n \rceil$ and to $\lceil \frac{8}{31}n \rceil$, respectively.

Problem 8   *(Erdős) Find the largest c such that any set of sufficiently many points in the plane contains a subset consisting of at least cn points in which the smallest distance is strictly larger than in the original set.*

[**Be02**]   K. BEZDEK: On the maximum number of touching pairs in a finite packing of translates of a convex body, *J. Combinatorial Theory Ser. A* **98** (2002) 192–200.

[**BeCK87**]   K. BEZDEK, R. CONNELLY, G. KERTÉSZ: On the average number of neighbors in a spherical packing of congruent circles, in *Intuitive Geometry 1985*, K. Böröczky et al., eds., *Colloq. Math. Soc. János Bolyai* **48** (1987) 37–52.

[**Br98**]   P. BRASS: On point sets with many unit distances in few
               directions, *Discrete Comput. Geom.* **19** (1998) 355–366.

[**Br96**]   P. BRASS: Erdős distance problems in normed spaces, *Comput. Geom. Theory App.* **6** (1996) 195–214.

[**Br92a**]   P. BRASS: The maximum number of second smallest distances in finite planar sets, *Discrete Comput. Geom.* **7** (1992) 371–379.

[**Br92b**]   P. BRASS: On the maximum densities of *j*-th smallest distances, in: *Sets, Graphs and Numbers*, G. Halász et al., eds., *Colloq. Math. Soc. János Bolyai* **60** (1992) 119–126.

[**Br92c**]   P. BRASS: *Beweis einer Vermutung von Erdős und Pach aus der kombinatorischen Geometrie*, Dissertation, Techn. Univ. Braunschweig 1992.

[**Cs99**]   G. CSIZMADIA: The multiplicity of the two smallest distances among points, *Discrete Math.* **194** (1999) 67–86.

[**Cs98**]   G. CSIZMADIA: On the independence number of minimum distance graphs, *Discrete Comput. Geom.* **20** (1998) 179–187.

[**EdS91**]   H. EDELSBRUNNER, M. SHARIR: A hyperplane incidence problem with applications to counting distances, in: *The Victor Klee Festschrift*, P. Gritzmann et al., eds., *DIMACS Ser. Discret. Math. Theor. Comput. Sci.* **4** (1991) 253–263.

[**Er75**]   P. ERDŐS: On some problems of elementary and combinatorial geometry, *Ann. Mat. Pura Appl. Ser. IV* **103** (1975) 99–108.

[**ErP90**]   P. ERDŐS, J. PACH: Variations on the theme of repeated distances, *Combinatorica* **10** (1990) 261–269.

[**GuPS94**]   L.J. GUIBAS, J. PACH, M. SHARIR: Sphere-of-influence graphs in higher dimensions, in *Intuitive Geometry* (Szeged, 1991), K. Böröczky et al., eds., *Colloq. Math. Soc. János Bolyai* **63** (1994) 131–137.

[**HaRM96**]   F. HARARY, W. RANDOLPH, P.G. MEZEY: A study of maximum unit-circle caterpillars—tools for the study of the shape of absorption patterns, *Discrete Appl. Math.* **67** (1996) 127–135.

[**Ha94**]   H. HARBORTH: Matchsticks in the plane, in: *The Lighter Side of Mathematics*, R.K. Guy et al., eds., MAA Spectrum, Washington D.C. 1994, 281–288.

[**Ha74**]   H. HARBORTH: Lösung zu Problem 664A, *Elemente Math.* **29** (1974) 14–15.

[HaH76]   H. HARBORTH, F. HARARY: Extremal animals, *J. Comb. Inf. Syst. Sci.* **1** (1976) 1–8.

[HaKS01] H. HARBORTH, M. KOCH, L. SZABÓ: Newton numbers for overlapping circular disks, *Studia Sci. Math. Hungar.* **37** (2001) 119–130.

[Ku94]   Y.S. KUPITZ: On the maximal number of appearances of the minimal distance among n points in the plane, in: *Intuitive Geometry* (Szeged, 1991), K. Böröczky et al., eds., *Colloq. Math. Soc. János Bolyai* **63** (1994) 217–244.

[MiQ94]   T.S. MICHAEL, T. QUINT: Sphere of influence graphs: A survey, *Congressus Numerantium* **105** (1994) 153–160.

[PaT96]   J. PACH, G. TÓTH: On the independence number of coin graphs, *Geombinatorics* **6** (1996) 30–33.

[Po85]   R. POLLACK: Increasing the minimum distance of a set of points, *J. Combinatorial Theory Ser. A* **40** (1985) p. 450.

[Re72]   O. REUTTER: Problem 664A, *Elemente Math.* **27** (1972) p. 19.

[So99]   M.A. SOSS: On the size of the Euclidean sphere of influence graph, in *CCCG 1999, Proc. Canadian Conf. Comput. Geom.* (Vancouver, 1999), 43–46.

[Sw02]   K.J. SWANEPOEL: Independence numbers of planar contact graphs, *Discrete Comput. Geom.* **28** (2002) 649–670.

[Tó97]   G. TÓTH: The shortest distance among points in general position, *Comput. Geom. Theory Appl.* **8** (1997), 33–38.

[Ve87]   K. VESZTERGOMBI: Bounds on the number of small distances in a finite planar set, *Stud. Sci. Math. Hungar.* **22** (1987) 95–101.

5.8 Frequent Large Distances

One of the oldest distance-related problems is the following: What is the maximum number of occurrences of the diameter, that is, the largest distance, in an n-element point set S? The planar version of this question was first posed by Hopf and Pannwitz in [HoP34] and answered by Sutherland [Su35] and others. In the plane, connecting two points of S by a line segment if their distance is maximum, we obtain a graph whose any two edges either cross each other or have an endpoint in common. By an observation of Erdős, this property implies that the number of edges (segments) is at most $|S| = n$ (see also Section 9.5).

Thus, the maximum number $l(n)$ of times that the largest distance occurs among n points in the plane is $l(n) = n$. In the extremal configuration, all points lie on a Reuleaux-type polygon, and the graph of largest distances consists of an odd cycle, possibly with a few additional vertices of degree one. Martini and Kupitz [MaK01] attempted to generalize this structure.

n POINTS WITH
n LARGEST DISTANCES

It was conjectured by Vázsonyi and proved by Grünbaum [Gr56], Heppes [He56], and Straszewicz [St57] that in three-dimensional space, the maximum number of largest distances among n points is $2n - 2$. The structure of the extremal sets is not completely known. All three proofs are based on the idea of placing around each point a ball whose radius is equal to the diameter of the point set and applying Euler's formula to the "spherical polytope" obtained by taking the intersection of these balls. The maximum number of largest distances is attained if and only if each vertex of the spherical polytope itself is one of the n points. This implies that any point set all of whose elements have degree at least two in the graph of largest distances can be extended to a set in which the number of largest distances is maximum. Thus, there is a large variety of extremal sets. The simplest example is the vertex set of a tetrahedron, with the remaining $n - 4$ points added on an edge of the spherical polytope. Further constructions were discussed by Neaderhouser and Purdy [NeP82]. It is known [Do00] that any two odd cycles in the graph of largest distances have a vertex in common. This suggests that the graph should not be far from being bipartite, although the chromatic number can be as large as four. A counterexample shows that at least three of the color classes can be large [Al03].

Conjecture 1   *(Brass) From any finite point set in three-dimensional Eu-*
clidean space, one can remove $O(1)$ points with the prop-
erty that the subgraph of the largest-distance graph in-
duced by the remaining points is three-colorable.

In higher dimensions, the problem becomes less interesting, since a
Lenz-type construction (see Section 5.1) is asymptotically extremal. The
exact maximum is not known for $d \geq 4$, but perhaps it is not very difficult
to determine this value. The problem can also be discussed in other normed
spaces. It was completely solved by Brass [Br96] in the planar case. Any
strictly convex norm behaves just like the Euclidean norm and for norms
that are not strictly convex, the exact maximum depends on the length
of the longest line segment along the unit circle. In higher-dimensional
normed spaces, the problem is open and similar to the problem of unit
distances.

The directed version of the problem has also received some attention.
Connect by a directed edge every point p of a given point set S to all other
elements whose distance from p is maximal. The resulting graph is called
the "furthest-neighbor graph" of S. The problem is to find the maximum
number of furthest-neighbor pairs, i.e., the maximum number of edges of
this graph over all n-element sets S. In the plane, this number is known
to be $3n - 3$ if n is odd and $3n - 4$ if n is even [EdS89], [Av84]. In three-
dimensional space, the maximum number of furthest-neighbor pairs deter-
mined by n points is $\lfloor \frac{1}{4}n^2 + \frac{3}{2}n + 3 \rfloor$ if $n \not\equiv 3 \bmod 4$ and $\lfloor \frac{1}{4}n^2 + \frac{3}{2}n + 4 \rfloor$
otherwise [Cs96], [AvEP88]. In higher dimensions, Lenz-type constructions
are asymptotically extremal.

Edelsbrunner and Skiena [EdS89] studied the maximum number of
furthest-neighbor pairs among n points in convex position. They proved
that in the plane this number is at most $2n$, which is attained, e.g., for
every set of n points with n diameters, but these are not the only extremal
configurations. The corresponding function in three-dimensional space is
still linear in n [ChSV03].

The maximum numbers of second-
largest and jth-largest distances were
studied by Vesztergombi et al. in several papers [ErLV88], [ErLV89], [Ve85],
[Ve87], [Ve96]. In particular, Vesztergombi [Ve87] proved that the second-
largest distance among n points in the
plane can occur at most $\frac{3}{2}n$ times.

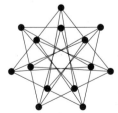

$\frac{3}{2}n$ SECOND-LARGEST DISTANCES

This bound cannot be improved. She also proved that the combined number
of occurrences of the largest and second-largest distances is at most $2n$, and
characterized all homogeneous inequalities involving these numbers and n.

In case the points are in convex position, she proved that the second-largest distance can occur at most $\frac{4}{3}n$ times. This bound is sharp apart from an additive constant [Ve85].

Problem 2   *(Vesztergombi [Ve85]) What is the maximum number of occurrences of the jth-largest distance among n points in the plane, for $j \geq 3$?*

[Al03]   N. ALON: Problems and results in extremal combinatorics I, *Discrete Math.* **273** (2003) 31–53.

[Av84]   D. AVIS: The number of furthest neighbor pairs of a finite planar set, *Amer. Math. Monthly* **91** (1984) 417–420.

[AvEP88]   D. AVIS, P. ERDŐS, J. PACH: Repeated distances in space, *Graphs Combinatorics* **4** (1988) 207–217.

[Br96]   P. BRASS: Erdős distance problems in normed spaces, *Comput. Geom. Theory Appl.* **6** (1996) 195–214.

[ChSV03]   O. CHEONG, C.S. SHIN, A. VIGNERON: Computing farthest neighbors on a convex polytope, *Theoretical Computer Science* **296** (2003) 47–58.

[Cs96]   G. CSIZMADIA: Furthest neighbors in space, *Discrete Math.* **150** (1996) 81–88.

[Do00]   V.L. DOL'NIKOV: Some properties of graphs of diameters, *Discrete Comput. Geom.* **24** (2000) 293–299.

[EdS89]   H. EDELSBRUNNER, S. SKIENA: On the number of furthest neighbor pairs in a point set, *Amer. Math. Monthly* **96** (1989) 614–618.

[ErLV89]   P. ERDŐS, L. LOVÁSZ, K. VESZTERGOMBI: On the graph of large distances, *Discrete Comput. Geom.* **4** (1989) 541–549.

[ErLV88]   P. ERDŐS, L. LOVÁSZ, K. VESZTERGOMBI: The chromatic number of the graph of large distances, in: *Combinatorics, Colloq. Math. Soc. János Bolyai* **52** (1988) 547–551.

[Gr56]   B. GRÜNBAUM: A proof of Vázsonyi's conjecture, *Bull. Res. Council Israel, Sect. A* **6** (1956) 77–78.

[He56]   A. HEPPES: Beweis einer Vermutung von A. Vázsonyi, *Acta Math. Acad. Sci. Hungar.* **7** (1956) 463–466.

[HoP34]   H. HOPF, E. PANNWITZ: Aufgabe Nr. 167, *Jahresbericht Deutsch. Math.-Verein.* **43** (1934) p. 114.

[MaK01]   H. MARTINI, Y.S. KUPITZ: From intersectors to successors, *Graphs Combinatorics* **17** (2001) 99-111.

[NeP82]   C.C. NEADERHOUSER, G. PURDY: On finite sets in E^k in which the diameter is frequently achieved, *Period. Math. Hungar.* **13** (1982) 253-257.

[St57]   S. STRASZEWICZ: Sur un problème géométrique de P. Erdős, *Bull. Acad. Pol. Sci., Cl. III* **5** (1957) 39–40.

[Su35]   J.W. SUTHERLAND: Lösung der Aufgabe 167, *Jahresbericht Deutsch. Math.-Verein.* **45** (1935) 33–35.

[Ve96]   K. VESZTERGOMBI: The two largest distances in finite planar sets, *Discrete Math.* **150** (1996) 379–386.

[Ve87]   K. VESZTERGOMBI: On large distances in planar sets, *Discrete Math.* **67** (1987) 191–198.

[Ve85]   K. VESZTERGOMBI: On the distribution of distances in finite sets in the plane, *Discrete Math.* **57** (1985) 129–145.

5.9 Chromatic Number of Unit-Distance Graphs

What is the minimum number of colors needed to color all points of the plane so that no two points at distance one receive the same color? In other words, what is the chromatic number of the (infinite) unit-distance graph induced by all points of the plane? It follows by compactness that this number, which is often refereed to as the *chromatic number of the plane*, is equal to the maximum chromatic number of finite unit-distance graphs [ErdB51]. (This argument is based on the axiom of choice. Shelah and Soifer showed that assuming other equally consistent axioms, the chromatic number can be strictly larger than the maximum chromatic number of its finite induced subgraphs [ShS03a], [ShS03b], [ShS04].)

   The history of this old and very popular problem, although obscure, was traced by Soifer [So03]. It is known as the Hadwiger–Nelson problem and appears in print in [Ha61], but is certainly older. In 1950, Edward Nelson (a student in Chicago at the time) formulated the question. A density version of this coloring problem had already been studied by Hadwiger in 1944 [Ha44].

Problem 1   *(Hadwiger–Nelson problem) What is the minimum number of colors for coloring the points of the plane so that no two points at unit distance receive the same color?*

Despite many publications on this problem, little progress has been made in the planar case. There are known examples of four-chromatic unit-distance graphs in the plane, the simplest being the Moser spindle [MoM61] and the rotated triangle. A seven-coloring of the whole plane shows that seven is an upper bound for the chromatic number of any unit-distance graph.

Many further four-chromatic unit-distance graphs have been found, starting with graphs without triangles [Wo79], [O'D94], [O'D95], [Ch95], [HoO96], and culminating in four-chromatic unit-distance graphs without cycles of length at most k, for arbitrarily large values of k [O'D00a], [O'D00b]. It has also been shown that a

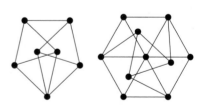

FOUR-CHROMATIC
UNIT-DISTANCE GRAPHS

unit-distance graph with chromatic number seven, if it exists, must have many vertices [Pr98] and must contain large odd cycles [ChFP95]. The maximum chromatic number of a unit-distance graph in the plane is probably four (otherwise, there must be some finite counterexample, which is hopefully not too large). On the other hand, there are numerous results that a proper four-coloring of the plane (if it exists) cannot have "nice" color classes (e.g., they cannot be open, closed, measurable).

The classical seven-coloring that avoids unit distances has very simple color classes (lattice-like collections of cells in a hexagonal tiling), and there are several six-colorings of the plane in which each color class avoids some distance (but not all the same) [So92a], [So94a], [So94b], [HoS94], [HoS96]. One can introduce several variants of the chromatic number, for which the color classes have to be measurable or satisfy some other property of being well-behaved, and try to prove better lower bounds for the number of colors required in such a coloring. One can also try to determine the maximum density of a measurable set that avoids some fixed distance or several distances.

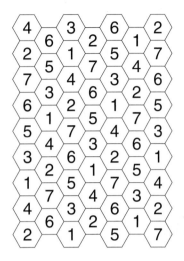

Seven-colored plane

If a point may belong to several color classes, and each class is a closed set, the lower bound was improved to six by Woodall [Wo73]. In the same situation, if each class is an open set, the lower bound was improved to five by Brown et al. [BrDP94]. If each color class is measurable, then Falconer [Fa81] obtained a lower bound of five. In an equivalent formulation, we can ask for the maximum number k with the property that in any coloring of the plane with k colors, one of the color classes determines the unit distance. We may require the stronger property that one of the color classes determines all distances. This version of the question was first studied by Hadwiger [Ha44], [Ha45], and later by Raiskii [Ra70], Woodall [Wo73], and Falconer [Fa81].

The Hadwiger–Nelson problem has also been investigated in other metric spaces. In other normed planes (often called Minkowski-planes), the situation is essentially the same as in the Euclidean plane: a similar four-chromatic example provides the lower bound four, while seven colors suffice. If the unit disk is a parallelogram or a hexagon, the chromatic number is known to be four [Ch91]. The situation is more interesting in higher dimensions. Larman and Rogers [LaR72], Cantwell [Ca96], and Raĭgorodskiĭ [Ra01] gave lower bounds for the chromatic numbers of d-dimensional Euclidean spaces:

d	1	2	3	4	5	6	7	8–10	11–13	14–22	23	24-76
$\chi(\mathbb{R}^d) \geq$	2	4	5	7	9	10	15	16	20	$\frac{4}{3}d$	100	102

Larman and Rogers [LaR72] also exhibited two general constructions, one providing the lower bound $\frac{4}{3}d$, and the other $\Omega(\frac{d^2}{\log^3 d})$, which was later improved to $\Omega(d^3)$ [La78a]. Upper bounds were established only for small dimensions. In three-dimensional space, Coulson [Co02] and Radoičić and Tóth [RaT03] defined a 15-coloring of the whole space that avoids distance one (cf. [Co97] for a weaker result). The best known lower bound, six, was established by Nechushtan [Ne02]. For larger values of d, Larman and Rogers [LaR72] proved that the chromatic number of $\mathbb{R}^d$ is at most $(3 + o(1))^d$. The first exponential lower bound, $(1 + o(1))(1.2)^d$, was found by Frankl and Wilson [FrW81]. This was improved to $(1.239 + o(1))^d$ by Raĭgorodskiĭ [Ra00], [Ra03] (see also [We02]).

Problem 2   (Larman–Rogers [LaR72]) Give good asymptotic bounds on the chromatic number of the unit-distance graph of d-dimensional Euclidean space.

The lower bound of Frankl and Wilson [FrW81] follows by letting k be a prime power, $d = 4k$, and considering the unit-distance graph induced by all points in d-dimensional Euclidean space that have $\frac{d}{2}$ coordinates equal to $\sqrt{\frac{2}{d}}$ and $\frac{d}{2}$ coordinates equal to 0. Two such points are at distance one if and only if their sets of nonzero coordinates have precisely $\frac{d}{4}$ elements in common. They show that the size of the largest independent set in this graph is at most $2\binom{d-1}{d/4-1}$, so that the chromatic number is at least

$$\frac{\binom{d}{d/2}}{2\binom{d-1}{d/4-1}} \geq (1 + o(1))\,(1.2)^d.$$

Independently of the original question, it would be interesting to determine the chromatic number of this particular unit-distance graph.

A "density version" of the Hadwiger–Nelson problem (Problem 1) is the following. What is the maximum density of a measurable subset of the plane (or of d-dimensional space) that does not contain a unit-distance pair? A simple set of density $\frac{\pi}{8\sqrt{3}} \approx 0.227$ satisfying this condition can be obtained by taking the union of all disks of radius $\frac{1}{2}$ centered at the points of a triangular lattice of edge length 2. This lower bound was improved to 0.229 by Croft [Cr67]. On the other hand, Székely [Sz84] showed that any set of density $\frac{12}{43} \approx 0.279$ contains a unit-distance pair.

Problem 3   (Croft [Cr67]) What is the maximum density of a measurable set in the plane that does not contain a unit-distance pair?

The standard tools for establishing upper bounds for this quantity are unit-distance graphs with small independence numbers. For higher dimensions, the same unit-distance graphs come into play as the ones used by Székely and Wormald [SzW89] and, previously, by Larman and Rogers [LaR72]. Once we have an upper bound on the size of an independent set in these specific graphs, we can essentially "average" over all of their translates or congruent copies to obtain a bound on the density of an independent set in the whole space. For d-dimensional unit spheres (in $(d+1)$-dimensional space), a similar theorem was obtained by Frankl and Wilson [FrW81]. Any set covering more than the fraction $(1+o(1))(1.13)^{-d}$ of the sphere contains an orthogonal pair, that is, two points at Euclidean distance $\sqrt{2}$. On the other hand, the intersection of the sphere with two antipodal open orthants is a set that has the required property, and its measure is 2^{-d} times the measure of the whole sphere.

It is easy to see that the maximum chromatic number of unit-distance graphs in L_∞^d is 2^d, but we do not know the chromatic number of any other normed space. Since L_∞^d is the only d-dimensional normed space in which the complete graph K_{2^d} is realizable as a unit-distance graph, and for this space the clique number and the chromatic number coincide, it might be the case that in every d-dimensional normed space the chromatic number of unit-distance graphs is at most as large as in L_∞^d. This would be the strongest bound one could hope for, but it probably fails even in the plane.

Problem 4   (Brass) Does there exist a normed plane whose unit-distance graph has chromatic number five?

The clique number and the chromatic number for any other normed space remain unknown. What are they for L_1^d? Füredi and Kang [FüK04] gave an $\Omega(1.139^d)$ lower bound for L_p^d, when $d = 4p - 1$ for an odd prime p and $1 \le p < \infty$, but this is probably not tight.

A related classical conjecture, due to Borsuk, can be rephrased as follows. What is the maximum chromatic number of largest-distance graphs in d-dimensional Euclidean space? Borsuk conjectured this to be $d+1$. This is true for $d = 2, 3$, but false in high dimensions. The first counterexamples were found by Kalai and Kahn [KaK93] (see also [Ni94], [Ra97], [We00]), and they provide a lower bound of $c(1.2)^{\sqrt{d}}$. It is not known what the smallest dimension is for which the conjecture fails. It does not hold for $d = 298$ [HiR03], but probably it fails already for much smaller values of d.

Problem 5   Is every largest-distance graph induced by a finite set of points in Euclidean four-dimensional space five-colorable?

The same question can be asked for smallest-distance graphs. In the plane, the chromatic number of any smallest-distance graph induced by a finite point set is four, since any such graph has a vertex of degree at most three. For dimension three, the corresponding number is still not known.

Problem 6   *What is the maximum chromatic number of a smallest-distance graph induced by a finite set of points in d-dimensional Euclidean space?*

Equivalently, we can ask what is the largest chromatic number of a graph whose vertices can be represented by nonoverlapping congruent balls such that two vertices are adjacent if and only if the corresponding balls touch each other. What happens if the balls do not have to be congruent? Koebe's coin graph theorem (see also Section 9.2) states that a graph is planar if and only if its vertices can be represented by not necessarily congruent nonoverlapping disks in the plane such that every pair of disks corresponding to adjacent vertices must touch. In three and higher dimensions the problem becomes more interesting.

Problem 7   *(Maehara–Noha [MaN97]) Let G be a graph whose vertices can be represented by nonoverlapping balls in 3-dimensional space that all touch a common plane and such that for any pair of adjacent vertices, the corresponding two spheres are tangent to each other. Is it true that G is 5-colorable?*

Maehara and Noha showed that the chromatic number of any such graph is at most six.

Erdős, Lovász, and Vesztergombi [ErLV89], [ErLV88] studied the chromatic number of the graph of k largest distances determined by an n-element point set. For example, in the plane, this number is at most seven, if $n > ck^2$ for a sufficiently large constant c. If the points are in convex position, this number drops to at most three.

There are many further spaces where unit-distance graphs have surprising structural properties, and it is interesting to study their chromatic numbers. On every $(d-1)$-dimensional Euclidean sphere of radius $r > \frac{1}{2}$ (in d-dimensional space), there are many $(d+1)$-chromatic unit-distance graphs. If the diameter is sufficiently close to one, the well-known partition of the sphere into $d+1$ parts of smaller diameter defines a $(d+1)$-coloring [Lo83]. The problems take on a number-theoretic flavor if one chooses rational spaces (the points with rational coordinates) as underlying sets [Wo73], [Za89], [Za92], [Man01]. All unit-distance graphs in rational two- and three-dimensional spaces are bipartite; in four-dimensional space the chromatic number is at most four; and in higher dimensions only some bounds are known.

Several other variants of the chromatic number, such as the fractional chromatic number and the choice number of unit-distance graphs, have also been considered [So92b], [Jo94], [Sch95], [JeT95], [Ka00].

It is a natural idea to define the "dimension" of a graph G as the smallest d for which G can be realized as a unit-distance graph in d-dimensional Euclidean space satisfying some conditions. Depending on these special requirements, one can define several closely related but not equivalent concepts of graph dimension. For instance, the *Euclidean dimension, faithful dimension*, and *contact dimension*, proposed in [ErHT65], [ErS80], and [Pa80], are defined as the minimum d such that G is a subgraph of a unit-distance graph, an induced subgraph of a unit-distance graph, and a subgraph of a smallest-distance graph in d-dimensional space, respectively. For any graph of n vertices, each of these parameters is at most $n - 1$, with equality only for the complete graphs. By the Lenz construction, the Euclidean dimension of any graph is at most twice its chromatic number. However, this is not true for the faithful dimension and for the contact dimension. Several interesting estimates and precise values of these parameters have been found for a number of graph classes [FrM88], [Ma88], [Ma89], [MaR90], [BuH88], including complete multipartite graphs and wheels. The number of edges that have to be subdivided to convert a graph into a planar unit-distance graph has also been studied [GeM00].

Conjecture 8   *(Erdős–Simonovits [ErS80]) The faithful dimension of a graph cannot exceed the maximum degree of its vertices. Equality holds for the complete graph.*

Erdős and Simonovits proved that the faithful dimension never exceeds twice the maximum degree plus one. See [MoP93] and [LoV02] for surveys.

[**BrDP94**]   N. Brown, N. Dunfield, G. Perry: Colorings of the plane III, *Geombinatorics* **3** (1994) 110–114.

[**BuH88**]   F. Buckley, F. Harary: On the Euclidean dimension of a wheel, *Graphs Combinatorics* **4** (1988) 23–30.

[**Ca96**]   K. Cantwell: Finite Euclidean Ramsey theory, *J. Combinatorial Theory, Ser. A* **73** (1996) 273–285.

[**Ch95**]   K.B. Chilakamarri: A 4-chromatic unit-distance graph with no triangles, *Geombinatorics* **4** (1995) 64–76.

[**Ch91**]   K.B. Chilakamarri: Unit-distance graphs in Minkowski metric spaces, *Geometriae Dedicata* **37** (1991) 345–356.

[**ChFP95**]   K.B. Chilakamarri, G.H. Fricke, M. Perkel: Packing an equilateral polygon in a thin strip, *Geometriae Dedicata* **54** (1995) 45–55.

[**Co02**]   D. Coulson: A 15-colouring of 3-space omitting distance one, *Discrete Math.* **256** (2002) 83–90.

[Co97]   D. COULSON: An 18-colouring of 3-space omitting distance one, *Discrete Math.* **170** (1997) 241–247.

[Cr67]   H.T. CROFT: Incidence incidents, *Eureka* (Cambridge) **30** (1967) 22–26.

[ErdB51] P. ERDŐS, N.G. DE BRUIJN: A colour problem for infinite graphs and a problem in the theory of relations, *Indag. Math.* **13** (1951) 371–373.

[ErHT65] P. ERDŐS, F. HARARY, W.T. TUTTE: On the dimension of a graph, *Mathematika* **12** (1965) 118–122.

[ErLV89] P. ERDŐS, L. LOVÁSZ, K. VESZTERGOMBI: On the graph of large distances, *Discrete Comput. Geom.* **4** (1989) 541–549.

[ErLV88] P. ERDŐS, L. LOVÁSZ, K. VESZTERGOMBI: The chromatic number of the graph of large distances, in: *Combinatorics*, A. Hajnal et al., eds., *Colloq. Math. Soc. János Bolyai* **52** (1988) 547–551.

[ErS80]   P. ERDŐS, M. SIMONOVITS: On the chromatic number of geometric graphs, *Ars Combinatoria* **9** (1980) 229–246.

[Fa81]   K.J. FALCONER: The realization of distances in measurable subsets covering $\mathbb{R}^n$, *J. Combinatorial Theory Ser. A* **31** (1981) 184–189.

[FrM88]   P. FRANKL, H. MAEHARA: On the contact dimension of graphs, *Discrete Comput. Geom.* **3** (1988) 89–96.

[FrW81]   P. FRANKL, R.M. WILSON: Intersection theorems with geometric consequences, *Combinatorica* **1** (1981) 357–368.

[FüK04]   Z. FÜREDI, J.-H. KANG: Distance graph on the integer grid with ℓ_1 norm, *Theoretical Computer Science* **319** (2004) 357–366.

[GeM00]   S.V. GERVACIO, H. MAEHARA: Subdividing a graph toward a unit-distance graph in the plane, *Europ. J. Comb.* **21** (2000) 223–229.

[GrJ92]   D. GREENWELL, P.D. JOHNSON JR.:   Forbidding prescribed distances for designated colors, *Geombinatorics* **2** (1992) 13–16.

[Ha61]   H. HADWIGER: Ungelöste Probleme Nr. 10, *Elemente Math.* **16** (1961) 103–104.

[Ha45]   H. HADWIGER: Überdeckung des Euklidischen Raumes durch kongruente Mengen, *Portugaliae Math.* **4** (1945) 238–242.

[Ha44]   H. HADWIGER: Ein Überdeckungssatz für den Euklidischen Raum, *Portugaliae Math.* **4** (1944) 140–144.

[HiR03]   A. Hinrichs, C. Richter: New sets with large Borsuk numbers, *Discrete Math.* **270** (2003) 137–147.

[HoO96]   R. Hochberg, P. O'Donnell: Some 4-chromatic unit-distance graphs without small cycles, *Geombinatorics* **5** (1996) 137–141.

[HoS96]   I. Hoffman, A. Soifer: Another six-coloring of the plane, *Discrete Math.* **150** (1996) 427–429.

[HoS94]   I. Hoffman, A. Soifer: Almost chromatic number of the plane, *Geombinatorics* **3** (1993) 38–40, see also correction in *Geombinatorics* **3** (1994) p. 102.

[JeT95]   T.R. Jensen, B. Toft: Choosability versus chromaticity: the plane unit distance graph has a 2-chromatic subgraph of infinite list-chromatic number, *Geombinatorics* **5** (1995) 45–64.

[Jo94]   P.D. Johnson Jr.:   The choice number of the plane, *Geombinatorics* **3** (1994) 122–128.

[Ka00]   M. Kahle: A generalization of the chromatic number of the plane, *Geombinatorics* **10** (2000) 69–74.

[KaK93]   J. Kahn, G. Kalai: A counterexample to Borsuk's conjecture, *Bull. Amer. Math. Soc., New Ser.* **29** (1993) 60–62.

[La78a]   D.G. Larman: A note on the realization of distances within sets in Euclidean space, *Comment. Math. Helvetici* **53** (1978) 529–535.

[La78b]   D.G. Larman: A triangle-free graph which cannot be $\sqrt{3}$-embedded in any Euclidean unit sphere, *J. Combinatorial Theory Ser. A* **24** (1978) 162–169.

[LaR72]   D.G. Larman, C.A. Rogers: The realization of distances within sets in Euclidean space, *Mathematika* **19** (1972) 1–24.

[Lo83]   L. Lovász: Self-dual polytopes and the chromatic number of distance graphs on the sphere, *Acta Sci. Math.* **45** (1983) 317–323.

[LoV02]   L. Lovász, K. Vesztergombi: Geometric representations of graphs, in: *Paul Erdős and His Mathematics, II* (Budapest, 1999), *Bolyai Soc. Math. Stud.* **11**, J. Bolyai Math. Soc., Budapest, 2002, 471–498.

[Ma89]   H. Maehara: Note on induced subgraphs of the unit distance graph E^n, *Discrete Comput. Geom.* **4** (1989) 15–18.

[Ma88]   H. Maehara: On the Euclidean dimension of a complete multipartite graph, *Discrete Math.* **72** (1988) 285–289.

[MaR90]   H. MAEHARA, V. RÖDL: On the dimension to represent
          a graph by a unit distance graph, *Graphs Combinatorics* **6**
          (1990) 365–367.

[MaN97]   H. MAEHARA, H. NOHA: On the graph represented by a
          family of solid balls on a table, *Ryukyu Math. J.* **10** (1997)
          51–64.

[Man01]   M. MANN: A new bound for the rational chromatic number
          of the rational five-space, *Geombinatorics* **11** (2001) 49–53.

[MoM61]   L. MOSER, W.O.J. MOSER: Solution to Problem 10, *Canad.
          Math. Bull.* **4** (1961) 187–189.

[MoP93]   W.O.J. MOSER, J. PACH: Recent developments in combi-
          natorial geometry, in: *New Trends in Discrete and Computa-
          tional Geometry*, J. Pach, ed., *Algorithms Combin. Ser.* **10**,
          Springer, Berlin, 1993, 281–302.

[Ne02]   O. NECHUSHTAN: On the space chromatic number, *Discrete
          Math.* **256** (2002) 499–507.

[Ni94]   A. NILLI: On Borsuk's problem, in: *Jerusalem combinatorics
          '93*, H. Barcelo et al., eds., *Contemporary Mathematics* **178**,
          AMS 1994, 209–210.

[O'D00a] P. O'DONNELL: Arbitrary girth, 4-chromatic unit distance
          graphs in the plane I: Graph description, *Geombinatorics* **9**
          (2000) 145–152.

[O'D00b] P. O'DONNELL: Arbitrary girth, 4-chromatic unit distance
          graphs in the plane II: Graph embedding, *Geombinatorics* **9**
          (2000) 180–193.

[O'D95]   P. O'DONNELL: A 40-vertex 4-chromatic triangle-free unit
          distance graph, *Geombinatorics* **5** (1995) 31–34.

[O'D94]   P. O'DONNELL: A triangle-free 4-chromatic graph in the
          plane, *Geombinatorics* **4** (1994) 23–29.

[Pa80]   J. PACH: Decomposition of multiple packing and covering, in:
          Diskrete Geometrie – 2. Kolloq., Math. Inst. Univ. Salzburg,
          1980, 169–178.

[Pr98]   D. PRITIKIN: All unit-distance graphs of order 6197 are 6-
          colorable, *J. Combinatorial Theory Ser. B* **73** (1998) 159–163.

[RaT03]   R. RADOIČIĆ, G. TÓTH: A note on the chromatic number
          of the space, in: *Discrete and Computational Geometry—The
          Goodman-Pollack Festschrift*, B. Aronov et al., eds., Springer,
          Berlin, 2003, 695–698.

[Ra70] D.E. RAISKII: The realization of all distances in a decomposition of the space $\mathbb{R}^n$ in $n+1$ parts, *Math. Notes* **7** (1970) 194–196.

[Ra03] A.M. RAĬGORODSKIĬ: The Erdős–Hadwiger problem, and the chromatic number of finite geometric graphs, *Doklady Russian Acad. Sci.* **392** (2003) 313–317.

[Ra01] A.M. RAĬGORODSKIĬ: Borsuk's problem and the chromatic numbers of some metric spaces, *Russ. Math. Surveys* **56** (2001) 103–139.

[Ra00] A.M. RAĬGORODSKIĬ: On the chromatic number of a space, *Russ. Math. Surveys* **55** (2000) 351–352.

[Ra97] A.M. RAĬGORODSKIĬ: On dimensionality in the Borsuk problem, *Russ. Math. Surveys* **52** (1997) 1324–1325.

[Sch95] J.H. SCHMERL: List chromatic number of Euclidean space, *Geombinatorics* **5** (1995) 65–68.

[ShS04] S. SHELAH, A. SOIFER: Axiom of choice and chromatic number: examples on the plane, *J. Combinatorial Theory Ser. A* **105** (2004) 359–364.

[ShS03a] S. SHELAH, A. SOIFER: Axiom of choice and chromatic number of the plane, *J. Combinatorial Theory Ser. A* **103** (2003) 387–391.

[ShS03b] S. SHELAH, A. SOIFER: Chromatic number of the plane. III. Its future, *Geombinatorics* **13** (2003) 41–46.

[So03] A. SOIFER: Chromatic number of the plane & its relatives. Part I: the problem & its history, *Geombinatorics* **12** (2003) 131–148.

[So94a] A. SOIFER: An infinite class of six-colorings of the plane, *Congr. Numerantium* **101** (1994) 83–86.

[So94b] A. SOIFER: Six-realizable set X_6, *Geombinatorics* **3** (1994) 140–145.

[So92a] A. SOIFER: A six-coloring of the plane, *J. Combinatorial Theory Ser. A* **61** (1992) 292–294.

[So92b] A. SOIFER: Relatives of chromatic number of the plane. I., *Geombinatorics* **1** (1992) 13–17.

[Sz84] L.A. SZÉKELY: Measurable chromatic number of geometric graphs and sets without some distances in Euclidean space, *Combinatorica* **4** (1984) 213–218.

[Sz83] L.A. SZÉKELY: Remarks on the chromatic number of geometric graphs, in: *Graphs and other combinatorial topics,*

M. Fiedler, ed., *Teubner-Texte Math.* **59** (1983) 312–315.

[SzW89]   L.A. SZÉKELY, N.C. WORMALD: Bounds on the measurable chromatic number of $\mathbb{R}^n$. *Discrete Math.* **75** (1989) 343–372.

[We02]   B. WEISSBACH: On the chromatic number of $\mathbb{R}^d$, Preprint 02-06, Mathematische Fakultät, Univ. Magdeburg 2002.

[We00]   B. WEISSBACH: Sets with large Borsuk numbers, *Beiträge Algebra Geom.* **41** (2000) 417–423.

[Wo73]   D.R. WOODALL: Distances realized by sets covering the plane, *J. Combinatorial Theory Ser. A* **14** (1973) 187–200.

[Wo79]   N.C. WORMALD: A 4-chromatic graph with a special plane drawing, *J. Austral. Math. Soc. Ser. A* **28** (1979) 1–8.

[Za92]   J. ZAKS: On the chromatic number of some rational spaces, *Ars Combinatoria* **33** (1992) 253–256.

[Za89]   J. ZAKS: On four-colourings of the rational four-space, *Aequationes Mathematicae* **37** (1989) 259–266.

5.10 Further Problems on Repeated Distances

In 1990, Erdős and Pach [ErP90] posed the following problem: Is it possible that for a set of $n > 4$ points in the plane all distances but the diameter occur more than n times? In a regular polygon all distances occur exactly n times, and in any set of n points the diameter always occurs at most n times; still, the existence of sets with all distances but the diameter occurring at least $n + 1$ times seems improbable.

Problem 1   *[ErP90] Is it true that for $n \geq 5$ points in the plane there is always one further distance besides the diameter that occurs at most n times?*

Another problem of Erdős on the realizability of given distance multiplicities is the following.

Problem 2   *(Erdős [Er86]) Do there exist, for arbitrarily large n, sets of n points in the plane, no three on a line, no four on a circle, that determine $n - 1$ distinct distances that can be ordered in such a way that the ith distance occurs i times $(1 \leq i \leq n - 1)$?*

Such examples were found for up to eight points in the plane [Pa89b], [Pa89c], [HaP94], and up to six points in three-dimensional space [Pa89a]. If we do not require the points to be in "general position," the above distance multiplicity distribution is realized by sets of n equidistant points on a line or on a circle. This problem is sometimes called the "seven points problem" of Erdős, which seems to be an especially ill-chosen name [Pa87], [Li86].

Another old problem stated several times by Erdős is to find the minimum diameter of a set of n points in the plane such that any two distinct distances determined by them differ by at least one. Obviously, any set can be blown up to satisfy this condition, but its diameter may become too large. The best known construction in the plane is a set of n points on a line such that any two consecutive points are at unit distance. The diameter of such a set is $n - 1$.

Conjecture 3   *(Erdős) The minimum diameter of a set of n points in the plane in which any two distinct distances differ by at least one is at least $n - 1$, provided that n is sufficiently large.*

There are many constructions whose diameters are asymptotically optimal, that is, bounded from above by some constant times n. For instance, a $\sqrt{n} \times \sqrt{n}$ section of the integer lattice has to be scaled by a factor of $c\sqrt{n}$ to satisfy the distance separation condition, and then its diameter becomes

$O(n)$. For $n \leq 9$, Brass [Br96] gave some examples for point sets whose diameters are much smaller than $n-1$, and proved the conjecture for point sets of a special type.

Clearly, the number of distinct distances determined by a point set meeting the requirements is a lower bound on its diameter. Kanold's [Ka81] original lower bound has been much superseded by the best known general lower bounds on the number of distinct distances determined by n points (see Section 5.3). In higher dimensions d, equidistant collinear point sets are certainly not extremal. Indeed, the diameter of a properly scaled $n^{\frac{1}{d}} \times \ldots \times n^{\frac{1}{d}}$ section of the d-dimensional integer lattice is $\Theta(n^{\frac{2}{d}})$.

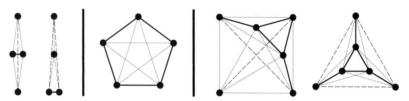

MINIMUM DIAMETER SETS WITH SEPARATED DISTANCES FOR ≤ 6 POINTS

Problem 4   (*Beck et al. [BeB*95]*) *Characterize all finite point sets S in the plane for which the sum of distances from every point $p \in S$ to all other elements of the set is the same for every point p.*

Define the distance count $\mathrm{dd}_S(p)$ of a point $p \in S$ as the number of distinct distances from p to all other points of S. In a set of n collinear equidistant points, the number of distinct distance counts is $\frac{n}{2}$. Improving some previous results of Erdős and Fishburn [ErF97], Csizmadia and Ismailescu [CsI97] constructed examples of n-element sets with $0.7n$ distinct distance counts, but no nontrivial upper bound is known.

Conjecture 5   (*Csizmadia–Ismailescu [CsI97]*) *There exists a constant $c < 1$ such that the number of distinct distance counts among n points in the plane is at most cn, provided that n is sufficiently large.*

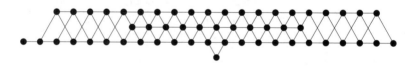

SET OF 55 POINTS WITH 31 DISTINCT DISTANCE COUNTS

[Ba90]   G. BARON: On point sets with differences of distances not less than the minimum distance, in: *Number-Theoretic Analysis*, Springer *Lect. Notes Math.* **1452** (1990) 1–5.

[BeB*95]   I. BECK, N. BEJLEGAARD, P. ERDŐS, P. FISHBURN: Equal distance sums in the plane, *Normat* **43** (1995) 150–161.

[Br96]   P. BRASS: On the Erdős-diameter of sets, *Discrete Math.* **150** (1996) 415–419.

[CsI97]   G. CSIZMADIA, D. ISMAILESCU: Maximum number of different distance counts, in: *Intuitive Geometry* (Budapest, 1995), I. Bárány et al., eds., *Bolyai Soc. Math. Stud.* **6** (1997) 301–309.

[Er86]   P. ERDŐS: On some metric and combinatorial geometric problems, *Discrete Math.* **60** (1986) 147–153.

[ErF97]   P. ERDŐS, P. FISHBURN: Distinct distances in finite planar sets, *Discrete Math.* **175** (1997) 97–132.

[ErP90]   P. ERDŐS, J. PACH: Variations on the theme of repeated distances, *Combinatorica* **10** (1990) 261–269.

[HaP94]   H. HARBORTH, L. PIEPMEYER: On special integral Erdős point sets, in: *Intuitive Geometry* (Szeged, 1991), K. Böröczky et al., eds., *Colloq. Math. Soc. János Bolyai* **63** (1994) 139–149.

[Ka81]   H.J. KANOLD: Über Punktmengen im k-dimensionalen euklidischen Raum, *Abh. Braunschw. Wiss. Ges.* **32** (1981) 55–65.

[Li86]   A. LIU: On the "Seven points problem" of P. Erdős, *Math. Chron.* **15** (1986) 29–33.

[Pa89a]   I. PALÁSTI: On some distance properties of sets of points in general position in space, *Studia Sci. Math. Hungar.* **24** (1989) 187–190.

[Pa89b]   I. PALÁSTI: Lattice-point examples for a question of Erdős, *Period. Math. Hungar.* **20** (1989) 231–235.

[Pa89c]   I. PALÁSTI: A distance problem of P. Erdős with some further restrictions, *Discrete Math.* **76** (1989) 155–156.

[Pa87]   I. PALÁSTI: On the seven points problem of P. Erdős, *Studia Sci. Math. Hungar.* **22** (1987) 447–448.

5.11 Integral or Rational Distances

In this section, we discuss some problems about point sets with integral or rational pairwise distances. Such sets are unexpectedly difficult to construct and are possibly quite rare. In fact, the essence of the following questions is if they are indeed as rare as they appear to be. The first two problems illustrate the extent of our ignorance.

Problem 1   (Ulam [Ul60]) Does there exist a subset everywhere dense in the plane such that all pairwise distances between its points are rational?

Problem 2   (Erdős) Does there exist a set of seven points in the plane, no three on a line, no four on a circle, such that all pairwise distances between them are rational (integral)?

A positive answer to the first question would imply a positive answer to almost all problems in this area, and a negative answer to the second problem would imply a negative answer to the same problems.

In the few known general constructions of arbitrarily large point sets with rational (integral) interpoint distances, the elements are in very special positions. Obviously, there are many such sets all of whose points lie on a line. We also know several constructions with all points on a circle. For instance, one can take all points of the form $(\cos i\alpha, \sin i\alpha)_{i=0}^{n}$, where $\sin \alpha, \cos \alpha$ are rational, or use Ptolemy's theorem, which implies that if four points lie on the same circle and five of the six distances determined by them are rational, then the sixth distance must also be rational. [Alt52], [Ann15], [AnnE45], [Mü53], [Si59]. One can even construct $n-2$ points on a line and two points off that line. For instance, if m is an odd number with at least t divisors, then $m^2 = y^2 - x^2$ has at least t distinct solutions $(x_i, y_i)_{i=1}^{t}$, and all distances between the points $(0, m), (0, -m), (x_1, 0), \ldots, (x_t, 0)$ are integers [AnnE45]. A different method to construct an integral-distance set with only $n - 2$ collinear points is given in [St63]. Huff [Hu48] and Peeples [Pe54] found n-element examples with only $n - 4$ points on a line and the four remaining points on a perpendicular line, and Solymosi used these constructions to obtain such sets with only $n - 3$ points on a circle. (See also [Ha70].) But that is all we know; we are not aware of any general construction of n-element point sets with no $n-4$ points on a line, no $n-3$ points on a circle, and all pairwise distances rational (integral).

There are many interesting questions between the two extremities represented by Problems 1 and 2. Perhaps the best known of them is the following.

Problem 3   *(Schoenberg) Can one approximate with arbitrary preci-sion every n-element point set in the plane by sets of n points, all of whose pairwise distances are rational?*

This is known as "Schoenberg's rational polygon problem," and is usu-ally formulated as an approximation problem for n-gons. In fact, the ques-tion has nothing to do with the actual order in which the points (vertices of the polygon) are listed, so it is really a problem for finite point sets. Evidently, the answer to Schoenberg's question is positive for $n = 3$, be-cause the only condition for the realizability of three given distances among three points is the triangle inequality ("rational triangles are dense in the set of all triangles"). Besicovitch [Bes59] and Mordell [Mo60] proved that the answer is also positive for $n = 4$: "rational quadrilaterals are dense in the set of all quadrilaterals." In fact, the set of possible rational extension points of a fixed nondegenerate rational triangle to a rational quadrilat-eral is everywhere dense in the plane [Alm63], [Ke90a]. Rational-distance approximations are also known to exist in some further special cases, e.g., for sets of six points with an axis of reflectional symmetry through two of them [Day63], and for approximation by triangles with some further ra-tionality conditions like rational area (Heron triangles) [Le899], rational medians [vWi50], rational coordinates [Ho48], [La51]. As building blocks for larger integral-distance sets, all triangles with integral sides and with a specified angle of rational cosine have been classified [Joh15], [Has77], [RoY77], [Wa53]. Rational quadrilaterals have been investigated in number theory for centuries by Brahmagupta, Kummer, and many others [Hae14], [Kum848]. However, Problem 3 is still open for $n = 5$.

The same questions can also be asked in higher dimensions [Fr02], in algebraic variants [Bel47], and for various special classes of point sets.

There are many similar problems combining the property that all in-terpoint distances are rational with some other combinatorial or geometric condition.

Problem 4   *Does there exist a point in the unit square at rational distances from all four vertices?*

See [DoSY77] for some comments and for the analogous question for arbitrary rectangles.

Problem 5   *Does there exist a rectangular parallelepiped such that all distances between its vertices are rational?*

Such a parallelepiped is known as a *perfect box*. A partial solution to this problem was given by Peterson and Jordan [PeJ95] and by Harborth and Möller [HaM98], who constructed some polyhedra, combinatorially iso-morphic to a cube, all of whose interpoint distances are rational. However, these polyhedra are not rectangular parallelepipeds.

There are many similar problems in which only some of the distances are required to be rational, and this condition is combined with various assumptions on the relative positions of the points [Ha94], [Ha98], [HaK90], [HaK*87], [HaM94], [Mö90]. A representation of a given graph G in the plane is called an "integral drawing" if every edge corresponds to a straight-line segment whose length is an integer. The main problem of this type is the following one [HaK*87], [HaM94], [Ha98], [KeH01].

Problem 6   *(Harborth) Does every planar graph admit an integral drawing in which no two edges share an interior point?*

A positive answer to this question would not even follow from the existence of arbitrarily large general-position point sets S all of whose interpoint distances are integral. Indeed, it is possible that the plane drawing condition forces a position incompatible with the order type of S. One can further restrict the allowable edge lengths by considering only special subsets of the integers, such as Fibonacci numbers [HaK91], [HaKR96]. In the extreme case in which there is only one admissible edge length, we again arrive at the concept of unit-distance graphs.

A mapping of the vertices of a graph G into the plane is called a "faithful representation" of G by rational distances when the distance between the images of any two vertices $u, v \in V(G)$ is rational if and only if u and v are adjacent in G. The restriction of "faithfulness," i.e., the condition that nonedges correspond to irrational distances, is essential here; otherwise, $V(G)$ could be mapped into any sufficiently large point set determining only rational distances.

Problem 7   *(Erdős) Does every finite graph admit a faithful representation by rational distances?*

Replacing "rational" by "integral" makes the problem easier: Maehara, Ota, and Tokushige [MaOT97] answered the corresponding question in the affirmative. That is, every finite graph admits a representation in which the distance between the images of two vertices is an integer if and only if they are adjacent.

Of course, if we have a (faithful) representation of a graph by rational distances, then by properly scaling it, we can remove the common denominator of the distances to obtain a (faithful) representation by integers.

The situation is quite different for infinite graphs, where the existence of a representation by rational numbers does not necessarily imply representability by integers. One of the oldest discoveries on this subject was made by Anning and Erdős [AnnE45], [Er45], who noticed that every infinite set of points in the plane that determines only integer distances is collinear. This was also shown by Choquet and Kreweras (see [Tr51]), and has since been rediscovered several times [Gle81]. The statement also holds in higher dimensions and has been partially extended to strictly convex metric spaces by Fullerton [Fu49]. Improving Erdős's [Er45] argument, Solymosi [So03] established a stronger quantitative result. He proved that if d denotes the distance between two elements p and q of a planar point set S, all of whose interpoint distances are integers, then either S is small, $|S| < d^3$ or, with the possible exception of at most $2d-1$ points, all elements of S belong to the line pq. Several variants of this question, in which the integers are replaced by some other sufficiently sparse set of allowable distances, were studied by Kuzminykh [Kuz80], [Kuz82], [Kuz98]. The most important observation used in the proofs of these theorems is that in an integer-distance set S, for any two points $a, b \in S$, all remaining elements $x \in S$ satisfy $d(x,a) - d(x,b) \in \mathbb{Z} \cap [-d(a,b), d(a,b)]$. Thus, the remaining points are contained in $2\lfloor d(a,b) \rfloor + 1$ hyperbolas determined by a and b.

In the framework of integral distances, the following question naturally arises: What is the minimum size of an integer-distance set of n elements measured, for example, by the diameter? (See Section 5.10, Conjecture 3.) The example of n points equally spaced at distance one along a line has diameter $n-1$. Solymosi [So03] gave an $\Omega(n)$ lower bound for the diameter of any integer-distance set of n points in the plane. Restricting our attention to noncollinear integral-distance sets, a construction of Harborth et al. [HaKM93] provided an $O(e^{c \log n \log \log n})$ upper bound for the minimum diameter, which improved their earlier bound [HaK85]. The minimum diameters of n-element integer-distance sets in d-dimensional space were determined by Harborth and Piepmeyer [HaP91], for several small values of n. Small examples with up to six points in general position (no three points collinear, no four points on a circle) in the plane were constructed in [Ke88], [Ke90b], [La83], [NoB89].

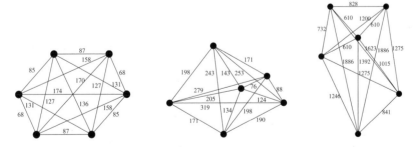

EXAMPLES OF SIX-POINT GENERAL-POSITION INTEGER DISTANCE SETS

Problem 8   *Give good bounds for the minimum diameter of a set of n points in the plane, not all on a line, such that all distances determined by them are integers.*

What is the maximum number of pairs among n points that can determine a distance belonging to a given set K of numbers? Several variants of this problem can be successfully settled by the results and methods of Turán-type extremal graph theory. A prototype of this question is the special case of K is the set of all odd integers. It was shown by Graham, Rothschild, and Straus [GrRS74] (see also [Ro97]) that $d + 2$ points with pairwise odd integral distances are realizable in Euclidean d-dimensional space only when $16 \mid (d + 2)$. In particular, in the plane, K_4 is not realizable with odd integral distances. Consequently, the maximum number of pairs among n points in the plane that determine odd distances is bounded from above by $\mathrm{ex}(n, K_4)$, the maximum number of edges of a K_4-free graph with n vertices. This bound cannot be improved. Piepmeyer [Pi96] and, by a different construction, Rosenfeld [Ro96] proved that the existence of a complete subgraph of four vertices is the only obstruction. That is, every finite K_4-free graph can be represented by odd distances in the plane. The result generalizes to three-dimensional space, where K_5 is unrealizable, and the maximum number of odd integral distances that can occur among n points is $\mathrm{ex}(n, K_5)$ [Br96]. However, the construction does not generalize to four-dimensional space. Graham et al. [GrRS74] sketched similar results for other divisibility classes. Realizability of graphs with other distance sets was also discussed in several other papers; see [Ma04] for a survey.

[Alm63]   J.H.J. Almering: Rational quadrilaterals, *Indagationes Math.* **25** (1963) 192–199. (= *Nederl. Akad. Wet., Proc., Ser. A* **66**)

[Alt52]   M. Altwegg: Ein Satz über Mengen von Punkten mit ganzzahliger Entfernung, *Elemente Math.* **7** (1952) 56–58.

[AnDS69]   D.D. Ang, D.E. Daykin, T.K. Sheng: On Schoenberg's rational polygon problem, *J. Austral. Math. Soc.* **9** (1969) 337–344.

[Ann15]   D. Anning: Relating to a geometric representation of integral solutions of certain quadratic equations, *Amer. Math. Monthly* **22** (1915) p. 321.

[AnnE45]   D. Anning, P. Erdős: Integral distances, *Bull. Amer. Math. Soc.* **51** (1945) 598–600.

[Bel47]   E.T. Bell: Diophantine equations suggested by elementary geometry, *Annals Math., Ser. 2* **48** (1947) 43–50.

[Bes59]   A.S. Besicovitch: Rational polygons, *Mathematika* **6** (1959) p. 98.

[Br96]   P. BRASS: Extremale Konstruktionen in der kombinato-
           rischen Geometrie, Habilitationsschrift, Universität Greifswald
           1996.

[Day63]   D.E. DAYKIN: Rational polygons, Mathematika **10** (1963)
           125–131.

[DoSY77]   C.W. DODGE, G. SHUTE, K.L. YOCOM: Seven integral
           distances. Solution to Problem 966, Math. Mag. **50** (1977)
           166–167. Problem 966 posed vol. **49** (1976) p. 43, one further
           comment appeared in vol. **59** (1986) p. 52.

[Er45]   P. ERDŐS: Integral distances, Bull. Amer. Math. Soc. **51**
           (1945) p. 996.

[Fr02]   J. FRICKE: On Heron-simplices and integer embedding,
           manuscript 2002, Universität Greifswald, Preprint 1/2002.

[Fu49]   R.E. FULLERTON: Integral distances in Banach spaces, Bull.
           Amer. Math. Soc. **55** (1949) 901–905.

[Gle81]   B. GLEIJESES: Points at mutual integral distances in S^n,
           Fibonacci Quart. **19** (1981) 153–160.

[GrRS74]   R.L. GRAHAM, B.L. ROTHSCHILD, E.G. STRAUS: Are there
           $n + 2$ points in E^n with odd integral distances? Amer. Math.
           Monthly **81** (1974) 21–25.

[Hae14]   E. HAENTZSCHEL: Die rationalen Vierecke des Inders Brah-
           magupta, Sitzungsber. Berliner Math. Ges. **14** (1914/15)
           23–31.

[Ha98]   H. HARBORTH: Integral distances in point sets, in: Karl
           der Grosse und sein Nachwirken. 1200 Jahre Kultur und Wis-
           senschaft in Europa. Band 2: Mathematisches Wissen,
           P.L. Butzer et al., eds., Brepols Verlag 1998, 213–224.

[Ha94]   H. HARBORTH: Match sticks in the plane, in: The Lighter
           Side of Mathematics, R.K. Guy et al., eds., MAA Spectrum,
           Washington D.C. 1994, 281–288.

[Ha71]   H. HARBORTH: Antwort auf eine Frage von P. Erdős nach
           fünf Punkten mit ganzzahligen Abständen, Elemente Math.
           26 (1971) 112–113.

[Ha70]   H. HARBORTH: On the problem of P. Erdős concerning points
           with integral distances, Annals New York Acad. Sci. **175**
           (1970) 206–207.

[HaK91]   H. HARBORTH, A. KEMNITZ: Fibonacci representations of
           graphs, in: Applications of Fibonacci numbers. Vol. 4,
           G.E. Bergum et al., eds., Kluwer (1991) 133–138.

[HaK90]   H. Harborth, A. Kemnitz: Integral representations of
           graphs, in: *Contemporary methods in graph theory. In honour
           of Prof. Dr. K. Wagner*, R. Bodendiek, ed., BI-Verlag 1990,
           359–367.

[HaK85]   H. Harborth, A. Kemnitz: Diameters of integral point
           sets, in: *Intuitive Geometry* (Siófok, 1985), K. Böröczky et
           al., eds., *Colloq. Math. Soc. János Bolyai* **48** (1987) 255–266.

[HaKM93]   H. Harborth, A. Kemnitz, M. Möller: An upper bound
           for the minimum diameter of integral point sets, *Discrete Com-
           put. Geom.* **9** (1993) 427–432.

[HaK*87]   H. Harborth, A. Kemnitz, M. Möller, A. Süssenbach:
           Ganzzahlige planare Darstellungen der platonischen Körper,
           Elemente Math. **42** (1987) 118–122.

[HaKR96]   H. Harborth, A. Kemnitz, N. Robbins: Non-existence
           of Fibonacci triangles, *Congressus Numerantium* **114** (1996)
           29–31.

[HaM98]   H. Harborth, M. Möller: Smallest integral combinato-
           rial box, in: *Applications of Fibonacci numbers. Volume 7*,
           G.E. Bergum et al., eds., Kluwer 1998, 153–159.

[HaM94]   H. Harborth, M. Möller: Minimum integral drawings of
           the Platonic graphs, *Math. Mag.* **67** (1994) 355–358.

[HaP94]   H. Harborth, L. Piepmeyer: On special integral Erdős
           point sets, in: *Intuitive Geometry* (Szeged, 1991), K. Böröczky
           et al., eds., *Colloq. Math. Soc. János Bolyai* **63** (1994) 139–149.

[HaP91]   H. Harborth, L. Piepmeyer: Points sets with small inte-
           gral distances, in: *Applied geometry and discrete mathemat-
           ics, Festschr. 65th Birthday Victor Klee, DIMACS Ser. Dis-
           cret. Math. Theor. Comput. Sci.* **4** (1991) 319–324.

[Has77]   H. Hasse: Ein Analogon zu den ganzzahligen pythagoreis-
           chen Dreiecken, *Elemente Math.* **32** (1977) 1–6.

[Ho48]   L. Holzer: Über Dreiecke mit ganzzahligen Koordinaten
           und ganzzahligen Seiten, *Elemente Math.* **3** (1948) 114–115.

[Hu48]   G.B. Huff: Diophantine problems in geometry and elliptic
           ternary equations, *Duke Math. J.* **15** (1948) 443–453.

[Joh15]   R.A. Johnson: Solution to Problem 195, *Amer. Math.
           Monthly* **22** (1915) 27–30.

[JoWW79]   J.H. Jordan, R. Walch, R.J. Wisner: Triangles with
           integer sides, *Amer. Math. Monthly* **86** (1979) 686–689.

[Ke94]   A. KEMNITZ: Rational triangles, in: *Intuitive Geometry* (Szeged, 1991), K. Böröczky et al., eds., *Colloq. Math. Soc. János Bolyai* **63** (1994) 181–187.

[Ke90a]   A. KEMNITZ: Rational quadrangles, *Congressus Numerantium* 76 (1990) 193–199.

[Ke90b]   A. KEMNITZ: Integral drawings of the complete graph K_6, in: *Topics in Combinatorics and Graph Theory. Essays in honour of Gerhard Ringel*, R. Bodendiek et al., eds., Physica-Verlag 1990, 421–429.

[Ke88]   A. KEMNITZ: *Punktmengen mit ganzzahligen Abständen*, Habilitationsschrift, TU Braunschweig 1988.

[KeH01]   A. KEMNITZ, H. HARBORTH: Plane integral drawings of planar graphs, *Discrete Math.* **236** (2001) 191–195.

[Kl79]   V. KLEE: Some unsolved problems in plane geometry, *Math. Mag.* **52** (1979) 131–145.

[Kum848]   E.E. KUMMER: Über die Vierecke, deren Seiten und Diagonalen rational sind, *J. Reine Angew. Math.* **37** (1848) 1–20.

[Kuz98]   A.V. KUZMINYKH: Sets with integral distances and a coloring of edges of an infinite graph, *Geombinatorics* **7** (1998) 132–138.

[Kuz82]   A.V. KUZMINYKH: Sets with almost integral distances, *Siberian Math. J.* **23** (1982) 527–530.

[Kuz80]   A.V. KUZMINYKH: Sets with nearly integral distances, *Sov. Math., Dokl.* **22** (1980) 593–595.

[La83]   J.L. LAGRANGE: Points du plan dont le distances mutuelles sont rationelles, *Séminaire de Théorie des Nombres de Bordeaux* 1982-1983, Exposé no. 27 (1983).

[La51]   R. LAUFFER: Rationale Dreiecke mit rationalen Koordinaten, *Elemente Math.* **6** (1951) p. 58.

[Le899]   D.N. LEHMER: Rational triangles, *Annals Math. Ser. 2* **1** (1899/1900) 97–102.

[Ma04]   H. MAEHARA: Distance graphs and rigidity, in: *Towards a Theory of Geometric Graphs*, J. Pach, ed., *Contemporary Mathematics* **342**, AMS 2004, 149–168.

[MaOT97]   H. MAEHARA, K. OTA, N. TOKUSHIGE: Every graph is an integral distance graph in the plane, *J. Comb. Theory, Ser. A* **80** (1997) 290–294.

[Mö90]   M. MÖLLER: *Ganzzahlige Darstellungen von Graphen in der Ebene*, Dissertation, TU Braunschweig 1990.

[**Mo60**] L.J. Mordell: Rational quadrilaterals, *J. London Math. Soc.* **35** (1960) 277–282.

[**Mü53**] A. Müller: Auf einem Kreis liegende Punktmengen ganzzahliger Entfernungen, *Elemente Math.* **8** (1953) 37–38.

[**NoB89**] L.C. Noll, D.I. Bell: n-clusters for $1 < n < 7$, *Math. Comp.* **53** (1989) 439–444.

[**Pe54**] W.D. Peeples Jr.: Elliptic curves and rational distance sets, *Proc. Amer. Math. Soc.* **5** (1954) 29–33.

[**PeJ95**] B.E. Peterson, J.H. Jordan: Integer hexahedra equivalent to perfect boxes, *Amer. Math. Monthly* **102** (1995) 41–45.

[**Pi96**] L. Piepmeyer: The maximum number of odd integral distances between points in the plane, *Discrete Comput. Geom.* **16** (1996) 113–115.

[**Ra87**] S. Rabinowitz: Problem 1261, *Math. Mag.* **60** (1987) p. 40.

[**Ro97**] M. Rosenfeld: In praise of the Gram matrix, in: *The mathematics of Paul Erdős.* Vol. 2, R.L. Graham et al., eds., Springer, *Algorithms Combinatorics Ser.* **14** (1997) 318–323.

[**Ro96**] M. Rosenfeld: Odd integral distances among points in the plane, *Geombinatorics* **5** (1996) 156–159.

[**RoY77**] D.P. Robbins, K.L. Yocom: Obtuse Pythagorean triplets. Solution to Problem E2566, *Amer. Math. Monthly* **84** (1977) 220–221.

[**Sh66**] T.K. Sheng: Rational polygons, *J. Australian Math. Soc.* **6** (1966) 452–459.

[**ShD66**] T.K. Sheng, D.E. Daykin: On approximating polygons by rational polygons, *Math. Mag.* **39** (1966) 299–300.

[**Si59**] W. Sierpiński: Sur les ensembles de points aux distances rationnelles situés sur un cercle, *Elemente Math.* **14** (1959) 25–27.

[**So03**] J. Solymosi: Note on integral distances, *Discrete Comput. Geom* **30** (2003) 337–342.

[**St63**] F. Steiger: Punkte mit ganzzahligen Abständen, *Elemente Math.* **18** (1963) p. 137.

[**St53**] F. Steiger: Zu einer Frage über Mengen von Punkten mit ganzzahliger Entfernung, *Elemente Math.* **8** (1953) 66–67.

[**Tr51**] E. Trost: Bemerkungen zu einem Satz über Mengen von Punkten mit ganzzahliger Entfernung, *Elemente Math.* **6** (1951) 59–60.

[Ul60]   S.M. ULAM: *A Collection of Mathematical Problems*, Inter-
            science 1960.

[Wa53]   E. WAAGE: Nahezu gleichseitige rationale und nahezu gleich-
            schenklige pythagoreische Dreiecke, *Elemente Math.* **8** (1953)
            111–113.

[vWi50]   A. VAN WIJNGARDEN: A table of partitions into two squares
            with an application to rational triangles, *Indagationes Math.*
            12 (1950) 313–325.

6. Problems on Repeated Subconfigurations

6.1 Repeated Simplices and Other Patterns

The set of unit-distance pairs determined by an n-element point set X can be regarded as an equivalence class of all point pairs under *congruence* as the equivalence relation. The basic questions discussed in Chapter 5 are to determine the size of the largest equivalence class and the number of distinct equivalence classes. Erdős and Purdy [ErP71], [ErP76] started the investigation of the same questions for k-dimensional simplices in $\mathbb{R}^d$, that is, for $(k + 1)$-tuples rather than point pairs. (In [ErP71] some of these problems are attributed to A. Oppenheim.) Let $u_{k,d}(n)$ denote the maximum number of mutually congruent k-dimensional simplices determined by n points in d-dimensional Euclidean space. Using the notation in the previous chapter, $u_{1,d}(n) = u_d(n)$ is the maximum number of unit distances determined by n points in $\mathbb{R}^d$.

For $k = 2$ and $d = 2$, for example, we would like to estimate $u_{2,2}(n)$, the maximum number of mutually congruent triangles in an n-element planar set. Fixing one side of a triangle T in a point set X, and marking the corresponding edge in each triangle congruent to T, we see that there are at most four triangles induced by X that share the same marked edge. Thus, $u_{2,2}(n) \leq 4u_2(n) = 4u(n) \leq O(n^{\frac{4}{3}})$. Any improvement of the upper bound for the number of unit distances in the plane will carry over to a bound for the number of congruent triangles. If Erdős's conjecture is true and the order of magnitude of $u_2(n)$ is attained for a section of the triangular lattice (see Section 5.1), then we have $u_{2,2}(n) = \Omega(u(n))$. Indeed, the rotational symmetry of this lattice guarantees that a positive fraction of all unit distance pairs can be extended to equilateral triangles. However, if Erdős's conjecture is wrong, it is possible that $u(n)$ is substantially larger than $u_{2,2}(n)$. Whatever the truth is, we have

$$ne^{\Omega\left(\frac{\log n}{\log \log n}\right)} \leq u_{2,2}(n) \leq O(n^{\frac{4}{3}}).$$

The lower bound remains true not only for equilateral triangles, but for any other congruence class of triangles. We sketch the proof of the lower bound. Fix a triangle T whose vertices are represented by complex numbers $0, 1, z$. Consider the set X of all points of the form $a + bi + (c + di)z$, where a, b, c, d are nonnegative integers not exceeding m. Let $n = |X|$ and pick a number r^2 that can be written as the sum of two squares $a^2 + b^2$ $(a, b \leq \frac{m}{|z|})$ in at least $e^{c\frac{\log n}{\log \log n}}$ different ways. For each of these choices, the triple $0, a + bi, (a + bi)z$ belongs to X and spans a triangle congruent to rT. Clearly, the set X looks essentially the same from each of its points. Thus, scaling down X by a factor of r, we obtain an n-element set containing at least $ne^{c\frac{\log n}{\log\log n}}$ congruent copies of T, as required. Note that if z is "generic," then $n = m^4$ and the constant c appearing in the exponent is roughly half of what we get for equilateral triangles.

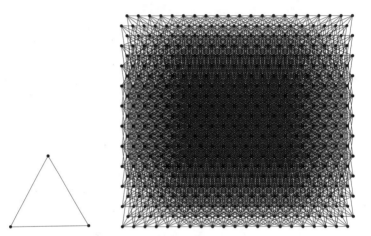

A UNIT EQUILATERAL TRIANGLE AND A LATTICE SECTION
CONTAINING MANY CONGRUENT COPIES OF THE TRIANGLE

For a given triangle T, let $u(T, n)$ denote the maximum number of triples induced by a set of n points in $\mathbb{R}^2$ that determine a triangle congruent to T. Clearly, we have $u_{2,2}(n) = \max_T u(T, n)$. One is tempted to make the following conjecture.

Conjecture 1 *(Pach) Let T_0 be an equilateral triangle. There exists a triangle T such that $\lim_{n\to\infty}\frac{u(T,n)}{u(T_0,n)} = 0$.*

The next interesting question occurs for $k = 2$ and $d = 3$. What is the maximum number of congruent triangles among n points in three-dimensional space? This question of Erdős and Purdy was treated in [AkTT98], [Br00], [AgS02], and [ÁbF02]. The current best upper bound, $u_{2,3}(n) \leq O(n^{\frac{5}{3}}\beta(n))$, is due to Agarwal and Sharir [AgS02]. Here $\beta(n)$ is an extremely slowly growing function related to the inverse Ackermann function $(\beta(n) = 2^{O(\alpha^2(n))}$, where α is the inverse Ackermann function). The best known lower bound, $u_{2,3}(n) \geq \Omega(n^{\frac{4}{3}})$, is based on a construction from [ErHP89] (see also [ClE*90]). Erdős et al. first lifted a planar config-uration of n points and n lines with $\Theta(n^{\frac{4}{3}})$ incidences to a configuration of points and great circles on the sphere. After replacing each point by the line connecting it with the center of the sphere, and replacing each great circle by its axis, they obtained two n-element sets of "directions" (i.e., lines through the origin), $\mathcal{L}_1$ and $\mathcal{L}_2$, with $\Theta(n^{\frac{4}{3}})$ orthogonal pairs (ℓ_1, ℓ_2),

$\ell_1 \in \mathcal{L}_1, \ell_2 \in \mathcal{L}_2$. Ábrego and Fernández-Merchant [ÁbF02] observed that these sets of directions can be used to construct many congruent copies of any given triangle ABC, as follows. Let AB be a side of maximum length, let F be the foot of the altitude from C to AB, and let r_A, r_B, r_C denote the distances between F and the vertices of the triangle. Along each line $\ell_1 \in \mathcal{L}_1$, choose two points at distances r_A and r_B from the center of the sphere, in opposite directions. Along each line $\ell_2 \in \mathcal{L}_2$, select a point at distance r_C from the center. If $\ell_1 \in \mathcal{L}_1$, $\ell_2 \in \mathcal{L}_2$ are orthogonal, then the points selected on them generate a congruent copy of the triangle ABC, so that its point corresponding to F is mapped to the center. In this way, we have obtained at most $3n$ points in $\mathbb{R}^3$ with $\Omega(n^{\frac{4}{3}})$ triples that span congruent copies of a prescribed triangle ABC. We can ensure that $\mathcal{L}_1$ and $\mathcal{L}_2$ are disjoint and that the selected points are in convex position.

Problem 2   *Is it true that the maximum number of mutually congru-ent triangles determined by n points in three-dimensional space is $O(n^{\frac{4}{3}})$?*

The problem of determining the maximum number of mutually congruent nondegenerate tetrahedra in three-dimensional space ($k = d = 3$) also seems to be quite interesting. The upper bound for the number of triangles carries over to this case, since each triangle can be extended to a congruent tetrahedron in at most six different ways. Thus, we have $u_{3,3}(n) \leq O(u_{2,3}(n))$. However, the lower bound construction does not seem to generalize because we cannot construct a set of n directions with many orthogonal triples. It is not even clear how many mutually congruent simplices can be found in a suitable lattice section. The iterated Minkowski sum construction (see Section 5.1) shows that $u_{3,3}(n) \geq \Omega(n \log n)$. One can do slightly better by taking a lattice section in the plane with many unit equilateral triangles and a section of the same lattice scaled by a factor of $\frac{1}{3}$ (so that it contains the centroids of all triangles determined by the original lattice section) in a parallel plane at the correct distance above the first lattice. This gives $\Omega\left(ne^{c\frac{\log n}{\log \log n}}\right) \leq u_{3,3}(n)$, which is the best known lower bound.

Problem 3   *(Erdős and Purdy [ErP76]) At most how many mutually congruent full-dimensional tetrahedra can be determined by n points in three-dimensional space?*

Problem 4   *(Brass) At most how many mutually congruent full-dimen-sional simplices can be selected from a three-dimensional lattice section of n points?*

Problem 5   *(Brass) At most how many orthogonal bases can be se-lected from n unit vectors in three-dimensional space?*

In dimensions $d \geq 4$, we have a Lenz-type construction (see Section 5.2). Arrange n points as evenly as possible on $\frac{d}{2}$ concentric circles lying in mutually orthogonal planes if d is even, and on $\frac{d-3}{2}$ such circles and a sphere in a three-dimensional subspace orthogonal to all of them if d is odd. These sets are known to contain asymptotically the largest possible number of mutually congruent one-dimensional simplices (unit distances). Moreover, it is easy to see that they are also optimal in the sense that the order of magnitude of the number of mutually congruent k-dimensional simplices generated by them is as large as possible for any $k < \left\lfloor \frac{d}{2} \right\rfloor$ [Er75]. Indeed, in these cases the order of magnitude of the trivial upper bound $u_{k,d}(n) \leq \binom{n}{k+1}$ is attained for the above construction. Erdős suspected that for any $k \geq \left\lfloor \frac{d}{2} \right\rfloor$, the order of magnitude of $u_{k,d}(n)$ is reached for a Lenz-type construction.

Conjecture 6   *(Erdős)* $u_{k,d}(n) = \Theta(n^{\frac{d}{2}})$ if $k \geq \frac{d}{2}$ and d is even,

and $u_{k,d}(n) = \Theta(n^{\frac{d}{2} - \frac{1}{6}})$ if $k \geq \frac{d-1}{2}$ and d is odd.

Moreover, the extremal sets can be covered by $\left\lfloor \frac{d}{2} \right\rfloor$ mutually orthogonal two- or three-dimensional subspaces.

Agarwal and Sharir [AgS02] obtained several bounds for $\frac{d}{2} < k \leq d-2$ and $d \leq 7$. In particular, they proved that in these cases we have $u_{k,d}(n) \leq O(n^{\frac{d}{2} + \varepsilon})$ for every $\varepsilon > 0$.

As for the problems on repeated distances discussed in Chapter 5, we have a "dual" set of questions. What is the minimum number of distinct equivalence classes of k-dimensional simplices determined by n points in d-dimensional space, under congruence as the equivalence relation? Very little seems to be known about this problem stated in [ErP76]. We denote this number by $v_{k,d}(n)$. The function $v_{1,d}(n) = v_d(n)$, the minimum number of distinct distances among n points in d-dimensional space, was discussed in detail in Section 5.4. If the (multi)sets of distances determined by two simplices do not coincide, they cannot be congruent. Therefore, up to a multiplicative constant, the number of distinct distances is a lower bound for the number of congruence classes of k-simplices. In general, the functions $v_{k,d}(n)$ are increasing in k. We obtain a better lower bound for $v_{k,d}(n)$ using the pigeonhole principle:

$$v_{k,d}(n) \geq \frac{\binom{n}{k+1}}{u_{k,d}(n)}.$$

For triangles in the plane, this bound can be improved by counting classes of congruent triangles with one marked edge, so that each congruence class is counted at most three times. Then, for a fixed edge ab taken as a marked edge, there are at most four distinct points c_1, c_2, c_3, c_4 in the set such that the triangles abc_i are congruent. Thus, every marked edge ab generates at least $\frac{1}{4}(n-2)$ distinct equivalence classes under the congruence relation on the set of triangles with one marked edge. Two marked edges of different lengths generate distinct equivalence classes, so the number of congruence classes of edge-marked triangles is at least $\frac{1}{4}(n-2)v_2(n) = \frac{1}{4}(n-2)v(n)$. Hence, we have $v_{2,2}(n) \geq \Omega(nv(n))$. The true value of $v_{2,2}(n)$ is most likely quadratic.

Problem 7   *(Brass) Is it true that the number of distinct congruence classes of triangles determined by a set of n points in the plane attains its minimum for a regular n-gon?*

In higher dimensions, the above-mentioned lower bound on $v_{k,d}(n)$ that follows from the pigeonhole principle is often too weak. In the case $k < \left\lfloor \frac{d}{2} \right\rfloor$, where the Lenz construction shows that a positive fraction of all simplices may be congruent, the bound is useless because it becomes a constant (while $v_{k,d}(n) \geq v_{1,d}(n) \to \infty$ as $n \to \infty$). For $k \geq \left\lfloor \frac{d}{2} \right\rfloor$ it becomes interesting again.

Problem 8   *(Erdős and Purdy) What is the minimum number $v_{k,d}(n)$ of distinct congruence classes of k-dimensional simplices determined by n points in d-dimensional Euclidean space?*

One not entirely obvious difficulty here is that the answer to the above question depends on whether we consider only nondegenerate simplices. If we also count congruence classes of lower-dimensional subsets, then the Lenz construction will not be optimal. Indeed, it has $\frac{2n}{d}$ points lying on a two-dimensional circle, and these points already determine $\Omega\left(\left(\frac{2n}{d}\right)^k\right)$ distinct congruence classes of k-element subsets. However, if we consider only subsets in affinely independent position (proper simplices), then they can have at most two points per circle (apart from one circle where we have three), which gives for d even and $k = d + 1$ only $\Omega(n^{\frac{d}{2}+1})$ distinct congruence classes.

We can formulate many other interesting problems, asking for the maximum number of copies of a given point set ("pattern") that can occur among n points in d-dimensional space, where we allow copies by translations, homotheties, similarities, or affine maps. These problems are also interesting algorithmically, where they can be interpreted as problems of geometric pattern matching in the exact point pattern matching model [Br02], [BrP05]. How do we find all occurrences of a given point pattern in a given background, and what is the largest number of such occurrences?

The case of translations is simple, and we can even give a good bound for the lower-order term. Define the *rational affine dimension* of a pattern (point set) A as the dimension of the affine subspace generated by A over the rationals. This is the same as the minimum number of vectors belonging to the difference set $A + (-A) = \{a_1 - a_2 \mid a_1, a_2 \in A\}$ such that any other element of the set can be obtained as their linear combination with rational coefficients. Obviously, the rational affine dimension of a pattern is at least as large as its affine dimension over the reals, but it can be larger. The rational dimension of $A \subset \mathbb{R}^d$ is the smallest dimension of a lattice whose projection to $\mathbb{R}^d$ contains a translate of A. The maximum number of translates of a given set A that can be contained in an n-element set in $\mathbb{R}^d$ is $n - \Theta(n^{1-\frac{1}{k}})$, where k is the rational dimension of A [Br02]. Hence, for translates the relevant dimension is not the real dimension of the embedding space but the rational dimension of the pattern.

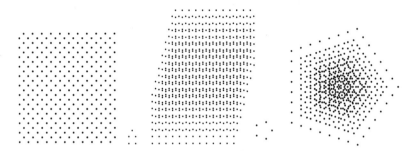

FIVE-POINT PATTERNS OF DIFFERENT RATIONAL DIMENSIONS
AND SETS WITH MANY TRANSLATES OF THESE PATTERNS

The minimum number of distinct translation equivalence classes of simplices determined by a set of n points in d-dimensional space can be estimated from below by the pigeonhole principle. Since each equivalence class has at most roughly n elements, the number of equivalence classes is at least $\Omega(n^d)$. This is asymptotically best possible, as shown by any lattice section.

The maximum number of homothetic copies of a fixed pattern was first treated by van Kreveld and de Berg [vKrB89] for the special case of hypercubes, and then solved for arbitrary sets by Elekes and Erdős [ElE94] (see also [Br02]). They proved that the maximum number of homothetic copies of a given full-dimensional set A that can occur as subsets of a set X of n points in d-dimensional space is $O(n^{1+\frac{1}{d}})$. This bound is asymptotically tight if all coordinates of the elements of A with respect to some basis are algebraic numbers. Otherwise, we have the slightly smaller lower bound of $\Omega(n^{1+\frac{1}{d}-b(\log n)^{-a}})$, for some $a, b > 0$ depending on the pattern A. Here the assumption that A is full-dimensional is important. The maximum decreases with the dimension of A, and X may have $\Theta(n^2)$ homothetic copies of A if we allow one-dimensional patterns. The above upper bound on the size of the homothety equivalence classes, combined with the pigeonhole principle, again yields a lower bound on the number of distinct homothety equivalence classes. The order of magnitude of this bound can probably be improved.

The case of similar copies is essentially solved in the plane, but is completely open in higher dimensions. In the plane, an n-element set can have at most $O(n^2)$ similar copies of a pattern (e.g., similar triangles), since a similarity is determined up to orientation by the image of a pair of points. This asymptotic bound is attained in a section of a triangular lattice. We lose some similar copies along the boundary, but a positive fraction of all pairs of points can be extended to an equilateral triangle. In fact, one can obtain a cn^2 lower bound with a fairly good constant c [ÁbF00]. We get similar results for any pattern A of a fixed size that is a subset of the triangular lattice. For general patterns, however, the situation is more complicated. The maximum number of similar copies of A is still quadratic or almost quadratic, as shown by a construction of Elekes and Erdős [ElE94]. In the planar case, one has a big technical advantage: the points of the plane can be interpreted as complex numbers. Given a triangle, whose vertices are represented by the complex numbers $0, 1, z$, consider all points of the form $\frac{i_1}{n}z$, $\frac{i_2}{n} + (1 - \frac{i_2}{n})z$, and $\frac{i_3}{n}z + (1 - \frac{i_3}{n})z^2$, where $0 < i_1, i_2, i_3 \leq \frac{n}{3}$. Any triangle $(\beta - \alpha)z$, $\alpha + (1-\alpha)z$, $\beta z + (1-\beta)z^2$ is similar to $0, 1, z$, which can be checked by computing the ratios of the sides. Thus, choosing $\alpha = \frac{i_2}{n}$, $\beta = \frac{i_3}{n}$, we obtain a quadratic number of similar copies of the triangle $0, 1, z$.

The *cross ratio* of an ordered four-tuple of points (a, b, c, d) in the complex plane is defined as

$$\frac{a - c}{b - c} : \frac{a - d}{b - d}.$$

Laczkovich and Ruzsa [LaR97] proved that the maximum number of similar copies of a given planar pattern among n points is quadratic if and only if the cross ratio of every four points is an algebraic number. Otherwise, the maximum is very slightly subquadratic [ElE94].

This fails to generalize to higher dimensions. Akutsu, Tamaki, and Tokuyama [AkTT98] gave an upper bound of $O(n^{2.2})$ for the maximum number of mutually similar full-dimensional tetrahedra induced by n points in three-dimensional space, but the true order of magnitude of this function is probably subquadratic. Thus, as in the case of homothetic copies, the maximum is probably smaller in three-dimensional space than in the plane. The only lower bound we have is $\Omega(n^{\frac{4}{3}})$, which can already be reached by homothetic copies without any use of rotations.

Conjecture 9 *[Br02] The maximum number of mutually similar full-dimensional tetrahedra generated by n points in $\mathbb{R}^3$ is $o(n^2)$.*

In higher dimensions, we are not aware of any nontrivial results concerning the number of similar copies. The only general lower bounds follow from the Lenz construction, just as in the case of congruent copies. Almost nothing is known about the number of distinct equivalence classes of simplices under similarity.

Problem 10   *(Erdős and Purdy) What is the maximum number of mutually similar k-dimensional simplices among n points in d-dimensional space?*

Problem 11   *(Erdős and Purdy) What is the minimum number of distinct similarity equivalence classes of k-dimensional simplices among n points in d-dimensional space?*

For affine copies, the situation becomes simpler. A set of n points in d-dimensional space contains at most $\binom{n}{d+1}$ affine copies of any fixed pattern. This bound cannot be improved: the $n^{\frac{1}{d}} \times \cdots \times n^{\frac{1}{d}}$ integer lattice contains $\Omega(n^{d+1})$ copies of any fixed pattern set that is itself a subset of a (smaller) lattice cube. However, it seems likely that the $\Omega(n^{d+1})$ bound cannot be attained for every pattern A. The true order of magnitude again may depend on some algebraic properties of A.

All the above questions can be raised under the assumption that the underlying set is in general or in convex position. In the case of triangles in a set of points in convex position in the plane, we obtain a nice variant of the problem on the maximum number of unit distances among the vertices of a convex n-gon. This question was answered by Pach and Pinchasi [PaP03] for equilateral triangles, and by Xu and Ding [XuD04] for isosceles right-angled triangles. In three-dimensional space, the construction in [ÁbF02] described after Conjecture 1 is already in convex position, so convexity is not a serious restriction, and the same holds for the Lenz construction in higher dimensions.

An interesting variant, at least in the plane, might be the question concerning the maximum number of empty congruent triangles. This problem was raised in an algorithmic context ("window matching") by Brass [Br02]. Although the number of congruent triangles among n points in the plane can certainly be superlinear, and the number of empty triangles can even be cubic (in a convex polygon all triangles are empty), it seems probable that the maximum number of empty congruent triangles is only linear.

Conjecture 12 *(Brass [Br02]) The largest number of mutually congruent empty triangles determined by n points in the plane is $O(n)$.*

If instead of empty congruent triangles we count empty triangles that are mutually similar, we have to assume that these triangles are nondegenerate. The lattice section construction contains $\Omega(n^2)$ empty similar copies of the degenerate triangle consisting of a segment and its midpoint.

To state the next set of related problems, we need some definitions. Given an underlying set X, we say that a pattern A is *repeated* under translation if X contains at least two translates of A. That is, there exist t_1, t_2 such that $A + t_i \subset X$, $i = 1, 2$. A repeated pattern A is *maximally repeated* under translation if there exists no proper superset $B \supset A$ whose translates appear at exactly the same places in X as the translates of A. More precisely, there is no $B \supset A$ such that $B + t \subset X$ whenever $A + t \subset X$. (Instead of translations, one can also consider any other group of transformations.)

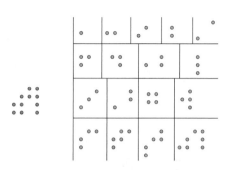

A SET AND ALL ITS MAXIMALLY
REPEATED SUBPATTERNS
UNDER TRANSLATION

These concepts were introduced by Brass [Br02]. They stem from some algorithmic applications, but seem to be of independent interest. We want to estimate the largest number of distinct maximally repeated patterns that may occur in a set of n points in the plane under translation, or in any other space with respect to any other group of geometric transformations. Two repeated patterns are considered *distinct* if they do not belong to the same equivalence class under the given group of transformations.

Conjecture 13 *(Brass [Br02]) The maximum number of distinct maximally repeated patterns under congruence, determined by n points in d-dimensional space, is $O(n^d)$.*

For translation, this problem reduces by projection to the one-dimensional case.

Problem 14   *(Brass [Br02]) Is it true that n real numbers determine at most $O(n)$ distinct maximally repeated patterns under translation?*

All of these problems are wide open. We have only a trivial $\Omega(n)$ lower bound, and an equally trivial exponential upper bound. The true order of magnitude is probably a low-degree polynomial for any usual geometric equivalence relation.

[AgS02]   P.K. Agarwal, M. Sharir: On the number of congruent simplices in a point set, *Discrete Comput. Geom.* **28** (2002) 123–150.

[AkTT98]   T. Akutsu, H. Tamaki, T. Tokuyama: Distribution of distances and triangles in a point set and algorithms for computing the largest common point sets, *Discrete Comput. Geom.* **20** (1998) 307–331.

[ÁbF02]   B.M. Ábrego, S. Fernández-Merchant: Convex polyhedra in R^3 spanning $\Omega(n^{4/3})$ congruent triangles, *J. Combinatorial Theory Ser. A* **98** (2002) 406–409.

[ÁbF00]   B.M. Ábrego, S. Fernández-Merchant: On the maximum number of equilateral triangles I, *Discrete Comput. Geom.* **23** (2000) 129–135.

[Br02]   P. Brass: Combinatorial geometry problems in pattern recognition, *Discrete Comput. Geom.* **28** (2002) 495–510.

[Br00]   P. Brass: Exact point pattern matching and the number of congruent triangles in a three-dimensional pointset, *Algorithms – ESA 2000*, M. Paterson, ed., Springer *LNCS* **1879** (2000) 112–119.

[BrP05]   P. Brass, J. Pach: Problems and results on geometric patterns, in: *Graph Theory and Combinatorial Optimization*, D. Avis et al., eds., Kluwer Academic Publishers, to appear.

[ClE*90]   K.L. Clarkson, H. Edelsbrunner, L.J. Guibas, M. Sharir, E. Welzl: Combinatorial complexity bounds for arrangements of curves and spheres, *Discrete Comput. Geom.* **5** (1990) 99–160.

[ElE94]   G. Elekes, P. Erdős: Similar configurations and pseudo grids, in: *Intuitive Geometry* (Szeged, 1991), K. Böröczky et. al., eds., *Colloq. Math. Soc. János Bolyai* **63** (1994) 85–104.

[Er75]   P. Erdős: On some problems of elementary and combinatorial geometry, *Ann. Mat. Pura Appl. Ser. IV* **103** (1975) 99–108.

[ErHP89]   P. Erdős, D. Hickerson, J. Pach: A problem of Leo Moser about repeated distances on the sphere, *Amer. Math. Monthly* **96** (1989) 569–575.

[ErP76]   P. Erdős, G. Purdy: Some extremal problems in geometry IV, *Congressus Numerantium* **17** (Proc. 7th South-Eastern Conf. Combinatorics, Graph Theory, and Computing, 1976) 307–322.

[ErP71] P. Erdős, G. Purdy: Some extremal problems in geometry, *J. Combinatorial Theory Ser. A* **10** (1971) 246–252.

[vKrB89] M.J. van Kreveld, M. de Berg: Finding squares and rectangles in sets of points, in: *WG 1989* (Graph-Theoretic Concepts in Comp. Sci.), M. Nagl, ed., Springer-Verlag *LNCS* **411** (1989) 341–355.

[LaR97] M. Laczkovich, I.Z. Ruzsa: The number of homothetic subsets, in: *The Mathematics of Paul Erdős, Vol. II*, R.L. Graham et al., eds., Springer-Verlag 1997 *Algorithms and Combinatorics Ser.* **14** 294–302.

[PaP03] J. Pach, R. Pinchasi: How many unit equilateral triangles can be generated by n points in convex position? *Amer. Math. Monthly* **110** (2003) 400–406.

[XuD04] C. Xu, R. Ding: The number of isosceles right triangles determined by n points in convex position in the plane, *Discrete Comput. Geom.* **31** (2004) 491–499.

6.2 Repeated Directions, Angles, Areas

For any equivalence relation defined on k-tuples of points, one can ask the following two natural questions:

(1) What is the maximum number of mutually equivalent k-tuples that can occur among n points?

(2) What is the minimum number of distinct equivalence classes?

In the previous section, we discussed these questions in the special case in which two k-tuples are considered equivalent if one can be carried to the other by translation, by rigid motion (congruence), by similarity, or by some other geometric transformation. In this section, we consider some variants where two equivalent k-tuples determine the same direction ($k = 2$), the same angle ($k = 3$), or simplices of the same volume. Most of these questions were first raised by Erdős and Purdy [ErP71], [ErP75], [ErP76], [ErP77].

Perhaps the oldest problem of this kind was studied by Scott [Sc70]. He conjectured that the smallest number of distinct directions determined by n points in the plane, not all on the same line, is n or $n-1$ depending on whether n is even or odd. Equality is attained for the regular n-gon and for the regular $(n-1)$-gon together with its center, respectively. (Two point pairs determine the same *direction* if their connecting lines are parallel.) After some preliminary results [BuP79], this conjecture was settled by Ungar [Un82], whose ingenious argument was based on allowable sequences.

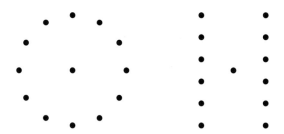

n POINTS, NOT ALL COLLINEAR, DETERMINING ONLY n DIRECTIONS

Surprisingly, there is a multitude of extremal configurations. For instance, for even n, two congruent sets consisting of $\frac{n}{2}$ equidistant points lying on two parallel lines determine the same number of directions. For odd n, we can add the center of symmetry of the above configuration. Two other infinite families of extremal examples and about one hundred sporadic configurations have been found by Jamison and Hill [JaH83], [Ja84a], [Ja84b], [Ja85], [Ja86a]. Two closed segments are called "convergent" if their supporting lines are parallel or meet at a point not belonging to either of the segments. Pach, Pinchasi, and Sharir [PaPS04a] generalized Ungar's theorem by proving that from all closed segments spanned by n points on a plane, not all on a line, one can always select n or $n-1$ elements, depending on whether n is even or odd, such that no two of them are convergent. There are also some structural results on the distribution of directions determined by n noncollinear points. Jamison [Ja86b], [Ja87], [Ja04] proved that one can select a spanning tree with straight-line edges, all of whose edges have different slopes. Moreover, Kleitman and Pinchasi [KlP05] showed that there also exists a straight-line Hamilton path with this property, provided that no three points are collinear.

In three-dimensional space, Scott's problem has been recently solved by Pach, Pinchasi, and Sharir [PaPS04b]. They showed that n points in three-dimensional space, not all in a plane, determine at least $2n-7$ directions if n is even and at least $2n-5$ directions if n is odd. For odd n, this bound is attained for any centrally symmetric double pyramid whose base is a centrally symmetric planar configuration of $n-2$ points that determine $n-3$ distinct directions. The special case of no three points collinear was settled in [PaPS04a], following some weaker results of Blokhuis and Seress [BlS02].

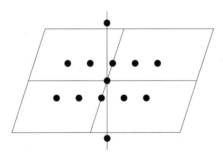

n POINTS IN THREE-DIMENSIONAL SPACE, NOT ALL COPLANAR,
WITH MINIMUM NUMBER OF DIRECTIONS

Conjecture 1 *(Blokhuis–Seress [BlS02]) Any set of n points in $\mathbb{R}^3$, not all coplanar and no three collinear, determines a direction assumed by at most two pairs of points.*

Very little is known about the analogous problems in higher dimensions.

Conjecture 2 *(Jamison [Ja85], Blokhuis–Seress [BlS02]) For every $d \geq 4$, any set of n points in $\mathbb{R}^d$, not all in a hyperplane and no three collinear, determines at least $(d-1)n - d(d-2)$ distinct directions.*

This conjecture, if true, is sharp, as shown by several higher-dimensional pyramids constructed in [Ja85] (Theorem 6.2). It seems likely that for any fixed d, up to an additive constant depending on d, the same upper bound remains valid even if we drop the condition that no three points are collinear. The case $d = 4$ was addressed in [PaPS04b].

The dual problem of finding the maximum number of point pairs that determine the same direction is trivial. If we assume that no three points are on a line, the regular n-gon is clearly optimal in the plane, and can be augmented to (almost) optimal higher-dimensional configurations by adding $d - 2$ other points. If collinear triples are not excluded, we can take $n - d + 1$ points on a line and add $d - 1$ other points in general position to obtain an extremal configuration.

One can also define "higher-dimensional directions" determined by a set of points $X \subset \mathbb{R}^d$. For any $k \geq 2$, two k-dimensional affine subspaces generated by subsets of X determine the same *k-dimensional direction* if they are parallel. Pach, Pinchasi, and Sharir (personal communication) proved that any set of n points in $\mathbb{R}^3$, not all on a plane, determine at least $\frac{n}{2}$ two-dimensional directions. They conjecture that the minimum number of two-dimensional directions is at least $n - 2$.

Conjecture 3   *What is the minimum number of k-dimensional directions determined by n points in d-dimensional space, not all in the same hyperplane?*

The k-dimensional directions determined by a set X are, in fact, the same as the equivalence classes of all k-dimensional affine subspaces induced by X under translation. Instead of affine subspaces, one can consider other types of spanned objects. For instance, two circles induced by points of X can be regarded as equivalent if their radii are the same. It is an old problem of Erdős to determine the maximum size of an equivalence class in this case.

Problem 4   *(Erdős) What is the maximum number of distinct unit circles spanned by a set of n points in the plane?*

Elekes [El84] gave an example of n points in the plane that determine at least $\Omega(n^{\frac{3}{2}})$ distinct unit circles. Let $v_1, \ldots, v_m$ be distinct unit vectors. For any $1 \leq i \leq j \leq k \leq m$, the three points $v_i + v_j$, $v_i + v_k$, $v_j + v_k$ lie on a unit circle centered at $v_i + v_j + v_k$. Therefore, the set of pairwise sums $\{v_i + v_j \mid 1 \leq i < j \leq m\}$ consists of $\binom{m}{2}$ points and determines at least $\binom{m}{3}$ unit circles. This construction is certainly not extremal. For instance, one can add the point 0 to create further unit circles centered at $v_1, \ldots, v_m$. Moreover, by the proper choice of the generating vectors $v_1 \ldots, v_m$, one can create even more unit circles, but this will not affect the asymptotic behavior of the lower bound. The only upper estimate we have on the number of unit circles is the trivial bound $\frac{1}{3}n^2 - O(n)$. Any subquadratic upper bound or any improvement of the lower bound would be interesting. Exact values were determined for $n \leq 8$ by Harborth and Mengersen [Ha85], [HaM86]. They suggest that the true value should be closer to the upper than to the lower bound.

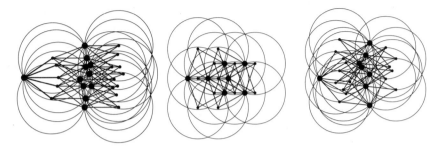

SETS SPANNING MANY UNIT CIRCLES: ELEKES'S CONSTRUCTION,
AND THE EXTREMAL SETS FOR 7 AND 8 POINTS

The corresponding question about the minimum number of distinct equivalence classes is the following.

Problem 5 *(Balog) What is the minimum number of distinct radii of circles spanned by n points in the plane, not all on a circle or a line?*

The set consisting of the vertices and the centroid of a regular $(n-1)$-gon shows that this number is $O(n)$. It is probably optimal or nearly optimal for every n. Elekes [El03] proved that the number of distinct radii is $\Omega(n)$.

The two analogous questions concerning the distribution of angles also appear to be interesting. What is the maximum number of ordered triples of n points that determine the same angle? At least how many distinct angles must occur? Pach and Sharir [PaS92] proved that for any $\gamma \in (0, \pi)$, there are at most $O(n^2 \log n)$ triples among n points in the plane that determine angle γ. Moreover, this order of magnitude is attained for a dense set of angles [PaS92]. We do not know whether this order can indeed be reached for every γ.

Problem 6 *(Pach–Sharir) Is it true that for every $\gamma \in (0, \pi)$ the maximum number of triples among n points in the plane that span angle γ is $\Omega(n^2 \log n)$?*

In three-dimensional space, Apfelbaum and Sharir [ApS05] showed that the same angle can occur among n points at most $O(n^{\frac{7}{3}})$ times and that for right angles this bound can be attained. (See also [CoC*79] for a weaker upper estimate.)

Problem 7   *(Apfelbaum–Sharir) Does there exist an angle $0 < \gamma < \pi$ different from $\frac{\pi}{2}$ that can occur among n points in three-dimensional space $\Omega(n^{\frac{7}{3}})$ times?*

Purdy [Pu88] noticed that in four-dimensional space the right angle can occur $\Theta(n^3)$ times, since the points $x_\varphi = (\cos\varphi, \sin\varphi, 0, 0)$, $y_s = (1, 0, s, 0)$, and $z_t = (-1, 0, 0, t)$ always determine a right angle at x_φ. For all other angles, there is an upper bound of $O\big(n^{\frac{5}{2}}\beta(n)\big)$ [ApS05]. Here $\beta(n)$ is an extremely slowly growing function related to the inverse Ackermann function $(\beta(n) = 2^{O(\alpha^2(n))})$, where α is the inverse Ackermann function). However, the best known lower bound for angles different from $\frac{\pi}{2}$ is the same as in the plane: $\Omega(n^2)$ and $\Omega(n^2 \log n)$ for some special values.

Problem 8   *(Purdy [Pu88]) At most how many times can the same angle $0 < \gamma < \pi$ different from $\frac{\pi}{2}$ occur among n points in four-dimensional space?*

In spaces of dimensions six and higher, any given angle can be represented by $\Theta(n^3)$ triples taken from an n-element set. This follows from the fact that in the Lenz construction (see Section 5.1) the number of mutually congruent triangles with an angle γ can be $\Omega(n^3)$. The analogous statement in five-dimensional space is not known to be true.

Problem 9   *Can every angle $0 < \gamma < \pi$ different from $\frac{\pi}{2}$ occur $\Omega(n^3)$ times among n points in five-dimensional space?*

The corresponding question about the minimum number of distinct equivalence classes was raised by Corrádi, Erdős, and Hajnal.

Conjecture 10 *(Corrádi–Erdős–Hajnal) Given n points in the plane, not all on a line, they always determine at least $n - 2$ distinct angles in $[0, \pi)$.*

It seems that this conjecture mentioned in [ErP95] has not been seriously studied up to now. The number of distinct angles determined by a regular n-gon is precisely $n - 2$, but there are several other configurations for which the conjectured lower bound is tight. It easily follows from the "weak Dirac conjecture" [Be83] that there is a constant $c > 0$ such that any noncollinear set of n points in the plane determines at least cn distinct angles.

Nothing is known about the higher-dimensional analogue of this question.

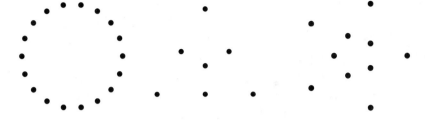

SETS OF n POINTS WITH $n - 2$ DISTINCT ANGLES

Problem 11   What is the minimum number of distinct angles determined by n points in $\mathbb{R}^d$, not all on a line?

There are many further ways to define equivalence classes of triangles or k-tuples of points selected from an n-element point set in the plane, and in each case one can pose similar questions. Several problems of this type that can be regarded as extensions or generalizations of the problem of unit distances have been proposed by Oppenheim (see [ErP95]), and Erdős and Purdy [ErP71], [ErP75], [ErP76], [ErP77].

Problem 12   (A. Oppenheim, 1967) What is the maximum number of triangles of unit area that can be determined by n points in the plane?

This function is bounded from above by $O(n^{\frac{7}{3}})$ [PaS92], while the lattice section construction gives an $\Omega(n^2 \log \log n)$ lower bound [ErP71].

Problem 13   (Pach–Sharir [PaS92]) What is the maximum number of triangles of unit perimeter determined by n points in the plane?

Pach and Sharir [PaS04] established the upper bound $O(n^{\frac{16}{7}})$. The best known lower bound, $\Omega(ne^{c\frac{\log n}{\log \log n}})$, follows from the fact that a section of the properly scaled triangular lattice spans this many equilateral triangles of edge length $\frac{1}{3}$. It is very likely that this bound can be substantially improved, because it can already be attained by congruent triangles (which automatically have the same perimeter).

As in the case of distances, it is much easier to estimate the number of occurrences of the maximum-area or the minimum-area (nondegenerate) triangles than the triangles of unit area, and the same holds for the perimeter. Brass, Rote, and Swanepoel [BrRS01] proved that there are at most n maximum-area or maximum-perimeter triangles and at most $\Theta(n^2)$ minimum-area and $\Theta(n)$ minimum-perimeter triangles.

It is conjectured by Erdős and Purdy [ErP77], [St78] that the minimum number of distinct areas of all triangles determined by n noncollinear points in the plane is $\lfloor \frac{n-1}{2} \rfloor$. This bound, if valid, is attained by the union of two sets of equidistant points lying on two parallel lines. The best known lower bound, $0.4142n - O(1)$, follows from the results in [BuP79], using Ungar's theorem [Un82] on the number of distinct directions determined by n noncollinear points in the plane.

Conjecture 14 *(Erdős–Purdy [ErP77]) The minimum number of triangles of distinct areas that must occur among n points in the plane, not all on a line, is $\left\lfloor \frac{n-1}{2} \right\rfloor$.*

For higher-dimensional spaces, Erdős and Purdy [ErP71], [Pu74] proved that among n points in d-dimensional space there are at most $O(n^{2+\frac{2}{3}})$ unit-area triangles if $d = 3$, and at most $O(n^{3-\varepsilon})$ unit-area triangles for some positive ε if $d = 4$ or 5. For $d \geq 6$, it follows from the Lenz construction that the maximum is $\Theta(n^3)$.

Problem 15 *(Erdős–Purdy [ErP71]) What is the maximum number of unit-area triangles determined by n points in $\mathbb{R}^d$ for $d = 3, 4$, and 5?*

Problem 16 *(Erdős–Purdy [ErP71]) What is the maximum number of unit-volume k-dimensional simplices determined by n points in d-dimensional space?*

Erdős and Purdy [ErP75] studied the maximum number of isosceles triangles spanned by a set X of n points in the plane. This question is closely related to the problem of determining the minimum number of distinct distances among n points [PaT03]. Counting the incidences between the elements of X and the perpendicular bisectors of its point pairs, it follows by the Szemerédi–Trotter theorem [SzT83] that n points in the plane determine at most $O(n^{2+\frac{1}{3}})$ isosceles triangles. Pach and Tardos improved this bound to $O(n^{2.13586})$ [PaT03]. More precisely, using the results of Katz and Tardos [KaT04], they proved that this number cannot exceed $O(n^{\frac{234-68e}{110-32e}+\varepsilon}) = O(n^{2.136})$, for any $\varepsilon > 0$, where e is the basis of the natural logarithm. On the other hand, counting isosceles triangles in a lattice section, we obtain an $\Omega(n^2 \log n)$ lower bound.

Problem 17 *(Erdős and Purdy [ErP75]) What is the maximum number of isosceles triangles determined by a set X of n points in the plane?*

Under the restriction that no three elements of X are collinear, one can almost precisely determine the maximum: it must be between $(n-2)(n-4)$ and $n(n-1)$. In higher dimensions $d \geq 3$, the maximum becomes $\Theta(n^3)$. Brass [Br03] used the estimates for the maximum number of isosceles triangles to bound the number of symmetric subsets.

[ApS05]   R. Apfelbaum, M. Sharir: Repeated angles in three and four dimensions, *SIAM J. Discrete Math.*, to appear.

[BlS02]   A. Blokhuis, Á. Seress: The number of directions determined by points in the three-dimensional euclidean space, *Discrete Comput. Geom.* **28** (2002) 491–494.

[Be83]   J. Beck: On the lattice property of the plane and some problems of Dirac, Motzkin and Erdős in combinatorial geometry, *Combinatorica* **3** (1983) 281–297.

[Br03]   P. Brass: On finding maximum-cardinality symmetric subsets, *Comput. Geom. Theory Appl.* **24** (2003) 19–25.

[BrRS01]   P. Brass, G. Rote, K.J. Swanepoel: Triangles of extremal area or perimeter in a finite planar pointset, *Discrete Comput. Geom.* **26** (2001) 51–58.

[BuP79]   G.R. Burton, G. Purdy: The directions determined by n points in the plane, *J. London Math. Soc. 2. Ser.* **20** (1979) 109–114.

[CoC*79]   J.H. Conway, H.T. Croft, P. Erdős, M.J.T. Guy: On the distribution of values of angles determined by coplanar points, *J. London Math. Soc. 2. Ser.* **19** (1979) 137–143.

[Cr61]   H.T. Croft: On 6-point configurations on 3-space, *J. London Math. Soc.* **36** (1961) 289–306.

[El03]   G. Elekes: On the number of distinct radii of circles determined by triplets and on parameters of other curves, *Studia Sci. Math. Hungar.* **40** (2003) 195–203.

[El84]   G. Elekes: n points in the plane can determine $n^{\frac{3}{2}}$ unit circles, *Combinatorica* **4** (1984) p. 131.

[Er75]   P. Erdős: Some problems on elementary geometry, *Australian Math. Soc. Gaz.* **2** (1975) 2–3.

[ErP95]   P. Erdős, G. Purdy: Extremal problems in combinatorial geometry, in: *Handbook of Combinatorics, Vol. 1*, R.L. Graham et al., eds., Elsevier 1995, 809–874.

[ErP77]   P. Erdős, G. Purdy: Some extremal problems in geometry V, *Congressus Numerantium* **19** (Proc. 8th South-Eastern Conf. Combinatorics, Graph Theory, and Computing, 1977) 569–578.

[ErP76]   P. Erdős, G. Purdy: Some extremal problems in geometry IV, *Congressus Numerantium* **17** (Proc. 7th South-Eastern Conf. Combinatorics, Graph Theory, and Computing, 1976) 307–322.

[ErP75]   P. Erdős, G. Purdy: Some extremal problems in geometry III, *Congressus Numerantium* **14** (Proc. 6th South-Eastern Conf. Combinatorics, Graph Theory, and Computing, 1975) 291–308.

[ErP71]   P. Erdős, G. Purdy: Some extremal problems in geometry, *J. Combinatorial Theory Ser. A* **10** (1971) 246–252.

[Ha85]   H. Harborth: Einheitskreise in ebenen Punktmengen, *3. Kolloquium über Diskrete Geometrie*, Universität Salzburg (1985) 163–168.

[HaM86]   H. Harborth, I. Mengersen: Point sets with many unit circles, *Discrete Math.* **60** (1986) 193–197.

[Ja04]   R.E. Jamison: Direction trees in centered polygons, in: *Towards a Theory of Geometric Graphs*, J. Pach, ed., *Contemporary Mathematics* **342**, AMS 2004, 87–98.

[Ja87]   R.E. Jamison: Direction trees, *Discrete Comput. Geom.* **2** (1987) 249–254.

[Ja86a]   R.E. Jamison: Few slopes without collinearity, *Discrete Math.* **60** (1986) 199–206.

[Ja86b]   R.E. Jamison: Direction paths, *Congressus Numerantium* **54** (Proc 17th South-Eastern Conf. Combinatorics, Graph Theory, and Computing 1986) 145–156.

[Ja85]   R.E. Jamison: A survey of the slope problem, in: *Discrete Geometry and Convexity*, J.E. Goodman et al., eds., *Annals New York Acad. Sci.* **440** (1985) 34–51.

[Ja84a]   R.E. Jamison: Planar configurations which determine few slopes, *Geometriae Dedicata* **16** (1984) 17–34.

[Ja84b]   R.E. Jamison: Structure of slope-critical configurations, *Geometriae Dedicata* **16** (1984) 249–277.

[JaH83]   R.E. Jamison, D. Hill: A catalogue of sporadic slope-critical configurations, *Congressus Numerantium* **40** (Proc. 14th South-Eastern Conf. Combinatorics, Graph Theory and Computing 1983) 101–125.

[KaT04]   N.H. Katz, G. Tardos: A new entropy inequality for the Erdős distance problem, in: *Towards a Theory of Geometric Graphs*, J. Pach, ed., *Contemporary Mathematics* **342**, AMS 2004, 119–126.

[KlP05]   D.J. Kleitman, R. Pinchasi: A note on caterpillar-embeddings with no two parallel edges, *Discrete Comput. Geom.*, to appear.

[PaPS04a] J. Pach, R. Pinchasi, M. Sharir: On the number of directions determined by a three-dimensional points set, *J. Combinatorial Theory Ser. A* **108** (2004) 1–16.

[PaPS04b] J. Pach, R. Pinchasi, M. Sharir: Solution of Scott's problem on the number of directions determined by a point set in 3-space, in: *SCG 04* (20th ACM Symp. Comput. Geom. 2004) 76–85.

[PaS04] J. Pach, M. Sharir: Geometric incidences, in: *Towards a Theory of Geometric Graphs*, J. Pach, ed., *Contemporary Mathematics* **342**, AMS 2004, 185–223.

[PaS92] J. Pach, M. Sharir: Repeated angles in the plane and related problems, *J. Combinatorial Theory Ser. A* **59** (1992) 12–22.

[PaT03] J. Pach, G. Tardos: Isosceles triangles determined by a planar point set, *Graphs Combinatorics* **18** (2002) 769–779.

[Pu88] G. Purdy: Repeated angles in E_4, *Discrete Comput. Geom.* **3** (1988) 73–75.

[Pu74] G. Purdy: Some extremal problems in geometry, *Discrete Math.* **7** (1974) 305–313.

[Sc70] P.R. Scott: On the sets of directions determined by n points, *Amer. Math. Monthly* **77** (1970) 502–505.

[St78] E.G. Straus: Some extremal problems in combinatorial geometry, in: *Proc. Internat. Conf. Combinatorial Theory*, (Canberra 1977), D.A. Holten et al., eds., *Springer Lecture Notes in Math.* **686** (1978) 308–312.

[SzT83] E. Szemerédi, W.T. Trotter: Extremal problems in discrete geometry, *Combinatorica* **3** (1983) 381–392.

[Un82] P. Ungar: $2N$ noncollinear points determine at least $2N$ directions, *J. Combinatorial Theory Ser. A* **33** (1982) 343–347.

6.3 Euclidean Ramsey Problems

In this section, we discuss several geometric problems concerning "unavoidable" patterns. These configurations are hard to avoid because either their monochromatic copies occur in all colorings of space with a certain number of colors, or they show up in every sufficiently "dense" set. This is the object of Euclidean Ramsey theory, a field started in a series of articles by Erdős, Graham, Montgomery, Rothschild, Spencer, Straus [ErG*73], [ErG*75a], [ErG*75b]. The starting point of these investigations is the Hadwiger-Nelson problem on the *chromatic number* of the plane (Section 5.9, Problem 1). What is the largest number r such that for any coloring of the plane with r colors, one can always find two points of the same color at unit distance from each other? The answer is known to be at least three and at most six. Erdős et al. generalized this question to patterns $A \subset \mathbb{R}^d$ consisting of more than two points and for different groups G of geometric transformations of $\mathbb{R}^d$, such as translations, similarities, isometries. Is it true that for any r-coloring of the points of $\mathbb{R}^d$, there is a monochromatic set A' that is the image of A under some transformation in G?

For instance, for three-element patterns in the plane under isometries, we have the following problem. For which triangles $\Delta = \{a, b, c\}$ is it true that for any two-coloring of the plane there is a monochromatic congruent copy of Δ? Several classes of triangles are known to have this property [ErG*75b], including all right triangles [Sh76]. Equilateral triangles, however, do not belong to this group, as is shown by the following coloring. Divide the plane into half-open vertical strips whose widths are equal to the height of the forbidden equilateral triangle, and color the strips alternately red and blue.

Conjecture 1 *(Erdős et al. [ErG*75b]) Any two-coloring of the plane with no monochromatic unit equilateral triangle consists of alternately colored parallel strips of width $\frac{\sqrt{3}}{2}$, with some freedom in coloring their boundary lines.*

Conjecture 2 *(Erdős et al. [ErG*75b]) For any nonequilateral triangle Δ and for any two-coloring of the plane, one can find a monochromatic congruent copy of Δ.*

It would be interesting to prove this conjecture at least in the special case that Δ is the degenerate (one-dimensional) triangle $\{1, 2, 3\}$ consisting of (the endpoints of) a segment and its midpoint.

One can also consider colorings with more than two colors. Since the plane can be colored by seven colors without having two points of the same color at unit distance, in a seven-colored plane we cannot guarantee the existence of a monochromatic congruent copy of any pattern A consisting of at least two elements. But what is the situation if we color the plane with fewer colors? Erdős et al. [ErG*73] constructed a four-coloring of the plane (and of all higher-dimensional spaces) that does not have a monochromatic congruent copy of $\{1, 2, 3\}$. They also found a three-coloring with no monochromatic copy of $\{1, 2, 3, 4\}$ and a two-coloring with no monochromatic copy of $\{1, 2, 3, 4, 5, 6\}$. These constructions raise a number of interesting questions.

Problem 3   (Erdős et al. [ErG*73]) Does there exist a two-coloring of the plane with no monochromatic copy of $\{1, 2, 3, 4\}$ (or $\{1, 2, 3, 4, 5\}$)?

Actually, there is no four-point pattern A for which it is known that any two-coloring of the plane contains a monochromatic congruent copy of A. Moreover, the few negative examples are not enlightening. The above-mentioned two-coloring that avoids an equilateral triangle Δ also shows that any superset of Δ and any square can be avoided. In the latter case, we can choose the width of the strips to be equal to the side length of the square.

Problem 4   (Erdős et al. [ErG*73]) Does there exist a three-coloring of the plane with no monochromatic congruent copy of $\{1, 2, 3\}$?

We know that in a three-colored plane one can always find many monochromatic unit-distance pairs. Perhaps some monochromatic three-point configurations must also occur.

Problem 5   Does there exist a triangle $\Delta = \{a, b, c\}$ such that in any three-coloring of the plane one can find a monochromatic congruent copy of Δ?

Erdős et al. [ErG*75a] have also studied some asymmetric variants of the above questions, where the forbidden patterns are different in different color classes. For instance, one can consider two-colorings of the plane with no unit-distance pair in the first color class and no congruent copy of a given set A in the second. Juhász [Ju79] proved that in any two-coloring of the plane, if the first color class contains no unit-distance pair, then the second color class must contain a congruent copy of every four-point set A. On the other hand, Csizmadia and Tóth [CsT94] constructed an eight-element set A that does not satisfy this condition.

Problem 6   (Juhász [Ju79], Csizmadia and Tóth [CsT94]) Do there exist a set A of five (or six, or seven) points in the plane and a two-coloring of $\mathbb{R}^2$ such that the first color class

> contains no unit-distance pair and the second color class
> contains no congruent copy of A?

Next we discuss some similar problems in three-dimensional space. It is known that for any triangle Δ, no matter how we two-color $\mathbb{R}^3$, there always exists a monochromatic congruent copy of Δ [ErG*73]. For four-point configurations, however, we have hardly any results of this kind. We know only a single planar configuration, $\{(0,0),(1,0),(2,0),(2,1)\}$, whose monochromatic congruent copy must occur in every two-coloring of $\mathbb{R}^3$.

Problem 7   *Is it true that for any set A of four points and for any two-coloring of $\mathbb{R}^3$, at least one of the color classes contains a congruent copy of A? Is this true at least for every full-dimensional set A?*

The special case that A is the unit square is particularly interesting. We have mentioned before that a parallel-strip coloring of $\mathbb{R}^2$ with two colors avoids monochromatic unit squares. On the other hand, Cantwell [Ca96] proved that any two-coloring of four-dimensional space contains a monochromatic unit square.

Problem 8   *(Erdős et al. [ErG*73]) Is it true that for any two-coloring of three-dimensional space, at least one of the color classes contains the vertex set of a unit square?*

In some sense, three-colorings of $\mathbb{R}^3$ seem to be similar to two-colorings of $\mathbb{R}^2$. As mentioned before, Shader [Sh76] proved that a congruent copy of a right triangle Δ can be found in at least one of the color classes of any two-colored plane. Bóna and Tóth [BóT96] showed that the same is true for any three-coloring of three-dimensional space.

Problem 9   *Given any triangle Δ, is it true that for any three-coloring of $\mathbb{R}^3$ one can find a monochromatic congruent copy of Δ?*

We cannot even rule out the possibility that a much stronger statement holds: any coloring of $\mathbb{R}^3$ with eleven colors must contain a monochromatic congruent copy of every triangle Δ. Bóna [Bó93] proved that this is not true with twelve colors; in this case a triangle with angles 30, 60, and 90 degrees can be avoided.

Most of the above questions also make sense in higher dimensions. Erdős et al. [ErG*73] called a finite point set A *Ramsey* if it has the following property. For any number of colors r, there is a dimension $d = d(r)$ such that in any r-coloring of $\mathbb{R}^d$ there is a monochromatic congruent copy of A. It is easy to see that every two-point configuration is Ramsey. To see this, let A denote a set of two points at distance a, set $d = r$, and consider a regular simplex of side length a in $\mathbb{R}^d$. For any r-coloring of its $r+1$ vertices, two vertices have the same color, and they induce a monochromatic copy of A. In [ErG*73], it was also shown that the Cartesian product of two Ramsey sets is also Ramsey. Thus, the vertex sets of boxes (rectangular parallelepipeds), and all of their subsets, are Ramsey. In particular, all acute and right triangles are Ramsey. Frankl and Rödl [FR90] proved that obtuse triangles are also Ramsey. In the other direction, Erdős et al. [ErG*73] proved that if a set is Ramsey, it must be *spherical*, that is, its vertices must lie on some sphere. Moreover, they conjectured that the converse statement is also true. Graham [Gr94] has offered $1000 for a proof of this conjecture, but even the planar version is unsolved.

Conjecture 10 *Every spherical set is Ramsey.*

Kříž [Kř91] proved that the vertex set of every regular polygon and three-dimensional regular polytope is Ramsey. Perhaps it is easier to settle the following problem.

Problem 11   *Is every quadrilateral inscribed in a circle Ramsey?*

Kříž [Kř92] showed that the answer is yes for trapezoids. The simplest non-Ramsey set consists of three collinear points. For more problems and results of this kind, consult [Gr80], [Gr83], [Gr85], [Gr90], [Gr94], [Gr97], and [MaR95].

To formulate a prototype of "density" problems, concerning the existence of "unavoidable" patterns in "dense" sets, we need a definition. For any Lebesgue measurable set $X \subseteq \mathbb{R}^d$, define its *upper density* as

$$\overline{d}(X, \mathbb{R}^d) = \limsup_{r \to \infty} \frac{\lambda(X \cap B(r))}{\lambda(B(r))},$$

where λ stands for the Lebesgue measure and $B(r)$ denotes the d-dimensional ball around the origin. One can raise the following general problem. Determine $\sup_X \overline{d}(X, \mathbb{R}^d)$ over all subsets $X \subset \mathbb{R}^d$ that do not contain any image A' of the forbidden pattern A under an element of the given transformation group.

For two-element patterns ("segments") in the plane, the question is to find the supremum $D(\mathbb{R}^2)$ of the upper densities of all measurable sets that do not contain two points at unit distance (see Section 5.9). The best known lower and upper bounds are due to L. Moser [Cr67] (see also [ScU97]) and Székely and Wormald [SzW89], [Sz84]. We have $0.2293 < D(\mathbb{R}^2) < \frac{12}{43} \approx 0.2791$. Erdős (see [Sz02]) conjectured that $D(\mathbb{R}^2) < \frac{1}{4}$. This conjecture, if true, immediately implies that the so-called *measurable chromatic number* of the plane is at least five. This means that for any four-coloring of the plane such that all color classes are Lebesgue measurable, there are two points of the same color whose distance is one. The last statement was proved by Falconer [Fa86]. In fact, Falconer showed that for every $d \geq 2$, the measurable chromatic number of $\mathbb{R}^d$ is at least $d + 3$. Székely and Wormald [SzW89] obtained some better estimates, using computer search. In particular, they improved Falconer's bound by 1 in four dimensions. According to Larman and Rogers [LaR72], the following conjecture was first made by L. Moser. It would yield a 2^d lower bound on the measurable chromatic number of $\mathbb{R}^d$, for every d.

Conjecture 12 (*L. Moser, Larman and Rogers [LaR72]*) $D(\mathbb{R}^2) \leq 2^{-d}$ *for every* $d \geq 1$.

Analogously, one can try to determine the supremum of the upper densities of all sets that do not contain any copy of A.

Problem 13   *What is the maximum upper density of a subset of the plane that does not contain a congruent copy of a given triangle?*

One can also ask density questions for geometric equivalence relations other than congruences. For equivalence under translation, the problem is trivial: any space can be two-colored without creating a monochromatic translate of a given set of two or more points. However, even in this case we can ask meaningful asymmetric questions.

Problem 14   *Determine the largest number r such that in any two-coloring of the plane, if the first color class contains no unit-distance pair, then the second color class must contain a translate of every planar r-element set.*

Szlam [Sz01] proved that for the largest r with the required property we have $1 \leq r \leq 6$. For homotheties and similarities, a powerful theorem of Furstenberg and Katznelson [FuK91] states that any set of positive density contains homothetic copies of all finite sets.

Most upper bounds on the density of point sets avoiding a fixed forbidden pattern A are based on estimates for finite sets. For instance, let $f(d, n)$ be the largest number such that any set of n points in $\mathbb{R}^d$ has an $f(d, n)$-element subset with no two points distance one apart (i.e., every unit-distance graph with n vertices has an independent set of this size). It is not hard to see that $D(\mathbb{R}^d) \leq \frac{f(d,n)}{n}$, which implies that the measurable chromatic number of $\mathbb{R}^d$ is at least $\frac{n}{f(d,n)}$. In fact, it also implies that the same lower bound holds for the usual *chromatic number* of $\mathbb{R}^d$. That is, for any coloring of $\mathbb{R}^d$ with fewer than this many colors, there are two points whose distance is one. (See Section 5.9.)

In this spirit, we can ask the following general questions.

Problem 15   *For any forbidden pattern A, determine the largest number $f_A(d, n)$ such that any set of n points in $\mathbb{R}^d$ has an $f_A(d, n)$-element subset containing no congruent (or similar, or homothetic) copy of A.*

Obviously, many such problems can be raised, not only for single patterns A, but also for various classes of forbidden objects. In Sections 5.3 and 5.7, for example, we have already mentioned some questions of this type, where isosceles triangles and occurrences of the minimum distance are forbidden.

Problem 16   *What is the largest number $f^{\mathrm{equi}}(n)$ such that from any set of n points in the plane one can select $f^{\mathrm{equi}}(n)$ elements such that no three of them induce an equilateral triangle?*

Problem 17   *What is the largest number $f^{\mathrm{mid}}(n)$ such that from any set of n points in the plane one can select $f^{\mathrm{mid}}(n)$ elements such that none of them is the midpoint of two others?*

The latter problem first appeared in the 1986 edition of *Research Problems in Discrete Geometry*, by Moser and Pach, and it was studied later by Bálint et al. [BáB*95]. Pach [Pa03] established the bounds

$$n^{1-c/\sqrt{\log n}} \le f^{\mathrm{mid}}(n) \le n/\log^{c'} n,$$

with suitable constants $c, c' > 0$.

Abbott [Ab80] noticed that any set of n points in the plane contains a subset of size at least $\lceil \sqrt{n}\, \rceil$ with no three points that form a right triangle. For example, such a subset can be obtained by selecting a monotone subsequence of the coordinate pairs. On the other hand, the $\lfloor \sqrt{n} \rfloor \times \lfloor \sqrt{n} \rfloor$ section of the integer lattice has at most $2\lfloor \sqrt{n} \rfloor - 2$ points with the required property.

Problem 18   *Let $f^{\mathrm{right}}(n)$ denote the largest number such that from any set of n points in the plane one can select $f^{\mathrm{right}}(n)$ elements such that no three of them determine a right angle. What is the largest constant c such that $f^{\mathrm{right}}(n) \ge (c + o(1))\sqrt{n}$?*

Clearly, the same question can be asked for any other angle α different from $\frac{\pi}{2}$.

[**Ab80**]   H.L. ABBOTT: On a conjecture of Erdős and Silverman in combinatorial geometry, *J. Combinatorial Theory Ser. A* **29** (1980) 380–381.

[BáB*95] V. BÁLINT, M. BRANICKÁ, P. GREŠÁK, I. HRINKO, P. NO-
          VOTNÝ, M. STACHO: Several remarks about midpoint-free
          subsets, Studies of University Transport and Communication
          in Žilina, Math.-Phys. Series 10 (1995) 3–10.

[Bó93]   M. BÓNA: A Euclidean Ramsey theorem, Discrete Math.
          122 (1993) 349–352.

[BóT96]   M. BÓNA, G. TÓTH: A Ramsey-type problem on right-
          angled triangles in space, Discrete Math. 150 (1996) 61–67.

[Ca96]   K. CANTWELL: Finite Euclidean Ramsey theory, J. Combi-
          natorial Theory Ser. A 73 (1996) 273–285.

[Cr67]   H.T. CROFT: Incidence incidents, Eureka (Cambridge) 30
          (1967) 22–26.

[CsT94]   G. CSIZMADIA, G. TÓTH: Note on a Ramsey-type problem
          in geometry, J. Combinatorial Theory Ser. A 65 (1994) 302–
          306.

[ErG*75a] P. ERDŐS, R.L. GRAHAM, P. MONTGOMERY, B.L. ROTH-
          SCHILD, J. SPENCER, E.G. STRAUS: Euclidean Ramsey theo-
          rems. II., in: Infinite and Finite Sets (Keszthely, 1973), A. Ha-
          jnal et al., eds., Colloq. Math. Soc. János Bolyai 10 (1975)
          559–584.

[ErG*75b] P. ERDŐS, R.L. GRAHAM, P. MONTGOMERY, B.L. ROTH-
          SCHILD, J. SPENCER, E.G. STRAUS: Euclidean Ramsey the-
          orems. III., in: Infinite and Finite Sets (Keszthely, 1973),
          A. Hajnal et al., eds., Colloq. Math. Soc. János Bolyai 10
          (1975) 559–584.

[ErG*73] P. ERDŐS, R.L. GRAHAM, P. MONTGOMERY, B.L. ROTH-
          SCHILD, J. SPENCER, E.G. STRAUS: Euclidean Ramsey the-
          orems. I. J. Combinatorial Theory 14 (1973) 341–363.

[Ex03]   G. EXOO: A Euclidean Ramsey problem, Discrete Comput.
          Geom. 29 (2003) 223–227.

[Fa86]   K.J. FALCONER: The realization of small distances in plane
          sets of positive measure, Bull. London Math. Soc. 18 (1986)
          471–474.

[FrR90]   P. FRANKL, V. RÖDL: A partition property of simplices in
          Euclidean space, J. Amer. Math. Soc. 3 (1990) 1–7.

[FuK91]   H. FURSTENBERG, Y. KATZNELSON: A density version of
          the Hales–Jewett theorem, J. Anal. Math. 57 (1991) 64–119.

[Gr97]   R.L. GRAHAM: Euclidean Ramsey theory, in: Handbook of
          Discrete and Computational Geometry, J.E. Goodman et al.,
          eds., CRC Press 1997, 153–166.

[**Gr94**]   R.L. GRAHAM: Recent trends in Euclidean Ramsey theory, *Discrete Math.* **136** (1994) 119–127.

[**Gr90**]   R.L. GRAHAM: Topics in Euclidean Ramsey theory, in: *Mathematics of Ramsey Theory*, J. Nešetřil, V. Rödl, eds., *Graphs and Combinatorics* **5**, Springer 1990, 200–213.

[**Gr85**]   R.L. GRAHAM: Old and new Euclidean Ramsey theorems, in: *Discrete Geometry and Convexity*, J.E. Goodman et al., eds., *Annals New York Acad. Sci.* **440** (1985) 20–30.

[**Gr83**]   R.L. GRAHAM: Euclidean Ramsey theorems on the n-sphere, *J. Graph Theory* **7** (1983) 105–114.

[**Gr80**]   R.L. GRAHAM: On partitions of E^n, *J. Combinatorial Theory Ser. A* **28** (1980) 89–97.

[**Ju79**]   R. JUHÁSZ: Ramsey type theorems in the plane, *J. Combinatorial Theory Ser. A* **27** (1979) 152–160.

[**Křž92**]   I. KŘIŽ: All trapezoids are Ramsey, *Discrete Math.* **108** (1992) 59–62.

[**Křž91**]   I. KŘIŽ: Permutation groups in Ramsey theory, *Proc. Amer. Math. Soc.* **112** (1991) 899–907.

[**LaR72**]   D.G. LARMAN, C.A. ROGERS: The realization of distances within sets in Euclidean space, *Mathematika* **19** (1972) 1–24.

[**MaR95**]   J. MATOUŠEK, V. RÖDL: On Ramsey sets on spheres, *J. Combinatorial Theory Ser. A* **70** (1995) 30–44.

[**Pa03**]   J. PACH: Midpoints of segments induced by a point set, *Geombinatorics* **13** (2003) 98–105.

[**ScU97**]   E.R. SCHEINERMAN, D.H. ULLMAN: *Fractional Graph Theory: A Rational Approach*, Wiley 1997.

[**Sh76**]   L. SHADER: All right triangles are Ramsey in E^2, *J. Combinatorial Theory Ser. A* **20** (1976) 385–389.

[**Sz01**]   A.D. SZLAM: Monochromatic translates of configurations in the plane, *J. Combinatorial Theory Ser. A* **93** (2001) 173–176.

[**Sz02**]   L.A. SZÉKELY: Erdős on unit distances and the Szemerédi–Trotter theorems, in: *Paul Erdős and His Mathematics, Vol. II*, G. Halász et al., eds., *Bolyai Society Mathematical Studies* **11** (2002) 649–666.

[**Sz84**]   L.A. SZÉKELY: Measurable chromatic number of geometric graphs and sets without some distances in Euclidean space, *Combinatorica* **4** (1984) 213–218.

[**SzW89**]   L.A. SZÉKELY, N.C. WORMALD: Bounds on the measurable chromatic number of $\mathbb{R}^n$. *Discrete Math.* **75** (1989) 343–372.

[**Tó96**]   G. TÓTH: A Ramsey-type bound for rectangles, *J. Graph Theory* **23** (1996) 53–56.

7. Incidence and Arrangement Problems

7.1 The Maximum Number of Incidences

Algebraic and combinatorial properties of incidence structures have been extensively studied in enumerative (algebraic) geometry, in the theory of combinatorial designs and finite projective spaces, etc. Among the oldest problems in combinatorial geometry, we find several questions on incidences between points and lines. For a general reference, see the classical monograph of Grünbaum [Gr72] or, for more recent developments, [ErP95] and [PaS04]. Edelsbrunner's monograph [Ed87] has put geometric arrangements and incidences in the focus of research in computational geometry (see also [AgS00]). Several interesting connections between incidence geometry, Fourier analysis, and measure theory are discussed in Iosevich's survey [Io01]. In particular, Wolff [Wo99] discovered that Kakeya's problem on the Hausdorff dimension of sets that contain a unit segment in every direction is related to incidence problems. The planar version of the problem was solved by Davies [Da71] and, in a stronger form, by Córdoba [Có77] and Bourgain [Bo99]. See the survey of Katz and Tao [KaT02]. Another related problem is Falconer's conjecture [Fa85] on distance sets, extensively studied by Iosevich, Bourgain, Wolff, and others. For some number-theoretic and algebraic aspects of incidence geometry, see [BoKT04], [El02].

Perhaps the most influential question on incidences was raised by Erdős: What is the maximum number of incidences $\mathrm{inc}(n, m)$ between n distinct points and m distinct lines in the plane? Since incidences are preserved by projection in a "generic" direction, $\mathrm{inc}(n, m)$ is also equal to the maximum number of incidences between n points and m lines in higher-dimensional Euclidean spaces. Furthermore, by duality we have $\mathrm{inc}(m, n) = \mathrm{inc}(n, m)$.

Using the facts that any two points are incident to at most one line and any two lines are incident to at most one point, we obtain the easy upper bounds $\mathrm{inc}(n, m) \leq O(n\sqrt{m} + m)$ and $\mathrm{inc}(n, m) \leq O(m\sqrt{n} + n)$. Together they easily imply

$$\mathrm{inc}(n, m) \leq O(n^{\frac{3}{4}} m^{\frac{3}{4}} + n + m).$$

A $n^{\frac{1}{2}} \times n^{\frac{1}{2}}$ square section of the integer lattice and m "rich" lattice lines that pass through as many points as possible show that the number of incidences can be $\Omega(n^{\frac{2}{3}} m^{\frac{2}{3}} + n + m)$. Erdős conjectured and, by a rather involved argument, Szemerédi and Trotter [SzT83] proved that this construction is asymptotically optimal, i.e., the maximum number of incidences satisfies

$$\mathrm{inc}(n, m) \leq O(n^{\frac{2}{3}} m^{\frac{2}{3}} + n + m).$$

Because of its wide range of applications and numerous generalizations, this result has received much attention. As a consequence, several different proofs have been found [ClE*90], [PaA95]. The simplest proof is due to Székely [Sz97] and is based on the so-called crossing lemma, providing a lower bound on the number of edge crossings in a graph with many edges drawn in the plane (see Section 9.3). The Szemerédi–Trotter theorem is an important prototype of a large class of incidence results, including the equivalent forms of several estimates concerning the distribution of distances among n points (see Chapter 5). It is one of the very few results in this area that are asymptotically tight.

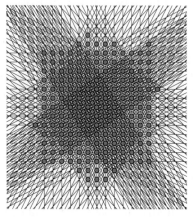

LATTICE SET WITH
MANY INCIDENCES

Refinements of the crossing lemma led to reasonable upper bounds on the constant hidden in the $O(.)$ notation. The currently best bounds are given by Pach, Radoičić, Tardos, and Tóth [PaR*04], [PaT97]. The maximum number of incidences between m lines and n points satisfies

$$0.42 m^{\frac{2}{3}} n^{\frac{2}{3}} + m + n \leq \text{inc}(n, m) \leq 2.5 m^{\frac{2}{3}} n^{\frac{2}{3}} + m + n.$$

Perhaps it is not hopeless to find the right multiplicative constant.

Problem 1   *What is the infimum of the values c for which*

$$\text{inc}(n, m) \leq c n^{\frac{2}{3}} m^{\frac{2}{3}} + O(n + m)$$

holds?

It should be noted that the extremal sets are certainly not the simple lattice sections mentioned above. Indeed, the "richest" lines in a $n^{\frac{1}{2}} \times n^{\frac{1}{2}}$ lattice section pass through $\Theta(n^{\frac{1}{2}})$ points and form bundles consisting of $\Theta(n^{\frac{1}{2}})$ parallel lines. For instance, for $m = n$ the average number of lines incident to a point is $\Theta(n^{\frac{1}{3}})$, so adding the "point at infinity" along each bundle of "rich" lines, we obtain a much better configuration. Applying a projective transformation, the "line at infinity" can be transformed into a real "limit" line, resulting in the "projective lattice" shown in the next figure [Br96]. Nevertheless, it is possible that the extremal sets are "nearly" lattice-like.

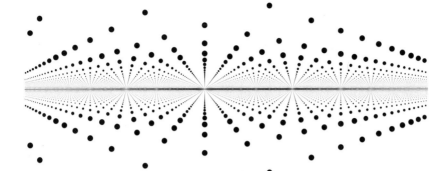

PROJECTIVE IMAGE OF LATTICE, FURTHER INCIDENCES ON "LIMIT" LINE

It is an old question of Erdős to decide whether the extremal configurations maximizing the number of incidences between n points and, say, n lines show any kind of lattice structure. In particular, do they necessarily contain $\Omega(n^{\frac{1}{2}})$ collinear points?

Let X and X' (respectively $\mathcal{L}$ and $\mathcal{L}'$) be two sets of points (respectively lines) in the plane. We say that the pairs $(X, \mathcal{L})$ and $(X', \mathcal{L}')$ are *isomorphic* if their incidence structures are the same.

Conjecture 2 *(Solymosi) For any set of points X and for any set of lines $\mathcal{L}$ in the plane, the maximum number of incidences between n points and n lines in the plane containing no subconfiguration isomorphic to $(X, \mathcal{L})$ is $o(n^{\frac{4}{3}})$.*

Solymosi (personal communication) managed to prove this conjecture in the special case that X is a fixed set of points in the plane, no three of which are on a line, and $\mathcal{L}$ consists of all of their connecting lines. It seems likely that in this case the exponent $\frac{4}{3}$ can also be lowered. In the first interesting unsolved case, X consists of four points, three of which are collinear, and $\mathcal{L}$ is the set of their four connecting lines. This special case of the result would already imply the existence of arithmetic progressions in a "dense" set of integers, which is a classical result.

Most proof techniques for the Szemerédi–Trotter theorem are quite flexible. In particular, they carry over to the case in which instead of straight lines, we consider "pseudo-lines" (that is, families $\mathcal{L}$ of two-way infinite curves in the plane, any pair of which intersect at most once), unit circles, or "pseudo-unit-circles." The number of incidences between n points and m members of $\mathcal{L}$ is $O(n^{\frac{2}{3}}m^{\frac{2}{3}} + n + m)$. More generally, we say that a family $\mathcal{L}$ of simple curves in the plane has d *degrees of freedom* if there exists a constant c such that

(1) for any d points, there are at most c curves in $\mathcal{L}$ passing through all of them, and

(2) any pair of curves belonging to $\mathcal{L}$ intersect in at most c points.

For example, the families of all lines, all unit circles, all circles, and all graphs of the form $y = p(x)$, where p is a polynomial of degree k, have two, two, three, and $k + 1$ degrees of freedom, respectively. Pach and Sharir [PaS98], [PaS92] generalized the Szemerédi–Trotter theorem by showing that the number of incidences between n points and m members of a family of curves with d degrees of freedom is at most

$$O\left(n^{\frac{d}{2d-1}} m^{\frac{2d-2}{2d-1}} + n + m\right).$$

Here the constant hidden in the O-notation depends on d and c.

Problem 3   *Is it true that for any family $\mathcal{L}$ of plane curves with a bounded degree of freedom, the number of incidences between n points and n members of $\mathcal{L}$ is at most $O(n^{\frac{4}{3}})$?*

As mentioned before, $O(n^{\frac{2}{3}} m^{\frac{2}{3}} + n + m)$ is also a trivial upper bound on the number of incidences between n points and m lines in $\mathbb{R}^3$. The problem can be made more interesting if we count only "truly three-dimensional" incidences. A point is called a *joint* of an arrangement of m lines in $\mathbb{R}^3$ if it is incident to at least three noncoplanar lines. Sharir and Welzl [ShW04] proved that the total number of incidences between a set of m lines in $\mathbb{R}^3$ and all joints determined by them is at most $O(m^{\frac{5}{3}})$. The best known lower bound, $\Omega(m^{3/2})$, is given by the set of lines parallel to the coordinate axes, passing through the lattice points in a cubic section of the integer lattice of side length $\sqrt{m/3}$.

Problem 4   *(Chazelle et al. [ChE*92]) Is it true that the maximum number of joints determined by m lines in $\mathbb{R}^3$ is at most $O\left(m^{\frac{3}{2}}\right)$?*

The best known upper bound for the number of joints, established by Feldman and Sharir [FeS05], is $O(m^{\frac{112}{69}} \log^{\frac{6}{23}} m) \leq O(m^{1.6232})$. This question is related to important problems in computer graphics (hidden surface removal). Let $\mathcal{L}$ be a family of lines in *general position* in three-dimensional space, that is, assume that

(1) no element of $\mathcal{L}$ is vertical (parallel to the z-axis),

(2) no two elements of $\mathcal{L}$ are parallel or have a point in common, and

(3) no vertical line passes through three elements of $\mathcal{L}$.

Let $\mathcal{L}'$ denote the orthogonal projection of $\mathcal{L}$ onto the xy-plane. Consider a bounded cell of the arrangement determined by $\mathcal{L}'$, whose sides belong to the lines $\ell'_1, \ldots, \ell'_k \in \mathcal{L}$, in cyclic order. We say that the corresponding lines $\ell_1, \ldots, \ell_k \in \mathcal{L}$ form an *elementary cycle* if the line ℓ_i passes above ℓ_{i+1} for all $1 \leq i \leq k$ (here we set $\ell_{k+1} := \ell_1$).

Conjecture 5 *(Chazelle et al. [ChE*92]) The number of elementary cycles determined by m lines in general position in $\mathbb{R}^3$ is $o(m^2)$.*

It is easy to see that any upper bound for the function appearing in the last conjecture is also an upper bound for the maximum number of joints determined by m lines in $\mathbb{R}^3$. It is possible that the best upper bound for both functions is $O(n^{\frac{3}{2}})$.

The most natural higher-dimensional extension of point–line incidence counting is to study point–hyperplane incidences. Taking n points along a line or some other proper affine subspace $U \subset \mathbb{R}^d$ and choosing m hyperplanes that pass through U, we have that the number of incidences between them is nm. If we want to ask nontrivial questions, we either have to restrict the class of admissible configurations or measure a parameter different from the number of incidences. In the literature, various restrictions have been considered. A minor but natural restriction is that our points must be selected from the set of vertices of the arrangement of hyperplanes. That is, each point must belong to d hyperplanes with linearly independent normal vectors. Of course, in this case, the inequality $n \leq \binom{m}{d}$ must hold. For $n \geq \binom{m-3}{d-2}$, Agarwal and Aronov [AgA92] proved that the maximum number of incidences between m hyperplanes in $\mathbb{R}^d$ and n vertices of their arrangement is $\Theta(n^{\frac{2}{3}}m^{\frac{d}{3}} + m^{d-1})$, while for $n \leq \binom{m-3}{d-2}$ the maximum is $\Theta(nm)$.

Conjecture 6 *(Brass) Let $d \geq 3$, $n \leq \binom{m-3}{d-2}$, and let μ denote the smallest integer such that $n \leq \binom{\mu}{d-2}$. Then any set $\mathcal{H}$ of m hyperplanes in $\mathbb{R}^d$ and any set X of n vertices of their arrangement that maximize the number of point–hyperplane incidences have the following structure:*

(1) $m - \mu$ elements of $\mathcal{H}$ intersect in a common $(d-2)$-dimensional affine subspace U,

(2) μ elements of $\mathcal{H}$ properly intersect U,

(3) the elements of X are vertices of the arrangement of the intersection of the μ hyperplanes with U.

For $d = 3$, such an arrangement consists of $m - n$ planes that intersect in a common line, and n planes that intersect that line and generate n vertices. Thus, the number of incidences between the vertices and the planes is $n(m - n + 1)$.

A stronger restriction is to require the hyperplanes and the points to be in "convex position." Suppose that $\mathcal{H}$ is a collection of m hyperplanes supporting the *facets* (i.e., $(d-1)$-dimensional faces) of a convex polytope $P \subset \mathbb{R}^d$, and let X be a set of n vertices of P. Under these assumptions, it is still possible that almost all facets share a common $(d - 3)$-dimensional face with n vertices, provided that $d > 3$ and m is large enough. Thus, almost all vertex–facet pairs can be incident, and perhaps it is possible to find the exact maximum.

Problem 7   *Determine the maximum number of incidences between m facets and n vertices of a d-dimensional convex polytope, for $d > 3$.*

By imposing much stronger restrictions on the admissible configurations, one can obtain asymptotically better, $o(mn)$, upper bounds on the number of point–hyperplane incidences. A result of this type was found by Edelsbrunner and Sharir [EdS91], who proved that the number of incidences between n points and m hyperplanes in $\mathbb{R}^4$ is $O(m^{\frac{2}{3}} n^{\frac{2}{3}} + m + n)$, provided that no three points are collinear, no three hyperplanes intersect in a two-dimensional plane, and each hyperplane bounds a closed half-space that contains all points.

Conjecture 8   *[EdS91] There exist sets of n points and n hyperplanes in $\mathbb{R}^4$ with a superlinear number of incidences between them such that*

(1) no three points are collinear;

(2) no three hyperplanes intersect in a two-dimensional plane;

(3) each hyperplane bounds a closed half-space that contains all points.

Given a family $\mathcal{S}$ of surfaces (curves or any geometric objects) and a set X of points, define their *incidence graph* as a bipartite graph on the vertex set $\mathcal{S} \cup X$, where two vertices are connected by an edge if and only if the corresponding elements of $\mathcal{S}$ and X are incident. For several algorithmic problems, such as counting or reporting incidences between large sets of geometric objects and points or counting intersections between two large sets of objects, it is easier to decompose the edge set of the corresponding incidence graph (respectively intersection graph) into few complete bipartite graphs than to list all edges. This enables us to give a compact representation of these graphs, by listing the vertex sets of the complete bipartite subgraphs that form a minimal decomposition. This technique is called *bipartite subgraph compression*. Erickson [Eri96] discovered that the maximum size of the smallest bipartite-subgraph decomposition of a point–hyperplane incidence relation serves as a lower bound on the running time of any algorithm (belonging to a certain class) for counting incidences.

Problem 9   *(Erickson [Eri96]) What is the smallest number $\mathrm{bsc}_d(n, m)$ such that the incidence graph of any set of n points and m*

> hyperplanes in $\mathbb{R}^d$ can be covered by $\mathrm{bsc}_d(n, m)$ complete bipartite subgraphs?

Improving Erickson's original bounds, Brass and Knauer [BrK03] showed that

$$\mathrm{bsc}_d(n, m) \geq \Omega\left((nm)^{1 - \frac{2}{d+3} - \varepsilon}\right),$$

for odd d (for $d = 3$, the result is true with $\varepsilon = 0$). If d is even, we have

$$\mathrm{bsc}_d(n, m) \geq \Omega\left((nm)^{1 - \frac{2(d+1)}{(d+2)^2} - \varepsilon}\right),$$

for any $\varepsilon > 0$. The best known upper bound,

$$\mathrm{bsc}_d(n, m) \leq O\left((nm)^{1 - \frac{1}{d+1}} + n + m\right),$$

is probably nearer to the true value [dBeS95], [BrK03], [ApS05].

The above questions have also been studied for several other types of higher-dimensional arrangements. In most cases, one can obtain asymptotically tight trivial answers, because the corresponding incidence graphs can be arbitrarily large complete bipartite graphs. Again, these problems become more interesting if we exclude certain subconfigurations. For example, the number of incidences between n points and m spheres can be mn. However, Clarkson et al. [ClE*90] proved an $O\left((mn)^{\frac{3}{4}} \beta(mn) + m + n\right)$ upper bound for the number of incidences between n points and m spheres in three-dimensional space, under the assumption that no three spheres have three points in common. This condition on the spheres is satisfied by arrangements of unit spheres. Hence, for $m = n$, the last result gives an upper bound for the maximum number of unit distances determined by n points in three-dimensional space (see Section 5.2). Here β is again a very slow growing function related to the inverse Ackermann function α, with $\beta(mn) = 2^{O(\alpha(mn)^2)}$. Furthermore, if the n points are vertices of the arrangement of m spheres, Clarkson et al. obtained an upper bound of $O\left(m^{\frac{9}{7}} n^{\frac{4}{7}} \beta(mn) + m^2\right)$ on the number of incidences. For d-dimensional point-sphere arrangements, Chung [Chu89] established the upper bound of $O\left((mn)^{1 - \frac{1}{d+2}} + mn^{1 - \frac{1}{d-1}} + n\right)$, under the assumption that the intersection of any i spheres is a unique $(d - i)$-dimensional sphere for all $i < d$. All of these upper bounds are probably far from tight. It would be interesting to extend these results, for example, to arrangements of (properly defined) "pseudo-hyperplanes" or "pseudo-spheres."

Returning to points and lines in the plane, a well-known equivalent formulation of the Szemerédi–Trotter theorem states that any set of n points in the plane spans at most $O(\frac{n^2}{k^3})$ lines that are incident to at least k points.

For $k < \sqrt{n}$, this bound is best possible: the set of integer points $\{(i, j) \mid 1 \leq i \leq \frac{n}{k}, 1 \leq j \leq k\}$ generates at least $\frac{1}{k-1} \lfloor \frac{n}{k} \rfloor^2$ lines containing exactly k points. It is possible to obtain reasonable bounds on the multiplicative constants involved in the $O(.)$ statement.

Problem 10   (Erdős [Er84]) For a fixed $k \geq 4$, what is the smallest constant c_k such that any set of n points in the plane spans at most $(c_k + o(1))n^2$ lines passing through precisely k points?

Clearly, we have $c_2 = 1$, and it was proved by Burr, Grünbaum, and Sloane [BuG*74] that $c_3 = \frac{1}{6}$. For other values $k \leq 14$, Palásti [Pal86] gave some constructions showing that $c_4 \geq \frac{1}{30}$, $c_5 \geq \frac{2}{135}$, $c_6 \geq \frac{47}{4860}$, etc. Of course, asymptotically, the relation $c_k = \Theta(k^{-3})$ holds.

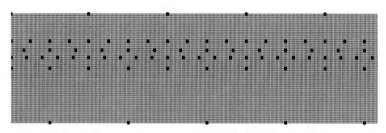

SET WITH MANY FOUR-POINT LINES

De Caen and Székely suggested an alternative proof of the Szemerédi–Trotter theorem that would rely on the validity of the following conjecture.

Conjecture 11   [dCaS97] The number of cycles of length six in the point-line incidence graph associated with any sets of n points and m lines in the plane is $O(mn)$.

It is likely that this statement, if true, can be proved by a simple counting argument. Note that a cycle of length six in the incidence graph corresponds to a "triangle" consisting of three noncollinear points and their connecting lines.

A popular class of incidence problems in projective geometry studied at the end of the nineteenth century is related to arrangements of n points and n lines in the plane, each point incident to k lines and each line incident to k points. Such an arrangement was called an (n_k)-configuration. The main problem was to enumerate all (n_3)-configurations; see Reye [Re882], Steinitz [St894], and Gropp [Gro97] for some historical remarks. Such a configuration, viewed as an abstract incidence structure, is also a natural combinatorial object. Since it is simpler to enumerate combinatorial objects than their geometric realizations, this problem led to the systematic study of combinatorial incidence structures and combinatorial designs. The realizability of several configurations discovered a long time ago was checked later. A nice result of this type is that every abstract (n_3)-configuration can be realized so that all but at most one of the "combinatorial lines" become straight lines [St894]. However, until recently no general method had been known to decide whether a given (n_k)-configuration is realizable. An important special aspect of this problem is treated under the heading of "stretchability of pseudo-line arrangements."

Problem 12   (Reye [Re882]) Given an integer k, for which values of n do there exist (n_k)-configurations?

Grünbaum studied (n_4)-configurations, especially those with maximal rotational symmetry [Gr02], [Gr00a], [Gr00b], [Gr00c], [Gr93], [GrR90], [Ber01]. He showed that for $n \neq 21$ and at most ten further exceptional values, connected (n_4)-configurations do exist [Gr00a]. It remains to settle these open cases. As far as we know, no work has been done concerning (n_k)-configurations for $k \geq 5$. Similar questions could be asked for higher dimensions.

Richter and Thomassen [RiT95], Salazar [Sa99], and Mubayi [Mu02] studied the following problem. What is the minimum number of intersection points in a family of n pairwise intersecting closed curves such that no three of them pass through the same point, and every tangency is counted as an intersection? If any two curves cross twice, as in an arrangement of pairwise crossing circles, the number of intersection points is $2\binom{n}{2}$. By making some pairs of curves meet tangentially instead of crossing each other, this number can be slightly reduced, but probably not by much. The best known lower bound is $\frac{8}{5}\binom{n}{2}$ [Mu02].

Problem 13   [RiT95] Is it true that n pairwise intersecting closed curves, no three of which pass through the same point, generate at least $\binom{n}{2} - O(n)$ points of intersection or tangency?

Another old question of Erdős and Grünbaum is intimately related to the problem of bounding the number of incidences between systems of curves and points, and appears in the analysis of several geometric algorithms. A set of closed curves in the plane, no three of which pass through the same point, and any two of which have at most two points in common, is called a family of *pseudo-circles*.

Problem 14   [ErG73] What is the maximum number of tangencies in a family $\mathcal{C}$ of n pseudo-circles, no three of which pass through the same point?

Erdős and Grünbaum proved that a complete bipartite graph $K_{3,9}$ cannot occur as a subgraph of a graph whose vertices are the elements of $\mathcal{C}$, and two vertices are joined by an edge if and only if the corresponding pseudo-circles touch each other. This fact alone implies an upper bound of $O(n^{\frac{5}{3}})$ on the number of tangencies (see also [TaT98]). This was improved by Marcus and Tardos [MaT05] to $O(n^{\frac{3}{2}} \log n)$. Previously, Agarwal et al. [AgN*04] established a similar result for special families of pseudo-circles. Furthermore, they proved a linear upper bound for the number of tangencies between n pairwise intersecting pseudo-circles. On the other hand, it is not hard to construct a family of n circles with $\Omega(n^{\frac{4}{3}})$ touching pairs. Start with an arrangement of $\frac{n}{2}$ lines and $\frac{n}{2}$ points having $\Omega(n^{\frac{4}{3}})$ incidences. Apply an inversion of the plane converting the lines into circles. Reducing the radius of each circle by some small $\varepsilon > 0$ and replacing each point by a circle of radius ε, one can achieve that each original incidence gives rise to a touching pair of circles.

[**AgA92**]   P.K. Agarwal, B. Aronov: Counting facets and incidences, *Discrete Comput. Geom.* **7** (1992) 359–369.

[**AgN*04**]   P.K. Agarwal, E. Nevo, J. Pach, R. Pinchasi, M. Sharir, S. Smorodinsky: Lenses in arrangements of pseudocircles and their applications, *J. ACM* **51** (2004) 139–186.

[**AgS00**]   P.K. Agarwal, M. Sharir: Arrangements and their applications, in: *Handbook of Computational Geometry*, J.-R. Sack et al., eds., North-Holland/Elsevier 2000, 49–119.

[**ApS05**]   R. Apfelbaum, M. Sharir: Incidences between points and hyperplanes, manuscript, 2005.

[**Ber01**]   L.W. Berman: A characterization of astral (n_4) configurations, *Discrete Comput. Geom.*, **26** (2001) 603–612.

[**Br96**]   P. Brass: *Extremale Konstruktionen in der kombinatorischen Geometrie*, Habilitationsschrift, Universität Greifswald 1996.

[**BrK03**]   P. Brass, C. Knauer: On counting point–hyperplane incidences, *Comput. Geom. Theory Appl.* **25** (2003) 13–20.

[**dBeS95**]   M. de Berg, O. Schwarzkopf: Cuttings and applications, *Internat. J. Comput. Geom. Appl.* **5** (1995) 343–355.

[**Bo99**]   J. Bourgain: On the dimension of Kakeya sets and related maximal inequalities, *Geom. Funct. Anal.* **9** (1999) 256–282.

[**BoKT04**]   J. Bourgain, N.H. Katz, T. Tao: A sum–product estimate in finite fields, and applications, *Geom. Funct. Anal.* **14** (2004) 27–57.

[**BuG*74**]   S.A. Burr, B. Grünbaum, N.J.A. Sloane: The orchard problem, *Geometriae Dedicata* **2** (1974) 397–424.

[dCaS97]   D. DE CAEN, L.A. SZÉKELY: On dense bipartite graphs of girth eight and upper bounds on certain point-line configurations., *J. Combinatorial Theory Ser. A* **77** (1997) 268–278.

[ChE*92]   B. CHAZELLE, H. EDELSBRUNNER, L.J. GUIBAS, R. POLLACK, R. SEIDEL, J. SNOEYINK: Counting and cutting cycles of lines and rods in space, *Comput. Geom. Theory Appl.* **1** (1992) 305–323.

[Chu89]   F.R.K. CHUNG: Sphere-and-point incidence relations in high dimensions with applications to unit distances and furthest-neighbor pairs, *Discrete Comput. Geom.* **4** (1989) 183–190.

[ClE*90]   K.L. CLARKSON, H. EDELSBRUNNER, L.J. GUIBAS, M. SHARIR, E. WELZL: Combinatorial complexity bounds for arrangements of curves and spheres, *Discrete Comput. Geom.* **5** (1990) 99–160.

[Có77]   A. CÓRDOBA: The Kakeya maximal function and the spherical summation multipliers, *Amer. J. Math.* **99** (1977) 1–22.

[Da71]   R.O. DAVIES: Some remarks on the Kakeya problem, *Proc. Cambridge Philos. Soc.* **69** (1971) 417–421.

[Ed87]   H. EDELSBRUNNER: *Algorithms in Combinatorial Geometry*, Springer-Verlag 1987.

[EdS91]   H. EDELSBRUNNER, M. SHARIR: A hyperplane incidence problem with applications to counting distances, in: *The Victor Klee Festschrift, DIMACS Ser. Discret. Math. Theor. Comput. Sci.* **4** (1991) 253–263.

[El02]   G. ELEKES: SUMS versus PRODUCTS in number theory, algebra and Erdős geometry, in: *Paul Erdős and His Mathematics. Vol. II* (Budapest, 1999), G. Halász et al., eds., *Bolyai Soc. Math. Stud.* **11** (2002) 241–290.

[Er84]   P. ERDŐS: Some old and new problems in combinatorial geometry, *Annals Discrete Math.* **20** (1984) 129–136.

[ErG73]   P. ERDŐS, B. GRÜNBAUM: Osculation vertices in arrangements of curves, *Geometriae Dedicata* **1** (1973) 322–333.

[ErP95]   P. ERDŐS, G. PURDY: Extremal problems in combinatorial geometry, in: *Handbook of Combinatorics. Vol. 1*, R.L. Graham et al., eds., North-Holland/Elsevier 1995, 809–874.

[Eri96]   J. ERICKSON: New lower bounds for Hopcroft's problem, *Discrete Comput. Geom.* **16** (1996) 389–418.

[Fa85]   K.J. FALCONER: On the Hausdorff dimension of distance sets, *Mathematika* **32** (1985) 206–212.

[FeS05]   S. FELDMAN, M. SHARIR: An improved bound for joints in arrangements of lines in space, *Discrete Comput. Geom.*, to appear.

[Gro97]   H. GROPP: Configurations and their realization, *Discrete Math.* **174** (1997) 137–151.

[Gr02]   B. GRÜNBAUM: Connected (n_4) configurations exist for almost all n — an update, *Geombinatorics* **12** (2002) 15–23.

[Gr00a]   B. GRÜNBAUM: Connected (n_4) configurations exist for almost all n, *Geombinatorics* **10** (2000) 24–29.

[Gr00b]   B. GRÜNBAUM: Which (n_4) configurations exist? *Geombinatorics* **9** (2000) 164–169.

[Gr00c]   B. GRÜNBAUM: Astral (n_4) configurations, *Geombinatorics* **9** (2000) 127–134.

[Gr93]   B. GRÜNBAUM: Astral (n_k) configurations, *Geombinatorics* **3** (1993) 32–37.

[Gr72]   B. GRÜNBAUM: *Arrangements and Spreads*, CBMS Regional Conference Series in Mathematics, No.10. AMS 1972, reprinted 1980.

[GrR90]   B. GRÜNBAUM, J.F. RIGBY: The real configuration (21_4), *J. London Math. Soc.* **41** (1990) 336–346.

[Io01]   A. IOSEVICH: Curvature, combinatorics, and the Fourier transform, *Notices Amer. Math. Soc.* **48** (2001) 577–583.

[KaT02]   N.H. KATZ, T. TAO: New bounds for Kakeya problems, *J. Anal. Math.* **87** (2002) 231–263.

[MaT05]   A. MARCUS, G. TARDOS: On topological graphs without self-intersecting 4-cycles, in: *Graph Drawing* (Proc. GD 2004, New York), J. Pach, ed., Springer *LNCS* **3383** (2005) 349–359.

[Mu02]   D. MUBAYI: Intersecting curves in the plane, *Graphs Combinatorics* **18** (2002) 583–589.

[PaA95]   J. PACH, P.K. AGARWAL: *Combinatorial Geometry*, Wiley, New York 1995.

[PaR*04]   J. PACH, R. RADOIČIĆ, G. TARDOS, G. TÓTH: Improving the Crossing Lemma by finding more crossings in sparse graphs, in: *SCG 04, Proc. 20th ACM Symp. Comput. Geom.* 2004, 68–75.

[PaS04]   J. PACH, M. SHARIR: Geometric incidences, in: *Towards a Theory of Geometric Graphs* (J. Pach, ed., *Contemporary Math.* **342** (2004) 185–223.

[PaS98]   J. PACH, M. SHARIR: On the number of incidences between points and curves, *Comb. Probab. Comput.* **7** (1998) 121–127.

[PaS92]   J. PACH, M. SHARIR: Repeated angles in the plane and related problems, *J. Combinatorial Theory Ser. A* **67** (1992) 12–22.

[PaT97]   J. PACH, G. TÓTH: Graphs drawn with few crossings per edge, *Combinatorica* **17** (1997) 427–439.

[Pal86]   I. PALÁSTI: A construction for arrangements of lines with vertices of large multiplicity, *Stud. Sci. Math. Hungar.* **21** (1986) 67–78.

[Re882]   TH. REYE: Das Problem der Configurationen, *Acta Math.* **1** (1882) 93–96.

[RiT95]   R.B. RICHTER, C. THOMASSEN: Intersections of curve systems and the crossing number of $C_5 \times C_5$, *Discrete Comput. Geom.* **13** (1995) 149–159.

[Sa99]   G. SALAZAR: On the intersections of systems of curves, *J. Combinatorial Theory Ser. B* **75** (1999) 56–60.

[ShW04]   M. SHARIR, E. WELZL: Point–line incidences in space, *Combin. Probab. Comput.* **13** (2004) 203–220.

[St894]   E. STEINITZ: *Über die Construction der Configurationen* n_3, Dissertation, Univ. Breslau 1894.

[Sz97]   L.A. SZÉKELY: Crossing numbers and hard Erdős problems in discrete geometry, *Combin. Probab. Comput.* **6** (1997) 353–358.

[SzT83]   E. SZEMERÉDI, W.T. TROTTER: Extremal problems in discrete geometry, *Combinatorica* **3** (1983) 381–392.

[TaT98]   H. TAMAKI, T. TOKUYAMA: How to cut pseudoparabolas into segments, *Discrete Comput. Geom.* **19** (1998) 265–290.

[Wo99]   T. WOLFF: Recent work related to the Kakeya problem, in: *Prospects in Mathematics* (Princeton, NJ, 1999), Amer. Math. Soc. 1999, 129–162.

7.2 Sylvester–Gallai-Type Problems

Perhaps the most famous problem concerning the set of lines spanned by finitely many points is Sylvester's following question [Sy893]. Is it true that any finite set of points in the Euclidean plane, not all on a line, has two elements whose connecting line does not pass through a third? Such a connecting line is called an *ordinary* line. The history of this problem is somewhat complicated; many proofs appeared backdated or were ascribed to someone else. Much of the confusion can be explained by interruptions to academic life caused by World War II. In the 1930s, the question was redis-covered by Erdős, and shortly thereafter, an affirmative answer was given by T. Grünwald (alias Gallai). In 1943, Erdős [Er43] posed the problem in the *American Mathematical Monthly*. In the following year, it was solved by Steinberg [St44] and others.[*] However, the oldest published proof is due to Melchior [Me41], who established the dual statement, as a corollary to a more general inequality (discussed in Section 7.3): any finite family of lines in the plane, not all of which pass through the same point, deter-mines a *simple* intersection point, i.e., a point that belongs to precisely two lines. Both the primal and dual forms of the result have become known as the *Sylvester–Gallai theorem*. Many alternative proofs and generaliza-tions have been found by de Bruijn and Erdős [dBrE48], Coxeter [Co48], Motzkin [Mo51], Lang [La55], Williams [Wi68], Lin [Li88], Edelstein et al. [EdHK63], [Ed70], P. Borwein [Bo84], [Bo83a], Giering [Gi95], Herzog and Kelly [HeK60], Kupitz [Ku92], Watson [Wa80]. A survey of these results was given by P. Borwein and W. Moser [BoM90].

The next natural task is to find the minimum number $ol(n)$ of ordinary lines (passing through precisely two points) determined by n noncollinear points in the plane.

Conjecture 1 *(Dirac [Di51], Motzkin) For every $n \neq 7, 13$, the number of ordinary lines determined by n noncollinear points in the plane is at least $\left\lceil \frac{1}{2}n \right\rceil$.*

Kelly and W. Moser [KeM58] proved that $ol(n) \geq \frac{3}{7}n$. The dual state-ment also holds for pseudo-line arrangements [KeR72]. The best known lower bound, $ol(n) \geq \frac{6}{13}n$, was found by Csima and Sawyer [CsS93], [CsS95]. The following exact values were determined by Crowe and Mc-

---

[*] The frequently cited note, "T. Gallai: Solution to problem 4065, *Amer. Math. Monthly* **51** (1944) 169–171" does not exist. At the cited pages, one finds [St44] followed by an editorial remark containing Gallai's proof.

Kee [CrM68], and by Brakke [Br72]:

n	3	4	5	6	7	8	9	10	11	12	13	14	15	16	17	18	19–21	22
$ol(n)$	3	3	4	3	3	4	6	5	6	6	6	7	?	8	?	9	?–?	11

Observe that the function $ol(n)$ is not monotonically increasing. There are two configurations with 7 and 13 points with exceptionally few ordinary lines. Apart from these values, the general upper bound is $ol(n) \leq \frac{n}{2}$ if n is even and $ol(n) \leq 3 \left\lfloor \frac{n}{4} \right\rfloor$ if n is odd.

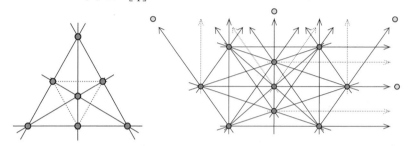

EXCEPTIONAL SETS WITH 7 AND 13 POINTS AND FEW ORDINARY LINES

For even n, the construction consists of a regular $\frac{n}{2}$-gon, which determines $\frac{n}{2}$ directions, and the $\frac{n}{2}$ projective points corresponding to these directions. If $n \equiv 1 \pmod 4$, the best known example can be obtained by adding the center of the polygon to the construction for $n-1$. If $n \equiv 3 \pmod 4$, one of the projective points has to be deleted from the construction for $n+1$. The substantial difference between the cases n even and n odd is quite unusual. It is especially strange in view of the fact that there are several other configurations of both parities that determine very few different directions and may serve as bases for constructions of point sets with few ordinary lines.

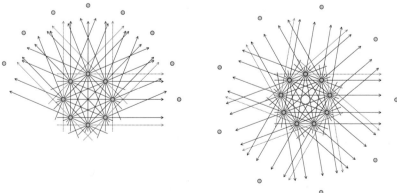

SETS OF 16 AND 18 POINTS WITH 8 AND 9 ORDINARY LINES

Problem 2   *For odd values of n, find sets of n points in the plane that determine only $\frac{1}{2}n + O(1)$ ordinary lines.*

A ghost reference related to the Dirac–Motzkin conjecture is the dissertation of S. Hansen from 1981. It claims to prove Conjecture 1, but one of its key lemmas turned out to be false [CsS93].

The Sylvester–Gallai theorem can be rephrased as follows. Let X be a finite set of points in the plane. If for any two elements $x_1, x_2 \in X$, there is a third point $x_3 \in X$ such that x_1, x_2, x_3 are collinear, then all points of X are collinear. This can be regarded as a strengthening of the trivial Helly-type statement: if any three points of X are collinear, then all of them are collinear. In the same spirit, we can try to strengthen any Helly-type theorem. Let Π be a property of finite sets, for which the following Helly-type statement holds:

> If any $(k + 1)$-element subset of X has property Π, then X has property Π.

We can ask whether the following stronger result is true:

> If every k-element subset of X can be extended to a $(k+1)$-element subset of X that has property Π, then X has property Π.

Results of the latter type are often called Sylvester–Gallai-type theorems. A few examples (in addition to the classical Sylvester–Gallai theorem) are listed below.

(a) Let $\mathcal{L}$ be a finite set of lines in the plane. If any two elements of $\mathcal{L}$ cross at a point that belongs to a third element, then all elements of $\mathcal{L}$ pass through the same point. (This is the dual version of the Sylvester–Gallai theorem.)

(b) Let $\mathcal{L}$ be a finite set of pseudo-lines in the plane, i.e., two-way infinite curves, any two of which cross precisely once. If any two elements of $\mathcal{L}$ cross at a point that belongs to a third element, then all elements of $\mathcal{L}$ pass through the same point [KeR72].

(c) If any three elements of a finite set X of points in the plane determine a circle that passes through a fourth element, then all elements of X lie on a circle [El67].

(d) If any five elements of a finite set X of points in the plane determine a conic that passes through a sixth element, then all elements of X lie on a conic [WiW88].

(e) Let $f : \mathbb{R} \to \mathbb{R}$ be a real function, and let $X = \{x_1, \ldots, x_n\}$ be a set of reals. We say that a polynomial *interpolates* f at x if its value at x is equal to $f(x)$. Suppose that for any $d+1$ elements $x_{i_0}, \ldots, x_{i_d} \in X$, the polynomial p of degree d that interpolates f at $x_{i_0}, \ldots, x_{i_d}$ also interpolates it at another place $x_j \in X$. Then the function f restricted to X is a polynomial of degree d [To89], [Bo83b].

The last three results suggest that it may be possible to prove a similar theorem for algebraic curves of degree k.

Problem 3   *Is it true that if a finite set X of points in the plane has the property that every $\frac{1}{2}k(k+3)$-element subset of X determines an algebraic curve of degree k that passes through another element of X, then all points of X lie on an algebraic curve of degree k?*

Not every Helly-type theorem can be strengthened in this way. For example, the "straightforward" Sylvester–Gallai-type generalization fails in the following two situations.

(a) There are finite point sets X in $\mathbb{R}^3$ (points distributed on two skew lines) such that for any three elements of X there is a plane passing through a fourth element, but X is not coplanar [Mo51].

(b) There are finite point sets X in the complex plane such that the complex line determined by any two elements of X passes through a third element, but not all elements of X lie in the same complex line.

To formulate a higher-dimensional analogue of the Sylvester–Gallai theorem, we have to define an *ordinary hyperplane* (with respect to a set $X \subset \mathbb{R}^d$) as a hyperplane in which all but exactly one point are contained in a $(d-2)$-dimensional affine subspace. With this definition, Motzkin [Mo51] and Hansen [Ha65] proved that any noncoplanar finite set of points in d-dimensional space spans an ordinary hyperplane (see also [BaBS66]). Similarly, Kelly [Ke86] established a weaker Sylvester–Gallai theorem for complex lines. The fact that the Sylvester–Gallai property fails in the complex plane shows that there does not exist a proof based purely on linear algebra that would work for any field. This prompted the study of abstract geometries in which the Sylvester–Gallai property holds; for a recent survey see [PrS05].

Let $oh_d(n)$ stand for the minimum number of ordinary hyperplanes spanned by a full-dimensional set of n points in $\mathbb{R}^d$. Improving some previous bounds from [Mo51], [Ha65], [BoK71], Hansen [Ha80] proved that $oh_3(n) \geq \frac{2}{5}n$. For higher dimensions, no lower bound is known apart from $oh_d(n) \geq 1$ [Ha65]. Neither do we have any reasonable conjecture for the exact value of $oh_d(n)$ for $d \geq 3$.

Problem 4   *(Motzkin [Mo51]) What is the minimum number $oh_d(n)$ of ordinary hyperplanes spanned by a full-dimensional set of n points in $\mathbb{R}^d$?*

The Sylvester–Gallai theorem has some colored variants, where the role of ordinary lines is played by special bichromatic lines or by monochromatic lines.

Pach and Pinchasi [PaP00] proved that given n black and n white points in the plane, there is a bichromatic line containing at most two black and two white points; but there need not be a bichromatic ordinary line. Motzkin [Mo67], Rabin, and Chakerian [Ch70] independently showed that any set of finitely many noncollinear black and white points in the plane spans a monochromatic line. (See also [ErP95] p. 819, [Gr99], and [PrS04].) However, there does not necessarily exist a monochromatic ordinary line (e.g., if the black points and white points lie on two separate lines).

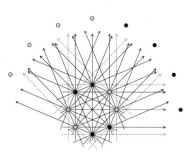

8 BLACK AND 8 WHITE POINTS
NO BICHROMATIC ORDINARY LINE

Chakerian's theorem does not extend to $\mathbb{R}^3$, where a two-colored noncoplanar point set need not span a monochromatic plane [ErP95].

Problem 5   *Given a set of finitely many black and white points in three-dimensional Euclidean space with full-dimensional color classes, does there always exist a monochromatic plane?*

See [BoM90] for many further colored variants of the Sylvester–Gallai theorem.

There are many interesting questions for types of curves other than straight lines. Any three points determine a (possibly degenerate) circle. A circle is called *ordinary* if it passes through precisely three elements of the underlying set. Elliott [El67] proved that any set of n points in the plane, not all on a circle, determine at least $\frac{4}{63}\binom{n}{2}$ ordinary circles. Bálintová and Bálint [BáB94] improved this bound to $\frac{22}{247}\binom{n}{2}$, but this result is surely far from optimal.

Problem 6   *(Elliott [El67]) Determine the supremum of all values c satisfying the condition that any set of n points in the plane, not all on a circle, determines at least $(c + o(1))n^2$ ordinary circles.*

It would be interesting to prove some analogues of the above results for conics [WiW88].

Problem 7   *(Brass) What is the largest $f(n)$ such that any set of n points in the plane, not all on a conic, determine at least $f(n)$ conics, each containing exactly five points?*

Similar quantitative questions can be asked in connection with the dual forms of the Sylvester–Gallai theorem and its generalizations.

Problem 8   *What is the minimum number of simple intersection points in an arrangement of n pairwise crossing pseudo-lines that are not all concurrent?*

As mentioned before, Kelly and Rottenberg [KeR72] established a lower bound of $\frac{3}{7}n$ for this number. There is an example of seven lines with three simple intersection points, dual to the exceptional seven-point configuration of the original Sylvester–Gallai theorem. It is not clear how to combine several copies of this construction to obtain larger examples without violating the condition that the curves are pairwise crossing.

Meyer [Me74] proved that any arrangement of n pairwise crossing simple closed curves in the plane that does not determine a *digon* (i.e., a two-sided cell) has at least $\frac{1}{2}n$ simple crossings. Here two curves are allowed to cross any finite number of times, but touching is not permitted.

Conjecture 9   *(Grünbaum [Gr72]) Any digon-free arrangement of n pairwise crossing simple closed curves in the plane, no two of which touch each other, has at least $n-1$ simple crossings. Equality holds only for $n \equiv 1 \pmod 3$.*

A typical family of pairwise intersecting unit circles has many digons. Nevertheless, A. Bezdek [Be92], [Be99] conjectured and Pinchasi [Pi02] proved that any set of at least five pairwise intersecting unit circles in the plane determines at least one simple crossing. Similarly, Agarwal et al. [AgN*04], [AlL*01] showed that any sufficiently large set of pairwise crossing pseudo-circles in the plane, not all of which pass through the same pair of points, determines an intersection point incident to at most three pseudo-circles.

A weakly related concept, now going under the name "Sylvester–Gallai configurations," is a geometric realization of block designs by lines and subspaces [BoRG92], [BoFK89].

[AgN*04]   P.K. AGARWAL, E. NEVO, J. PACH, R. PINCHASI, M. SHARIR, S. SMORODINSKY: Lenses in arrangements of pseudo-circles and their applications, *J. ACM* **51** (2004) 139–186.

[AlL*01]   N. ALON, H. LAST, R. PINCHASI, M. SHARIR: On the complexity of arrangements of circles in the plane, *Discrete Comput. Geom.* **26** (2001) 465–492.

[BáB94]   A. BÁLINTOVÁ, V. BÁLINT: On the number of circles determined by n points in the Euclidean plane, *Acta Math. Hungar.* **63** (1994) 283–289.

[BaBS66] R.H. Balomenos, W.E. Bonnice, R.J. Silverman: Extensions of Sylvester's theorem, *Canadian Math. Bull.* **9** (1966) 1–14.

[Be99] A. Bezdek: Incidence problems for points and unit circles, in: *Paul Erdős and His Mathematics*, A. Sali et al., eds., Budapest 1999, 33–36.

[Be92] A. Bezdek: On the intersection points of unit circles, *Amer. Math. Monthly* **99** (1992) 779-780.

[BoRG92] J. Bokowski, J. Richter-Gebert: A new Sylvester–Gallai configuration representing the 13-point projective plane in R^4, *J. Combinatorial Theory Ser. B* **54** (1992) 161–165.

[BoK71] W.E. Bonnice, L.M. Kelly: On the number of ordinary lines, *J. Combinatorial Theory* **11** (1971) 45–53.

[BoFK89] E. Boros, Z. Füredi, L.M. Kelly: On representing Sylvester-Gallai designs, *Discrete Comput. Geom.* **4** (1989) 345–348.

[Bo84] P.B. Borwein: Sylvester's problem and Motzkin's theorem for countable and compact sets, *Proc. Amer. Math. Soc.* **90** (1984) 580–584.

[Bo83a] P.B. Borwein: A conjecture related to Sylvester's problem, *Amer. Math. Monthly* **90** (1983) 389–390.

[Bo83b] P.B. Borwein: On Sylvester's problem and Haar spaces, *Pacific J. Math.* **109** (1983) 275–278.

[BoM90] P.B. Borwein, W.O.J. Moser: A survey of Sylvester's problem and its generalizations, *Aequationes Math.* **40** (1990) 111–135.

[Br72] K.A. Brakke: Some new values for Sylvester's function for n noncollinear points, *J. Undergrad. Math.* **4** (1972) 11-14.

[dBrE48] N.G. de Bruijn, P. Erdős: On a combinatorial problem, *Proc. Akad. Wet. Amsterdam* **51** (1948) 1277–1279 = *Indagationes Math.* **10** (1948) 421–423.

[Ch70] G.D. Chakerian: Sylvester's problem on collinear points and a relative, *Amer. Math. Monthly* **77** (1970) 164–167.

[Co48] H.S.M. Coxeter: A problem of collinear points, *Amer. Math. Monthly* **55** (1948) 26–28.

[CrM68] D.W. Crowe, T.A. McKee: Sylvester's problem on collinear points, *Math. Mag.* **41** (1968) 30–34.

[CsS95] J. Csima, E.T. Sawyer: The 6n/13 theorem revisited, in: *Graph Theory, Combinatorics, Algorithms and Applications.*

Vol. 1, Y. Alavi et al., eds, Wiley, New York, 235–249.

[CsS93]   J. CSIMA, E.T. SAWYER: There exist $6n/13$ ordinary points, *Discrete Comput. Geom.* **9** (1993) 187–202.

[Di51]   G.A. DIRAC: Collinearity properties of sets of points, *Quarterly J. Math.* **2** (1951) 221–227.

[Ed70]   M. EDELSTEIN: Generalizations of the Sylvester problem, *Math. Mag.* **43** (1970) 250–254.

[EdHK63]   M. EDELSTEIN, F. HERZOG, L.M. KELLY: A further theorem of the Sylvester type, *Proc. Amer. Math. Soc.* **14** (1963) 359–363.

[El67]   P.D.T.A. ELLIOTT: On the number of circles determined by n points, *Acta Math. Acad. Sci. Hungar.* **18** (1967) 181–188.

[Er43]   P. ERDŐS: Problem 4065, *Amer. Math. Monthly* **50** (1943) p. 65.

[ErP95]   P. ERDŐS, G. PURDY: Extremal problems in combinatorial geometry, in: *Handbook of Combinatorics. Vol. 1*, R.L. Graham et al., eds., North-Holland/Elsevier 1995, 809–874.

[Gi95]   O. GIERING: Zum Problem von Sylvester Punktmengen mit k-Tripeln, *Sitzungsber., Abt. II, Österr. Akad. Wiss., Math.-Naturwiss. Kl.* **204** (1995) 119–143.

[Gr99]   B. GRÜNBAUM: Monochromatic intersection points in families of colored lines, *Geombinatorics* **9** (1999) 3–9.

[Gr72]   B. GRÜNBAUM: *Arrangements and Spreads*, CBMS Regional Conference Series in Mathematics, No.10. AMS 1972, reprinted 1980.

[Ha80]   S. HANSEN: On configurations in 3-space without elementary planes and on the number of ordinary planes, *Math. Scand.* **47** (1980) 181–194.

[Ha65]   S. HANSEN: A generalization of a theorem of Sylvester on the lines determined by a finite point set, *Math. Scand.* **16** (1965) 175–180.

[HeK60]   F. HERZOG, L.M. KELLY: A generalization of the theorem of Sylvester, *Proc. Am. Math. Soc.* **11** (1960) 327–331.

[Ke86]   L.M. KELLY: A resolution of the Sylvester–Gallai problem of J.-P. Serre, *Discrete Comput. Geom.* **1** (1986) 101–104.

[KeM58]   L.M. KELLY, W.O.J. MOSER: On the number of ordinary lines determined by n points, *Canadian J. Math.* **10** (1958) 210–219.

[KeR72]　L.M. Kelly, R. Rottenberg: Simple points in pseudoline arrangements, *Pacific J. Math.* **40** (1972) 617–622.

[Ku92]　Y.S. Kupitz: On a generalization of the Gallai-Sylvester theorem, *Discrete Comput. Geom.* **7** (1992) 87–103.

[La55]　G.D.W. Lang: The dual of a well-known theorem, *Math. Gazette* **39** (1955) p. 314.

[Li88]　X.B. Lin: Another brief proof of the Sylvester theorem, *Amer. Math. Monthly* **95** (1988) 932–933.

[Me41]　E. Melchior: Über Vielseite der projektiven Ebene, *Deutsche Mathematik* **5** (1941) 461–475.

[Me74]　W. Meyer: On ordinary points in arrangements, *Israel J. Math.* **17** (1974) 124–135.

[Mo67]　T.S. Motzkin: Nonmixed connecting lines, Abstract 67T 605, *Notices Amer. Math. Soc.* **14** (1967) p. 837.

[Mo51]　T.S. Motzkin: The lines and planes connecting the points of a finite set, *Trans. Amer. Math. Soc.* **70** (1951) 451–464.

[PaP00]　J. Pach, R. Pinchasi: Bichromatic lines with few points, *J. Combinatorial Theory Ser. A* **90** (2000) 326–335.

[Pi02]　R. Pinchasi: Gallai-Sylvester theorem for pairwise intersecting unit circles, *Discrete Comput. Geom.* **28** (2002) 607–624.

[PrS04]　L.M. Pretorius, K.J. Swanepoel: An algorithmic proof of the Motzkin–Rabin theorem, *Amer. Math. Monthly* **111** (2004) 245-251.

[PrS05]　L.M. Pretorius, K.J. Swanepoel: The Sylvester–Gallai theorem, colourings and algebra, manuscript.

[St44]　R. Steinberg: Solution to Problem 4065, *Amer. Math. Monthly* **51** (1944) 169–171.

[Sy893]　J.J. Sylvester: Mathematical question 11851, *Educational Times* **46** (1893) 156.

[To89]　F.A. Toranzos: A combinatorial approach to the generalized Sylvester's problem, *Rev. Union Mat. Argent.* **35** (1989/1990) 101–103.

[Wa80]　K.S. Watson: Sylvester's problem for spreads of curves, *Canadian J. Math.* **32** (1980) 219–239.

[Wi68]　V.C. Williams: A proof of Sylvester's theorem on collinear points, *Amer. Math. Monthly* **75** (1968) 980–982.

[WiW88]　J.A. Wiseman, P.R. Williams: A Sylvester theorem for conic sections, *Discrete Comput. Geom.* **3** (1988) 295–305.

7.3 Line Arrangements Spanned by a Point Set

We say that a line is *spanned* by a point set X if it passes through at least two elements of X. Erdős [Er43] noticed that the Sylvester–Gallai theorem (see Section 7.1) easily implies, by induction, that any planar set of n points, not all on a line, spans at least n distinct lines. Later, de Bruijn and Erdős [dBrE48] found the following purely combinatorial generalization of this result, whose stronger form (Fisher inequality) plays an important role in the theory of combinatorial designs. Any family $\mathcal{L}$ of proper subsets of an n-element set X with the property that every pair $\{x_1, x_2\} \in X$ occurs in exactly one member of $\mathcal{L}$ has at least n members. Equality holds here only for finite projective planes or when $\mathcal{L}$ consists of an $(n-1)$-element set $X' \subset X$ and all two-element sets containing the remaining point of $X \setminus X'$. In the first case, $(X, \mathcal{L})$ cannot be realized as a point–line incidence structure in the Euclidean plane. The second case corresponds to a "near-pencil," that is, to a family of n lines generated by n points, all but one of which belong to the same line. What is the minimum number of lines determined by a planar point set that is not a near-pencil? Elliott [El67] and Kelly and W. Moser [KeM58] proved that the answer is $2n - 4$ for $n \geq 11$. What other numbers of lines, larger than $2n - 4$, are realizable? The situation is somewhat diffuse: there are increasingly many small numbers of lines that are realizable. On the other hand, Erdős [Er72] showed that all values from $cn^{\frac{3}{2}}$ to $\binom{n}{2}$ are realizable, with the exception of $\binom{n}{2} - 1$ and $\binom{n}{2} - 3$. A complete description was attempted by Salamon and Erdős [SaE88]. For the realizable numbers of circles spanned by an n-element point set, similar results were found by Beck [Bec85].

Problem 1   *[KeM58] For a given $k \geq 2$, what is the minimum number of lines determined by n points in the plane, no $n - k$ of which are collinear?*

As mentioned in the previous section, Motzkin [Mo51] and Hansen [Ha65] proved that any set X of n points in $\mathbb{R}^d$, not all on a hyperplane, spans an "ordinary hyperplane," i.e., a hyperplane whose intersection with X, with the exception of precisely one element, lies in a $(d-2)$-dimensional affine subspace. This implies, by induction, that X determines at least n distinct hyperplanes. This bound is sharp, as shown by the configuration of $n - d + 1$ collinear points together with $d - 1$ points in "general position," not belonging to this line. In this case, it is hard to imagine any purely combinatorial generalization, because large groups of points can be collinear and thus occur together in several hyperplanes.

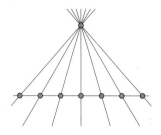
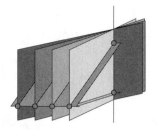

n POINTS GENERATING n LINES IN THE PLANE AND n PLANES IN SPACE

With additional assumptions on the point set, one can obtain stronger lower bounds for the number of spanned hyperplanes. Erdős and Purdy [ErP77] showed that n points in convex position (vertices of a convex polyhedron) in three-dimensional space determine at least $\binom{n-1}{2} + 1$ planes. This bound is attained for a pyramid over an $(n-1)$-gon. In d-dimensional space, n points, not all of which lie in a hyperplane and no d of which are in a $(d-2)$-dimensional affine subspace, determine at least $c_d n^{d-1}$ hyperplanes, each containing exactly d points. No work has been done on the realizable numbers of hyperplanes in higher dimensions.

Problem 2　　*For which values of k does there exist an n-element set in $\mathbb{R}^d$ that spans exactly k hyperplanes?*

One can ask similar questions for many other structures. Simple duality shows that in the projective plane, n lines, not all of which pass through the same point, determine at least n distinct intersection points. (This is true only in the projective plane, where every pair of lines intersect.) For circles, Elliott [El67] proved that any set of $n \geq 394$ points in the plane, not all on the same circle, determines at least $\binom{n-1}{2} + 1$ distinct circles. Apart from the assumption on n, this bound is clearly best possible. Furthermore, A. Bezdek [Bez92], [Bez99] showed that any connected set of n unit circles, no two of which are tangent, determines at least n intersection points. This result is also best possible.

A conjecture of Purdy [Pu86], [ErP75] attempts to connect the numbers of lines and planes spanned by n points in three-dimensional space.

Conjecture 3　　*(Purdy [Pu86]) If n is sufficiently large, then any set of n points in three-dimensional space that cannot be covered by two lines, or by a plane and a point, spans at least as many planes as lines.*

Counterexamples are known for up to 16 points, but they do not seem to generalize. Beck [Bec83] proved that n points in $\mathbb{R}^d$ span at least $c_d n^d$ hyperplanes, unless at least $c'_d n$ of them lie in a hyperplane.

Another famous problem related to the system of lines spanned by a finite point set in the plane has become known as the "strong Dirac conjecture" [Di51].

Conjecture 4 *(Dirac [Di51]) There is a constant c such that any set X of n points, not all on a line, has an element incident to at least $\frac{n}{2} - c$ lines spanned by X.*

If X is equally distributed on two lines, then this bound is tight with $c = 0$. Many small examples listed by Grünbaum [Gr72] show that the conjecture is false with $c = 0$. An infinite family of counterexamples was constructed by Felsner (personal communication): $6k+7$ points, each of them incident to at most $3k+2$ lines. The "weak Dirac conjecture," proved by Beck [Bec83], states that there exists $\epsilon > 0$ such that one can always find a point incident to at least ϵn lines spanned by X. This statement also follows from the Szemerédi–Trotter theorem on the number of point–line incidences [SzT83], [PaT97] (see Section 7.1).

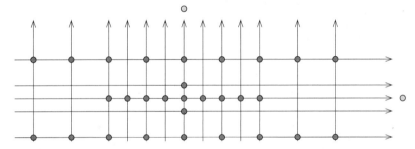

COUNTEREXAMPLE TO THE STRONG DIRAC CONJECTURE WITH $c = 0$

Jung and Melchior [JuM36] made an attempt to classify, for small values of k, all planar point sets whose every element is incident to the same number k of spanned lines.

Artes et al. [ArGL97], [ArGL98] noticed that the arrangement of invariant lines of a polynomial differential system $x' = p_1(x, y)$, $y' = p_2(x, y)$ has some interesting combinatorial properties, depending on the degree n of p_1 and p_2. They made the conjecture that every arrangement satisfying these properties consists of at most $2n + O(1)$ lines.

Conjecture 5 *(Artes, Grünbaum, Llibre [ArGL97]) Let $\mathcal{L}$ be an arrangement of straight lines in the plane satisfying the following four conditions:*
(1) Each line $\ell \in \mathcal{L}$ has at most n distinct intersection points with other lines in $\mathcal{L}$.
(2) The number of distinct intersection points between the lines in $\mathcal{L}$ is at most $O(n^2)$.
(3) $\mathcal{L}$ has at most n pairwise parallel lines.
(4) Every point is incident to at most $n + 1$ lines in $\mathcal{L}$.
Then $\mathcal{L}$ consists of at most $2n + O(1)$ lines.

An arrangement of n horizontal and n vertical lines meets the above requirements, and it is easy to construct many other examples. Note that the validity of this conjecture would immediately follow from the "strong Dirac conjecture" (Conjecture 4).

A similar question was raised by de Berg et al. [dBB*98], in a model of *geometric probing* of lines by lines. Suppose there is a "hidden" set $\mathcal{L}$ of lines in the plane. We are allowed to draw k parallel *query lines* across the plane, and we obtain all intersection points of these lines with the elements of $\mathcal{L}$. How large can the hidden set be if each query line contains at most n intersection points? Let $f^{\mathrm{hl}}(n, k)$ denote this maximum.

Trivially, we have $f^{\mathrm{hl}}(n, 2) = n^2$.
One can show that

$$f^{\mathrm{hl}}(n, 3) = \left\lceil \tfrac{3}{4} n^2 \right\rceil.$$

For a fixed k, the $k \times n$ integer lattice gives

$$f^{\mathrm{hl}}(n, k) \geq \frac{1}{k-1} n^2,$$

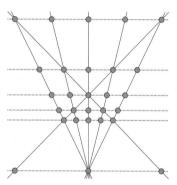

and the point–line incidence bound (Szemerédi–Trotter theorem [SzT83]) implies $f^{\mathrm{hl}}(n, k) = O(\frac{n^2}{k} + n)$. For large values of k, the situation is not so clear. There is a construction for $n \geq 3$ that shows

$$f^{\mathrm{hl}}(n, n + 1) \geq n + 2$$

$n + 2$ LINES INTERSECTING
EACH OF $n + 1$ PARALLEL LINES
ONLY IN AT MOST n POINTS

(and not equal to n, as claimed in [dBB*98]).

Problem 6   *(de Berg et al. [dBB*98]) Given a set $\mathcal{Q}$ of k horizontal ("query") lines in the plane, at most how many nonhorizontal lines can be selected with the property that they intersect each element of $\mathcal{Q}$ in at most n distinct points?*

Consider all lines spanned by a given configuration of n points in the plane. For any $k \geq 2$, let t_k denote the number of lines passing through precisely k points. Clearly, we have $\sum_k \binom{k}{2} t_k = \binom{n}{2}$. There are several other important inequalities involving these parameters. For instance, Euler's polyhedral formula easily implies that

$$t_2 \geq 3 + t_4 + 2t_5 + 3t_6 + \cdots, \quad \text{if } t_n = 0.$$

This simple result is known as *Melchior's inequality* [Me41], and it yields the existence of at least three ordinary lines in any set of points, not all on a line.

Another useful result is *Hirzebruch's inequality* [Hi86]:

$$t_2 + \tfrac{3}{4} t_3 \geq n + t_5 + 3t_6 + 5t_7 + \cdots, \quad \text{if } t_n = t_{n-1} = t_{n-2} = 0.$$

This inequality was derived using algebraic geometry, and no elementary proof is known.

Problem 7   *Find an elementary proof for Hirzebruch's inequality.*

The assumption that no more than $n-3$ points are collinear is necessary in Hirzebruch's inequality. The inequality fails for an arrangement of $n-2$ points on a line and two points off that line, for $n \geq 8$. Apart from this fact, there is no reason to believe that the inequality is sharp.

The *Erdős–Purdy inequalities* [ErP78] are similar. If $t_n = 0$, then

$$\max(t_2, t_3) \geq n - 1 \quad \text{and} \quad \max(t_2, t_3) \geq t_i, \text{ for all } i.$$

They also showed that if $t_2 \leq n - 2$ and $t_n = 0$, then $t_3 \geq cn^2$ for some absolute constant c. Also, if $t_3 \geq \alpha t_2$ holds for some $\alpha \in [0, 1]$ and $t_n = 0$, then $t_3 \geq c'\alpha^2 n^2$ and the number of spanned lines is at least $c''\alpha n^2$, for some constants $c', c'' > 0$.

Beck's result [Bec83], mentioned already in the context of Dirac's conjecture, can be stated as follows. If $t_k = 0$ for all $k \geq \frac{1}{100}n$, then the total number of spanned lines satisfies $\sum_k t_k \geq cn^2$, for a suitable constant $c > 0$.

Problem 8   *For a fixed r, find further linear inequalities for the numbers t_k of k-point lines spanned by a set of n points in the plane with no $n - r$ collinear elements.*

It follows by a simple application of the Szemerédi–Trotter theorem [SzT83] on the number of incidences between n points and m lines in the plane (see [PaT97], [PaR*04]) that

$$t_k + t_{k+1} + t_{k+2} + \cdots \leq 15.5\frac{n^2}{(k - 1)^3}, \quad \text{for } k < n^{\frac{1}{3}}, n > n_0(k).$$

The order of magnitude of this bound cannot be improved, as mentioned already in Section 7.1. All known examples for which this inequality is tight apart from a constant factor have the property that many lines pass through a large number (more than $c\frac{n}{k}$) of points. This is probably not a coincidence, but it does not seem to follow from any known result.

Problem 9   *For a fixed $r > k$, what is the maximum number t_k of k-point lines spanned by a set of n points in the plane with no r collinear points?*

An extreme case of this question (for $r = 4$) is Sylvester's famous "orchard problem": what is the maximum number of collinear triples determined by n points in the plane, provided that no four points are collinear? For the interesting and rather complicated history of this problem, see

[BuGS74]. In general, let $t_k^{\mathrm{orchard}}(n)$ denote the maximum of t_k over all n-element point sets in the plane containing no $k+1$ collinear points.

Problem 10　　What is the maximum number $t_k^{\mathrm{orchard}}(n)$ of k-point lines in a set of n points in the plane containing no $k+1$ collinear elements?

Sylvester gave a construction with $\frac{n^2}{8} + O(n)$ collinear triples, using the fact that the points (x, x^3), (y, y^3) and (z, z^3) are collinear if and only if $x+y+z = 0$. This was improved by Burr, Grünbaum, and Sloane [BuGS74] and by Füredi and Palásti [FüP84] to

$$t_3^{\mathrm{orchard}}(n) \geq \lfloor \tfrac{1}{6}n^2 - \tfrac{1}{2}n + 1 \rfloor,$$

using other cubic curves and their associated group structures. The best current construction uses the points

$$p(\alpha) = \left(\frac{\sin\alpha - \sin 2\alpha}{\cos\alpha + \cos 2\alpha}, \frac{\sin 3\alpha}{\cos\alpha + \cos 2\alpha} \right),$$
$$\alpha \in \{\tfrac{1}{n}\pi, \tfrac{3}{n}\pi, \ldots, \tfrac{2n-1}{n}\pi\}.$$

The points $p(\alpha_1)$, $p(\alpha_2)$, $p(\alpha_3)$ are col-
linear if and only if $\alpha_1 + \alpha_2 = \alpha_3 \equiv 0(\mathrm{mod}\,4\pi)$. The above lower bound is
very close to the trivial upper bound
$t_3^{\mathrm{orchard}}(n) \leq \frac{1}{3}\binom{n}{2}$, which follows from
the fact that each pair of points occurs
in at most one collinear triple.

POINTS ON A CUBIC CURVE
WITH MANY COLLINEAR TRIPLES

　　Since each ordinary line uses a pair of points that does not occur in a collinear triple, the upper bound can be improved to $t_3^{\mathrm{orchard}}(n) \leq \lfloor \frac{1}{6}n^2 - \frac{25}{78}n \rfloor$, using the lower bound of Csima and Sawyer [CsS93], [CsS95] on the number of ordinary lines. It should be possible to close the remaining small gap and perhaps even prove that any extremal set lies on a cubic curve. Some exact values were determined in [BuGS74] and again in [Gi95]. For $n \leq 12$ and $n = 16$, there are sets with a minimum number of ordinary lines in which all other lines pass through collinear triples.

n	3	4	5	6	7	8	9	10	11	12	13	14	15	16
$t_3^{\mathrm{orchard}}(n)$	1	1	2	4	6	8	10	13	16	20	?	?	?	37

Problem 11　　Is it true that for n sufficiently large, any set of n points in the plane with the maximum number of collinear triples, but with no collinear four-tuple, can be covered by a cubic curve?

　　For higher values of k ($k \geq 4$), the problem of bounding $t_k^{\mathrm{orchard}}(n)$ becomes difficult, interesting, and wide open. To improve the trivial upper bound $t_4^{\text{orchard}} \leq \frac{1}{12}n^2$, even to $t_4^{\text{orchard}} \leq \left(\frac{1}{12} - \varepsilon\right)n^2$, was often stated by Erdős [Er84b], [ErP95b] as an important open problem. A simple improvement to $t_4^{\text{orchard}} < \frac{1}{14}n^2$, using Melchior's inequality, was noticed by Brass [Br03]. The real problem here is to reduce the upper bound to a subquadratic one.

Conjecture 12 *For any fixed $k \geq 4$, we have $t_k^{\text{orchard}} = o(n^2)$.*

For a long time, the best known lower bound was $t_k^{\text{orchard}}(n) \geq c_k n^{1+\frac{1}{k-2}}$. In his paper in which the generalized orchard problem was first stated, Grünbaum [Gr76] described the following ingenious recursive construction. Let X be a set of n points, no $k+1$ on a line, that spans m lines with k points each, and let q be an integer parameter. Using X, Grünbaum constructed a set of $qn + m$ points, no $k+2$ on a line, that spans qm lines with $k+1$ points each. For this purpose, he chose an affine map φ with no invariant directions (e.g., a rotation), and defined a family of affine maps $\varphi_1, \ldots, \varphi_q$ by $\varphi_i = \frac{q-1-i}{q-1}I + \frac{i}{q-1}\varphi$, where I stands for the identity. Let the new set consist of the union of all affine copies $\varphi_i(X)$ and, for each line g spanned by X, the intersection point of the affine pencil $\varphi_1(g), \ldots, \varphi_q(g)$. Thus, we have $t_{k+1}^{\text{orchard}}\left(qn + t_k^{\text{orchard}}(n)\right) \geq q t_k^{\text{orchard}}(n)$. Starting with $t_3^{\text{orchard}}(n) \geq \frac{n^2}{6}$ and choosing $q = n^{\frac{1}{k-2}}$, we obtain Grünbaum's lower bound $t_k^{\text{orchard}}(n) \geq c_k n^{1+\frac{1}{k-2}}$.

Using a tricky combination of homothetic lattice cubes, Ismailescu [Is02] showed that $t_k^{\text{orchard}}(n) \geq c_k n^{\frac{\log(k+4)}{\log k}}$. For $5 \leq k \leq 35$, this is better than Grünbaum's construction. It can be used as the starting point of Grünbaum's recursive construction to obtain $t_k^{\text{orchard}}(n) \geq c_k n^{1+\frac{1}{k-3.59}}$ for all $k \geq 18$, which is currently the best bound. Another construction with the same asymptotic behavior was presented by Brass [Br03]. It deserves some attention, because of its simplicity and relation to the problem of "probing lines by lines" (Problem 6). Brass showed that if $f^{\text{hl}}(k, k) \geq k + a$, then $t_k^{\text{orchard}}(n) \geq c_k n^{\frac{\log(a+k)}{\log k}}$.

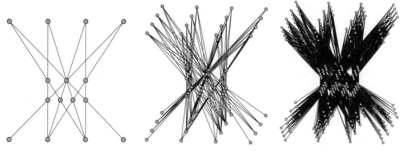

SETS WITH NO 5 POINTS COLLINEAR, MANY COLLINEAR 4-TUPLES

Problem 13   *Improve the lower bound on $t_k^{\mathrm{orchard}}(n)$ for $k \geq 4$.*

It would be especially interesting to improve the lower bound in the first open case $k = 4$, where all three known constructions [Gr76], [Is02], [Br03] provide the same bound $t_4^{\mathrm{orchard}}(n) \geq \Omega(n^{\frac{3}{2}})$.

The following closely related problem, also often mentioned by Erdős, is open for every $k \geq 3$.

Problem 14   *(Erdős) What is the largest number $f(n, k)$ such that any set of n points in the plane, no $k+1$ of which are collinear, has an $f(n, k)$-element subset with no k collinear points?*

Füredi [Fü91] showed that $\lim_{n \to \infty} \frac{f(n,k)}{n} = 0$ for a fixed k, but his proof uses a powerful tool: the density version of the Hales–Jewett theorem, due to Furstenberg and Katznelson [FuK91]. If we take any projection of the d-dimensional lattice cube $\{0, 1, \ldots, k-1\}^d$, we obtain a set of $n = k^d$ points in the plane with many collinear k-tuples, and for $d > d_0(\varepsilon)$, any subset of εn points will contain a collinear k-tuple. Hence, we cannot achieve a positive fraction of the points in the subset without a collinear k-tuple.

As for a lower bound, we can find a subset of $cn^{1 - \frac{1}{k-1}}$ points, with no k elements collinear, using greedy or randomized selection. This bound can be slightly improved to $f(n, k) \geq c_k n^{1 - \frac{1}{k-1}} (\log n)^{\frac{1}{k-1}}$ using better probabilistic tools [Fü91]. Füredi's argument already makes optimal use of the combinatorial information that the collinear k-tuples form a partial Steiner system. Phelps and Rödl [PhR86] and Rödl and Šinajová [RöS94] constructed partial Steiner systems with no independent sets of size larger than a constant times the above value. To make any further progress, one needs to explore the geometric structure of the problem.

One can ask several interesting questions on how to cut a point set into as equal parts as possible. Of course, if no three elements of a finite point set X in the plane are collinear, then one can find a line spanned by X with $\lfloor \frac{n-2}{2} \rfloor$ points on one of its sides and $\lceil \frac{n-2}{2} \rceil$ points on the other. In general, such exactly balanced partitions need not exist if each "nearly halving" line contains many elements of X.

One way to measure the "balancing" of a partition realized by a line ℓ is to determine the smaller of the two numbers of points lying on the left-hand side and on the right-hand side of ℓ.

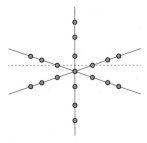

There is an example of $6k + 1$ points such that any spanned line has at most $3k - 2$ points on each of its sides. (The case $k = 3$ is depicted in the figure.) On the other hand, Pinchasi [Pi03] showed that any set of n points, not all collinear, spans a line with at least $\frac{n}{2} - c \log \log n$ points on each side, for some constant $c > 0$.

19 POINTS SUCH THAT EACH SPANNED LINE HAS ON ONE SIDE AT MOST 7 POINTS

Problem 15   *(Pinchasi [Pi03]) Is it true that any set of n points in the plane, not all on a line, spans a line with at least $\frac{n}{2} - 3$ points on each side?*

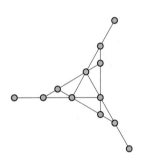

Another natural way to measure the "balancing" of a partition realized by a line ℓ is to determine the difference between the numbers of points on the two sides of ℓ. Alon [Al02] found a set X of $8k + 4$ points in the plane such that the numbers of points on the two sides of any line spanned by X differ by at least two. On the other hand, Pinchasi's theorem [Pi03] yields an upper bound of $O(\log \log n)$.

EACH DIVISION BY A SPANNED LINE IS UNBALANCED BY ≥ 2 POINTS

Problem 16   *(Kupitz [Ku79], see also [Er84a]) Is it true that any set of n points in the plane spans a line for which the numbers of points on its two sides differ by at most two?*

In their effort to solve the last problem, Perles and Pinchasi [PeP04] raised the following question. Is it true that for any $k \leq \frac{1}{2}(n - 2)$, every set X of n noncollinear points in the plane spans a line bounding an open half-plane that contains precisely k elements of X? The answer is no. However, a slightly weaker statement is true: one can always find a line spanned by X with either k or $k + 2$ points on one of its sides. Unfortunately, this result does not settle the previous problem, because the line with the required property (for $k = \left\lceil \frac{1}{2}(n - 2) \right\rceil$) is not necessarily ordinary: it may pass through many points.

The concept of an omissible point was introduced by Koutský and Polák [KoP60] and rediscovered in [Gr99b] (under the name "omittable").

An element x of a finite point set X in the plane is *omissible* if $X \setminus \{x\}$ and X span the same sets of lines.

Problem 17   *What is the maximum number $\psi(n)$ of omissible points in any set of n points in the plane, not all on a line?*

Grünbaum [Gr99b] proved that $\psi(n) \geq \lfloor \frac{1}{3}n \rfloor$, which had been previously known only in the special case that all omissible points are collinear [KoP60]. Grünbaum [Gr99b] also found several examples with more than $\lfloor \frac{1}{3}n \rfloor$ omissible points, but he made the following conjecture.

Conjecture 18 *(Grünbaum [Gr99b]) The maximum number $\psi(n)$ of omissible points that a noncollinear set of n points in the plane can have satisfies $\limsup\limits_{n \to \infty} \frac{\psi(n)}{n} = \frac{1}{3}$.*

Conjecture 19 *(Grünbaum [Gr99b]) If n is sufficiently large, then in any set of n points in the plane, all but at most one omissible points are collinear.*

Motzkin, Grünbaum [Gr75], [Gr99], Erdős and Purdy [ErP78], [ErP95a] independently asked the following question. What is the smallest number $m(n)$ of points necessary to represent (i.e., block) all lines spanned by n noncollinear points in the plane if the generating points cannot be used? Equivalently, what is the smallest number $m(n)$ for which there is a configuration of $m(n)$ blue and n red points in the plane, the red points not all collinear, that does not span a monochromatic red line? The red and blue points have to be distinct; otherwise, the problem becomes trivial. An $\Omega(n)$ lower bound follows from the "weak Dirac conjecture," (Beck's theorem [Bec83], [SzT83]) that guarantees the existence of a point that lies on $\Omega(n)$ connecting lines. Clearly, each of these lines requires a separate blue point to block it. A set of $n - 1$ collinear red points and one point off that line shows that $m(n) \leq n$, but Grünbaum [Gr99a] constructed several configurations with a slightly smaller number of blue points. He obtained $m(n) \leq n - 4$, provided that n is sufficiently large, and gives one example showing that $m(16) \leq 10$. Up to now it is not known whether

$$\lim_{n \to \infty} \frac{m(n)}{n} = 1.$$

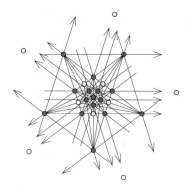

16 DARK AND 10 WHITE POINTS, NO MONOCHROMATIC DARK LINE

Problem 20   *(Grünbaum) Is it true that there is a constant C with the property that all lines spanned by a set X of n noncollinear points in the plane can never be represented by fewer than $n - C$ points not belonging to X?*

[Al02]   N. ALON: Problem 365: Splitting lines for planar pointsets, *Discrete Math.* **257** (2002) 601–602.

[ArGL98]   J.C. ARTES, B. GRÜNBAUM, J. LLIBRE: On the number of invariant straight lines for polynomial differential systems *Pacific J. Math.* **184** (1998) 207–230.

[ArGL97]   J.C. ARTES, B. GRÜNBAUM, J. LLIBRE: On the invariant straight lines of polynomial differential systems, *Differ. Equ. Dyn. Syst.* **5** (1997) 317–327.

[Bec85]   J. BECK: Remarks on combinatorial geometry I, *Stud. Sci. Math. Hungar.* **20** (1985) 249–254.

[Bec83]   J. BECK: On the lattice property of the plane and some problems of Dirac, Motzkin and Erdős in combinatorial geometry, *Combinatorica* **3** (1983) 281–297.

[dBB*98]   M. DE BERG, P. BOSE, D. BREMNER, W. EVANS, L. NARAYANAN: Recovering lines with fixed linear probes, in: *CCCG 1998, Proc. Tenth Canadian Conf. Comput. Geom.*, see extended version at
           http://www.cs.arizona.edu/people/will/papers/probe.ps.gz

[Bez99]   A. BEZDEK: Incidence problems for points and unit circles, in: *Paul Erdős and His Mathematics* (extended abstracts of papers from the International Conference held in memory of Paul Erdős in Budapest, July 4–11, 1999) A. Sali et al., eds., Budapest 1999, 33–36. [Math. Reviews 1901861]

[Bez92]   A. BEZDEK: On the intersection points of unit circles, *Amer. Math. Monthly* **99** (1992) 779-780.

[Br03]   P. BRASS: On point sets without k collinear points, in: *Discrete Geometry—in Honor of W. Kuperberg's 60th Birthday*, A. Bezdek, ed., Marcel Dekker, 2003, 185–192.

[dBrE48]   N.G. DE BRUIJN, P. ERDŐS: On a combinatorial problem, *Proc. Akad. Wet. Amsterdam* **51** (1948) 1277–1279. Also in: *Indagationes Math.* **10** (1948) 421–423.

[BuGS74]   S.A. BURR, B. GRÜNBAUM, N.J.A. SLOANE: The orchard problem, *Geometriae Dedicata* **2** (1974) 397–424.

[CsS95]   J. Csima, E.T. Sawyer: The $6n/13$ theorem revisited, in:
           Graph Theory, Combinatorics, Algorithms and Applications
           Vol. 1, Y. Alavi et al., eds, Wiley, 1995, 235–249.

[CsS93]   J. Csima, E.T. Sawyer: There exist $6n/13$ ordinary points,
           Discrete Comput. Geom. **9** (1993) 187–202.

[Di51]   G.A. Dirac: Collinearity properties of sets of points, Quar-
           terly J. Math. **2** (1951) 221–227.

[El67]   P.D.T.A. Elliott: On the number of circles determined by
           n points, Acta Math. Acad. Sci. Hungar. **18** (1967) 181–188.

[Er84a]   P. Erdős: Some old and new problems in combinatorial
           geometry, in: Convexity and Graph Theory, M. Rosenfeld et
           al., eds., Annals Discrete Math. **20** (1984) 129–136.

[Er84b]   P. Erdős: Research Problems 36, Periodica Math. Hungar-
           ica **15** (1984) 101–103.

[Er72]   P. Erdős: On a problem of Grünbaum, Canadian Math.
           Bull. **15** (1972) 23–25.

[Er43]   P. Erdős: Problem 4065, Amer. Math. Monthly **50** (1943)
           p. 65.

[ErP95a]   P. Erdős, G. Purdy: Two combinatorial problems in the
           plane, Discrete Comput. Geom. **13** (1995) 441–443.

[ErP95b]   P. Erdős, G. Purdy: Extremal problems in combinatorial
           geometry, in: Handbook of Combinatorics. Vol. 1, R.L. Gra-
           ham et al., eds., North-Holland/Elsevier 1995, 809–874.

[ErP78]   P. Erdős, G. Purdy: Some combinatorial problems in the
           plane, J. Combinatorial Theory Ser. A **25** (1978) 205–210.

[ErP77]   P. Erdős, G. Purdy: Some extremal problems in geometry
           V, Congressus Numerantium **19** (1977)   (Proc. 8th South-
           Eastern Conf. Combinatorics, Graph Theory, and Computing)
           569–578.

[ErP75]   P. Erdős, G. Purdy: Some extremal problems in geometry
           III, Congressus Numerantium **14** (1975)   (Proc. 6th South-
           Eastern Conf. Combinatorics, Graph Theory, and Computing)
           291–308.

[Fü91]   Z. Füredi: Maximal independent subsets in Steiner systems
           and in planar sets, SIAM J. Discrete Math. **4** (1991) 196–199.

[FuK91]   H. Furstenberg, Y. Katznelson: A density version of
           the Hales–Jewett theorem, J. Anal. Math. **57** (1991) 64–119.

[FüP84]   Z. Füredi, I. Palásti: Arrangements of lines with large
           number of triangles, Proc. Amer. Math. Soc. **92** (1984)

561–566.

[Gi95]   O. Giering: Zum Problem von Sylvester Punktmengen mit
           k-Tripeln, *Sitzungsber., Abt. II, Österr. Akad. Wiss., Math.-
           Naturwiss. Kl.* **204** (1995) 119–143.

[Gr99a]   B. Grünbaum: Monochromatic intersection points in fami-
           lies of colored lines, *Geombinatorics* **9** (1999) 3–9.

[Gr99b]   B. Grünbaum: Omittable points, *Geombinatorics* **9** (1999)
           57–62.

[Gr76]   B. Grünbaum: New views of some old questions of combina-
           torial geometry, in: *Colloq. Int. Theorie Comb., Roma 1973,
           Tomo I* (1976) 451–468.

[Gr75]   B. Grünbaum: Arrangements of colored lines, Abstract 720-
           50-5, *Notices Amer. Math. Soc.* **22** (1975) A-200.

[Gr72]   B. Grünbaum: *Arrangements and Spreads*, CBMS Regional
           Conference Series in Mathematics, No.10. AMS 1972, re-
           printed 1980.

[Ha65]   S. Hansen: A generalization of a theorem of Sylvester on
           the lines determined by a finite point set, *Math. Scand.* **16**
           (1965) 175–180.

[Hi86]   F. Hirzebruch: Singularities of algebraic surfaces and char-
           acteristic numbers, in: *Proc. Lefschetz Centennial Conf.*
           (Mexico City, 1984), Part I, *Contemporary Mathematics* **58**,
           AMS 1986, 141–155.

[Is02]   D. Ismailescu: Restricted point configurations with many
           collinear k-tuplets, *Discrete Comput. Geom.* **28** (2002) 571–
           575.

[JuM36]   W. Jung, E. Melchior: Symmetrische Geradenkonfigura-
           tionen, *Deutsche Mathematik* **1** (1936) 239–255.

[KeM58]   L.M. Kelly, W.O.J. Moser: On the number of ordinary
           lines determined by n points, *Canadian J. Math.* **10** (1958)
           210–219.

[KoP60]   K. Koutský, V. Polák: Note on the omissible points
           in complete systems of points and straight lines in the plane
           (in Slovakian, with English abstract), *Časopis pro pěstování
           matematiky* **85** (1960) 60–69.

[Ku79]   Y.S. Kupitz: *Extremal Problems in Combinatorial Geo-
           metry*, Lecture Notes Ser. Vol. 53, Matematisk Institut, Aarhus
           Universitet 1979.

[Me41]   E. Melchior: Über Vielseite der projektiven Ebene,
           Deutsche Mathematik **5** (1941) 461–475.

[Mo51]　　T.S. Motzkin:　The lines and planes connecting the points of a finite set, *Trans. Amer. Math. Soc.* **70** (1951) 451–464.

[PaR*04]　J. Pach, R. Radoičić, G. Tardos, G. Tóth:　Improving the Crossing Lemma by finding more crossings in sparse graphs, in: *SCG 04 Proc. 20th ACM Symp. Comput. Geom.*, 2004, 68–75.

[PaT97]　J. Pach, G. Tóth:　Graphs drawn with few crossings per edge, *Combinatorica* **17** (1997) 427–439.

[PeP04]　M.A. Perles, R. Pinchasi:　Large sets must contain either a k-edge or a $(k+2)$-edge, in: *Towards a Theory of Geometric Graphs*, J. Pach, ed., *Contemporary Mathematics* **342**, AMS 2004, 225–232.

[PhR86]　K.T. Phelps, V. Rödl:　Steiner Triple Systems with minimum independence number, *Ars Combinatoria* **21** (1986) 167–172.

[Pi03]　R. Pinchasi:　Lines with many points on both sides, *Discrete Comput. Geom.* **30** (2003) 415–435.

[Pu86]　G. Purdy:　Two results about points, lines and planes, *Discrete Math.* **60** (1986) 215–218.

[RöS94]　V. Rödl, E. Šinajová:　Note on independent sets in Steiner systems, *Random Structures Algorithms* **5** (1994) 183–190.

[SaE88]　P. Salamon, P. Erdős:　The solution of a problem of Grünbaum, *Canadian Math. Bull.* **31** (1988) 129–138.

[SzT83]　E. Szemerédi, W.T. Trotter:　Extremal problems in discrete geometry, *Combinatorica* **3** (1983) 381–392.

8. Problems on Points in General Position

8.1 Structure of the Space of Order Types

In this chapter, we treat problems on points in "general position." Problems that are invariant under slight perturbation of the point sets and that require that no three elements be collinear can usually be described as problems on *order types*. The order type is an equivalence relation on sets of n points in the plane, no three collinear, in which two sets S, S' are equivalent if and only if there is a bijection $\beta \colon S \to S'$ between the points such that each triangle (ordered point triple) $abc \in S$ has the same orientation as its image $\beta(a)\beta(b)\beta(c)$. A similar definition can be given in higher dimensions: two n-element sets in general position in $\mathbb{R}^d$ have the same order type if there is a bijection between them that does not change the orientation of any $(d+1)$-tuple. The order type captures the relative position of the elements of a point set and preserves many special properties such as convex position.

ORDER TYPES OF 3, 4, AND 5 POINTS

The space of all order types has been studied extensively [GoP93], and it turns out to be unexpectedly complicated. One reason for this phenomenon is that the order types are defined as equivalence classes of n-point sets for each n independently, and the fact that two sets have the same order type does not mean that the possible order types of their extensions by one point are also the same. For any set P, the order types of the extensions of P by a point q are in one-to-one correspondence with the cells determined by the connecting lines of P, depending on where q falls.

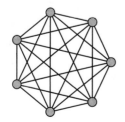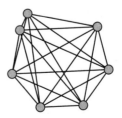

TWO CONVEX 7-GONS WITH THE SAME ORDER TYPE,
BUT DIFFERENT DIAGONAL ARRANGEMENTS

However, the combinatorial type of the arrangement of the connecting lines is not uniquely determined by the order type of P. In particular, for a fixed n, all convex n-gons are of the same order type, but the cell decompositions realized by their diagonals may be quite different. Therefore, the recursive enumeration of order types is a difficult task.

Goodman and Pollack determined the asymptotic behavior of the number of different order types of n-element point sets in d dimensions as $n \to \infty$ and d is fixed. They proved that this number is $\left(\frac{n}{d}\right)^{d^2 n (1 + \Theta(\frac{1}{\log(n/d)}))}$ [GoP93], [GoP91], [GoP86], [Al86]. Aichholzer, Aurenhammer, and Krasser [AiK01], [AiAK02], [Kr03] have managed to give a complete list of all planar order types with up to eleven points. They obtained the following exact numbers of planar order types:

n	3	4	5	6	7	8	9	10	11
#ordertype(n)	1	2	3	16	135	3315	158 817	14 309 547	2 334 512 907

Problem 1   *What is the number of different order types of n points in the plane?*

A further difficulty in describing the space of all order types is that some order types of n points allow only very "big" realizations. Goodman, Pollack, and Sturmfels [GoPS89] proved that if we want to realize them by points of integer coordinates, then we are forced to use doubly exponentially large numbers in terms of n. In a slightly different formulation, the ratio between the largest volume and the smallest volume of a triangle induced by a set of n points in the plane grows at least doubly exponentially with n [GoPS90]. There are probably many other ways to describe this phenomenon. For example, it would be interesting to find a result of this type based on distances. Although one can define "typical" point sets in several different ways, in most interpretations it is true that certain order types cannot occur in typical sets (e.g., in data given by integer coordinates of polynomial length). Thus, excluding sets that have no subsets of some "forbidden" type may dramatically change the solution of many extremal problems. Questions of this kind related to the Erdős–Szekeres problem and to the halving lines problem are discussed in Sections 8.2 and 8.3. Since many order types are extremely unlikely to occur in randomly chosen point sets, it is not possible to test certain geometric conjectures (or algorithms) by checking random examples.

Problem 2   *For a given constant $\alpha > 0$, what is the number of order types of n points that can be represented by integer coordinates smaller than 2^{n^α}?*

Problem 3   *For a given constant $\beta > 0$, what is the number of order types of n points that can be represented by integer coordinates smaller than n^β?*

Problem 4   *For a given constant $\gamma > 0$, what is the number of order types of n points that have a realization S with $\frac{\mathrm{diam}(S)}{\mathrm{mindist}(S)} \leq n^\gamma$?*

The concept of order types can be generalized in many different ways. We mention some of these without going into details. We start with the notion of *allowable sequences*, introduced by Goodman and Pollack [GoP80], [GoP81] and since studied in many papers (see [GoP93] for a survey). Consider a set P of n points in the plane, labelled by the integers $1, \ldots, n$. Perpendicularly projecting these points onto a line ℓ, we obtain a permutation of $1, \ldots, n$. The circular sequence of permutations obtained by rotating ℓ through an angle π is an allowable sequence. An important property of such a sequence is that each permutation can be obtained from the previous one by reversing some consecutive intervals: those corresponding to the connecting lines of P orthogonal to the position of ℓ. If no three points in P are on a line and no two connecting lines are parallel, then in every step we reverse a single pair of elements, and the corresponding allowable sequence is said to be *simple*. A related algorithmic notion, "sorting by reversals," has recently also received some attention. Another generalization of order types is the concept of "CC-systems." It was introduced by Knuth [Kn92], mainly with the algorithmic perspective to axiomatize the orientation test for point triples. By dualizing (simple) order types or allowable sequences, one obtains combinatorial types of (simple) line arrangements. Further generalizations include (simple) *pseudo-line arrangements* and *oriented matroids*.

An arrangement of *pseudo-lines* is a family of x-monotone curves in the plane, any pair of which intersect precisely once, at an ordinary crossing. If no three pseudo-lines pass through the same point, the arrangement is said to be *simple*. Felsner [Fe97] enumerated all simple pseudo-line arrangements with ten or fewer elements, and established the best known upper bound for the number of different simple arrangements of n pseudo-lines: roughly $2^{0.7n^2}$.

Conjecture 5   *(Knuth [Kn92]) The number of different simple arrangements of n pseudo-lines is at most $2^{\binom{n}{2}}$.*

It is known that this number is at least $2^{\frac{n^2}{6} - \frac{5n}{2}}$.

[AiK01]   O. AICHHOLZER, H. KRASSER: The point set order type data base: a collection of applications and results, in: *CCCG 01, 13th Canadian Conf. Comput. Geom.* (2001).

[AiAK02] O. Aichholzer, F. Aurenhammer, H. Krasser: Enumerating order types for small point sets with applications, *Order* **19** (2002) 265–281.

[Al86] N. Alon: The number of polytopes, configurations, and real matroids, *Mathematika* **33** (1986) 62–71.

[Fe97] S. Felsner: On the number of arrangements of pseudolines, *Discrete Comput. Geom.* **18** (1997) 257–267.

[GoP93] J.E. Goodman, R. Pollack: Allowable sequences and order types in discrete and computational geometry, in: *New Trends in Discrete and Computational Geometry*, J. Pach, ed., Springer *Algorithms and Combinatorics* **10** (1993) 103–134.

[GoP91] J.E. Goodman, R. Pollack: The complexity of point configurations, *Discrete Appl. Math.* **31** (1991) 167–180.

[GoP86] J.E. Goodman, R. Pollack: Upper bounds for configurations and polytopes in $\mathbb{R}^d$, *Discrete Comput. Geom.* **1** (1986) 219–227.

[GoP81] J.E. Goodman, R. Pollack: A combinatorial perspective on some problems in geometry, *Congressus Numerantium* **32** (1981) 383–394.

[GoP80] J.E. Goodman, R. Pollack: On the combinatorial classification of nondegenerate configurations in the plane, *J. Combinatorial Theory Ser. A* **29** (1980) 220–235.

[GoPS90] J.E. Goodman, R. Pollack, B. Sturmfels: The intrinsic spread of a configuration in $\mathbb{R}^d$, *J. Amer. Math. Soc.* **3** (1990) 639–651.

[GoPS89] J.E. Goodman, R. Pollack, B. Sturmfels: Coordinate representation of order types requires exponential storage, in: *SToC 89, 21st ACM Symp. Theory of Computing* (1989) 405–410.

[Kn92] D.E. Knuth: *Axioms and Hulls*, Springer-Verlag *LNCS* **606** (1992).

[Kr03] H. Krasser: *Order Types of Point Sets in the Plane*, PhD thesis, Institute for Theoretical Computer Science, Graz University of Technology, Austria, October 2003.

8.2 Convex Polygons and the Erdős–Szekeres Problem

In this section, just as in the rest of Chapter 8, all point sets are assumed to be in *general position*, that is, in the plane, they do not contain three points on a line, or, in d-dimensional space, $d + 1$ points in a hyperplane. These point sets can be described by their order types, as defined in Section 8.1.

Perhaps the most famous problem concerning point sets in general position is the "Erdős–Szekeres convex polygon problem," also known as Esther Klein's problem or the Happy End problem* (see [MoS00], [BáK01], for recent surveys). What is the smallest number $f^{\mathrm{ES}}(r)$ such that any set of $f^{\mathrm{ES}}(r)$ points in general position in the plane contains r points that form the vertex set of a convex r-gon? Instead of saying that a point set contains the "vertex set of a convex r-gon," in the sequel we simply say that it contains a "convex r-gon."

In their first joint paper, Erdős and Szekeres [ErS35] established the existence of this number by reducing the result to a Ramsey-type statement and rediscovering Ramsey's theorem for its proof. In a much later paper [ErS60], the same authors obtained the lower bound $f^{\mathrm{ES}}(r) \geq 2^{r-2} + 1$. The only known exact values of this function are $f^{\mathrm{ES}}(3) = 3$, $f^{\mathrm{ES}}(4) = 5$, and $f^{\mathrm{ES}}(5) = 9$ [Bo74], [KaKS70]. It is conjectured that the lower bound is sharp for all r.

Conjecture 1 *(Erdős–Szekeres [ErS35]) Any set of $2^{r-2} + 1$ points in general position in the plane contains a convex r-gon.*

Let $R_k(r, s)$ denote the smallest number n with the property that for any coloring of the k-tuples of an n-element set with red and blue, one can find either an r-element subset all of whose k-tuples are red or an s-element subset all of whose k-tuples are blue.

In their original paper [ErS35], Erdős and Szekeres first established the upper bound $f^{\mathrm{ES}}(r) \leq R_4(5, r)$ by coloring all four-tuples of points with two colors according to whether they are in convex position or not. According to Esther Klein's old observation, among any five points there are four in convex position. Furthermore, any set of r points, all of whose four-tuples are in convex position, is a convex r-gon. Thus, if a set of n points does not induce a convex r-gon, then its four-tuples can be colored by two colors so that one of its color classes does not contain a complete four-uniform subhypergraph $K_5^{(4)}$ on five vertices and the other color class does not contain a $K_r^{(4)}$.

---

* The problem proposer Esther Klein married the problem solver George Szekeres in 1937, four years after collaborating on this problem.

There are two other alternative reductions of this problem to Ramsey problems, both leading to the upper bound $f^{\mathrm{ES}}(r) \leq R_3(r, r)$. In both cases, we color the triples of points. In the first such argument, due to Johnson, we color a triple pqr according to the parity of the number of points in the interior of the triangle pqr [Jo86]. In the second, the color of a triple depends on its orientation relative to some fixed linear ordering of the vertex set [Le76]. In both cases, a monochromatic r-element subset is necessarily the vertex set of a convex r-gon.

Although Ramsey's theorem is a powerful tool for deriving existence results, it is often useless to obtain good bounds. As Erdős and Szekeres [ErS35] discovered, to get a better result one can use a more delicate Ramsey-type argument for "caps" and "cups," which was used in all subsequent (modest) improvements of their estimate. An r-*cap* or r-*cup* is an r-vertex polygonal arc that is concave or convex with respect to some preferred direction. Let $f^{\mathrm{cap/cup}}(r, s)$ denote the minimum number n such that every set of n points in general position, no two of which are on a line parallel to the preferred direction, contains either an r-cap or an s-cup. Obviously, we have $f^{\mathrm{ES}}(r) \leq f^{\mathrm{cap/cup}}(r, r)$, and the function $f^{\mathrm{cap/cup}}$ is easier to analyze. It satisfies the simple recurrence relation

$$f^{\mathrm{cap/cup}}(r, s) \leq f^{\mathrm{cap/cup}}(r - 1, s) + f^{\mathrm{cap/cup}}(r, s - 1) - 1,$$

based on the fact that if a point is simultaneously the left endpoint of an $(r - 1)$-cup and the right endpoint of an $(s - 1)$-cap, then one of these arcs can be extended by one extra segment. This observation together with the boundary condition $f^{\mathrm{cap/cup}}(r, 3) = f^{\mathrm{cap/cup}}(3, r) = r$ allows us to prove the bound

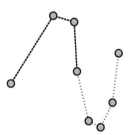

SET WITH 4-CAP AND 5-CUP
THE 5-CUP CAN BE EXTENDED
TO A 6-CUP

$$f^{\mathrm{cap/cup}}(r, s) \leq \binom{r + s - 4}{r - 2} + 1.$$

It is not hard to come up with a construction achieving this bound, which essentially mirrors the inductive decomposition [ErS35]. Thus, we have

$$f^{\mathrm{cap/cup}}(r, s) = \binom{r + s - 4}{r - 2} + 1 \quad \text{for } r, s \geq 3.$$

Since $f^{\mathrm{ES}}(r) \leq f^{\mathrm{cap/cup}}(r, r) = \Theta(4^r / \sqrt{r})$, this upper bound is still quite far from the lower bound $2^{r-2} + 1$ on $f^{\mathrm{ES}}(r)$. On the other hand, by arbitrarily selecting a "preferred direction," we "typically" lose half of the vertices of a convex r-gon. A "typical" r-gon is the union of an $\left(\frac{r}{2}+1\right)$-cap and an $\left(\frac{r}{2}+1\right)$-cup, and

$$f^{\mathrm{cap/cup}}\left(\frac{r}{2}+1,\frac{r}{2}+1\right)=\Theta\left(\frac{2^r}{\sqrt{r}}\right).$$

This may be regarded as further evidence for the conjecture that 2^r is the right order of magnitude of $f^{\mathrm{ES}}(r)$.

The upper bound $f^{\mathrm{ES}}(r)\le f^{\mathrm{cap/cup}}(r,r)$ stood unimproved for a long time. Then, in a short period of time, three better bounds were published in the same special issue of *Discrete & Computational Geometry* dedicated to the memory of Erdős [ChG98], [KlP98], [TóV98]. The proof of the strongest among them was based on the idea of using a projective transformation to move one point of the set to infinity, perpendicularly to the preferred direction, so that each $(r-1)$-cup in the transformed set (minus the point at infinity) corresponds to a convex r-gon in the original set [TóV98]. Thus, $f^{\mathrm{ES}}(r)\le f^{\mathrm{cap/cup}}(r,r-1)+1$. A further improvement by 1 was made by Tóth and Valtr [TóV05], so that we have

$$f^{\mathrm{ES}}(r)\le\binom{2r-5}{r-2}+1,$$

for $r>4$, which is the currently best upper bound. However, its order of magnitude is still $4^{r+o(r)}$, compared to the lower bound of order 2^r.

The construction of the lower bound in [ErS60] is somewhat complicated. It is pieced together from smaller configurations, for which the above upper bound on $f^{\mathrm{cap/cup}}(r,s)$ is attained (see Lovász's book [Lo79] for a simple presentation). The complexity of this construction is reflected by the fact that none of the numerous papers on the Erdős–Szekeres convex polygon problem includes a picture of the 16-point set without a convex hexagon. The problem is usually illustrated by the well-known example of an eight-point set with no convex pentagon, but it is not equivalent to the set resulting from the construction of Erdős and Szekeres. This eight-point set is the union of two concentric squares, but this example does not seem to generalize: any three concentric pentagons determine a convex hexagon. Kalbfleisch and Stanton [KaS95] gave explicit coordinates for the 2^{r-2} points in the Erdős–Szekeres construction. However, even in the case $r=6$ the coordinates are so large that they cannot be used for a reasonable illustration. The order type of the 16-point set shown here is different from that of the standard construction. Thus, even if the Erdős–Szekeres conjecture is true and $f^{\mathrm{ES}}(r)=2^{r-2}+1$ holds, the order type of the extremal configuration is certainly not uniquely determined.

SETS OF 4, 8, AND 16 POINTS NOT CONTAINING

A CONVEX QUADRILATERAL, PENTAGON, OR HEXAGON

The exponential blowup of the coordinates in the above lower bound constructions may be necessary. It is possible that all extremal configurations belong to the class of order types that have no small realizations. The answer to the Erdős–Szekeres problem becomes quite different if we restrict our attention to point sets with the property that the ratio of the largest distance to the smallest distance determined by them is only $O(\sqrt{n})$. Of course, this is the strongest possible restriction of this kind for an n-element point set in the plane. It was shown in [AlKP89], [Va92b] that every set S of n points in general position in the plane satisfying the condition

$$\frac{\mathrm{diam}(S)}{\mathrm{mindist}(S)} \leq \alpha \sqrt{n}$$

contains a convex polygon with $\Omega(n^{\frac{1}{3}})$ vertices. This is much larger than the $\Omega(\log n)$ bound that follows from the Erdős–Szekeres theorem for general point sets. On the other hand, there exists such a set S with no more than $O(n^{\frac{1}{3}})$ points in convex position. That is, the size of the largest forced convex polygon depends strongly on metric restrictions on the underlying set S. Therefore, it might be interesting to study the effect of other restrictions of this kind.

Problem 2   *Does there exist, for every $\beta \geq 1$, a suitable constant $\varepsilon(\beta) > 0$ with the following property: any set S of n points in general position in the plane with $\frac{\mathrm{diam}(S)}{\mathrm{mindist}(S)} < n^{\beta}$ contains a convex $n^{\varepsilon(\beta)}$-gon?*

In his thesis, Valtr showed that the answer to the last question is yes for $\beta < 1$.

Problem 3   *Does there exist, for every $\gamma \geq 1$, a suitable constant $\varepsilon(\gamma) > 0$ with the following property: any set of n points in general position in the plane with positive integer coordinates that do not exceed n^{γ} contains a convex $n^{\varepsilon(\gamma)}$-gon?*

For a fixed r and for large values of n, every n-element point set in the plane contains many convex r-gons. Moreover, for every set X of n points in general position in the plane, there are r subsets, $Y_1, \ldots, Y_r \subset X$, each containing a positive fraction of all points ($|Y_i| \geq c_r n$ for $i = 1, \ldots, r$), such that each of their *transversals* forms a convex r-gon [BáV98]. (A transversal of the sets $Y_1, \ldots, Y_r$ is a sequence of elements $y_1, \ldots, y_r$ where $y_i \in Y_i$ for every i.) The last statement follows from the so-called *Same Type Lemma* ([BáV98], also [PaS98]): given $m, d \geq 2$, there exists a constant $c(m, d) > 0$ such that any set X of sufficiently many points in general position in d-space has m subsets $Y_1, \ldots, Y_m$, $|Y_i| \geq c(m, d)|X|$, such that all transversals $y_1, \ldots, y_m$, $y_i \in Y_i$, are of the same order type. So, if $m \geq f^{\mathrm{ES}}(r)$, at least r of these subsets have the property that each of their transversals forms a convex polygon.

Problem 4   *What is the largest c_r with the property that any sufficiently large point set X in general position in the plane has r disjoint subsets $Y_1, \ldots, Y_r \subset X$ with $|Y_i| \geq c_r|X|$ for every i, such that all of their transversals form convex r-gons?*

The best known bounds for c_r are $r2^{-32r} \leq c_r \leq K_\varepsilon (\frac{1}{2} + \varepsilon)^r$ for any $\varepsilon > 0$, where K_ε is a constant depending on ε [PóV02]. Pór and Valtr have also proved that from any point set X in the plane with no three points collinear, one can delete at most $c_1(r)$ "exceptional" points so that the remainder X' satisfies the following condition: X' can be partitioned into $c_2(r)$ groups, each consisting of r subsets of equal sizes, such that all transversals within a group form convex r-gons [PóV02]. In the first nontrivial case, $r = 4$, they showed that 26 groups and four exceptional points are sufficient.

The following interesting variant of the Erdős–Szekeres problem was raised by Erdős and has received much attention: what happens if we want to find an *empty* convex r-gon, that is, r points that form the vertex set of a convex r-gon with no other points of the set in its interior? Obviously, every set of at least three points in general position determines an empty triangle, and it is easy to see that every set of at least five points in general position determines an empty convex quadrilateral.

To guarantee an empty convex pentagon we need ten points [Ha78]. This is the first case in which we obtain a different answer if we insist on empty r-gons: nine points are already enough if we are satisfied with any convex pentagon. It was a big surprise when Horton [Ho83] constructed sets of arbitrarily many elements with no empty convex heptagon.

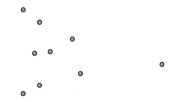

NINE POINTS WITHOUT
EMPTY CONVEX PENTAGON

The only case that remains undecided is $r = 6$: Do there exist arbitrarily large sets in general position in the plane that do not determine any empty convex hexagon? The largest currently known set with this property has 29 points [Ov03], [OvSV89].

Problem 5   *Does every sufficiently large set of points in general position in the plane contain an empty convex hexagon?*

29 POINTS WITHOUT EMPTY CONVEX HEXAGON

The diameters of the sets resulting from Horton's construction [Ho83], relative to the smallest distances, grow exponentially fast with the number of points. However, here the same phenomenon can be observed already for "small" order types. Valtr [Va92b] constructed sets of n points without empty convex heptagons that satisfy $\frac{\text{diam}(S)}{\text{mindist}(S)} \leq \alpha\sqrt{n}$ for arbitrarily large n. Some nice identities and inequalities involving the number of empty k-gons were obtained in [EdJ87], [AhGM99], [EdR00], [Kl99] and, using very elementary arguments, by Pinchasi, Radoičić, and Sharir [PiRS04].

Conjecture 6   *(Valtr [Va02]) For every $r \geq 3$, there exists an integer $f = f^{\text{cap/empty−cup}}(r)$ such that any set of at least f points in general position in the plane, no two of them on a vertical line, contains an r-cap or an empty r-cup.*

This conjecture, if true, would imply that for any r and s, every sufficiently large set of points in the plane with no s-cap contains an empty r-gon.

The empty convex polygon problem has also been studied under a restriction on the maximum number of points in the interior of a triangle. If every triangle spanned by an n-point set is empty, then the points form an empty convex n-gon. If every triangle spanned by the set contains at most q points of the set for some fixed q, then this is still sufficient to guarantee the existence of an empty convex r-gon in any sufficiently large set [KáPT01], [Va02]. The strongest result of this kind was proved by Kun and Lippner [KuL03]. Another variant of the problem in which the points are not required to be in general position was discussed in [BiFT03].

The empty convex polygon problem also has a colored version. Any two-colored set of n points in general position in the plane contains an empty monochromatic triangle. Indeed, for $n \geq 10$ this is almost immediate, because the set contains an empty pentagon, and at least three vertices of this pentagon have the same color. Empty monochromatic triangles in sets without the general position assumption have also been studied [BaB77], [Bo84]. Devillers, Hurtado, Károlyi, and Seara [DeH*03] suggested the following question.

Problem 7   *[DeH*03] Does every sufficiently large two-colored point set in general position in the plane contain an empty monochromatic convex quadrilateral?*

A two-colored sets of 20 points that does not contain a monochromatic empty quadrilateral was given in [Br04]. Friedman [Fri04] improved this lower bound to 30. Constructions of two- and three-colored sets with arbitrarily large numbers of points and without a monochromatic empty pentagon or triangle can be found in [DeH*03]. Thus, in two-colored sets only the existence of empty monochromatic quadrilaterals remains an open problem.

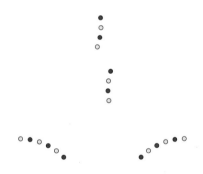

20 POINTS WITH NO EMPTY
MONOCHROMATIC QUADRILATERAL

The Erdős–Szekeres problem can also be extended to higher dimensions. We say that a set of points $X \subset \mathbb{R}^d$ is in *convex position* if X is the vertex set of a convex polytope. Let $f_d^{ES}(r)$ denote the smallest n such that among any n points in general position in d-space one can find r points in convex position. Projections into lower-dimensional spaces can be used to bound these functions, since almost all projections preserve general position, and the preimage of a set in convex position must itself be in convex position. This shows that

$$f_d^{ES}(r) \leq \cdots \leq f_3^{ES}(r) \leq f_2^{ES}(r),$$

so the classical Erdős–Szekeres problem in the plane gives an upper bound for all higher dimensions. However, it seems likely that already $f_3^{\mathrm{ES}}(r)$ is much smaller than $f_2^{\mathrm{ES}}(r)$.

Problem 8   *What is the smallest number $f_d^{\mathrm{ES}}(r)$ such that any set of $f_d^{\mathrm{ES}}(r)$ points in general position in d-space has r elements in convex position?*

The best known lower bound, proved by Károlyi and Valtr, is only $f_d^{\mathrm{ES}}(r) \geq c^{r^{1/(d-1)}}$ [KáV03]. The currently best upper bound can be obtained by a minor improvement of the projection argument [Ká01]. It yields the recurrence relation

$$f_d^{\mathrm{ES}}(r) \leq f_{d-1}^{\mathrm{ES}}(r-1) + 1,$$

which implies

$$f_d^{\mathrm{ES}}(r) \leq \binom{2r - 2d - 1}{r - d} + d.$$

Morris and Soltan [MoS00] made the very strong conjecture that

$$f_d^{\mathrm{ES}}(r) = 4f_d^{\mathrm{ES}}(r - d) - 3$$

for any $d \geq 2$ and $r \geq \lfloor \frac{1}{2}(3d+1) \rfloor$. For small r, more values have been determined. Clearly, we have $f_d^{\mathrm{ES}}(r) = r$ for $r \leq d + 1$. It is easy to see that $f_d^{\mathrm{ES}}(d + 2) = d + 3$. Furthermore, $f_d^{\mathrm{ES}}(r) = 2r - d - 1$ holds for all $d + 2 \leq r \leq \lfloor \frac{3}{2}d \rfloor + 1$ [BiS94], [BiH95]. Finally, Bisztriczky and V. Soltan [BiS94] proved that $f_3^{\mathrm{ES}}(6) = 9$.

One can even guarantee a specific combinatorial type of the subset in convex position: any sufficiently large set in general position in d-dimensional space contains the vertex set of polytope combinatorially equivalent to a cyclic polytope. This follows by a direct generalization of the planar proof of Lewin [Le76]. First color all $(d + 1)$-tuples of points according to their orientation with respect to a fixed numbering (see [Ma02], p. 99, Exercise 3), and then apply Ramsey's theorem to this coloring. However, the number of points that guarantee the existence of a cyclic polytope of r vertices will be much larger than $f_d^{\mathrm{ES}}(r)$.

The empty convex polygon problem also has a higher-dimensional analogue. As in the planar case, if $r > 12^{d-1}(P(d - 1) + 1)$, where $P(d - 1)$ stands for the product of the first $d - 1$ primes, there are arbitrarily large sets that induce no empty convex polytopes with r vertices [Va92a]. For small values of r, one can force the existence of empty convex polytopes with r vertices. Indeed, for every $r \leq 2d+1$, any set of $f_d^{\mathrm{ES}}(4d+1)$ points in general position induces an empty convex polytope with r vertices. Moreover, Bisztriczky, V. Soltan, and Harborth [BiS94],[BiH95] proved that for any $d + 1 \leq r \leq \lfloor \frac{3}{2}d \rfloor + 1$, every set of size $2r - d - 1$ induces an empty convex polytope of r vertices and that this result cannot be improved. In the three-dimensional case, Valtr [Va92a] constructed arbitrarily large sets with no empty convex polytopes of 22 vertices.

Problem 9 *What is the largest number $r \leq 21$ such that every sufficiently large set of points in general position in three-dimensional space induces an empty convex polytope with r vertices?*

The empty convex polygon problem also has a "modular" variant. Instead of looking for convex r-gons containing no other point of the set, one may try to find convex r-gons with the property that the number of points in their interiors is $0 \pmod{q}$, for some fixed q. The question is whether, for a fixed r and q, every sufficiently large set of n points in general position in the plane induces an r-gon with this property. One should note that there is no monotonicity here: it makes a big difference whether we would like to find an r-gon with the required property or whether we are satisfied with an *at least r-gon*, i.e., at least r points in convex position. This problem, the "modular Erdős–Szekeres convex polygon problem," was first studied by Bialostocki, Dierker, and Voxman [BiDV91]. They conjectured that for any r and q, every sufficiently large $n = n(r, q)$-element set induces an r-gon with the required property. This conjecture was proved for $r \geq q + 2$ and for $r \equiv 2 \pmod{q}$ [BiDV91]. The upper bound in [BiDV91] was improved to $n(r, q) \leq 2^{c(q)k}$ by Caro [Ca96]. The conjecture also holds for $r \geq \frac{5}{6}q + O(1)$ [KáPT01], and moreover, according to Valtr (unpublished), for $r \geq \frac{3}{4}q + O(1)$, but almost nothing is known when r is small relative to q.

Conjecture 10 *(Bialostocki–Dierker–Voxman [BiDV91]) For any $r \leq q + 1$, every sufficiently large set of points in general position in the plane contains a convex r-gon such that the number of points in its interior is divisible by q.*

Bisztriczky and G. Fejes Tóth [BiFT89a], [BiFT89b] initiated the generalization of the Erdős–Szekeres problem to families $\mathcal{F}$ of convex bodies, instead of points. It seems that in the plane the relevant analogue of the assumption that no three points are collinear is that there are no three sets $A, B, C \in \mathcal{F}$ such that $B \subset \text{conv}(A \cup C)$. If $\mathcal{F}$ satisfies this condition, we say that its members are in *general position*. Analogously, r members $A_1, \ldots, A_r \in \mathcal{F}$ form a *convex r-gon* [PaT98], [Tó00], [PaT00] or are *convexly independent* [BiFT90] if each of the sets contributes a piece to the boundary of the convex hull of their union; that is, we have for all i,

$$A_i \not\subset \text{conv}\left(\bigcup_{\substack{j=1,\ldots,r \\ j \neq i}} A_j \right).$$

Given any $r \geq 3$, let $f^{\mathrm{ES-sets}}(r)$ denote the smallest number n such that any family of n convex sets in general position in the plane has r members in convex position. Clearly, we have $f^{\mathrm{ES}}(r) \leq f^{\mathrm{ES-sets}}(r)$ for every r. The existence of $f^{\mathrm{ES-sets}}(r)$ was proved in [BiFT89a], [BiFT89b], where an astronomically large upper bound was established. This was reduced to a single exponential bound in [PaT98], while the best known lower bound is only the $2^{r-2} + 1$ already known for $f^{\mathrm{ES}}(r)$.

Problem 11   (Bisztriczky–G. Fejes Tóth [BiFT90])
    Is $f^{\mathrm{ES-sets}}(r) = f^{\mathrm{ES}}(r)$ true for every r?

This has been proved for $r \leq 5$ [BiFT89a].

According to our terminology, a family of convex bodies is in general position if any triple of its members are in convex position. The upper bounds on $f^{\mathrm{ES-sets}}(r)$ can be substantially improved under the stronger condition that any s members of the family are in convex position for some $s > 3$ [BiFT90]. If any four sets are in convex position, then r^3 disjoint convex sets are sufficient to guarantee that r of them are in convex position. If any five sets are in convex position, then r^2 sets are enough, while assuming that any six or more sets are in convex position, the bound becomes linear in r [PaT98], [Tó00]. Pach and Tóth [PaT00] studied how far the condition of disjointness on the underlying sets can be relaxed. One can also establish the following generalization: Every sufficiently large family $\mathcal{F}$ of pairwise disjoint convex bodies in the plane has r disjoint subfamilies, each containing a positive fraction of the members of $\mathcal{F}$, such that picking one element from each subfamily, they are always in convex position [PaS98].

Most of the above questions can be generalized to higher dimensions [FeTP*03]. A family $\mathcal{C}$ of convex bodies in d-space is called *separable* if every pair of its members can be separated from any third one by a hyperplane. A family $\mathcal{C}$ is said to be *nondegenerate* if no $d + 1$ of its members have a common supporting hyperplane. The elements of $\mathcal{C}$ are in *convex position* if the boundary of the convex hull of their union contains a piece of the boundary of every $C \in \mathcal{C}$.

Bisztriczky and G. Fejes Tóth (personal communication) showed that there exists a function $g(n)$ tending to infinity that satisfies the following condition. Every separable nondegenerate family of n convex bodies in three-dimensional space, any seven of which are in convex position, has at least $g(n)$ members that are in convex position.

Problem 12   [FeTP*03] Does there exist, for every fixed k and for every sufficiently large n, a separable nondegenerate family of convex bodies in three-dimensional space, any k of which are in convex position, but not all of them are in convex position?

Problem 13   *[FeTP*03] Do there exist, for every $d > 3$, a constant
$k = k(d)$ and a function $g_d(n)$ tending to infinity that satisfies the following condition: every separable nondegenerate family of n convex bodies in d-dimensional space,
any k of which are in convex position, has at least $g_d(n)$
members that are in convex position?*

Some of the Erdős–Szekeres-type questions also make sense in a "dual"
setting, exchanging the roles of points and lines. An arrangement of lines
or pseudolines in the projective plane is called *simple* if no three of its
elements pass through the same point. Simple line arrangements are the
dual of point sets in general position, and the Erdős–Szekeres theorem implies that any simple arrangement of $f^{ES}(r)$ lines contains a subarrangement with an r-gonal face. This generalizes to pseudolines. Harborth and
Möller [HaM94] proved that every simple arrangement of nine pseudolines
in the projective plane determines a subarrangement with a hexagonal face.
Moreover, for any fixed r, any simple arrangement consisting of sufficiently
many pseudolines has a subarrangement with an r-gonal face.

Problem 14   *(Harborth–Möller [HaM94]) What is the smallest number $f^{ps-line}(r)$ such that any arrangement of $f^{ps-line}(r)$
pseudolines has a subarrangement with an r-gonal face?*

Instead of studying the Erdős–Szekeres-type properties of a point set
X in the plane, one can extend the investigations to all segments connecting point pairs in X, that is, to the straight-line drawing of the complete
graph induced by X. A subset $Y \subseteq X$ in convex position corresponds to a
complete subgraph induced by Y whose edges are represented by pairwise
crossing segments. In other words, the elements of Y are in convex position
if the number of crossings between the edges induced by them is as large as
possible. Using this language, Harborth, Mengersen, and Schelp [HaMS95]
proved the following analogue of the Erdős–Szekeres theorem for general
(not necessarily straight-line) drawings of the complete graph K_n with n
vertices: given any r, there exists $n_0(r)$ such that for any $n > n_0(r)$, every
drawing of K_n induces a subdrawing of K_r that determines $\binom{r}{4}$ crossings.
Clearly, $\binom{r}{4}$ is the maximum possible number of crossings that any drawing
of K_r can have, provided that two edges can cross only at most once.

Let $f^{KT}(r, s)$ denote the minimum number of points such that any
set of $n \geq f^{KT}(r, s)$ points, no three on a line, contains a subset of s
points whose convex hull has at least r vertices. Clearly, we have $f^{ES}(r) \leq
f^{KT}(r, s) \leq f^{ES}(s)$ for all $r \leq s$. Károlyi and Tóth [KáT99] obtained better
bounds on this function, especially for small values of r.

Following Erdős et al. [ErTV96], Károlyi [Ká01] studied the following
question. Let n be a multiple of r. When is it possible to partition a set of

n points in general position into $\frac{n}{r}$ sets, each of which consists of r elements in convex position? He proved that in dimension $d \geq 3$, if n is sufficiently large, then one can always find such a partition. However, in the plane, we may be left with a remainder set, the so-called *Ramsey remainder*, that cannot be decomposed into convex r-gons. For $r = 4$, Károlyi characterized all sets of $4n$ points that cannot be partitioned into n convex quadrilaterals. According to a related result [Fr99], [GuS01], every set of nine points in general position in the plane induces two disjoint convex quadrilaterals.

The smallest number R such that every n-element set in general position in the plane can be split into convex r-gons and at most one "remainder" set of size at most R is denoted by $\mathrm{RR}(n, r)$. Let $f^{\mathrm{convex/cup}}(r, s)$ be the smallest number such that any set of this size contains a convex r-gon or an s-cup. Erdős, Tuza, and Valtr [ErTV96] proved that $\mathrm{RR}(n, r) = f^{\mathrm{convex/cup}}(r, r - 2) - 1$, provided that n is sufficiently large in terms of r. If the Erdős–Szekeres conjecture (Conjecture 1) is true, we have

$$f^{\mathrm{convex/cup}}(r, r - 2) = 2^{r-2} - r + 2.$$

Several algebraic variants of the convex polygon problem were established by Motzkin and O'Neil [MoO67]. Perhaps the most natural of them is the *comonotone* version: given d and r, any set of sufficiently many points in d-dimensional space has a subset of size r, all of whose elements can be connected by a path P such that every plane that does not contain an edge of P intersects it in at most d points.

[AhGM99] C. Ahrens, G. Gordon, E.W. McMahon: Convexity and the beta invariant, *Discrete Comput. Geom.* **22** (1999) 411–424.

[AlKP89] N. Alon, M. Katchalski, W.R. Pulleyblank: The maximum size of a convex polygon in a restricted set of points in the plane, *Discrete Comput. Geom.* **4** (1989) 245–251.

[BaB77] V.J. Baston, F.A. Bostock: Generalizations of a combinatorial lemma of Kelly, *J. Combinatorial Theory Ser. A* **22** (1977) 241–245.

[BáK01] I. Bárány, G. Károlyi: Problems and results around the Erdős–Szekeres convex polygon theorem, in: *JCDCG 00, Jap. Conf. Disc. Comput. Geom.*, J. Akiyama et al., eds., Springer *LNCS* **2098** (2001) 91–105.

[BáV98] I. Bárány, P. Valtr: A positive fraction Erdős–Szekeres theorem, *Discrete Comput. Geom.* **19** (1998) 335–342.

[BiDV91] A. Bialostocki, P. Dierker, B. Voxman: Some notes on the Erdős–Szekeres theorem, *Discrete Math.* **91** (1991) 231–238.

[BiFT03] T. Bisztriczky, G. Fejes Tóth: The Erdős–Szekeres problem for planar points in arbitrary position, in: *Discrete Geometry. In Honor of W. Kuperberg's 60th Birthday*, A. Bezdek, ed., Monogr. Textbooks Pure Appl. Math. **253**, Marcel Dekker 2003, 49–58.

[BiFT90] T. Bisztriczky, G. Fejes Tóth: Convexly independent sets, *Combinatorica* **10** (1990) 195–202.

[BiFT89a] T. Bisztriczky, G. Fejes Tóth: Nine convex sets determine a pentagon with convex sets as vertices, *Geometriae Dedicata* **31** (1989) 89–104.

[BiFT89b] T. Bisztriczky, G. Fejes Tóth: A generalization of the Erdős–Szekeres convex n-gon theorem, *J. Reine Angew. Math.* **395** (1989) 167–170.

[BiH95] T. Bisztriczky, H. Harborth: On empty convex polytopes, *J. Geometry* **52** (1995) 25–29.

[BiS94] T. Bisztriczky, V.P. Soltan: Some Erdős–Szekeres type results about points in space, *Monatshefte Math.* **118** (1994) 33–40.

[Bo74] W.E. Bonnice: On convex polygons determined by a finite planar set, *Amer. Math. Monthly* **81** (1974) 749–752.

[Bo84] P.B. Borwein: On monochromatic triangles, *J. Combinatorial Theory Ser. A* **37** (1984) 200–204.

[Br04] P. Brass: Empty monochromatic fourgons in two-colored point sets, *Geombinatorics* **14** (2004) 5–7.

[Ca96] Y. Caro: On the generalized Erdős–Szekeres conjecture — a new upper bound, *Discrete Math.* **160** (1996) 229–233.

[ChG98] F.R.K. Chung, R.L. Graham: Forced convex n-gons in the plane, *Discrete Comput. Geom.* **19** (1998) 367–371.

[DeH*03] O. Devillers, F. Hurtado, G. Károlyi, C. Seara: Chromatic variants of the Erdős–Szekeres theorem, *Comput. Geom. Theory Appl.* **26** (2003) 193–208.

[EdJ87] P.H. Edelman, R.E. Jamison: The theory of convex geometries, *Geometriae Dedicata* **19** (1987) 247–270.

[EdR00] P.H. Edelman, V. Reiner: Counting the interior of a point configuration, *Discrete Comput. Geom.* **23** (2000) 1–13.

[ErS60] P. Erdős, G. Szekeres: On some extremum problems in elementary geometry, *Ann. Univ. Sci. Budapest* **3-4** (1960–61) 53–62.

[ErS35]   P. Erdős, G. Szekeres: A combinatorial problem in geometry, *Compositio Math.* **2** (1935) 463–470.

[ErTV96]   P. Erdős, Z. Tuza, P. Valtr: Ramsey-remainder, *European J. Combinatorics* **17** (1996) 519–532.

[FeTP*03]   G. Fejes Tóth, J. Pach, G. Tóth: Problem 19, in: *Discrete Geometry. In Honor of W. Kuperberg's 60th Birthday,* A. Bezdek, ed., *Monogr. Textbooks Pure Appl. Math.* **253**, Marcel Dekker, 2003, 454–455.

[Fr99]   J. Francis: Some combinatorial geometry for convex quadrilaterals, *Period. Math. Hungar.* **39** (1999) 145–152.

[Fri04]   E. Friedman: 30 two-colored points with no empty monochromatic convex fourgons, *Geombinatorics* **14** (2004) 53–54.

[GuS01]   A. Gulyás, L. Szabó: Disjoint empty convex polygons in planar point sets, *Elemente Math.* **56** (2001) 62–70.

[Ha78]   H. Harborth: Konvexe Fünfecke in ebenen Punktmengen, *Elemente Math.* **33** (1978) 116–118.

[HaMS95]   H. Harborth, I. Mengersen, R.H. Schelp: The drawing Ramsey number $Dr(K_n)$, *Australasian J. Combinatorics* **11** (1995) 151–156.

[HaM94]   H. Harborth, M. Möller: The Esther Klein problem in the projective plane, *J. Combin. Math. Combin. Comput.* **15** (1994) 171–179.

[Ho83]   J.D. Horton: Sets with no empty convex 7-gon, *Canad. Math. Bull.* **26** (1983) 482–484.

[Jo86]   S. Johnson: A new proof of the Erdős–Szekeres convex k-gon result, *J. Combinatorial Theory Ser. A* **42** (1986) 318–319.

[KaKS70]   J.D. Kalbfleisch, J.G. Kalbfleisch, R.G. Stanton: A combinatorial problem on convex n-gons, in: *Proc. Louisiana Conf. Combinatorics, Graph Theory, and Computing,* Baton Rouge (1970) 180–188.

[KaS95]   J.G. Kalbfleisch, R.G. Stanton: On the maximum number of coplanar points containing no convex n-gons, *Utilitas Math.* **47** (1995) 235–245.

[Ká01]   G. Károlyi: Ramsey-remainder for convex sets and the Erdős–Szekeres theorem, *Discrete Appl. Math.* **109** (2001) 163–175.

[KáPT01]   G. Károlyi, J. Pach, G. Tóth: A modular version of the Erdős–Szekeres theorem, *Studia Sci. Math. Hungar.* **38** (2001) 245–259.

[KáT99]   G. KÁROLYI, G. TÓTH: An Erdős–Szekeres type problem in the plane, *Period. Math. Hungar.* **39** (1999) 153–159.

[KáV03]   G. KÁROLYI, P. VALTR: Point configurations in d-space without large subsets in convex position, *Discrete Comput. Geom.* **30** (2003) 277–286.

[Kl99]   D. KLAIN: An Euler relation for valuations of polytopes, *Adv. Math.* **147** (1999) 1–34.

[KlP98]   D.J. KLEITMAN, L. PACHTER: Finding convex sets among points in the plane, *Discrete Comput. Geom.* **19** (1998) 405–410.

[KuL03]   G. KUN, G. LIPPNER: Large empty convex polygons in k-convex sets, *Period. Math. Hungar.* **46** (2003) 81–88.

[Le76]   M. LEWIN: A new proof of a theorem of Erdős and Szekeres, *Math. Gazette* **60** (1976) 136–138.

[Lo79]   L. LOVÁSZ: *Combinatorial Problems and Exercises*, North-Holland 1979.

[Ma02]   J. MATOUŠEK: Lectures in Discrete Geometry, Springer 2002.

[MoS00]   W. MORRIS, V.P. SOLTAN: The Erdős–Szekeres problem on points in convex position – a survey, *Bull. Amer. Math. Soc., New Ser.* **37** (2000) 437–458.

[MoO67]   T.S. MOTZKIN, P.E. O'NEIL: Bounds assuring subsets in convex position, *J. Combinatorial Theory* **3** (1967) 252–255.

[Ov03]   M.H. OVERMARS: Finding sets of points without empty convex 6-gons, *Discrete Comput. Geom.* **29** (2003) 153–158.

[OvSV89]   M.H. OVERMARS, B. SCHOLTEN, I. VINCENT: Sets without empty convex 6-gons, *Bulletin of the EATCS* **37** (1989) 160–168.

[PaS98]   J. PACH, J. SOLYMOSI: Canonical theorems for convex sets, *Discrete Comput. Geom.* **19** (1998) 427–435.

[PaT00]   J. PACH, G. TÓTH: Erdős–Szekeres-type theorems for segments and non-crossing convex sets, *Geometriae Dedicata* **81** (2000) 1–12.

[PaT98]   J. PACH, G. TÓTH: A generalization of the Erdős–Szekeres theorem to disjoint convex sets, *Discrete Comput. Geom.* **19** (1998) 437–445.

[PiRS04]   R. PINCHASI, R. RADOIČIĆ, M. SHARIR: On empty convex polygons in a planar point set, *SCG'04, Proc. 20th ACM Symp. Comput. Geom.* (2004) 391–400.

[Pó03]   A. PÓR: A partitioned version of the Erdős–Szekeres theorem for quadrilaterals, *Discrete Comput. Geom.* **30** (2003) 321-336.

[PóV02]   A. PÓR, P. VALTR: The partitioned version of the Erdős–Szekeres theorem, *Discrete Comput. Geom.* **28** (2002) 625–637.

[Tó00]   G. TÓTH: Finding convex sets in convex position, *Combinatorica* **20** (2000) 589–596.

[TóV05]   G. TÓTH, P. VALTR: The Erdős–Szekeres theorem: upper bounds and related results, in: *Combinatorial and Computational Geometry*, J.E. Goodman et al., eds., Cambridge Univ. Press, *MSRI Publications* **52** (2005), to appear.

[TóV98]   G. TÓTH, P. VALTR: Note on the Erdős–Szekeres theorem, *Discrete Comput. Geom.* **19** (1998) 457–459.

[Va02]   P. VALTR: A sufficient condition for the existence of large empty convex polygons, *Discrete Comput. Geom.* **28** (2002) 671–682.

[Va92a]   P. VALTR: Sets in $\mathbb{R}^d$ with no large empty convex subsets, *Discrete Math.* **108** (1992) 115–124.

[Va92b]   P. VALTR: Convex independent sets and 7-holes in restricted planar point sets, *Discrete Comput. Geom.* **7** (1992) 135–152.

8.3 Halving Lines and Related Problems

The "halving lines problem" and the "k-set problem" were first formulated by Erdős, Lovász, Simmons, and Straus [Lo71], [ErL*73]. Due to their important algorithmic consequences, twelve years later they were rediscovered by Edelsbrunner and Welzl [EdW83], [EdW85] and have caught the attention of many computational geometers, especially in their dual version as k-levels in arrangements. By now there is a large literature related to the subject.

As everywhere in this chapter, all point sets X are finite and in general position. Using a not too illuminating but widely accepted term, a k-element subset $Y \subset X$ that can be separated from $Y \setminus X$ by a straight line (by a hyperplane in higher dimensions) not passing through any point of X is called a k-set. The main problem is to determine the maximum number of distinct k-sets that an n-element set can have. In the special case $k = \frac{n}{2}$ (n even), the separating lines are called halving lines. Two lines not passing through any element of X are considered equivalent if they realize the same partition of X, or, equivalently, if one can be continuously moved to the other while avoiding all elements of X. We are interested in the number of different partitions of X that can be realized by lines.

Problem 1 (Straus; see [Lo71]) For even n, what is the maximum number $f^{\text{halving}}(n)$ of distinct ways in which one can cut a set of n points in general position by a line into two halves, each of cardinality $\frac{n}{2}$?

Currently, the best known asymptotic bounds are $f^{\text{halving}}(n) \leq O(n^{\frac{4}{3}})$ [De98] and $f^{\text{halving}}(n) \geq n e^{\Omega(\sqrt{\log n})}$ [Tó01]. Exact values are known for up to 14 points [BeR02], [AnA*98], [AiK01]:

n	4	6	8	10	12	14
$f^{\text{halving}}(n)$	3	6	7	13	18	22

Problem 2 [ErL*73] What is the maximum number $f(n, k)$ of distinct ways in which one can cut off k points by a line from a set of n points in general position?

Again, the best known asymptotic bounds $n e^{\Omega(\sqrt{\log k})} \leq f(n, k) \leq O(nk^{\frac{1}{3}})$ are due to Tóth [Tó01] and Dey [De98]. Notice that for $k = \frac{n}{2}$ (n even), each halving line determines two $\frac{n}{2}$-sets, on both sides of the line, so that we have $f^{\text{halving}}(n) = \frac{1}{2} f(n, \frac{n}{2})$. The exact maximum numbers of k-sets are known for $k = 2$ and for $n \leq 10$. The equation $f(n, 2) = \lfloor \frac{3}{2}n \rfloor$

was established by Edelsbrunner and Welzl [EdW85]. Using their order type database, Aichholzer and Krasser [AiK01] found the values

n	3	4	5	6	7	8	9	10
$f(n,2)$	-	6	7	9	10	12	13	15
$f(n,3)$	-	-	-	12	12	14	16	18
$f(n,4)$	-	-	-	-	-	18	18	21
$f(n,5)$	-	-	-	-	-	-	-	26

There is a simple recursive construction leading to the lower bound

$$f^{\text{halving}}(n) \geq \Omega(n \log n).$$

For given a set X of n points with a maximum number of $f^{\text{halving}}(n)$ halving lines, let us flatten X by an affine transformation. Take three rotated copies, X_1, X_2, X_3, of the flattened set in such a way that each of the original halving lines of each copy X_i separates the other two copies, so that it is also a halving line of $X_1 \cup X_2 \cup X_3$. But there are $\Omega(n)$ further halving lines, cutting two sets X_i, X_{i+1} and leaving the third entirely to one side [Lo71], [ErL*73], [EdW83], [EdW85].

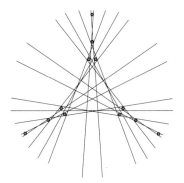

12 POINTS WITH
15 HALVING LINES

A more involved construction of Tóth [Tó01] yields

$$f^{\text{halving}}(n) \geq n e^{\Omega(\sqrt{\log n})}.$$

All known upper bounds on the number of k-sets are based on the analysis of a directed graph that can be constructed as follows. Given a set X of n points in general position in the plane, there are many equivalent directed separating lines that have the same k-set on their left sides.

Pick any one of them and rotate it in the clockwise direction until it hits a point on either side. By this operation, each k-set gives rise to a directed line that passes through two elements of X and that has precisely $k-1$ elements on its left side. This is a one-to-one correspondence. Thus, with the portion of the directed line between the two points of X being called a $(k-1)$-edge, the number of k-sets determined by X coincides with the number of $(k-1)$-edges.

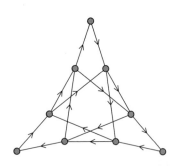

9 POINTS WITH 16 2-EDGES
CORRESPONDING TO 3-SETS

The basic geometric properties of the directed graph $\overrightarrow{G}_{k-1}$ formed by the $(k-1)$-edges were studied in detail by Erdős, Lovász, Simmons, and Straus [ErL*73]. Perhaps the most significant of them is the "local alternation" of the in-edges and out-edges around each point, which had been discovered earlier by Lovász [Lo71] and which was exploited in all other proofs. To formulate this property precisely, rotate a half-line h around any element $x \in X$. It will meet the edges of $\overrightarrow{G}_{k-1}$ in cyclic order. Let e and e' be two out-edges incident to x, consecutive in this cyclic order. Then the angular region antipodal to the region swept by h as it rotates from e to e' contains a unique in-edge at x. Another interesting property of the directed graph $\overrightarrow{G}_{k-1}$ is that any line that separates X into two sets, X_1 and X_2, crosses $\min(k+1, |X_1|, |X_2|)$ directed $(k-1)$-edges between X_1 and X_2 in both directions.

From these facts it is not hard to deduce that $f(n, k) \leq O(n\sqrt{k})$, $f^{\text{halving}}(n) \leq O(n^{\frac{3}{2}})$. Edelsbrunner and Welzl [EdW83], [EdW85] obtained the same bounds using a different technique, the method of allowable sequences. These upper bounds were slightly improved by Pach, Steiger, and Szemerédi [PaSS92]. Finally, Dey [De98] reduced them to $f(n, k) \leq 6.48n(k+1)^{\frac{1}{3}}$, $f^{\text{halving}}(n) \leq O(n^{\frac{4}{3}})$. Apart from the local alternation property, Dey's argument was based on the so-called crossing lemma (see Chapter 9.3). An alternative argument [AnA*98] gives the better numerical result $f(n, k) \leq 3.24nk^{\frac{1}{3}}$. The constant can be further improved by using a stronger version of the crossing lemma, due to Pach, Radoičić, Tardos, and Tóth [PaR*04].

Some modified versions of this problem have also attracted considerable interest. Welzl [We86] studied the total number of k-sets for several values of k. He proved that given a set $K \subset \{1, \ldots, \frac{n}{2}\}$, the total number of k-sets of an n-element set over all $k \in K$ is at most $\sqrt{8}n \left(\sum_{k \in K} k\right)^{\frac{1}{2}}$. This is, of course, better than the sum of the individual upper bounds, but it can certainly be improved using Dey's technique [De98]. Bárány and Steiger [BáS94] studied the expected number of k-sets for sets of n points independently chosen by some probability distribution. The answer depends on the distribution. They showed that for uniform distribution over a convex body or for distributions that have a rotational symmetry around a point, the expected numbers of halving lines and k-sets are linear in n. But this does not hold for every distribution. They have also constructed a distribution for which the expected number is $\Omega(n \log n)$. Alt, Felsner, Hurtado, Noy, and Welzl [AlF*00] proved that if a set X can be covered by a fixed number of convex curves, the number of its halving lines is linear in n. Edelsbrunner, Valtr, and Welzl [EdVW97] obtained a stronger bound on the number of halving lines of a "restricted" point set X, that is, for an n-element set with $\frac{\text{diam}(X)}{\text{mindist}(X)} \leq \alpha\sqrt{n}$, where $\alpha > 0$ is a fixed constant. They showed that such sets have at most $O(n^{7/6})$ halving lines. Although this bound is better than the best known upper bound for "unrestricted" point sets, it is still much larger than the general lower bound. Thus, in contrast to the Erdős–Szekeres problem, here it is not entirely clear whether the extremal configurations necessarily require "large" realizations and whether it changes the nature of the problem if we consider only restricted sets. ("Restricted" point sets are often called "dense" in the literature. We decided to avoid this term because of its multiple meanings.)

Problem 3   *Does there exist, for every $\alpha \geq (12/\pi^2)^{1/4}$, a constant $\beta = \beta(\alpha)$ such that any set of n points in general position in the plane that satisfies $\frac{\mathrm{diam}(S)}{\mathrm{mindist}(S)} \leq \alpha\sqrt{n}$ has at most $\beta n \log n$ halving lines?*

A construction in [EdVW97] shows that there are "restricted" n-element sets with $\Omega(n \log n)$ halving lines, so that if the answer to the last question is affirmative it is essentially best possible.

No matter how we restrict our point sets, we cannot hope for any upper bound better than $\frac{n}{2}$. Indeed, any set of n points in general position in the plane has at least one halving segment through each point. This lower bound is attained by regular n-gons. Pach and Solymosi [PaS99] showed that a set of n points has precisely $\frac{n}{2}$ halving lines if and only if it has a perfect *cross-matching*, i.e., a matching that consists of $\frac{n}{2}$ pairwise crossing straight-line segments.

Surprisingly, it is much easier to estimate the number of "*at most k-sets*" (for short, $\leq k$-*sets*), that is, all subsets of size *at most k* that can be separated from the remaining points by a line. Let $f(n, \leq k)$ denote the maximum number of $\leq k$-sets of a set of n points in general position in the plane. This function was introduced by Goodman and Pollack [GoP84]. It is known that $f(n, \leq k) = kn$ for every $k < \frac{n}{2}$, with equality for convex n-gons [AlG86], [Pe85].

The *minimum* number of k-sets and $\leq k$-sets, for a fixed k and for sufficiently large n, was determined by Lovász, Vesztergombi, Wagner, and Welzl [LoV*04] (following [EdH*89]). They have also found an elegant application of their results to obtain a lower bound for the rectilinear crossing number of a complete graph (see more about this in Chapter 9). The minimum number of k-sets is $2k+1$, which is attained by a regular $(2k+1)$-gon with all the remaining $n - 2k - 1$ points lying close to its center. The minimum number of $\leq k$-sets is $3\binom{k+1}{2}$. To see that this bound can be tight for $n \geq 3k$, choose three rays with a common origin such that the angle between any two of them is $2\pi/3$, and select $\frac{n}{3}$ points on each ray, slightly perturbed and well spaced. For $n \in \{2k, \ldots, 3k-1\}$ this set is not extremal.

Problem 4   *What is the minimum number of $\leq k$-sets determined by n points in the plane for $2k \leq n < 3k$?*

The best estimate in this range were obtained by Balogh and Salazar [BaS05].

The problem of determining the maximum number of halving hyperplanes determined by n points in general position in d-dimensional space appears to be at least as difficult as its planar version.

Problem 5   *[BáFL90] Let n be even. What is the maximum number of distinct ways to separate by a hyperplane a set of n points in general position in $\mathbb{R}^d$ into two parts of size $\frac{n}{2}$?*

This question for $d = 3$ was first studied by Bárány, Füredi, and Lovász [BáFL90], who established an $O(n^{3-\varepsilon})$ upper bound. This was improved in several steps ([ArC*91], [Ep93], [DeE94], [KáW01]) to $O(n^{\frac{5}{2}})$, which was obtained by Sharir, Smorodinsky, and Tardos [ShST01]. The only lower bound we have is only slightly superquadratic [Tó01].

Problem 6   *What is the maximum number of k-sets that a set of n points in general position in $\mathbb{R}^d$ can have?*

In four dimensions, Matoušek, Sharir, Smorodinsky, and Wagner [MaS*05] established the $O(n^{4-2/45})$ upper bound on the number k-sets, for any k. In higher dimensions, Živaljević and Vrećica [ŽiV92] proved that the answer to the last question is $O(n^{d-\varepsilon_d})$, for some small $\varepsilon_d > 0$. See also [AgA*98].

As in the planar case, the question becomes much easier and can be asymptotically answered if we consider $\leq k$-sets rather than k-sets. Clarkson and Shor [ClS89] showed that n points in general position in $\mathbb{R}^d$ have at most $O(n^{\lfloor \frac{d}{2} \rfloor} k^{\lceil \frac{d}{2} \rceil})$ subsets of cardinality at most k that can be cut off by a hyperplane. This order is attained when the points are selected from the moment curve $(t, t^2, \ldots, t^d)_{t \in \mathbb{R}}$. (See [AnW03] and [We01] for a simple proof.)

Obviously, the same questions make sense for the *minimum* numbers of k-sets and $\leq k$-sets in any dimension.

Problem 7   *For $d > 2$, what is the minimum number of k-sets that a set of n points in general position in $\mathbb{R}^d$ can have?*

Problem 8   *For $d > 2$, what is the minimum number of $\leq k$-sets that a set of n points in general position in $\mathbb{R}^d$ can have?*

The notion of k-sets can be "dualized" as follows. Consider an arrangement $\mathcal{A}$ of lines or curves in the plane, or hyperplanes, or hypersurfaces in higher dimensions, and fix a preferred direction. Suppose that the members of $\mathcal{A}$ have the property that every line in the preferred direction, say in the "vertical" direction of the x_d-axis, intersects each member of the arrangement in at most one point. Thus, the arrangement can be regarded as a collection of graphs of some (partially defined) functions. The 1-*level* of the arrangement is the lower envelope of the members of $\mathcal{A}$, that is, the pointwise minimum of the functions.

In general, the *k-level* consists of all points p belonging to at least one member of $\mathcal{A}$ that have the property that every open ray starting at p and pointing to the negative direction of the x_d-axis meets precisely $k - 1$ other members (that do not contain p). The number of k-sets corresponds to the "combinatorial complexity" of the k-level, that is, the total number of its faces of all dimensions. For instance, up to a multiplicative constant, the bound on the number of k-sets in the plane is equivalent to a bound on the number of vertices of the k-level of a simple arrangement of n lines. The k-levels of concave surfaces were considered in [KaT02].

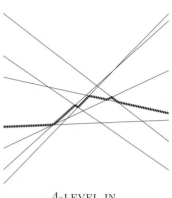

4-LEVEL IN
ARRANGEMENT OF 7 LINES

Levels in sets of curves and surfaces with restrictions on their pairwise intersection lead to problems of Davenport–Schinzel sequences [Sh91]. In general, levels in arrangements are important in many algorithmic applications.

[AgA*98] P.K. Agarwal, B. Aronov, T.M. Chan, M. Sharir: On levels in arrangements of lines, segments, planes, and triangles, *Discrete Comput. Geom.* **19** (1998) 315–331.

[AiK01] O. Aichholzer, H. Krasser: The point set order type data base: a collection of applications and results, in: *CCCG 01, 13th Canadian Conf. Comput. Geom.* (2001) 17–20.

[AlG86] N. Alon, E. Győri: The number of small semispaces of a finite set of points, *J. Combinatorial Theory Ser. A* **41** (1986) 154–157.

[AlF*00] H. Alt, S. Felsner, F. Hurtado, M. Noy, E. Welzl: A class of point sets with few k-sets, *Comput. Geom. Theory Appl.* **16** (2000) 95–101.

[AnA*98] A. Andrzejak, B. Aronov, S. Har-Peled, R. Seidel, E. Welzl: Results on k-sets and j-facets via continuous motion, in: *SCG 98, 14th ACM Symp. Comput. Geom.* (1998)

192–199.

[AnF99]   A. ANDRZEJAK, K. FUKUDA: Optimization over k-set poly-topes and efficient k-set enumeration, in: *WADS 99, Workshop Algorithms Data Structures*, Springer *LNCS* **1663** (1999) 1–12.

[AnW03]   A. ANDRZEJAK, E. WELZL: In between k-sets, j-facets, and i-faces: (i, j)-partitions, *Discrete Comput. Geom.* **29** (2003) 105–131.

[ArC*91]   B. ARONOV, B. CHAZELLE, H. EDELSBRUNNER, L.J. GUI-BAS, M. SHARIR, R. WENGER: Points and triangles in the plane and halving planes in space, *Discrete Comput. Geom.* **6** (1991) 435–442.

[BaS05]   J. BALOGH, G. SALAZAR: On k-sets, convex quadrilaterals, and the rectilinear crossing number of K_n, in: *GD 2004* (Proc. 12th Int. Symp. Graph Drawing, New York, 2004), J. Pach, ed., Springer *LNCS* **3383** (2005) 25–35.

[BáFL90]   I. BÁRÁNY, Z. FÜREDI, L. LOVÁSZ: On the number of halving planes, *Combinatorica* **10** (1990) 175–183.

[BáS94]   I. BÁRÁNY, W. STEIGER: On the expected number of k-sets, *Discrete Comput. Geom.* **11** (1994) 243–263.

[BeR02]   A. BEYGELZIMER, S.P. RADZISZOWSKI: On halving line arrangements, *Discrete Math.* **257** (2002) 267–283.

[ClS89]   K.L. CLARKSON, P.W. SHOR: Applications of random sampling in computational geometry, II, *Discrete Comput. Geom.* **4** (1989) 387–421.

[De98]   T.K. DEY: Improved bounds for planar k-sets and related problems, *Discrete Comput. Geom.* **19** (1998) 373–382.

[DeE94]   T.K. DEY, H. EDELSBRUNNER: Counting triangle crossings and halving planes, *Discrete Comput. Geom.* **12** (1994) 281–289.

[EdH*89]   H. EDELSBRUNNER, N. HASAN, R. SEIDEL, X.J. SHEN: Circles through two points that always enclose many points, *Geometriae Dedicata* **32** (1989) 1–12.

[EdW85]   H. EDELSBRUNNER, E. WELZL: On the number of line separations of a finite set in the plane, *J. Combinatorial Theory Ser. A* **38** (1985) 15–29.

[EdW83]   H. EDELSBRUNNER, E. WELZL: On the number of equal-sized semispaces of a set of points in the plane (extended abstract), in; *ICALP 83, 10th Colloq. Automata, Languages, and Programming*, Springer *LNCS* **154** (1983) 182–187.

[EdVW97] H. EDELSBRUNNER, P. VALTR, E. WELZL: Cutting dense point sets in half, *Discrete Comput. Geom.* **17** (1997) 243–255.

[Ep93] D. EPPSTEIN: Improved bounds for intersecting triangles and halving planes, *J. Combinatorial Theory Ser. A* **62** (1993) 176–182.

[ErL*73] P. ERDŐS, L. LOVÁSZ, A. SIMMONS, E.G. STRAUS: Dissection graphs of planar point sets, in: *A Survey of Combinatorial Theory*, J.N. Srivastava et al., eds., North-Holland 1973, 139–149.

[GoP84] J.E. GOODMAN, R. POLLACK: On the number of k-sets of a set of n points in the plane, *J. Combinatorial Theory Ser. A* **36** (1984) 101–104.

[KáW01] G. KÁROLYI, E. WELZL: Crossing-free segments and triangles in point configurations, *Discrete Appl. Math.* **115** (2001) 77–88.

[KaT02] N. KATOH, T. TOKUYAMA: K-levels of concave surfaces, *Discrete Comput. Geom.* **27** (2002) 567–589.

[Lo71] L. LOVÁSZ: On the number of halving lines, *Ann. Univ. Sci. Budapest Eötvös Sect. Math.* **14** (1971) 107–108.

[LoV*04] L. LOVÁSZ, K. VESZTERGOMBI, U. WAGNER, E. WELZL: Convex quadrilaterals and k-sets, in: *Towards a Theory of Geometric Graphs*, J. Pach, ed., *Contemporary Mathematics* **342**, AMS 2004, 139–148.

[MaS*05] J. MATOUŠEK, M. SHARIR, S. SMORODINSKY, U. WAGNER: On k-sets in four dimensions, *Discrete Comput. Geom.*, to appear.

[PaR*04] J. PACH, R. RADOIČIĆ, G. TARDOS, G. TÓTH: Improving the Crossing Lemma by finding more crossings in sparse graphs, in: *SCG 04, 20th ACM Symp. Comput. Geom.*, (2004) 68–75.

[PaS99] J. PACH, J. SOLYMOSI: Halving lines and perfect cross-matchings, in: *Advances in Discrete and Computational Geometry*, B. Chazelle et al., eds., *Contemporary Mathematics* **223**, AMS 1999, 245–249.

[PaSS92] J. PACH, W. STEIGER, E. SZEMERÉDI: An upper bound on the number of planar k-sets, *Discrete Comput. Geom.* **7** (1992) 109–123.

[Pe85] G.W. PECK: On 'k-sets' in the plane, *Discrete Math.* **56** (1985) 73–74.

[**Sh91**]   M. SHARIR: On k-sets in arrangements of curves and surfaces, *Discrete Comput. Geom.* **6** (1991) 593–613.

[**ShST01**]   M. SHARIR, S. SMORODINSKY, G. TARDOS: An improved bound for k-sets in three dimensions, *Discrete Comput. Geom.* **26** (2001) 195–204.

[**Tó01**]   G. TÓTH: Point sets with many k-sets, *Discrete Comput. Geom.* **26** (2001) 187–194.

[**We01**]   E. WELZL: Entering and leaving j-facets, *Discrete Comput. Geom.* **25** (2001) 351–364.

[**We86**]   E. WELZL: More on k-sets of finite sets in the plane, *Discrete Comput. Geom.* **1** (1986) 95–100.

[**ŽiV92**]   R. ŽIVALJEVIĆ, S. VREĆICA: The colored Tverberg's problem and complexes of injective functions, *J. Combinatorial Theory Ser. A* **61** (1992) 309–318.

8.4 Extremal Number of Special Subconfigurations

According to Esther Klein's observation mentioned in Section 8.2, among any five or more points in general position in the plane there are four in convex position. By a simple counting argument, this implies that every set of n points in general position has at least $\frac{1}{5}\binom{n}{4}$ four-tuples that are in convex position.

Problem 1   (*Erdős [Er84]*) *What is the minimum number* $\mathrm{conv}_k(n)$ *of convex k-gons determined by n points in general position in the plane?*

It is easy to see that $\lim_{n\to\infty}\frac{\mathrm{conv}_k(n)}{\binom{n}{k}} = c_k$ for some constant $c_k \in (0, 1)$. For $k = 4$, this problem is equivalent to the following: what is the minimum number $\mathrm{lin\text{-}cr}(K_n)$ of crossing edge pairs in a straight-line drawing of the complete graph with n vertices? This parameter, the *rectilinear crossing number* of K_n, has been extensively studied for a long time [HaH62], [ErG73], [Gu81]. See more about this and many similar topics in Chapter 9.

For small values of n, this number was determined "by hand" [Je71], [BrDG01], [BrDG03]. The results were verified by computer, using a database for order types created by Aichholzer, Aurenhammer, and Krasser [AiAK02b], [AiAK01], [AiK01], [Kr03]. For $n \leq 16$, the minimum numbers of convex quadrilaterals determined by n points are as follows:

n	4	5	6	7	8	9	10	11	12	13	14	15	16
$\mathrm{conv}_4(n)$	0	1	3	9	19	36	62	102	153	229	324	447	603

The best known lower bound on the asymptotic behavior of $\mathrm{conv}_4(n)$ was established by Balogh and Salazar [BaS05], who improved on earlier results by Wagner [Wa03], Lovász et al. [LoV*04], and Ábrego and Fernández-Merchant [ÁbF05]. The strongest upper bounds were found by Aichholzer, Aurenhammer, and Krasser [AiAK02a], who improved some earlier constructions of Brodsky, Durocher, and Gethner [BrDG01], [BrDG03]. We have

$$0.3755 \leq \lim_{n\to\infty}\frac{\mathrm{conv}_4(n)}{\binom{n}{4}} \leq 0.3807.$$

12 POINTS WITH MINIMUM NUMBER OF CONVEX QUADRILATERALS

This limit can also be interpreted as the infimum over all open sets X in the plane, with finite measure, of the probability that four points chosen independently and uniformly at random from X are in convex position [ScW94]. If the points are chosen instead on a sphere, in uniform distribution, the expected number of crossing geodesics was determined by Moon [Mo65].

As in Section 8.2, we can also count the number of *empty* convex polygons determined by a point set X, that is, closed convex polygons that do not contain any element of X other than their vertices. The maximum number of empty convex k-gons determined by an n-point set is obviously $\binom{n}{k}$, since all k-gons determined by the vertices of a convex n-gon are empty. The problem of finding the minimum number of empty convex k-gons determined by n points in general position was addressed in a number of papers [KaM88], [BáF87], [De87], [Va95], [Du00].

Problem 2   *What is the minimum number $\text{empty}_k(n)$ of empty convex k-gons determined by n points in general position in the plane?*

Horton's examples [Ho83] discussed in Section 8.2 show that $\text{empty}_k(n)$ is equal to zero for $k \geq 7$ and all sufficiently large n. It is an open problem to decide whether $\text{empty}_6(n) = 0$ for all n, but we know that $\text{empty}_6(29) = 0$ [Ov03], [OvSV89]. Thus we are left with the problems of estimating the numbers of empty triangles, quadrilaterals, and pentagons. The presently known exact values and best lower bounds, due to Harborth [Har78] and Dehnhardt [De87], are as follows:

n	3	4	5	6	7	8	9	10	11	12	≥ 13
$\text{empty}_3(n)$	0	3	7	13	21	31	43	58	75	94–95	$\geq n^2 - 5n + 10$
$\text{empty}_4(n)$	0	0	1	3	6	10	15	23	32	42–44	$\geq \binom{n-3}{2} + 6$
$\text{empty}_5(n)$	0	0	0	0	0	0	0	1	2	3–4	$\geq 3 \lfloor \frac{1}{12} n \rfloor$

The upper bounds from [BáF87] have been improved subsequently by Valtr [Va95], Dumitrescu [Du00], and Bárány and Valtr [BáV04]. The current records are, for n sufficiently large,

$$\text{empty}_3(n) \leq 1.6196n^2 \qquad\qquad \text{empty}_4(n) \leq 1.9397n^2$$
$$\text{empty}_5(n) \leq 1.0206n^2 \qquad\qquad \text{empty}_6(n) \leq 0.2006n^2.$$

In this context perhaps the most interesting unsolved question is the following.

Problem 3   *Does there exist a constant $\varepsilon > 0$ such that $\text{empty}_3(n) \geq (1+\varepsilon)n^2$ for every $n > n_0$?*

As a first step towards resolving Problem 3, Bárány suggested the following question.

Problem 4   *(Bárány [BáK01]) For any finite set X of points in general position in the plane, let $\Delta(X)$ denote the maximum number of empty triangles induced by X that share an edge. Is it true that*

$$\lim_{n \to \infty} \min_{|X|=n} \Delta(X) = \infty\,?$$

We can also ask whether there exists a positive ε such that
(a) $\text{empty}_4(n) \geq \left(\frac{1}{2} + \varepsilon\right) n^2$ or
(b) $\text{empty}_5(n) \geq \varepsilon n^2$.
These two questions are equivalent to Problem 3 in the sense that a positive answer to any of them would yield positive answers to the other two [PiRS04]. As for (b), even the following problem is open.

Problem 5   *Is it true that*

$$\lim_{n \to \infty} \frac{\text{empty}_5(n)}{n} = \infty\,?$$

Empty triangles can also be defined in a somewhat different context: for drawings of complete graphs [Har87], [Har98]. Consider a drawing of K_n with not necessarily straight-line edges satisfying the property that any two edges have at most one point in common, which is either an endpoint or a proper crossing. An empty triangle is a region bounded by three edges that does not contain any element of the vertex set in its interior. It turns out that the minimum number of empty triangles in a drawing of K_n is no longer quadratic in n. Harborth [Har98] constructed a drawing of K_n with $n - 2$ empty triangles, but we have no nontrivial lower bound. See more about this and many related questions in Section 9.6.

Most of the above questions can also be asked in higher dimensions, but only the case of empty simplices determined by a set of n points in d-dimensional space has been studied in detail. Bárány and Füredi [BáF87] proved that the minimum number of empty simplices induced by n points in general position in d-space is less than $d^{d^2} \binom{n}{d}$.

Problem 6   *What is the minimum number $\text{empty}^{(d)}(n)$ of empty simplices determined by n points in general position in d-dimensional space?*

Problem 7   *What is the minimum number $\text{conv}_k^{(d)}(n)$ of k-vertex convex polytopes determined by n points in general position in d-dimensional space?*

The number of convex k-gons induced by n points in the plane attains its maximum for point sets in convex position. In this sense, Problems 1 and 7 address the question how far an n-element point set can be from being convex. Another interesting and extensively studied measure of non-convexity is the following [Ak79], [NeM80], [AjC*82], [Hay87], [GaT98], [GaNT00]: What is the maximum number of simple closed n-gons (that is, noncrossing Hamiltonian cycles) induced by a set X of n points in general position in the plane? Denoting this parameter by simple(X), we clearly have that simple(X) = 1 if and only if X is in convex position. The problem of maximizing simple(X) over all n-element point sets plays a role in random generation of simple polygons (in maximizing the probability that a random permutation of a given set of points defines a simple polygon).

Problem 8   *(Newborn, W. Moser [Ak79]) What is the maximum number* simple(n) *of simple closed n-gons that can be drawn on the same set of n points in general position in the plane?*

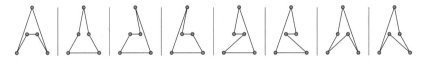

Five points admitting 8 distinct simple polygons

The function simple(n) is easily seen to satisfy the following inequality:

$$\text{simple}(a + b) \geq \text{simple}(a)\,\text{simple}(b),$$

so the limit $\lim_{n \to \infty} \left(\text{simple}(n)\right)^{1/n}$ exists (it is finite by [AjC*82]). The currently best bounds [GaNT00], [SaS03], [AlFK99] are

$$4.642 \leq \lim_{n \to \infty} \left(\text{simple}(n)\right)^{\frac{1}{n}} \leq 198.83 \,.$$

The upper bound follows by combining the bound on the number of triangulations of a set by Santos and Seidel [SaS03] with the bound on the number of simple cycles within a triangulation by Alt, Fuchs, and Kriegel [AlFK99]. With the help of computers, exact values have been determined for $n \leq 10$ [AiK01]:

n	3	4	5	6	7	8	9	10
simple(n)	1	3	8	29	92	339	1282	4994

Given a finite point set X in general position in the plane, let triang(X) denote the number of its distinct triangulations. In contrast to the previous questions, neither the maximization nor the minimization problem related to this parameter is trivial. Let

$$\text{max\_triang}(n) := \max_{|X|=n} \text{triang}(X), \qquad \text{min\_triang}(n) := \min_{|X|=n} \text{triang}(X),$$

where the maximum and minimum are taken over all n-element point sets X in general position in the plane.

Problem 9   *What is the maximum number of triangulations that a set of n points in general position in the plane can have?*

It is known [NeM80], [AjC*82], [DeS97], [Se98], [GaNT00], [SaS03] that

$$8^{n-\Theta(\log n)} \leq \max\_\text{triang}(n) \leq O\left(\frac{59^n}{n^6}\right).$$

Here the upper bound is due to Santos and Seidel [SaS03].

Problem 10   *What is the minimum number of triangulations that a set of n points in general position in the plane can have?*

The best known bounds, established by Aichholzer, Hurtado, and Noy [AiHN04], are the following:

$$\Omega\left((2.33)^n\right) \leq \min\_\text{triang}(n)$$
$$\leq (\sqrt{12})^{n-\Theta(\log n)}.$$

It is conjectured that the minimum is attained by "double-circle configurations," and not by point sets in convex position. Again, the exact values were determined for $n \leq 11$ by Aichholzer and Krasser [AiK01], [Kr03]:

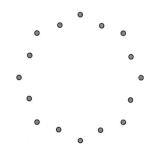

DOUBLE-CIRCLE
CONFIGURATION

n	3	4	5	6	7	8	9	10	11
max_triang(n)	1	2	5	14	42	150	780	4550	26888
min_triang(n)	1	1	2	4	11	30	89	250	776

Many other related extremal problems are discussed in [AiK01].

There are a number of interesting problems on triangulations that occur only in dimensions larger than two. By a *triangulation* of a finite set $X \subset \mathbb{R}^d$ we mean a decomposition of its convex hull into simplices with pairwise disjoint interiors, each of which is spanned by $d + 1$ elements of X. In the plane, every triangulation of a fixed set has the same number of triangles: a triangulation of a set of n points whose convex hull has k vertices contains $2n - k - 2$ triangles. This easily follows from the fact that the sum of the interior angles of each triangle is π. However, nothing similar holds for $d \geq 3$. Any triangulation of an n-element point set in d-space contains at least $n - d$ simplices, and this bound can be attained by gluing together simplices face to face, where each new simplex contributes a new vertex and all vertices are in convex position [RoS85]. It is also known that the maximum number of simplices in any triangulation of a set of n points has $f_d(n + 1, d + 1) - (d + 1) = \Theta(n^{\lceil \frac{1}{2} d \rceil})$, where $f_d(n + 1, d + 1)$ denotes the number of d-dimensional faces of a $(d + 1)$-dimensional cyclic polytope on $n + 1$ vertices [RoS85]. This follows from the upper bound theorem for simplicial spheres. For $d = 3$, stronger results have been obtained by Edelsbrunner, Preparata, and West [EdPW90]. These bounds take into account the number of points in the interior of the convex hull, and they can be attained only for special point sets. It is easy to see that every point set has a "small" triangulation with $O(n)$ simplices (at most $3n - 11$ simplices in three-dimensional space) [EdPW90]. But does every set have a "large" triangulation?

Problem 11   *What is the maximum number $R_d(n)$ such that every set of n points in general position in d-dimensional space has a triangulation consisting of at least $R_d(n)$ simplices?*

Brass [Br05] established the upper bounds $R_3(n) \leq O(n^{\frac{5}{3}})$ and $R_d(n) \leq O\left(n^{\frac{1}{d} + \frac{d-1}{d}\lceil \frac{1}{2} d \rceil}\right)$. Any set that results from the d-dimensional lattice cube $\{1, \ldots, n^{\frac{1}{d}}\}^d$ by a small perturbation cannot have a larger triangulation than this number. The first open problem is to decide whether

$$\lim_{n \to \infty} \frac{R_3(n)}{n} = \infty$$

holds. Urrutia [Ur03] proved that $R_3(n) \geq 3n$. This inequality has the interesting consequence that any four-colored point set in three-dimensional space contains an empty monochromatic simplex. If for some integer c there exist arbitrarily large c-colored sets with no empty monochromatic simplex, then the same argument yields that $R_3(n) = O(n)$.

[ÁbF05]   B.M. Ábrego, S. Fernández-Merchant: A lower bound for the rectilinear crossing number, *Graphs Combinatorics*, to appear.

[AiAK02a] O. Aichholzer, F. Aurenhammer, H. Krasser: On the crossing number of complete graphs, in: *SCG 02, 18th ACM Symp. Comput. Geom.* (2002) 19–24.

[AiAK02b] O. Aichholzer, F. Aurenhammer, H. Krasser: Enumerating order types for small point sets with applications, *Order* **19** (2002) 265–281.

[AiAK01]   O. Aichholzer, F. Aurenhammer, H. Krasser: Progress on rectilinear crossing numbers, Technical report, IGI-TU Graz 2001.

[AiHN04]   O. Aichholzer, F. Hurtado, M. Noy: A lower bound
           on the number of triangulations of planar point sets, *Comput.
           Geom. Theory Appl.* **29** (2004) 135–145.

[AiK01]   O. Aichholzer, H. Krasser: The point set order type
           data base: a collection of applications and results, in: *CCCG
           01, 13th Canadian Conf. Comput. Geom.* (2001) 17–20.

[AjC*82]   M. Ajtai, V. Chvátal, M.M. Newborn, E. Szemerédi:
           Crossing-free subgraphs, *Annals Discrete Math.* **12** (1982) 9–
           12.

[Ak79]   S. Akl: A lower bound on the maximum number of crossing-
           free Hamilton cycles in a rectilinear drawing of K_n, *Ars Com-
           binatoria* **7** (1979) 7–18.

[AlFK99]   H. Alt, U. Fuchs, K. Kriegel: On the number of simple
           cycles in planar graphs, *Combinatorics Probab. Comput.* **8**
           (1999) 397–405.

[BaS05]   J. Balogh, G. Salazar: On k-sets, convex quadrilaterals,
           and the rectilinear crossing number of $K(n)$, in: *Graph Draw-
           ing 2004*, J. Pach, ed., Springer *LNCS* **3383** (2005) 25–25.

[BáF87]   I. Bárány, Z. Füredi: Empty simplices in euclidean space,
           Canad. Math. Bull. **30** (1987) 436–445.

[BáK01]   I. Bárány, G. Károlyi: Problems and results around
           the Erdős-Szekeres convex polygon theorem, in: *JCDCG 2000*
           (Jap. Conf. Discrete Comput. Geom.) J. Akiyama et al., eds.,
           Springer *LNCS* **2098** (2001) 91–105.

[BáV04]   I. Bárány, P. Valtr: Planar point sets with a small number
           of empty convex polygons, *Studia Math. Hungar.* **41** (2004)
           243–266.

[Br05]   P. Brass: On the size of higher-dimensional triangulations,
           in: *Combinatorial and Computational Geometry*, J.E. Good-
           man et al., eds., Cambridge Univ. Press *MSRI Publications*
           52 (2005) 145–151.

[BrDG03]   A. Brodsky, S. Durocher, E. Gethner: Toward the rec-
           tilinear crossing number of K_n: New drawings, upper bounds,
           and asymptotics, *Discrete Math.* **262** (2003) 59–77.

[BrDG01]   A. Brodsky, S. Durocher, E. Gethner: The rectilinear
           crossing number of K_{10} is 62, *Electronic J. Combinatorics* **8**
           (2001) #R23.

[De87]   K. Dehnhardt: *Leere konvexe Vielecke in ebenen Punkt-
           mengen*, Dissertation, TU Braunschweig, 1987.

[DeS97]   M.O. DENNY, C.A. SOHLER: Encoding a triangulation as a permutation of its point set, in *CCCG 97, 9th Canadian Conf. Comput. Geom.* (1997) 39–43.

[Du00]   A. DUMITRESCU: Planar sets with few empty convex polygons, *Studia Sci. Math. Hungar.* **36** (2000) 93–109.

[EdPW90] H. EDELSBRUNNER, F.P. PREPARATA, D.B. WEST: Tetrahedralizing point sets in three dimensions, *J. Symbolic Computation* **10** (1990) 335–347.

[Er84]   P. ERDŐS: Some old and new problems in combinatorial geometry, in: *Convexity and Graph Theory*, M. Rosenfeld et al., eds., *Annals Discrete Math.* **20** (1984) 129–136.

[ErG73]   P. ERDŐS, R.K. GUY: Crossing number problems, *Amer. Math. Monthly* **80** (1973) 52–58.

[GaNT00] A. GARCIA, M. NOY, J. TEJEL: Lower bounds on the number of crossing-free subgraphs of K_N, *Comput. Geom. Theory Appl.* **16** (2000) 211–221.

[GaT98]   A. GARCIA, J. TEJEL: A lower bound on the number of polygonalizations of N points in the plane, *Ars Combinatoria* **49** (1998) 3–19.

[Gu81]   R.K. GUY: Unsolved problems, *Amer. Math. Monthly* **88** (1981) p.757.

[HaH62]   F. HARARY, A. HILL: On the number of crossings in a complete graph, *Proc. Edinburgh Math. Soc.* **13** (1962/63) 333–338.

[Har98]   H. HARBORTH: Empty triangles in drawings of the complete graph, *Discrete Math.* **191** (1998) 109–111.

[Har87]   H. HARBORTH: Konvexe Vielecke in Punktmengen, in: *3. Kolloquium Geometrie und Kombinatorik, 1987*, TU Karl-Marx-Stadt 1988, 71–73.

[Har78]   H. HARBORTH: Konvexe Fünfecke in ebenen Punktmengen, *Elemente Math.* **33** (1978) 116–118.

[Hay87]   R.B. HAYWARD: A lower bound for the optimal crossing-free Hamiltonian cycle problem, *Discrete Comput. Geom.* **2** (1987) 327–343.

[Ho83]   J.D. HORTON: Sets with no empty convex 7-gon, *Canad. Math. Bull.* **26** (1983) 482–484.

[Je71]   H.F. JENSEN: An upper bound for the rectilinear crossing number of a complete graph, *J. Combinatorial Theory* **11** (1971) 212–216.

[KaM88]   M. Katchalski, A. Meir: On empty triangles determined by points in the plane, *Acta Math. Hungar.* **51** (1988) 323–328.

[Kr03]   H. Krasser: *Order Types of Point Sets in the Plane*, PhD thesis, Institute for Theoretical Computer Science, Graz University of Technology, Austria, October 2003.

[LoV*04]   L. Lovász, K. Vesztergombi, U. Wagner, E. Welzl: Convex quadrilaterals and *k*-sets, in: *Towards a Theory of Geometric Graphs*, J. Pach, ed., *Contemporary Mathematics* **342**, AMS 2004, 139–148.

[Mo65]   J. Moon: On the distribution of crossings in random complete graphs, *J. Soc. Industr. Appl. Math.* **13** (1965) 506–510.

[NeM80]   M.M. Newborn, W.O.J. Moser: Optimal crossing-free Hamiltonian circuit drawings of K_n, *J. Combinatorial Theory Ser. B* **29** (1980) 12–26

[Ov03]   M.H. Overmars: Finding sets of points without empty convex 6-gons, *Discrete Comput. Geom.* **29** (2003) 153–158.

[OvSV89]   M.H. Overmars, B. Scholten, I. Vincent: Sets without empty convex 6-gons, *Bulletin of the EATCS* **37** (1989) 160–168

[PiRS04]   R. Pinchasi, R. Radoičić, M. Sharir: On empty convex polygons in a planar point set, in: *SCG 04, Proc. 20th ACM Symp. Comput. Geom.* (2004) 391–400.

[RoS85]   B.L. Rothschild, E.G. Straus: On triangulations of the convex hull of *n* points, *Combinatorica* **5** (1985) 167–179.

[SaS03]   F. Santos, R. Seidel: A better upper bound on the number of triangulations of a planar point set, *J. Combinatorial Theory Ser. A* **102** (2003) 186–193.

[ScW94]   E.R. Scheinerman, H.S. Wilf: The rectilinear crossing number of a complete graph and Sylvester's "Four point problem" of geometric probability, *Amer. Math. Monthly* **101** (1994) 939–943.

[Se98]   R. Seidel: Note on the number of triangulations of planar point sets, *Combinatorica* **18** (1998) 297–299.

[Ur03]   J. Urrutia: Colorings, tetrahedralization, and empty tetrahedra in colored point sets in $\mathbb{R}^3$ (in Spanish), in: *Proc. X Encuentros de Geometria Computacional* (Sevilla 2003), 95–100.

[**Va95**]   P. VALTR: On the minimum number of empty polygons in planar point sets, *Studia Sci. Math. Hungar.* **30** (1995) 155–163.

[**Wa03**]   U. WAGNER: On the rectilinear crossing number of complete graphs, in: *SoDA 03, 14th ACM-SIAM Symp. Discr. Algor.* (2003) 583–588.

8.5 Other Problems on Points in General Position

Avis, Hosono, and Urabe [AvHU00] asked the following innocent-looking question. For which r is it true that any finite set X in general position in the plane that has sufficiently many points in the interior of its convex hull has a subset whose convex hull has precisely r points of X in its interior? If such a subset exists, we say that X contains a "convex polygon with r interior points." Note that in contrast to the Erdős–Szekeres problem discussed in Section 8.2, we are not interested in the number of vertices of such a polygon.

Clearly, the answer to the above question is affirmative for $r = 0$, because every face in a triangulation of X is empty. The answer is also yes for $r = 1$. Indeed, if the interior of $\mathrm{conv}(X)$ is not empty, then there exists a triangle T induced by X that has at least one interior point. If T has precisely one interior point, we are done. Otherwise, picking any interior point and connecting it to all vertices of T, we see that at least one of the three resulting triangles will be nonempty and will contain fewer interior points than T. Continuing this procedure, we finally obtain a triangle with precisely one point in its interior. Avis et al. [AvHU01] also gave a positive answer to their question for $r = 2$, but the case $r = 3$ is still open [Fe03]. It is essential in this problem to insist on *precisely* r interior points; if we relax this property by requiring at most r interior points, the problem becomes much simpler [AvHU00].

For any set $S \subseteq \mathbb{R}^2$, let $\mathrm{int}(S)$ and $\mathrm{conv}(S)$ denote the interior of S and the convex hull of S, respectively.

Problem 1   (Avis–Hosono–Urabe [AvHU00], [AvHU01]) Does there exist for every $r \geq 3$ a smallest integer $I = I(r)$ such that any finite set X of points in general position in the plane with $\left| X \cap \mathrm{int}(\mathrm{conv}(X)) \right| \geq I(r)$ has a subset $Y \subseteq X$ with $\left| X \cap \mathrm{int}(\mathrm{conv}(Y)) \right| = r$?

It was proved by Avis et al. [AvHU01] that $I(1) = 1$ and $I(2) = 4$. They also gave an example showing that seven points in the interior of X are not sufficient to guarantee a convex polygon with exactly three interior points, so that we have $I(3) \geq 8$, if $I(3)$ exists at all.

Hosono, Károlyi, and Urabe [HoKU03]
noticed that it would be sufficient to establish the statement in Problem 1 for
sets X whose convex hull is a triangle.
Furthermore, Bisztriczky et al. [BiH*04]
proved that for any r and k, there exists a
number $I = I_k(r)$ such that any finite set
of points in general position in the plane
that has at least I interior points and satisfies the condition that each triple determines a triangle with at most k interior points contains a convex polygon
with precisely r points in its interior.

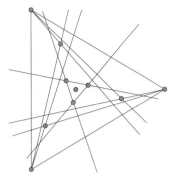

NO SUBSET WITH
EXACTLY 3 INTERIOR POINTS

In Section 8.2, we discussed some Erdős–Szekeres-type results that
guarantee the existence of r large subsets, $X_1, \ldots, X_r$, in any set of sufficiently many points in the plane such that all of their transversals have the
same order type. Instead of picking one point from each X_i, as in the case
of transversals, we can set a condition for the order type of certain small
subsets that may have more than one element from the same X_i. Two sets,
X and Y, in the plane are called *mutually avoiding* if no line spanned by
two points from one set separates two points of the other. Thus, for any
$x_1, x_2 \in X$, all points of Y lie on the same side of the line x_1x_2, which
implies that all triangles x_1x_2y ($y \in Y$) have the same orientation, and a
similar property holds for every pair $y_1, y_2 \in Y$.

It is known that any set of n points in general position in the plane
has two mutually avoiding subsets of size at least $c\sqrt{n}$ and that this order
of magnitude is best possible [ArE*94], [Va96], [Va97]. This problem can
be generalized to more than two sets and to dimensions larger than two.
Two sets in d-dimensional space are said to be *mutually avoiding* if no
hyperplane induced by d elements of one set intersects the convex hull of
the other set.

Problem 2   *(Aronov et al. [ArE*94]) Let $d, r \geq 2$ be fixed integers.
What is the largest number $f = f_d(r, n)$ with the property
that any set of n points in general position in $\mathbb{R}^d$ has r
pairwise disjoint mutually avoiding subsets, each consisting of at least f points?*

The convex hull of n points taken from the moment curve $\{(t, t^2, \ldots, t^d) \mid t \in \mathbb{R}\}$ is called a *cyclic d-polytope* with n vertices. It is well known (see,
e.g., p. 398 of [BjL*93] or Exercise 3 on p. 99 of [Ma02]) that the Erdős–
Szekeres theorem generalizes to higher dimensions, as follows. For any
$n > d \geq 2$, there exists a smallest integer $F_d(n)$ such that any set of at
least $F_d(n)$ points in general position in d-dimensional space has n elements

that form the vertex set of a polytope combinatorially equivalent to a cyclic d-polytope. In fact, it is not hard to see that this statement does not hold for any other d-dimensional polytope combinatorially different from the cyclic polytope. Clearly, we have

$$f_d(r, n) \geq \frac{1}{r} F_d^{-1}(n),$$

because by partitioning the vertices of a cyclic polytope into contiguous intervals we obtain mutually avoiding sets.

The problem of partitioning a set of n points into a small number of convex polygons was studied by Urabe [Ur96], [Ur99]. Without any further restriction on the partition classes, it follows from the Erdős–Szekeres convex polygon theorem that one can always partition a set X of n points in general position in the plane into $O(\frac{n}{\log n})$ classes, each class consisting of points in convex position. The order of magnitude of this bound cannot be improved.

If we want to partition X into (geometrically) *disjoint* convex polygons, then we obtain two different problems depending on the definition of "disjointness." Either we require that the convex hulls of the polygons be pairwise disjoint, or we are satisfied with the disjointness of their boundaries. Note that in the latter case a nested sequence of simple closed polygons is permitted, while in the former case it is forbidden.

Problem 3   *(Urabe [Ur96]) What is the smallest number $G(n)$ with the property that every set of $n \geq 3$ points in general position in the plane can be partitioned into $G(n)$ subsets, each of which is the vertex set of a convex polygon and the convex hulls of these polygons are pairwise disjoint?*

Since there are arbitrarily large sets that do not induce any empty heptagons, we have that $G(n) \geq \frac{1}{6}n$. The best known lower and upper bounds on $G(n)$,

$$\left\lceil \frac{1}{4}(n+1) \right\rceil \leq G(n) \leq \left\lceil \frac{5}{18}n \right\rceil,$$

are due to Xu and Ding [XuD02] and to Hosono and Urabe [HoU01], respectively (see also [DiH*03]). In fact, the upper bound can be improved to $\left\lceil \frac{3}{11}(n+1) \right\rceil$ for infinitely many n [HoU01].

There is another variant of this question: we can replace the assumption that the polygons do not cross with the requirement that they be *empty* and that their vertex sets be disjoint. In this case, the upper bound can be reduced to $\left\lceil \frac{4}{15}n \right\rceil$, while we still have the same lower bound [XuD02], [Ur96].

Problem 4 *(Urabe [Ur96]) What is the smallest number $g(n)$ with the property that any set of $n \geq 3$ points in general position in the plane can be partitioned into $g(n)$ disjoint subsets such that their convex hulls are bounded by pairwise noncrossing empty convex polygonal curves?*

Obviously, we have $g(n) \leq G(n)$ for every n.

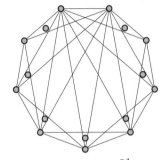

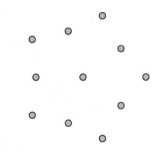

n POINTS REQUIRING $\left\lceil \frac{1}{4}(n+1) \right\rceil$ 11 POINTS CAN BE PARTITIONED INTO
EMPTY CONVEX POLYGONS   3 EMPTY OR 4 INTERIOR-DISJOINT
WITH DISJOINT INTERIORS   EMPTY CONVEX POLYGONS

The above problems can also be raised in higher dimensions [Ur99]. In three-dimensional space, the best known upper bound for the number of required classes in either of the variants is $\left\lceil \frac{2}{9}n \right\rceil$, and there is no nontrivial lower bound better than the general $\Omega(\frac{n}{\log n})$ bound, which follows from the Erdős–Szekeres theorem (without using the condition of disjointness).

Conjecture 5 *(Urabe [Ur99]) Given $d \geq 2$, let $G_d(n)$ denote the smallest number such that any set of n points in general position in $\mathbb{R}^d$ can be partitioned into $G_d(n)$ subsets whose convex hulls are pairwise disjoint empty polytopes. We have $G_d(n) = \frac{1}{2d}n + O(1)$ as n tends to infinity.*

A *convex decomposition* of a point set X in the plane is a subdivision of its convex hull into convex polygons with pairwise disjoint interiors whose vertices belong to X. The following conjecture is due to Rivera-Campo and Urrutia (personal communication).

Conjecture 6 *Let $U(n)$ denote the smallest integer such that every set of n points in general position in the plane permits a convex decomposition with at most $U(n)$ faces. Then we have $U(n) = n + O(1)$ as n tends to infinity.*

It was shown by Rivera-Campo and Urrutia (unpublished) that $U(n) \leq \frac{3}{2}n$ for every n.

Most of the functions introduced above express how far the worst possible n-element point sets are from being convex. Another interesting function of this kind was introduced by Arkin et al. [ArF*03]. A vertex of a simple closed polygon is said to be *reflex* if its interior angle is larger than π. The *reflexivity* $R(X)$ of a point set X is the smallest number of reflex vertices in a noncrossing simple closed polygon whose vertex set is X.

Conjecture 7   *(Arkin et al. [ArF*03]) Let $R(n)$ denote the maximum of $R(X)$ over all sets of n points in general position in the plane. Then we have $R(n) = (\frac{1}{4} + o(1))n$ as n tends to infinity.*

It was proved in [ArF*03] that $\lfloor \frac{n}{4} \rfloor < R(n) < \lceil \frac{n}{2} \rceil$ for every n.

Many interesting problems are concerned with colored order types. One typically tries to partition a colored point set into a small number of parts that either are monochromatic or have a "balanced" coloring. Many questions of this type are surveyed by Kaneko and Kano [KaK03].

Dumitrescu and Pach [DuP02] proved that any two-colored set of n points in general position in the plane can be partitioned into $\lceil \frac{1}{2}(n+1) \rceil$ monochromatic subsets with pairwise disjoint convex hulls, and this result is tight.

Problem 8   *[DuP02] What is the minimum number $\mathrm{part}_d(n,r)$ such that any r-colored set of n points in general position in d-space can be partitioned into $\mathrm{part}_d(n,r)$ monochromatic classes with disjoint convex hulls?*

It is known that

$$\left\lceil \left(1 - \frac{1}{r}\right)n + \frac{1}{r} \right\rceil \leq \mathrm{part}_2(n,r) \leq \left(1 - \frac{1}{r + \frac{1}{6}}\right)n + O(1).$$

Dumitrescu and Pach [DuP02] have also established a slightly sublinear lower bound on $\mathrm{part}_d(n,2)$ for any fixed $d \geq 3$. It is a challenging unsolved problem to decide whether

$$\lim_{n\to\infty} \frac{\mathrm{part}_3(n,2)}{n} = 0.$$

Conversely, does there exist a constant $\delta > 0$ such that there are arbitrarily large two-colored sets X in general position in $\mathbb{R}^3$ that cannot be decomposed into fewer than $\delta|X|$ monochromatic subsets whose convex hulls are pairwise disjoint?

Consider a set $X = B \cup W$ of n black and n white points in general position in the plane, $|B| = |W| = n$. A line connecting a black point $b \in B$ and a white point $w \in W$ is said to realize a *balanced partition* of X if the number of black points minus the number of white points is zero on both sides of the line bw. Pach and Pinchasi [PaP01] proved that the number of lines bw realizing balanced partitions of X is always at least n, and that this bound can be attained. Sharir and Welzl [ShW03] gave an alternative proof based on the generalized lower bound theorem for convex polytopes.

Suppose now that the coloring of $X = B \cup W$ is not necessarily balanced, and we have $|B| = rn$ and $|W| = sn$ for some positive integers r and s. Then it is possible to partition X into n groups whose convex hulls are pairwise disjoint so that each group consists of r black and s white points [ItUY00], [BeKS99].

Conjecture 9 (*Kaneko and Kano [KaK01]*) *Let $n = n_1 + n_2 + \ldots + n_g$, where $1 \leq n_i \leq n/3$ for $1 \leq i \leq g$. Suppose that $X = B \cup W$ is a set of points in general position in the plane with $|B| = rn$ and $|W| = sn$ for some positive integers r and s. Then the plane can be partitioned into g convex regions, $R_1 \cup R_2 \cup \ldots \cup R_g$, such that R_i contains precisely rn_i black and sn_i white points for every $1 \leq i \leq g$.*

By projection, all results of this kind generalize to higher dimensions.

[ArF*03]   E.M. Arkin, S.P. Fekete, F. Hurtado, J.S.B. Mitchell, M. Noy, V. Sacristan, S. Sethia: On the reflexivity of point sets, in: *Discrete and Computational Geometry — The Goodman–Pollack Festschrift*, B. Aronov et al., eds., Springer, 2003, 139–156.

[ArE*94]   B. Aronov, P. Erdős, W. Goddard, D.J. Kleitman, M. Klugerman, J. Pach, L.J. Schulman: Crossing families, *Combinatorica* **14** (1994) 127–134.

[AvHU01]   D. Avis, K. Hosono, M. Urabe: On the existence of a point subset with a specified number of interior points, *Discrete Math.* **241** (2001) 33–40.

[AvHU00]   D. Avis, K. Hosono, M. Urabe: On the existence of a point subset with 4 or 5 interior points, in: *JCDCG 98* (Jap. Conf. Disc. Comput. Geom. 1998), J. Akiyama et al., eds., Springer *LNCS* **1763** (2000) 57–64.

[BeKS99]   S. Bespamyatnikh, D. Kirkpatrick, J. Snoeyink: Generalizing ham sandwich cuts to equitable subdivisions, in: *SCG 99, 15th ACM Symp. Comput. Geom.* (1999) 49–58.

[BiH*04]   T. Bisztriczky, K. Hosono, G. Károlyi, M. Urabe: Constructions from empty polygons, *Period. Math. Hungar.* **49** (2004) 1–8.

[BjL*93]   A. Björner, M. Las Vergnas, B. Sturmfels, N. White, G.M. Ziegler: Oriented Matroids, Encyclopedia Math. Appl. Ser. **46**, Cambridge Univ. Press, 1993.

[DiH*03]   R. Ding, K. Hosono, M. Urabe, C. Xu: Partitioning a planar point set into empty convex polygons, in: JCDCG 2002 (Jap. Conf. Discr. Comput. Geom. 2002), J. Akiyama et al., eds., Springer LNCS **2866** (2003) 129–134.

[DuP02]   A. Dumitrescu, J. Pach: Partitioning colored point sets into monochromatic parts, Internat. J. Comput. Geom. Appl. **12** (2002) 401–412.

[Fe03]   T. Fevens: A note on point subsets with a specified number of interior points, in: JCDCG 2002 (Jap. Conf. Discr. Comput. Geom. 2002), J. Akiyama et al., eds., Springer LNCS **2866** (2003) 152–158.

[HoKU03]   K. Hosono, G. Károlyi, M. Urabe: On the existence of a convex polygon with a specified number of interior points, in: Discrete Geometry. In Honor of W. Kuperberg's 60th Birthday, A. Bezdek, ed., Monogr. Textbooks Pure Appl. Math. **253**, Marcel Dekker, 2003, 351–358.

[HoU01]   K. Hosono, M. Urabe: On the number of disjoint convex quadrilaterals for a planar point set, Comp. Geom. Theory Appl. **20** (2001) 97–104.

[ItUY00]   H. Ito, H. Uehara, M. Yokoyama: 2-Dimensional ham sandwich theorem for partitioning into three convex pieces, in: JCDCG 98 (Jap. Conf. Disc. Comput. Geom. 1998), J. Akiyama et al., eds., Springer LNCS **1763** (2000) 129–157.

[KaK03]   A. Kaneko, M. Kano: Discrete geometry on red and blue points in the plane – a survey, in: Discrete and Computational Geometry — The Goodman–Pollack Festschrift, B. Aronov et al., eds., Springer, 2003, 551–570.

[KaK01]   A. Kaneko, M. Kano: Generalized balanced partitions of two sets of points in the plane, in: JCDCG 2000 (Jap. Conf. Disc. Comput. Geom. 2000), J. Akiyama et al., eds., Springer LNCS **2098** (2001) 176–186.

[Ma02]   J. Matoušek: Lectures in Discrete Geometry, Grad. Texts in Math. 212, Springer 2002.

[PaP01]   J. Pach, R. Pinchasi: On the number of balanced lines, Discrete Comput. Geom. **25** (2001) 611–628.

[ShW03]   M. Sharir, E. Welzl: Balanced lines, halving triangles, and the Generalized Lower Bound Theorem,   in: Discrete and *Computational Geometry — The Goodman–Pollack Festschrift*, B. Aronov et al., eds., Springer, 2003, 789–797.

[**Ur99**]   M. URABE: Partitioning point sets into disjoint convex polytopes, *Comput. Geom. Theory Appl.* **13** (1999) 173–178.

[**Ur96**]   M. URABE: On a partition in convex polygons, *Discrete Appl. Math.* **64** (1996) 179–191.

[**Va97**]   P. VALTR: On mutually avoiding sets, in: *The Mathematics of Paul Erdős*, R.L. Graham, J. Nešetřil, eds., Springer-Verlag *Algorithms and Combinatorics* Series **14** (1997) 324–328.

[**Va96**]   P. VALTR: Lines, line-point incidences and crossing families in dense sets, *Combinatorica* **16** (1996) 269–294.

[**XuD02**]   C. XU, R. DING: On the empty convex partition of a finite set in the plane, *Chin. Ann. of Math.* **23B** (2002) 487–494.

9. Graph Drawings and Geometric Graphs

9.1 Graph Drawings

In the traditional areas of graph theory (Ramsey theory, extremal graph theory, random graphs, etc.), graphs are regarded as abstract binary relations. The relevant methods are often incapable of providing satisfactory answers to questions arising in geometric applications. Geometric graph theory focuses on combinatorial and geometric properties of graphs drawn in the plane by straight-line edges (or, more generally, by edges represented by simple Jordan arcs). It is a fairly new discipline abounding in open problems, and it has already yielded some striking results that led to the solution of several problems in combinatorial and computational geometry and number theory. Many interesting questions arise or are directly motivated by practical problems in network design (VLSI), cartography, geographic information systems (GIS), visualization in chemical and biological phenomena, etc. Pach and Agarwal devoted a chapter to geometric graph theory in their monograph [PaA95]. For more recent surveys, see [Pa99], [Pa04]. The algorithmic aspects of graph drawing are discussed in the monograph of di Battista et al. [dBE*98]. See also the proceedings of the annual conferences on graph drawing, published in the Lecture Notes in Computer Science Series of Springer [GrDr].

We say that a graph G is *drawn* in the plane if its vertices are represented by distinct points and its edges are represented by (possibly crossing) simple Jordan arcs connecting the corresponding points and satisfying the following properties:

1. No arc representing an edge e passes through any point representing a vertex different from the endpoints of e.
2. Any two arcs have only finitely many points in common.
3. If two arcs have a common interior point, then they *properly cross* at this point; i.e., the first arc passes from one side of the second arc to the other side.
4. No three arcs have a common interior point.

If the edges of the graph G are represented by segments, we call the drawing a *straight-line drawing*.

Unless it leads to confusion, we make no notational or terminological difference between the vertices (edges) of G and the points (arcs) representing them. If we place the drawings themselves at the center of our attention, then, in order to emphasize that they are not merely *secondary* constructs derived from the underlying abstract graphs, we call them *topo-*

logical graphs or, in the case of straight-line drawings, *geometric graphs*.

A common interior point of two edges of a topological or geometric graph is said to be a crossing. We always assume that the vertices of a geometric graph are in *general position*, in the sense that no three of them are on a line.

[**dBE*98**]   G. DI BATTISTA, P. EADES, R. TAMASSIA, I. TOLLIS: *Graph Drawing: Algorithms for the Visualization of Graphs*, Prentice Hall, 1998.

[**GrDr**]   GRAPH DRAWING: Conference Series, Springer *LNCS (Lecture Notes in Computer Science)* **894, 1027, 1190, 1353, 1547, 1731, 1984, 2265, 2528, 2912, 3383**, 1994–2004.

[**Pa04**]   J. PACH: Geometric Graph Theory, in: *Handbook of Discrete and Computational Geometry* (only in the 2nd ed.), J.E. Goodman et al., eds., CRC Press, 2004, 219–238.

[**Pa99**]   J. PACH: Geometric graph theory, in: *Surveys in Combinatorics, 1999* (Canterbury 1999), 167–200, *London Math. Soc. Lecture Note Ser.* **267** (1999) Cambridge Univ. Press.

[**PaA95**]   J. PACH, P.K. AGARWAL: *Combinatorial Geometry*, J. Wiley and Sons, 1995.

9.2 Drawing Planar Graphs

Not surprisingly, much of our knowledge of geometric and topological graphs is based on our understanding of the properties of *planar* graphs. Although Euler's polyhedral formula, Kuratowski's criterion for planarity, Whitney's theorem, the four color theorem, etc., provide important structural information about planar graphs, many simple questions remain open. A large proportion of the problems listed in this section have been raised fairly recently by computational geometers.

According to Fáry's theorem [Fá48], [Wa36], every planar graph G has a straight-line drawing with pairwise noncrossing edges. Moreover, if G is three-connected, it has a straight-line drawing with all bounded faces convex [St22], [St51], and G and its dual have simultaneous straight-line drawings in the plane such that only dual pairs of edges cross and every such pair is perpendicular [BrS93]. (Here the unbounded face must be represented by the "point at infinity." A graph is k-connected if it remains connected after the removal of any $k - 1$ of its vertices.)

We say that a planar graph G *permits* a straight-line drawing *on a point set* X if G can be drawn by pairwise noncrossing line segments such that every vertex is represented by an element of X. De Fraysseix, Pach, and Pollack [dFPP90] and Schnyder [Sch90] proved that every planar graph on n vertices permits a straight-line drawing on the $(n-1) \times (n-1)$ integer grid. Moreover, such a drawing can be constructed in $O(n)$ time (Chrobak and Payne [ChP95]).

Problem 1   *Does every planar graph on n vertices permit a straight-line drawing on the $\lceil 2n/3 \rceil \times \lceil 2n/3 \rceil$ integer grid?*

It is clear that if n is divisible by 3, drawing a nested sequence of $n/3$ triangles requires a grid of this size. It is possible that every four-connected planar graph with n vertices permits a straight-line drawing on the $\lceil n/2 \rceil \times \lceil n/2 \rceil$ integer grid. The slightly weaker statement that every such graph has a straight-line drawing on an $a \times b$ grid, for a suitable pair of integers a, b satisfying $a + b \le n$, was proved by He [He97].

A point set U_n is said to be *universal* for planar graphs of n vertices if every such graph permits a straight-line drawing on U_n. It follows from the above-mentioned results [dFPP90], [Sch90] that the minimum size $u(n)$ of a universal set for planar graphs of n vertices is $O(n^2)$. It is also known [dFPP90], [ChK89] that there exists an $\varepsilon > 0$ such that $u(n) > (1 + \varepsilon)n$ holds for every sufficiently large n, but Kobourov (private communication) has checked by exhaustive search that $u(n) = n$ for all $n \le 14$.

Problem 2   *Does the minimum size $u(n)$ of a universal set for planar graphs of n vertices satisfy $u(n) = \Theta(n)$?*

It is interesting to note that every *outerplanar* graph G (every triangulated cycle) of n vertices permits a straight-line drawing on *any* set P of n points in the plane, no three collinear [GrM*91] (see [KaW02] and [Bo02] for algorithmic versions). In fact, for any edge v_1v_2 of the outer cycle of G and for any two consecutive vertices, p_1 and p_2, of the convex hull of P, there is a drawing in which v_1 and v_2 are represented by p_1 and p_2, respectively.

Ikebe, Perles, Tamura, and Tokunaga [IkP*94] (see also [PaT93]) proved a stronger statement for trees: every rooted tree T of n vertices permits a straight-line drawing on any set P of n points in general position in the plane with an arbitrarily specified element $p \in P$ such that the root of T is mapped to p. Kaneko and Kano [KaK99] generalized this theorem to any *pair* of rooted trees T_1 and T_2, as follows. For any set P of $n = |V(T_1)| + |V(T_2)|$ points in general position in the plane, there is a crossing-free straight-line drawing of $T_1 \cup T_2$ on P in which the roots are mapped to arbitrarily specified elements of P. The analogous statement for *triples* of trees is false.

If we want to construct a crossing-free drawing of a fixed planar graph with n vertices such that the location of *every* vertex is specified, then we cannot insist on straight-line edges. Our only hope is that we can find a drawing in which the edges are represented by noncrossing polygonal paths, and the total number of extra vertices (bends) is not too large. Pach and Wenger [PaW01] showed that there always exists such a drawing with $O(n^2)$ bends, and that this bound is asymptotically tight. To prove the tightness, they showed that if $P = \{p_1, p_2, \ldots, p_{2n}\}$ is the vertex set of a convex polygon, and we want to connect the corresponding point pairs of a randomly selected perfect matching of P by pairwise disjoint polygonal paths, then as $n \to \infty$, almost surely every such drawing has at least $\Omega(n^2)$ bends. However, it is quite likely that the same result holds for arbitrary sets in general position.

Conjecture 3   *(Pach) Let $P = \{p_1, p_2, \ldots, p_{2n}\}$ be any point set in general position in the plane, and let $\pi : \{1, 2, \ldots, 2n\} \to \{1, 2, \ldots, 2n\}$ be a randomly and uniformly selected permutation. Then as $n \to \infty$, with probability tending to one, the minimum number of bends in a noncrossing drawing in which every $p_{\pi(2i)}$ is connected to $p_{\pi(2i-1)}$ by a polygonal path, for $1 \le i \le n$, is $\Omega(n^2)$.*

The following question raised by Aichholzer, Aurenhammer, Hurtado, and Krasser [AiA*03] was motivated by possible applications in image analysis and morphing.

Conjecture 4 *[AiA*03] Let P and Q be two finite sets of points in general position in the plane. If |P| = |Q| and their convex hulls also have the same number of vertices, then one can triangulate the convex hulls of P and Q using only the points of P and Q so that the resulting triangulations are combinatorially isomorphic.*

It is possible that two such isomorphic triangulations exist even if we arbitrarily specify one vertex of each convex hull and we insist that they should correspond to each other. Aichholzer et al. [AiA*03] established this stronger statement in the special case that P and Q have at most three interior points. They also exhibited, for every $n \geq m \geq 3$, an n-element point set $P_{n,m}$ in general position whose convex hull has m vertices, and whenever $P_{n,m}$ and Q meet the requirements in Conjecture 4, they have isomorphic triangulations. However, this does not settle the conjecture because the relation that P and Q permit isomorphic triangulations is not transitive.

Hanani (alias Chojnacki*) [Ch34] and Tutte [Tu70] found a remarkable characterization of planar graphs: a graph G is planar if and only if it can be drawn in the plane so that any two arcs representing edges of G cross an *even* number of times. We call such a drawing an *even drawing* of G. If we combine this result with Fáry's theorem, according to which every planar graph permits a straight-line drawing, we conclude that every even drawing can be "stretched."

A drawing is called *x-monotone* if every vertical line intersects every edge in at most one point. Pach and Tóth [PaT02] strengthened the above statement by showing that every x-monotone even drawing of a (planar) graph can be stretched without changing the x-coordinates of the vertices (see also Eades and Feng [EaF97]). However, the Hanani–Tutte theorem remains true under the weaker assumption that any two *nonadjacent* edges cross an even number of times, i.e., there is no need to make any assumption about edges incident to the same vertex.

Problem 5 *Suppose that G has an x-monotone drawing D such that every pair of nonadjacent edges cross an even number of times. Does it necessarily follow that G has a noncrossing straight-line drawing in which the x-coordinates of the vertices are the same as in D?*

---

* Chaim Chojnacki emigrated in 1935 to Israel and changed his name around 1950 to Haim Hanani. All his later papers in design theory appeared under that name.

In 1936, Koebe [Ko36] established a far-reaching generalization of Fáry's theorem. He proved that the vertices of any planar graph G can be represented by nonoverlapping circular disks in the plane such that any two of them are tangent to each other if and only if the corresponding vertices are joined by an edge in G. Moreover, for triangulated planar graphs this representation is uniquely determined up to inversion (conformal transformation of the plane). Connecting the centers of the disks by segments tangent to each other, we obtain a straight-line drawing of G. For Thurston's combinatorial proof of Koebe's theorem and for related questions and results consult Colin de Verdière [CdV89], p. 96 in [PaA95], [MaR90], Andreev [An70a], [An70b], de Fraysseix, de Mendez, Rosenstiehl [dFMR94]. The following question may have been first asked by Scheinerman [Sch84].

Problem 6   *Can the vertices of every planar graph G be represented by line segments so that two segments have a point in common if and only if the corresponding vertices are adjacent in G?*

For bipartite planar graphs, the answer is yes. Hartman, Newman, and Ziv [HaNZ91] and de Fraysseix, de Mendez, and Pach [dFMP94], [dFMP95] showed that in this case one can use only vertical and horizontal segments and achieve that two segments are tangent to each other if and only if the corresponding vertices are adjacent but no two segments cross properly. De Castro, Cobos, Dana, Márquez, and Noy [dCC*99] proved the existence of the required representation for triangle-free planar graphs. De Fraysseix and de Mendez (unpublished, see also [dFdM04]) generalized this result to three-colorable planar graphs. Surprisingly, in this case one cannot always find a representation by noncrossing segments of at most three different directions. West [We91] raised a much stronger form of Problem 6: is it true that the vertices of every planar graph are representable by line segments of four different slopes such that two segments have a point in common if and only if the corresponding vertices are adjacent? If such a representation exists with the additional property that every pair of parallel segments are disjoint, then this would imply the four color theorem.

Actually, we do not even know the answer to the much weaker question: is it true that the vertices of every planar graph G can be represented by continuous arcs ("strings") so that two of them intersect if and only if the corresponding vertices are adjacent, and (1) any two arcs intersect at most once, (2) if two arcs intersect at a point p, then they properly cross at p? (Without the latter assumptions, the statement immediately follows from Koebe's theorem.)

We close this subsection with two curious questions. The first one may have more to do with number theory than with geometry; the second may be more graph-theoretic.

Problem 7 *Does every planar graph have a straight-line drawing in which the length of every edge is an integer?*

This problem has already been mentioned in a different context in Section 5.11; Harborth and his collaborators [HaK*87], [HaK90], [HaM94a], [HaM94b] have found such representations for some special graphs, including the skeletons of Platonic solids.

Conjecture 8 *(Papadimitriou and Ratajczak [PaR04]) Any three-connected planar graph can be drawn in the plane with (possibly crossing) straight-line edges so that for any ordered pair of vertices (u, v), there is a path from u to v along which the Euclidean distance to v strictly decreases.*

Moreover, it is conjectured that there exists such an embedding even with pairwise noncrossing edges such that every face of the resulting drawing is a convex polygon. It is not hard to prove that the complete bipartite graphs $K_{2,11}$ and $K_{3,16}$ permit no embeddings that meet the requirements in Conjecture 8. However, $K_{2,11}$ is not three-connected and $K_{3,16}$ is not planar.

[AiA*03] O. Aichholzer, F. Aurenhammer, F. Hurtado, H. Krasser: Towards compatible triangulations, *Theoretical Computer Science* **296** (2003) 3–13.

[An70a] E.M. Andreev: Convex polyhedra of finite volume in Lobačevskiĭ space (in Russian), *Mat. Sb. (N.S.)* **83 (125)** (1970) 256–260; English transl. in *Math. USSR Sb.* **12** (1970) 255–259.

[An70b] E.M. Andreev: Convex polyhedra in Lobačevskiĭ spaces (in Russian), *Mat. Sb. (N.S.)* **81 (123)** (1970) 445–478; English transl. in *Math. USSR Sb.* **10** (1970) 413–440.

[dBE*98] G. Di Battista, P. Eades, R. Tamassia, I. Tollis: *Graph Drawing: Algorithms for the Visualization of Graphs*, Prentice Hall, 1998.

[Bo02] P. Bose: On embedding an outer-planar graph in a point set, *Comput. Geom. Theory Appl.* **23** (2002) 303–312.

[BrS93] G.R. Brightwell, E.R. Scheinerman: Representations of planar graphs, *SIAM J. Discrete Math.* **6** (1993) 214–229.

[dCC*99] N. de Castro, F. Javier Cobos, J. Carlos Dana, A. Márquez, M. Noy: Triangle-free planar graphs as segments intersection graphs, in: *Graph Drawing 99*, J. Kratochvíl, ed., *LNCS* **1731**, Springer-Verlag, Berlin, 1999, 341–350.

[Ch34] CH. CHOJNACKI: Über wesentlich unplättbare Kurven im dreidimensionalen Raume, *Fund. Math.* **23** (1934) 135–142 (Ch. Chojnacki = H. Hanani).

[ChK89] M. CHROBAK, H. KARLOFF: A lower bound on the size of universal sets for planar graphs, *SIGACT News* **20**, 4 (1989), 63–86.

[ChP95] M. CHROBAK, T.H. PAYNE: A linear-time algorithm for drawing a planar graph on a grid, *Inform. Process. Lett.* **54** (1995) 241–246.

[CdV89] Y. COLIN DE VERDIÈRE: Empilements de cercles: convergence d'une méthode de point fixe, *Forum Math.* **1** (1989) 395–402.

[EaF97] P. EADES, Q.-W. FENG: Drawing clustered graphs on an orthogonal grid, in: *Graph Drawing 97*, G. di Battista, ed., *Lecture Notes in Computer Science* **1353**, Springer 1997, 146–157.

[Fá48] I. FÁRY: On straight line representation of planar graphs, *Acta Univ. Szeged. Sect. Sci. Math.* **11** (1948) 229–233.

[dFdM04] H. DE FRAYSSEIX, P.O. DE MENDEZ: Stretching of Jordan arc contact systems, in: *Graph Drawing 2003*, G. Liotta, ed., Springer *LNCS* **2912** (2004) 71–85.

[dFMP95] H. DE FRAYSSEIX, P.O. DE MENDEZ, J. PACH: A left-first search algorithm for planar graphs, *Discrete Comput. Geom.* **13** (1995) 459–468.

[dFMP94] H. DE FRAYSSEIX, P.O. DE MENDEZ, J. PACH: Representation of planar graphs by segments, in: *Intuitive Geometry* (Szeged, 1991), K. Böröczky et al., eds., *Colloq. Math. Soc. János Bolyai* **63** (1994) 109–117.

[dFMR94] H. DE FRAYSSEIX, P.O. DE MENDEZ, P. ROSENSTIEHL: On triangle contact graphs, *Combin. Probab. Comput.* **3** (1994) 233–246.

[dFPP90] H. DE FRAYSSEIX, J. PACH, R. POLLACK: How to draw a planar graph on a grid, *Combinatorica* **10** (1990) 41–51.

[GrM*91] P. GRITZMANN, B. MOHAR, J. PACH, R. POLLACK: Embedding a planar triangulation with vertices at specified points (solution to Problem E3341), *Amer. Math. Monthly* **98** (1991) 165–166.

[HaK90]   H. Harborth, A. Kemnitz: Integral representations of
           graphs, in: Contemporary Methods in Graph Theory, R. Bo-
           dendiek, ed., BI Verlag, 1990, 359–367.

[HaK*87]   H. Harborth, A. Kemnitz, M. Möller, A. Süssenbach:
           Ganzzahlige planare Darstellungen der platonischen Körper,
           Elemente Math. 42 (1987) 118–122.

[HaM94a]   H. Harborth, M. Möller: Ebene geradlinige Darstellun-
           gen der platonischen Graphen mit wenigen verschiedenen Kan-
           tenlängen, Abh. Braunschweig. Wiss. Ges. 45 (1994) 7–20.

[HaM94b]   H. Harborth, M. Möller: Minimum integral drawings of
           the platonic graphs, Math. Mag. 67 (1994) 355–358.

[HaNZ91]   I.B.-A. Hartman, I. Newman, R. Ziv: On grid intersec-
           tion graphs, Discrete Math. 87 (1991) 41–52.

[He97]   X. He: Grid embeddings of 4-connected plane graphs, Dis-
           crete Comput. Geom. 17 (1997) 329–358.

[IkP*94]   Y. Ikebe, M.A. Perles, A. Tamura, S. Tokunaga: The
           rooted tree embedding problem into points in the plane, Dis-
           crete Comput. Geom. 11 (1994) 51–63.

[KaK99]   A. Kaneko, M. Kano: Straight-line embeddings of two
           rooted trees in the plane, Discrete Comput. Geom. 21 (1999)
           603–613.

[KaW02]   M. Kaufmann, R. Wiese: Embedding vertices at points:
           few bends suffice for planar graphs, J. Graph Algorithms Appl.
           6 (2002) 115–129.

[Ko36]   P. Koebe: Kontaktprobleme der konformen Abbildung, Ber.
           Sächs. Akad. Wiss. Leipzig, Math.-Phys. Kl. 88 (1936) 141–
           164.

[MaR90]   A. Marden, B. Rodin: On Thurston's formulation and
           proof of Andreev's theorem, in: Computational Methods and
           Function Theory, S. Ruscheweyh et al., eds., Springer Lecture
           Notes Math. 1435 (1990) 103–115.

[Pa04]   J. Pach: Geometric graph theory, in: Handbook of Discrete
           and Computational Geometry (2nd edition only), J.E. Good-
           man et al., eds., CRC Press, 2004, 219–238.

[Pa99]   J. Pach: Geometric graph theory, in: Surveys in Combina-
           torics, (Canterbury, 1999), 167–200, London Math. Soc. Lec-
           ture Note Ser. 267 (1999) Cambridge Univ. Press.

[PaA95]   J. Pach, P.K. Agarwal: Combinatorial Geometry, J. Wi-
           ley and Sons, 1995.

[PaT93]   J. Pach, J. Töröcsik: Layout of rooted trees, in: *Planar Graphs, DIMACS Ser. Discrete Math. Theoret. Comput. Sci.* **9**, Amer. Math. Soc., 1993, 131–137.

[PaT02]   J. Pach, G. Tóth: Monotone drawings of planar graphs, in: *Algorithms and Computation*, P. Bose, P. Morin, eds., Springer *LNCS* **2518** (2002) 647–653.

[PaW01]   J. Pach, R. Wenger: Embedding planar graphs at fixed vertex locations, *Graphs Combinatorics* **17** (2001) 717–728.

[PaR04]   C.H. Papadimitriou, D. Ratajczak: On a conjecture related to geometric routing, in: *Algosensors* 2004, S. Nikoletseas et al., eds., Springer *LNCS* **3121** (2004) 9–17.

[Sch84]   E.R. Scheinerman: *Intersection Classes and Multiple Intersection Parameters of Graphs*, Ph.D. thesis, Princeton University, 1984.

[Sch90]   W. Schnyder: Embedding planar graphs on the grid, in: *SoDA 90, Proc. 1st ACM-SIAM Sympos. Discrete Algorithms*, ACM, 1990, 138–148.

[St51]   S.K. Stein: Convex maps, *Proc. Amer. Math. Soc.* **2** (1951) 464–466.

[St22]   E. Steinitz: Polyeder und Raumeinteilungen, in: *Enzykl. Math. Wiss.*, Part 3AB12 (1922) 1–139.

[Tu70]   W.T. Tutte: Toward a theory of crossing numbers, *J. Combinatorial Theory* **8** (1970) 45–53.

[Tu60]   W.T. Tutte: Convex representation of graphs, *Proc. London Math. Soc.* **10** (1960) 304–320.

[Wa36]   K. Wagner: Bemerkungen zum Vierfarbenproblem, *Jahresbericht Deutsch. Math. Verein.* **46** (1936) 26–32.

[We91]   D.B. West: Open problems, *SIAM J. Discrete Math. Newsletter* **2** (1991) 10–12.

9.3 The Crossing Number

Turán [Tu77] defined the *crossing number* of G, $\mathrm{cr}(G)$, as the smallest number of edge crossings in any drawing of G. Clearly, $\mathrm{cr}(G) = 0$ if and only if G is planar.

Garey and Johnson [GaJ83] proved that the determination of the crossing number is an *NP-complete* problem. During the last twenty years of the twentieth century, it has turned out that crossing numbers play an important role in various fields of discrete and computational geometry, and they can also be used, for example, to obtain lower bounds on the chip area required for the VLSI circuit layout of a graph [Le83].

The following question is also known as "Turán's brick factory problem." It occurred to Turán during World War II, while he was working at forced labor in a brick factory pushing carts filled with bricks between the kilns and the storage places. He wondered whether a better network of railroad tracks would reduce the number of crossings and thus the amount of time and energy wasted at these places.

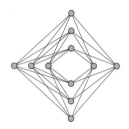

COMPLETE BIPARTITE GRAPH $K_{5,6}$ DRAWN WITH MINIMUM NUMBER OF CROSSINGS.

Conjecture 1 *(Zarankiewicz's conjecture [Gu69]) The crossing number of the complete bipartite graph $K_{n,m}$ with n and m vertices in its classes satisfies*

$$\mathrm{cr}(K_{n,m}) = \left\lfloor \frac{m}{2} \right\rfloor \cdot \left\lfloor \frac{m-1}{2} \right\rfloor \cdot \left\lfloor \frac{n}{2} \right\rfloor \cdot \left\lfloor \frac{n-1}{2} \right\rfloor .$$

Kleitman [Kl70] verified this conjecture in the special case $\min\{m,n\} \le 6$ and Woodall [Wo93] did so for $m = 7$, $n \le 10$.

Conjecture 2 *The crossing number of the complete graph K_n satisfies*

$$\mathrm{cr}(K_n) = \frac{1}{4} \left\lfloor \frac{n}{2} \right\rfloor \cdot \left\lfloor \frac{n-1}{2} \right\rfloor \cdot \left\lfloor \frac{n-2}{2} \right\rfloor \cdot \left\lfloor \frac{n-3}{2} \right\rfloor .$$

Of course, the best known upper bounds for both $\mathrm{cr}(K_{n,m})$ and $\mathrm{cr}(K_n)$ are the values in Conjectures 1 and 2. The best known lower bounds are

$$\mathrm{cr}(K_{m,n}) \ge n^2 m^2 \left(\frac{1}{20} - o(1) \right), \quad \mathrm{cr}(K_n) \ge n^4 \left(\frac{1}{80} - o(1) \right),$$

and they can be deduced from Kleitman's result [Kl70] by an easy counting argument.

Richter and Thomassen [RiT97] showed that if Zarankiewicz's conjecture is true (at least asymptotically, as $m = n \to \infty$), then, if we replace $\frac{1}{4}$ by $\frac{1}{4} + o(1)$, the expression in Problem 2 also gives the (asymptotically) correct value of $\operatorname{cr}(K_n)$.

Because of its computational complexity, the exact value of the crossing number is known or conjectured only for a few specific graphs. Let C_n denote a cycle with n vertices. For any two graphs G and H, let $G \times H$ denote the graph on the vertex set $V(G) \times V(H)$, where every vertex $(x, y), x \in V(G), y \in V(H)$ is connected to all vertices (x', y') such that $x = x'$ and $yy' \in E(H)$, or $xx' \in E(G)$ and $y = y'$.

Conjecture 3 *(Harary–Kainen–Schwenk [HaKS73]) For every $n \geq m \geq 3$, we have $\operatorname{cr}(C_n \times C_m) = n(m - 2)$.*

This conjecture was proved by Glebsky and Salazar [GlS04] for every $m \geq 3$ and for all $n \geq m(m + 1)$. Another not entirely hopeless problem may be to determine at least the asymptotic behavior of the crossing number of Q_n, the skeleton of the n-dimensional hypercube. We have $1/20 + o(1) \leq \operatorname{cr}(Q_n)/4^n \leq 163/1024$ [FaH00], [SýV93].

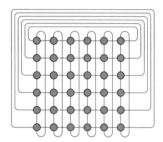

DRAWING OF $C_6 \times C_6$

WITH FEW CROSSINGS

It is easy to verify that the expressions in Conjectures 1 and 2 are within a constant factor from the true values of the corresponding crossing numbers. In particular, this is a direct corollary to the following general lower bound on crossing numbers, discovered by Ajtai, Chvátal, Newborn, and Szemerédi [AjC*82] and, independently, by Leighton [Le83]. For any graph G with n vertices and $e \geq 7n$ edges, we have

$$\operatorname{cr}(G) \geq \frac{1}{33.1} \frac{e^3}{n^2}. \tag{1}$$

This estimate, known as the *crossing lemma*, is tight up to a constant factor. The best known constant, $1024/31827 \approx 1/33.1 \approx 0.032$, in (1) is due to Pach, Radoičić, Tardos, and Tóth [PaR*04] (see also [PaT97]), and the result does not remain true if we replace 0.032 by roughly 0.09.

Let $f(n) \ll g(n)$ stand for $\lim_{n \to \infty} f(n)/g(n) = 0$. It was shown by Pach, Spencer, and Tóth [PaST00] that

$$\lim_{\substack{n \to \infty \\ n \ll e \ll n^2}} \frac{\min\{\operatorname{cr}(G) : |V(G)| = n, |E(G)| = e\}}{e^3/n^2} = K_0 \tag{2}$$

exists and is positive. It follows from what we have said before that

$$0.032 < 1024/31827 \le K_0 \le 0.09.$$

Problem 4   *[PaST00] Determine the precise value of K_0.*

Problem 5   *(Erdős–Guy [ErG73]) Do there exist positive constants C_1 and C_2 such that for any function $e = e(n)$ satisfying $C_1 n \le e \le C_2 n^2$, we have*

$$\lim_{\substack{n \to \infty \\ C_1 n \le e \le C_2 n^2}} \frac{\min\{cr(G) : |V(G)| = n, |E(G)| = e\}}{e^3/n^2} = K_0 ?$$

If the answer to the last question is yes, then clearly, $C_1 > 3$. We would also have that $C_2 < \frac{1}{2}$, because for $e = \binom{n}{2}$, $cr(K_n) > \left(\frac{1}{10} - \varepsilon\right) \frac{e^3}{n^2}$ holds for any $\varepsilon > 0$, provided that n is sufficiently large [Gu72].

Let G be a graph with vertex set $V(G)$ and edge set $E(G)$. The *bisection width* $b(G)$ of G is defined as the minimum number of edges whose removal splits the graph into two roughly equal subgraphs. More precisely, $b(G)$ is the minimum number of edges between V_1 and V_2 over all partitions of the vertex set of G into two disjoint parts $V_1 \cup V_2$ such that $|V_1|, |V_2| \ge |V(G)|/3$.

Leighton [Le84] observed that there is an intimate relationship between the bisection width and the crossing number of a graph based on the Lipton–Tarjan separator theorem for planar graphs [LiT79]. The following version of this relationship was obtained by Pach, Shahrokhi, and Szegedy [PaSS96] and by Sýkora and Vrťo [SýV94]. Let G be a graph of n vertices with degrees $d_1, d_2, \ldots, d_n$; then

$$b(G) \le 10\sqrt{cr(G)} + 2\sqrt{\sum_{i=1}^{n} d_i^2}. \tag{3}$$

Djidjev and Vrťo [DjV02] slightly strengthened (3) by replacing $b(G)$ with the *cut width* of G, which is defined as the minimum number $c(G)$ such that there is a drawing of G in which no two vertices have the same x-coordinate and every vertical line crosses (the interior of) at most $c(G)$ edges.

The example of a star (i.e., a tree consisting of a vertex connected to all other vertices) shows that (3) does not remain true if we remove the last term on its right-hand side. However, it is possible that the dependence of the bound on the degrees of the vertices can be improved.

Problem 6   *[PaT00] Does there exist a constant $t < 2$ such that*

$$b(G) = O\left(\sqrt{cr(G)} + \left(\sum_{i=1}^{n} d_i^t\right)^{1/t}\right)$$

holds for every graph of n vertices with degrees $d_1, \ldots, d_n$?

For every $g \geq 0$, one can define a new crossing number $\mathrm{cr}_g(G)$ as the minimum number of crossings in any drawing of G on a closed oriented surface of genus g. It was shown in [PaST00] that (2) remains true with exactly the same constant K_0 if we replace $\mathrm{cr}(G)$ by $\mathrm{cr}_g(G)$ and keep g fixed. What happens if g tends to infinity with n?

Problem 7　　　*Find a function $g = g(n)$ tending to infinity such that for any function $e = e(n)$ satisfying $n \ll e \ll n^2$, the limit*

$$\lim_{\substack{n \to \infty \\ n \ll e \ll n^2}} \frac{\min\{\mathrm{cr}_g(G) : |V(G)| = n, |E(G)| = e\}}{e^3/n^2}$$

exists and is positive.

The lower bound (1) for crossing numbers can be substantially improved if we restrict our attention to some special classes of graphs, for example, to graphs not containing some fixed so-called *forbidden* subgraph.

A graph property $\mathcal{P}$ is said to be *monotone* if
(a) for any graph G satisfying $\mathcal{P}$, every subgraph of G also satisfies $\mathcal{P}$; and
(b) if G_1 and G_2 satisfy $\mathcal{P}$, then their disjoint union also satisfies $\mathcal{P}$.
For any monotone property $\mathcal{P}$, let $\mathrm{ex}(n, \mathcal{P})$ denote the maximum number of edges that a graph of n vertices can have if it satisfies $\mathcal{P}$. In the special case that $\mathcal{P}$ is the property that the graph does not contain any (not necessarily induced) subgraph isomorphic to a fixed forbidden subgraph H, we write $\mathrm{ex}(n, H)$ for $\mathrm{ex}(n, \mathcal{P})$.

Let $\mathcal{P}$ be a monotone graph property with $\mathrm{ex}(n, \mathcal{P}) \leq O(n^{1+\alpha})$ for some $\alpha > 0$. Pach, Spencer, and Tóth [PaST00] proved that there exist two constants $c, c' > 0$ such that the crossing number of any graph G with property $\mathcal{P}$ that has n vertices and $e \geq cn \log^2 n$ edges satisfies

$$\mathrm{cr}(G) \geq c' \frac{e^{2+1/\alpha}}{n^{1+1/\alpha}}. \tag{4}$$

This bound is asymptotically tight, up to a constant factor.

In some interesting special cases, when we know the precise order of magnitude of the function $\mathrm{ex}(n, \mathcal{P})$, we can obtain a slightly stronger result: (4) is valid for every $e \geq 4n$ [PaST00]. For instance, if $\mathcal{P}$ is the property that G does not contain C_4 (a cycle of length four) as a subgraph, then $\mathrm{ex}(n, \mathcal{P}) = \mathrm{ex}(n, C_4) = \Theta(n^{3/2})$, and from this one can deduce that the crossing number of any graph with n vertices and $e \geq 4n$ edges satisfying this property is at least a constant times e^4/n^3. This bound is asymptotically tight. For an alternative proof, see Székely's paper [Sz04].

If the answer to the question formulated in the following problem turns out to be yes, one could extend this stronger result to many further graph properties $\mathcal{P}$.

Problem 8 Let H be a bipartite graph, and let H' be a graph that can be obtained from H by identifying two vertices whose distance is at least three. Is it true that

$$\mathrm{ex}(n, H) \leq O\left(\mathrm{ex}(n, H')\right)?$$

[AjC*82] M. AJTAI, V. CHVÁTAL, M.M. NEWBORN, E. SZEMERÉDI: Crossing-free subgraphs, *Annals Discrete Math.* **12** (1982) 9–12.

[DjV02] H. DJIDJEV, I. VRŤO: An improved lower bound for crossing numbers, in: *Graph Drawing 2001*, P. Mutzel et al., eds., Springer *LNCS* **2265** (2002) 96–101.

[ErG73] P. ERDŐS, R.K. GUY: Crossing number problems, *Amer. Math. Monthly* **80** (1973) 52–58.

[FaH00] L. FARIA, C.M.H. DE FIGUERADO: On Eggleton and Guy conjectured upper bounds for the crossing number of the n-cube, *Math. Slovaca* **50** (2000) 271–287.

[GaJ83] M.R. GAREY, D.S. JOHNSON: Crossing number is NP-complete, *SIAM J. Algebraic Discrete Methods* **4** (1983) 312–316.

[GlS04] L.Y. GLEBSKY, G. SALAZAR: The crossing number of $C_m \times C_n$ is as conjectured for $n \geq m(m+1)$, *J. Graph Theory* **47** (2004) 53–72.

[Gu72] R.K. GUY: Crossing numbers of graphs, in: *Graph Theory and Applications (Proc. Conf. Western Michigan Univ., Kalamazoo, Mich., 1972)*, Y. Alavi et al., eds., *Lecture Notes in Mathematics* **303**, Springer, Berlin, 111–124.

[Gu69] R.K. GUY: The decline and fall of Zarankiewicz's theorem, in: *Proof Techniques in Graph Theory*, Academic Press, 1969, 63–69.

[HaKS73] F. HARARY, P.C. KAINEN, A.J. SCHWENK: Toroidal graphs with arbitrarily high crossing numbers, *Nanta Math.* **6** (1973) 58–67.

[Kl70] D.J. KLEITMAN: The crossing number of $K_{5,n}$, *J. Combinatorial Theory* **9** (1970) 315–323.

[Le84] F.T. LEIGHTON: New lower bound techniques for VLSI, *Math. Systems Theory* **17** (1984) 47–70.

[Le83] F.T. LEIGHTON: *Complexity Issues in VLSI*, Foundations of Computing Series, MIT Press, 1983.

[LiT79]   R.J. LIPTON, R.E. TARJAN: A separator theorem for planar graphs, *SIAM J. Applied Mathematics* **36** (1979) 177–189.

[PaR*04]   J. PACH, R. RADOIČIĆ, G. TARDOS, G. TÓTH: Improving the Crossing Lemma by finding more crossings in sparse graphs, in: *SCG 04, 20th ACM Symposium on Computational Geometry*, ACM Press, New York, 2004, 68–75.

[PaSS96]   J. PACH, F. SHAHROKHI, M. SZEGEDY: Applications of the crossing number, *Algorithmica* **16** (1996) 111–117.

[PaST00]   J. PACH, J. SPENCER, G. TÓTH: New bounds on crossing numbers, *Discrete Comput. Geom.* **24** (2000) 623–644.

[PaT00]   J. PACH, G. TÓTH: Thirteen problems on crossing numbers, *Geombinatorics* **9** (2000) 194–207.

[PaT97]   J. PACH, G. TÓTH: Graphs drawn with few crossings per edge, *Combinatorica* **17** (1997) 427–439.

[RiT97]   R.B. RICHTER, C. THOMASSEN: Relations between crossing numbers of complete and complete bipartite graphs, *Amer. Math. Monthly* **104** (1997) 131–137.

[SýV94]   O. SÝKORA, I. VRŤO: On VLSI layouts of the star graph and related networks, *Integration, The VLSI Journal* **17** (1994) 83–93.

[SýV93]   O. SÝKORA, I. VRŤO: On the crossing number of the hypercube and the cube connected cycles, *BIT* **33** (1993) 232–237.

[Sz04]   L.A. SZÉKELY: A short proof of a theorem of Pach, Spencer, and Tóth, in: *Towards a Theory of Geometric Graphs*, J. Pach, ed., *Contemporary Mathematics* **342**, AMS 2004, 281–283.

[Tu77]   P. TURÁN: A note of welcome, *J. Graph Theory* **1** (1977) 7–9.

[Wo93]   D.R. WOODALL: Cyclic-order graphs and Zarankiewicz's crossing-number conjecture, *J. Graph Theory* **17** (1993) 657–671.

9.4 Other Crossing Numbers

There are three important variants of the notion of crossing number.

(1) The *rectilinear crossing number* lin-cr(G) of a graph G is the minimum number of crossings in a drawing of G in which every edge is represented by a line segment.

(2) The *pairwise crossing number* of G, pair-cr(G), is the minimum number of crossing pairs of edges over all drawings of G. (Here the edges can be represented by arbitrary continuous curves, so that two edges may cross more than once, but every pair of edges contributes at most one to pair-cr(G).)

(3) The *odd-crossing number* of G, odd-cr(G), is the minimum number of those pairs of edges that cross an odd number of times over all drawings of G.

It readily follows from the definitions that

$$\text{odd-cr}(G) \le \text{pair-cr}(G) \le \text{cr}(G) \le \text{lin-cr}(G).$$

Bienstock and Dean [BiD93] exhibited a series of graphs with crossing number four whose rectilinear crossing numbers are arbitrarily large. The following is perhaps the most exciting unsolved problem in the area.

Problem 1 *(Pach–Tóth [PaT00b]) Does*

$$\text{odd-cr}(G) = \text{pair-cr}(G) = \text{cr}(G)$$

hold for every graph G?

According to the theorem of Hanani [Ch34] and Tutte [Tu70] quoted in the previous subsection, odd-cr(G) = 0 implies that cr(G) = 0. Note that in this case, by Fáry's theorem [Fá48], we also have that lin-cr(G) = 0.

The main difficulty with this problem is that a graph has so many essentially different drawings that the computation of any of the above crossing numbers for a graph of only 15 vertices appears to be a hopelessly difficult task even for a computer [RiT97].

As we have already mentioned, the determination of the crossing number is an *NP-complete* problem [GaJ83]. The same is true for the odd-crossing number [PaT00b] and for the pairwise crossing number [ScSS02]. However, we know only that computing the rectilinear crossing number is *NP-hard* [Bi91].

Problem 2 *Given a graph G of n vertices and an integer K, can one check in polynomial time whether lin-cr(G) $\le K$? In other words, is the above problem in NP?*

Since presently the determination of the crossing number of the complete graph K_n seems to be out of reach, a lot of attention has been given to estimating lin-cr(K_n).

Problem 3   *Determine the value of*

$$\kappa := \lim_{n \to \infty} \frac{\text{lin-cr}(K_n)}{\binom{n}{4}}.$$

The lower bound $\kappa \geq 3/8 = 0.375$ was proved independently by Lovász, Vesztergombi, Wagner, Welzl [LoV*04] and by Ábrego, Fernández [ÁbFM05]. Actually, Lovász et al. proved strict inequality, but their result was slightly superseded by Balogh and Salazar [BaS05], who established the inequality $\kappa \geq 0.37533$. The best known upper bound, $\kappa \leq 0.381$, is due to Aichholzer, Aurenhammer, Krasser [AiAK02]. The known exact values of lin-cr(G) are listed below [BrDG01], [Kr03].

n	4	5	6	7	8	9	10	11	12	13	14	15	16
lin-cr(K_n)	0	1	3	9	19	36	62	102	153	229	324	447	603

Concerning Problem 1, it is known that the parameters cr(G), pair-cr(G), and odd-cr(G), are not completely unrelated. More precisely, it was proved in [PaT00b] that cr(G) $\leq 2($odd-cr$(G))^2$ for every graph G. Valtr [Va05] showed that

$$\text{cr}(G) \leq O\left(\frac{(\text{pair-cr}(G))^2}{\log(\text{pair-cr}(G))} \right).$$

Perhaps the next step would be to answer the following question.

Problem 4   *Does there exist a constant C such that*

$$\text{cr}(G) \leq C\,(\text{odd-cr}(G))$$

holds for every graph G?

A formula for pair-cr(G), analogous to but slightly weaker than bound (3) in Section 9.3 was established by Kolman and Matoušek [KoM04].

Problem 5   *[PaT00a] Does there exist a suitable constant C such that every graph G satisfies*

$$b(G) \leq C\left(\sqrt{\text{odd-cr}(G)} + \sqrt{\sum_{i=1}^{n} d_i^2} \right) ?$$

Here d_i denotes the degree of the ith vertex, $1 \leq i \leq n$.

Another natural approach toward Problem 1 is to decide whether the *expected values* of $\mathrm{cr}(G), \mathrm{pair\text{-}cr}(G)$, and $\mathrm{odd\text{-}cr}(G)$ coincide for randomly selected graphs.

Let $G = G(n, p)$ be a *random* graph with n vertices whose edges are chosen independently with probability $p = p(n)$. Let e denote the *expected number* of edges of G, i.e., $e = \binom{n}{2}p$. It is not difficult to see that if $e > 10n$, almost surely $b(G) \geq e/10$. Therefore, it follows from inequality (3) in Section 9.3 that almost surely we have $\mathrm{cr}(G) \geq e^2/4000$. Evidently, the order of magnitude of this bound cannot be improved.

We cannot always determine the precise order of magnitude of the expected value of $\mathrm{odd\text{-}cr}(G)$ and $\mathrm{pair\text{-}cr}(G)$ for a random graph $G = G(n, p)$.

Conjecture 6   *[PaT00a] Let $G = G(n, p)$ be a random graph with n vertices, with edge probability $0 < p < 1$, and let $e = \binom{n}{2}p > 4n$. Then there exist suitable positive constants c_1 and c_2 such that*

(1) $$\mathrm{E}\left[\mathrm{pair\text{-}cr}(G)\right] \geq c_1 e^2,$$

(2) $$\mathrm{E}\left[\mathrm{odd\text{-}cr}(G)\right] \geq c_2 e^2.$$

This conjecture was verified by Spencer and Tóth [SpT02] under the stronger condition that $e > n^{1+\epsilon}$ for some $\epsilon > 0$.

Each of the above crossing numbers with the exception of $\mathrm{lin\text{-}cr}(G)$ can be further modified by applying one of the following rules:

Rule $+$: Consider only those drawings of G in which two edges with a common endpoint do not cross each other.

Rule 0 : Two edges with a common endpoint are allowed to cross, and their crossing counts.

Rule $-$: Two edges with a common endpoint are allowed to cross, but their crossing does not count.

In the original definitions we have always used Rule 0. If we apply Rule $+$ or Rule $-$ in the definition of the crossing numbers, then we indicate this by using the corresponding subscript, as shown in the table below. This gives us an array of nine different crossing numbers. It is easy to see that in a drawing of a graph that minimizes the number of crossing points, any two edges have at most one point in common (see, for example, [RiT97]). Therefore, $\mathrm{cr}_+(G) = \mathrm{cr}(G)$, which slightly simplifies the picture.

Rule +	odd-cr$_+(G)$	pair-cr$_+(G)$	cr(G)
Rule 0	odd-cr(G)	pair-cr(G)	
Rule −	odd-cr$_-(G)$	pair-cr$_-(G)$	cr$_-(G)$

Moving from left to right or from bottom to top in this array, the numbers do not decrease. It is not hard to generalize the crossing number lower bound cr$(G) \geq \frac{1}{60.75} \frac{e^3}{n^2}$ to each of these crossing numbers. In particular, we obtain (as in [PaT00b]) that

$$\text{odd-cr}_-(G) \geq \frac{1}{60.75} \frac{e^3}{n^2}$$

for any graph G with n vertices and with $e \geq 4n$ edges. We are not aware of any other results about odd-cr$_-(G)$, pair-cr$_-(G)$, or cr$_-(G)$. It is conjectured in [PaT00a] that these values are very close to, if not the same as, odd-cr(G), pair-cr(G), and cr(G). That is, by letting pairs of *incident* edges cross an arbitrary number of times, one probably cannot substantially reduce the total number of crossings between *independent* pairs of edges. The simplest open questions are the following.

Problem 7   *Do there exist suitable functions f_1, f_2, f_3 such that every graph G satisfies*

(1)   odd-cr$(G) \leq f_1(\text{odd-cr}_-(G))$,
(2)   pair-cr$(G) \leq f_2(\text{pair-cr}_-(G))$,
(3)   cr$(G) \leq f_3(\text{cr}_-(G))$?

The *thickness* of a graph G is the minimum number of edge-disjoint planar graphs into which G can be partitioned. Equivalently, the thickness is the minimum number t of colors such that there exists a drawing of G in the plane with the property that its edges can be colored by t colors with no two edges of the same color crossing each other. The definition of the *geometric thickness* of G is the same, except that now we consider only straight-line drawings. Clearly, the thickness of a graph can never exceed its geometric thickness. Eppstein [Ep04] constructed a series of graphs of thickness three whose geometric thickness is arbitrarily large.

Problem 8   *Construct a series of graphs with thickness two whose geometric thickness is arbitrarily large.*

In practical applications, graphs of thickness two play a special role. This motivates the following definition. The *biplanar crossing number* of a graph is the minimum of $\mathrm{cr}(G_1) + \mathrm{cr}(G_2)$ over all partitions of the graph into two edge-disjoint subgraphs, G_1 and G_2. Thus, the biplanar crossing number of a graph is 0 if and only if its thickness is at most 2. It is known [SýSV05] that the biplanar crossing number of every graph is at most 3/8 times its crossing number. The best value of the constant may be as small as 7/24. Spencer [Sp04] proved that the analogue of Conjecture 6 is true for the biplanar crossing number, or, more generally, for the minimum total crossing number of the parts over all partitions of the graph into k edge-disjoint subgraphs, where $k \geq 2$ is fixed.

Place the vertices of a graph G on a circle and connect the corresponding points by line segments. The resulting drawing is said to be a *convex drawing* of G. The *convex crossing number* of G is the minimum number of crossings in a convex drawing of G. It was shown by Shahrokhi, Sýkora, Székely, and Vrťo [ShS*04] that the convex crossing number of a graph G with n vertices having degrees $d_1, \ldots, d_n$ can be bounded from above by a constant times

$$\left(\mathrm{cr}(G) + \sum_{i=1}^{n} d_i^2 \right) \log n.$$

Shahrokhi (personal communication) suggested that a much better result may hold for the *biplanar convex crossing number* of G, defined as the minimum number of crossings between edges of the same color for any coloring of the edges of G with two colors in any convex drawing of G. Clearly, the biplanar convex crossing number of a graph is at least as large as its biplanar crossing number and at most as large as its convex crossing number.

Problem 9   (Shahrokhi) Is it true that the biplanar convex crossing number of every graph G with n vertices having degrees $d_1, \ldots, d_n$ is at most

$$O\left(\mathrm{cr}(G) + \sum_{i=1}^{n} d_i^2 \right)?$$

An affirmative answer to this question would lead to efficient new approximation schemes for estimating $\mathrm{cr}(G)$.

[ÁbFM05] B.M. ÁBREGO, S. FERNÁNDEZ-MERCHANT: A lower bound for the rectilinear crossing number, *Graphs Combinatorics*, to appear.

[AiAK02] O. Aichholzer, F. Aurenhammer, H. Krasser: Enumerating order types for small point sets with applications, *Order* **19** (2002) 265–281.

[BaS05] J. Balogh, G. Salazar: Improved bounds for the number of k-sets, convex quadrilaterals, and the rectilinear crossing number of K_n, in: *Graph Drawing 2004*, J. Pach, ed., Springer *LNCS* **3383** (2005) 25–35.

[Bi91] D. Bienstock: Some provably hard crossing number problems, *Discrete Comput. Geom.* **6** (1991) 443–459.

[BiD93] D. Bienstock, N. Dean: Bounds for rectilinear crossing numbers, *J. Graph Theory* **17** (1993) 333–348.

[BrDG01] A. Brodsky, S. Durocher, E. Gethner: The rectilinear crossing number of K_{10} is 62, *Electron. J. Combinatorics* **8** (2001), no. 1, Research Paper 23.

[Ch34] Ch. Chojnacki: Über wesentlich unplättbare Kurven im dreidimensionalen Raume, *Fund. Math.* **23** (1934) 135–142 (Ch. Chojnacki = H. Hanani).

[Ep04] D. Eppstein: Separating thickness from geometric thickness, in: *Towards a Theory of Geometric Graphs*, J. Pach, ed., *Contemporary Mathematics* **342**, AMS 2004, 75–86.

[Fá48] I. Fáry: On straight line representation of planar graphs, *Acta Univ. Szeged. Sect. Sci. Math.* **11** (1948) 229–233.

[GaJ83] M.R. Garey, D.S. Johnson: Crossing number is NP-complete, *SIAM J. Algebraic Discrete Methods* **4** (1983) 312–316.

[Gu72] R.K. Guy: Crossing numbers of graphs, in: *Graph Theory and Applications* (Proc. Conf. Western Michigan Univ., Kalamazoo, Mich., 1972) *Lecture Notes in Mathematics* **303**, Springer, Berlin, 111-124.

[KoM04] P. Kolman, J. Matoušek: Crossing number, pair-crossing number, and expansion, *J. Combinatorial Theory Ser. B* **92** (2004) 99–113.

[Kr03] H. Krasser: *Order Types of Point Sets in the Plane* (PhD thesis), Institute for Theoretical Computer Science, Graz University of Technology, Austria, October 2003.

[LoV*04] L. Lovász, K. Vesztergombi, U. Wagner, E. Welzl: Convex quadrilaterals and k-sets, in: *Towards a Theory of Geometric Graphs*, J. Pach, ed., *Contemporary Mathematics* **342**, AMS 2004, 139–148.

[PaT00a]   J. PACH, G. TÓTH: Thirteen problems on crossing numbers, *Geombinatorics* **9** (2000) 194–207.

[PaT00b]   J. PACH, G. TÓTH: Which crossing number is it, anyway? *J. Combinatorial Theory Ser. B* **80** (2000) 225–246.

[RiT97]   R.B. RICHTER, C. THOMASSEN: Relations between crossing numbers of complete and complete bipartite graphs, *Amer. Math. Monthly* **104** (1997) 131–137.

[ScSS02]   M. SCHAEFER, E. SEDGWICK, D. ŠTEFANKOVIČ: Recognizing string graphs in NP, in: *SToC 2002, Proc. Thirty-fourth Annual ACM Symposium on the Theory of Computing*, ACM Press, New York, 2002, 1–6.

[ShS*04]   F. SHAHROKHI, O. SÝKORA, L.A. SZÉKELY, I. VRŤO: The gap between the crossing number and the convex crossing number, in: *Towards a Theory of Geometric Graphs*, J. Pach, ed., *Contemporary Mathematics* **342**, AMS 2004, 249–258.

[Sp04]   J. SPENCER: The biplanar crossing number of the random graph, in: *Towards a Theory of Geometric Graphs*, J. Pach, ed., *Contemporary Mathematics* **342**, AMS 2004, 269–271.

[SpT02]   J. SPENCER, G. TÓTH: Crossing numbers of random graphs, *Random Structures Algorithms* **21** (2002) 347–358.

[SýSV05]   O. SÝKORA, L.A. SZÉKELY, I. VRŤO: Crossing numbers and biplanar crossing numbers II: using the probabilistic method, manuscript.

[Tu70]   W.T. TUTTE: Toward a theory of crossing numbers, *J. Combinatorial Theory* **8** (1970) 45–53.

[Va05]   P. VALTR: On the pair-crossing number, in: *Combinatorial and Computational Geometry*, J.E. Goodman et al., eds., Cambridge Univ. Press, *MSRI Publications* **52** (2005), to appear.

9.5 From Thrackles to Forbidden Geometric Subgraphs

A *topological graph* is a graph drawn in the plane so that its vertices are represented by points and their edges by non-self-intersecting continuous arcs connecting the corresponding point pairs, but not passing through any other point representing a vertex. The edges of a topological graph are not allowed to have tangencies, i.e., if two edges share an interior point p, then they must properly cross at p. If the vertex set $V(G)$ of a topological graph G is in *general position*, i.e., no three points that represent vertices are collinear and the edges are drawn by straight-line segments, then G is called a *geometric graph*. If in addition, $V(G)$ is the vertex set of a convex polygon, then G is called a *convex geometric* graph.

Conway defined a *thrackle* as a topological graph with any pair of non-adjacent edges crossing precisely once, but with no two adjacent edges sharing an interior point [Wo71], [Wo72], [GrR95], [Ri96]. He offered a cash prize for the solution of the following problem.

CYCLES C_5 AND C_{10}
DRAWN AS THRACKLES

Conjecture 1 *(Conway's thrackle conjecture)* The number of edges of a thrackle cannot exceed the number of its vertices.*

Woodall [Wo71] has shown that if the conjecture is false, then there exists a counterexample consisting of two even cycles that share a vertex. It was shown by Lovász et al. [LoPS97] that a thrackle of n vertices has fewer than $2n$ edges. This bound was improved by Cairns and Nikolayevsky [CaN00] to $3(n-1)/2$. They also generalized Conway's conjecture as follows: any thrackle of n vertices on a closed oriented surface of genus g has at most $n + 2g$ edges. This statement has been verified only for graphs with at most five vertices.

Conway's conjecture is known to be true for straight-line thrackles [HoP34], [Su35]. In this special case, the statement can be reformulated as follows. Let G be a geometric graph with n vertices that does not contain two disjoint edges (i.e., two edges that do not cross and do not even share an endpoint). Then G has at most n edges.

Turán's classical graph theorem [Tu54] determines the maximum number of edges that an *abstract* graph with n vertices can have without containing a complete subgraph with k vertices. This suggests that many sim-

---

* "Thrackle" appears to be a word invented by Conway.

ilar problems can be raised for *geometric* graphs in which some geometric subconfiguration is forbidden. The systematic study of questions of this type was initiated by Avital and Hanani [AvH66], Erdős, Kupitz [Ku79], and Perles. Given a class $\mathcal{H}$ of so-called *forbidden geometric subgraphs*, what is the maximum number of edges that a geometric graph G of n vertices can have without containing a geometric subgraph belonging to $\mathcal{H}$? For instance, what is the maximum number of edges of a geometric graph with n vertices containing no k disjoint edges? For $k = 3$, Černý [Če03] proved that the answer is $2.5n + O(1)$ (see Goddard Katchalski, Kleitman [GoKK96] and Alon, Erdős [AlE89] for weaker upper bounds).

Problem 2   *What is the smallest constant c such that the maximum number of edges of a geometric graph of n vertices, containing no four disjoint edges, is $cn + o(n)$?*

For larger (but fixed) values of k, the first linear upper bound was found in [PaT94]. The best currently known result is due to Tóth [Tó00]: every geometric graph with n vertices and no k disjoint edges has at most $2^{10}(k-1)^2 n$ edges.

Problem 3   *Does there exist an absolute constant C such that every geometric graph with n vertices and no k disjoint edges has at most Ckn edges?*

For convex geometric graphs, i.e., when the vertices are in convex position, the answer to this question is clearly yes. It follows from an old result of Perles (see [KáP*98]) that if a convex geometric graph of $n \geq 2k$ vertices has more then $(k-1)n$ edges, it contains a noncrossing path of length $2k - 1$. This bound cannot be improved. Obviously, every noncrossing path of length $2k - 1$ has k disjoint edges. It is tempting to conjecture that Ckn edges are sufficient to guarantee the existence of a noncrossing path of length $2k - 1$ in any geometric graph with n vertices.

Another interesting variant of this question was studied by Valtr [Va98], [Va99]. We call two edges of a geometric graph *avoiding* if neither of the lines supporting one of them has a point in common with the other edge. Using the theory of generalized Davenport–Schinzel sequences [Va98], Valtr proved that for every k there exists a constant C_k such that every geometric graph with n vertices and no k pairwise avoiding edges has at most $C_k n$ edges.

If we are looking for noncrossing subconfigurations in topological graphs whose edges are not necessarily line segments, then we have to impose some further conditions on their intersections. Otherwise, it may happen that every pair of edges cross each other. We call a topological graph *simple* if any two of its edges meet at most once. According to this definition, every thrackle is a simple topological graph.

Problem 4   (Pach–Tóth [PaT05]) Is it true that, for every fixed $k \geq$ 3, the maximum number of edges of a simple topological graph with n vertices that contains no k disjoint edges is $O(n)$?

For $k = 2$, this follows from the partial results concerning the thrackle conjecture [LoPS97], [CaN00]. For $k > 2$, the best known upper bound, which is n times a polylogarithmic factor, was established in [PaT05].

It was proved in [PaST03] that every simple topological complete graph with n vertices has a noncrossing subgraph isomorphic to any fixed tree T with at most $c \log^{1/6} n$ vertices, where $c > 0$ is a suitable constant. In particular, it has at least $\lfloor (c/2) \log^{1/6} n \rfloor$ disjoint edges. It follows from the results in [PaT05] that the latter bound can be improved to $c' \log n / \log \log n$ for some $c' > 0$, but most likely this estimate is still far from being optimal.

Problem 5   Does there exist a constant $c > 0$ such that every simple topological complete graph with n vertices has at least n^c disjoint edges?

One can try to find many disjoint edges by the following greedy method. Select an edge that intersects the smallest number of other edges, delete these edges, and repeat the procedure. The problem is that we may quickly run out of edges.

Let $h = h(n)$ denote the smallest integer such that every simple topological complete graph with n vertices has an edge crossing fewer than h other edges. The following construction of Valtr, drawn on the surface of a cube, shows that $h(n) \geq \Omega(n^{3/2})$: Put a square grid of $n/6$ points on each face of the of the cube, and slightly perturb the points in order to achieve general position. Connect any two points by a geodesic (that is, a shortest path connecting them on the surface of the cube). To see this,

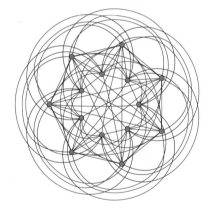

K_{14} drawn with each edge crossing at least three others

it is enough to notice that taking any row or column of the grid sitting on a face of the cube, the corresponding path is crossed by $\Omega(n^2)$ edges, and these crossings are approximately equally distributed among the edges of the path. (For a construction showing that $h(n) > \left(\frac{3}{4} + o(1)\right) n$, see [HaT94]). We do not have any reasonable upper bound on $h(n)$.

Conjecture 6 $h(n) = o(n^2)$.

[AlE89]   N. ALON, P. ERDŐS: Disjoint edges in geometric graphs,
           Discrete Comput. Geom. **4** (1989) 287–290.

[AvH66]   S. AVITAL, H. HANANI: Graphs (in Hebrew), *Gilyonot Le-
           matematika* **3** (1966) 2–8.

[CaN00]   G. CAIRNS, Y. NIKOLAYEVSKY: Bounds for generalized
           thrackles, *Discrete Comput. Geom.* **23** (2000) 191–206.

[Če03]   J. ČERNÝ: Geometric graphs with no three disjoint edges
           (preprint), *KAM-DIMATIA Series* 2003-649, Charles Univer-
           sity, Prague 2003.

[GoKK96]   W. GODDARD, M. KATCHALSKI, D.J. KLEITMAN: Forcing
           disjoint segments in the plane, *European J. Combinatorics* **17**
           (1996) 391–395.

[GrR95]   J.E. GREEN, R.D. RINGEISEN: Combinatorial drawings and
           thrackle surfaces, in: *Graph Theory, Combinatorics, and Algo-
           rithms, Vol. 1, 2* (Kalamazoo, 1992), Wiley-Interscience, 1995,
           999–1009.

[HaT94]   H. HARBORTH, C. THÜRMANN: Minimum number of edges
           with at most s crossings in drawings of the complete graph,
           Congressus Numerantium **102** (1994) 83–90.

[HoP34]   H. HOPF, E. PANNWITZ: Aufgabe Nr. 167, *Jahresbericht
           Deutsch. Math.-Verein.* **43** (1934) p. 114.

[KáP*98]   G. KÁROLYI, J. PACH, G. TÓTH, P. VALTR: Ramsey-type
           results for geometric graphs. II, *Discrete Comput. Geom.* **20**
           (1998) 375–388.

[Ku79]   Y.S. KUPITZ: *Extremal Problems in Combinatorial Geo-
           metry, Aarhus University Lecture Notes Series* **53**, Aarhus
           University, Denmark, 1979.

[LoPS97]   L. LOVÁSZ, J. PACH, M. SZEGEDY: On Conway's thrackle
           conjecture, *Discrete Comput. Geom.* **18** (1997) 369–376.

[PaST03]   J. PACH, J. SOLYMOSI, G. TÓTH: Unavoidable configura-
           tions in complete topological graphs, *Discrete Comput. Geom.*
           30 (2003) 311–320.

[PaT94]   J. PACH, J. TÖRŐCSIK: Some geometric applications of Dil-
           worth's theorem, *Discrete Comput. Geom.* **12** (1994) 1–7.

[PaT05]   J. PACH, G. TÓTH: Disjoint edges in topological graphs, in:
           *The Indonesia-Japan Conference on Combinatorial Geometry
           and Graph Theory (IJCCGGT 2003)*, J. Akiyama et al., eds.,
           Springer *LNCS* **3330** (2005) 133–140.

[**Ri96**]   R.D. RINGEISEN: Two old extremal graph drawing conjectures: progress and perspectives, *Congressus Numerantium* **115** (1996) 91–103.

[**Su35**]   J.W. SUTHERLAND: Lösung der Aufgabe 167, *Jahresbericht Deutsch. Math.-Verein.* **45** (1935) 33–35.

[**Tó00**]   G. TÓTH: Note on geometric graphs, *J. Combinatorial Theory Ser. A* **89** (2000) 126–132.

[**Tu54**]   P. TURÁN: On the theory of graphs, *Colloquium Math.* **3** (1954) 19–30.

[**Va99**]   P. VALTR: Generalizations of Davenport–Schinzel sequences, in: Contemporary Trends in Discrete Mathematics (Štiřin Castle, 1997) *DIMACS Ser. Discrete Math. Theoret. Comput. Sci.* **49**, AMS 1999, 349–389.

[**Va98**]   P. VALTR: On geometric graphs with no k pairwise parallel edges, *Discrete Comput. Geom.* **19** (1998) 461–469.

[**Wo72**]   D.R. WOODALL: Unsolved problems, in: *Combinatorics, Proc. Conf. Combinatorial Math.* (Inst. Math. Appl. Oxford, Southend-on-Sea, 1972), 351–363.

[**Wo71**]   D.R. WOODALL: Thrackles and deadlock, in: *Combinatorial Mathematics and Its Applications*, D.J.A. Welsh, ed., Academic Press, London, 1971, 335–347.

9.6 Further Turán-Type Problems

If a topological graph G with $n \geq 3$ vertices has no two crossing edges, by Euler's polyhedral formula its number of edges cannot exceed $3n-6$. It was shown by Agarwal, Aronov et al. [AgA*97] (and by a simpler argument in [PaRT03]) that under the weaker condition that no three edges are *pairwise crossing*, the number of edges of G is still $O(n)$. It is not known, even for geometric graphs, whether this statement remains true if we assume only that no *four* edges are pairwise crossing.

Problem 1   *Does there exist for every $k > 3$ a constant c_k such that any geometric graph of n vertices containing no k pairwise crossing edges has at most $c_k n$ edges?*

Slightly improving the first nearly linear upper bound of Pach, Shahrokhi, and Szegedy [PaSS96], Valtr [Va97] proved that the above maximum is $O(n \log n)$, where the constant hidden in the notation depends on k. It was shown in [PaP*05] that if a topological graph G with n vertices contains no two k-element sets of edges such that every edge in the first set crosses all edges in the second, then G has $O(n)$ edges. In other words, if a topological graph contains no large "gridlike" crossing pattern, then its number of edges is at most linear in n.

For *convex* geometric graphs G with $n \geq 2k$ vertices and no k pairwise crossing edges, Capoyleas and Pach [CaP92] proved that

$$|E(G)| \leq 2(k-1)n - \binom{2k-1}{2}.$$

This bound is tight. In fact, it is also known that the edges of any convex geometric graph G with the above property can be colored by at most 2^{k+5} colors so that no two edges of the same color cross each other [Ko88], [KoK97]. In such a coloring every color class induces a planar graph. Thus, for a fixed k, this immediately yields that $|E(G)| \leq O(n)$. However, even for $k = 3$ we do not know whether this type of coloring result is true if we drop the assumption that G is convex, i.e., that its vertices are in convex position.

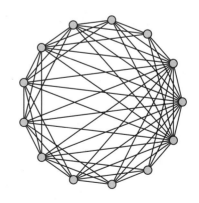

13 POINTS, 57 EDGES, WITHOUT
4 PAIRWISE CROSSING EDGES

Erdős (see [Gy87], [KoN98]) raised the following more general problem.

Problem 2   *Does there exist for every $k \geq 3$ a constant d_k with the property that the edges of every geometric (or topological) graph with no k pairwise crossing edges can be colored by d_k colors so that no two edges of the same color cross each other?*

McGuinness [Mc00] proved that if $k = 3$, the answer to this question is affirmative for every *simple* topological graph (i.e., any two edges cross at most once) with the property that there exists a closed Jordan curve crossing every edge of G precisely once.

It is worth mentioning that Karzanov and Lomonosov [KaL78] made a purely combinatorial conjecture that, if true, would imply that for any fixed k the number of edges of a convex geometric graph with n vertices and no k pairwise crossing edges is $O(n)$. Two subsets, A and B, of an n-element underlying set V are said to *cross* if none of the sets $A \cap B$, $A \setminus B$, $B \setminus A$, and $V \setminus (A \cup B)$ is empty.

Conjecture 3   *[KaL78] There exists a constant $C > 0$ such that the size of any family of subsets of an n-element set that contains no k pairwise crossing members is at most Ckn.*

For $k = 3$ this conjecture was proved by Pevzner [Pe94] (see [Fl01] for a simple alternative argument), but for $k > 3$ the best known general upper bound is $O(kn \log n)$. However, the coloring version of the Karzanov–Lomonosov conjecture is false: there are set systems on n-element underlying sets containing no three pairwise crossing members that cannot be partitioned into a bounded number of subsystems consisting of pairwise noncrossing members [Pe94].

No matter what the answer to the above problem is, it seems plausible that dense topological graphs must contain many pairwise crossing edges. In particular, for complete geometric graphs the following problem is open:

Problem 4   *[ArE*94] Does there exist a constant $c > 0$ such that every complete geometric graph with n vertices has at least cn pairwise crossing edges?*

It was shown by Aronov et al. [ArE*94] that one can always find at least $c\sqrt{n}$ pairwise crossing edges, but the argument requires the edges to be straight. The answer to the question in Problem 4 may be in the affirmative for all topological complete graphs, but in this case we can guarantee the existence of only $c\frac{\log n}{\log \log n}$ pairwise crossing edges [PaT05].

Here we mention another open question of Harborth [Ha98] for simple complete topological graphs K_n. The edges between a triple of vertices in K_n are said to form an *empty triangle* if the simple closed curve determined by them contains in its interior either all or none of the vertices of K_n.

Problem 5   *Is every vertex of a simple complete topological graph K_n a vertex of at least two empty triangles?*

If the answer to this question is positive, it would follow that every simple complete topological graph determines at least $2n/3$ empty triangles. The best known lower bound is two. On the other hand, there are examples of such graphs with at most $2n - 4$ empty triangles [Ha98].

Another relaxation of planarity was introduced in [PaP*02]: a topological graph is called *k-locally planar* if G has no self-intersecting path of length at most k. Roughly speaking, this means that the embedding of the graph is planar in a neighborhood of radius $k/2$ around any vertex. In [PaP*02], it was shown that the number of edges of a 3-locally planar geometric graph with n vertices is $O(n \log n)$ and that this result is asymptotically tight. The best known upper bound for topological graphs is only $O(n^{3/2})$.

Problem 6   *For every fixed $k \geq 3$ and for every $\varepsilon > 0$, is the number of edges of every k-locally planar topological graph of n vertices $O(n^{1+\varepsilon})$?*

For every $k > 3$, Tardos [Ta05] constructed a sequence of k-locally planar geometric graphs with n vertices and a superlinear number of edges (n times approximately the $\lfloor k/2 \rfloor$ times iterated logarithm of n). Moreover, these graphs are bipartite, and all of their edges can be stabbed by the same line.

For some geometric applications [AgN*04], one needs to exclude short self-intersecting cycles rather than paths. Clearly, every graph G has a bipartite subgraph containing at least half of its edges, and such a subgraph avoids all odd cycles. Therefore, if we want to establish only the order of magnitude of $|E(G)|$, then forbidding odd cycles makes almost no difference.

Problem 7   *Does there exist a constant $c > 0$ such that the number of edges of any geometric graph with n vertices that does not contain a self-intersecting cycle of length four is at most $cn^{3/2}$?*

Obviously, this result, if true, cannot be improved, because there exist C_4-free abstract graphs with $n^{3/2}$ edges. Pinchasi and Radoičić [PiR03] established an $O(n^{8/5})$ upper bound, which has been improved to $O(n^{3/2} \log n)$ by Marcus and Tardos [MaT05].

In the spirit of the result of Capoyleas and Pach [CaP92] mentioned in the beginning of this section, one can raise many other nontrivial Turán-type questions for convex geometric graphs. Given any convex geometric graph F, what is the maximum number of edges $\mathrm{ex}_{\circlearrowleft}(n, F)$ that a convex geometric graph with n vertices can have without containing a geometric subgraph isomorphic to F? Kupitz and Perles [KuP96] solved this problem in the special case that the forbidden configuration F is a "convex matching" of size $k \leq \frac{n}{2}$ (i.e., k pairwise disjoint edges such that the convex hull of their union is a convex $2k$-gon whose boundary contains these edges). Perles (unpublished) and ten years later Brass, Károlyi, and Valtr [BrKV03] carried out systematic studies of problems of this type. They defined the *cyclic chromatic number* $\chi_{\circlearrowleft}(F)$ of a convex geometric graph F as the minimum number of colors needed to color the vertices of F so that no two vertices of the same color are adjacent and each color class forms an interval in the cyclic order of the vertices. According to this definition, for example, the cyclic chromatic number of a noncrossing convex cycle of length four is equal to four. It can be proved by an application of the Erdős–Stone theorem on abstract graphs [ErS46] that $\mathrm{ex}_{\circlearrowleft}(n, F) = \left(1 - \frac{1}{\chi_{\circlearrowleft}(F)-1}\right)\binom{n}{2} + o(n^2)$. Thus, the maximum edge number is asymptotically determined by the cyclic chromatic number of the forbidden subgraph, provided that this number is at least three. For convex geometric graphs with cyclic chromatic number two the situation appears even more complicated than for abstract graphs. Surprising phenomena can be observed already when F is a "folded" path of length three [BrKV03]. We have

$$\mathrm{ex}_{\circlearrowleft}(n, \text{⋈}) = 2n - 3 \quad \text{and} \quad \mathrm{ex}_{\circlearrowleft}(n, \underbrace{\text{⋈} \cdots \text{⋈}}_{k \text{ times } \text{⋈}}) = \Theta(n) \text{ for any } k, \text{ but}$$

$$\mathrm{ex}_{\circlearrowleft}(n, \text{⋈⋈}) = \Theta(n \log n) \quad \text{and} \quad \mathrm{ex}_{\circlearrowleft}(n, \text{⋈⋈}) = \Theta(n \log n).$$

For abstract graphs of n vertices forbidding two disjoint copies of the same subgraph rather than just one, the maximum number of edges remains asymptotically the same. However, this is not the case for convex geometric graphs. Furthermore, although the growth rate of the maximum number of edges in a convex geometric graph of n vertices with no ⋈⋈ is the same as for geometric graphs with no ⋈⋈, the extremal graphs have completely different structures.

Another interesting difference is that for abstract graphs of n vertices avoiding a certain subgraph, the growth rate of the maximum number of edges cannot be $\Theta(n \log n)$. Perhaps it is not a hopeless task to describe the possible growth rates at the low end of the spectrum. We do not have any example of a convex geometric graph F with cyclic chromatic number two whose underlying graph is a tree and for which the number of edges of an F-free geometric graph with n vertices can grow strictly faster than $\Theta(n \log n)$. On the other hand, we have been unable to obtain even an $O(n^{1+\varepsilon})$ upper bound on these functions for every $\varepsilon > 0$.

Conjecture 8 *(Pach and Tardos, [BrKV03]) Let F be a tree drawn as a convex geometric graph, all of whose edges can be stabbed by a straight line, i.e., $\chi_{\oslash}(F) = 2$. Then the maximum number of edges that an F-free convex geometric graph with n vertices can have is $O(n^{1+\varepsilon})$ for every $\varepsilon > 0$.*

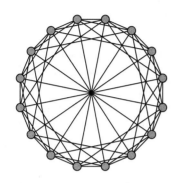 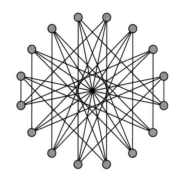

CONVEX GEOMETRIC GRAPHS WITHOUT ⨝⨝ AND WITHOUT ⨉⨉

[AgA*97]   P.K. AGARWAL, B. ARONOV, J. PACH, R. POLLACK, M. SHARIR: Quasi-planar graphs have a linear number of edges, *Combinatorica* **17** (1997) 1–9.

[AgN*04]   P.K. AGARWAL, E. NEVO, J. PACH, R. PINCHASI, M. SHARIR, S. SMORODINSKY: Lenses in arrangements of pseudo-circles and their applications, *J. ACM* **51** (2004) 139–186.

[ArE*94]   B. ARONOV, P. ERDŐS, W. GODDARD, D.J. KLEITMAN, M. KLUGERMAN, J. PACH, L.J. SCHULMAN: Crossing families, *Combinatorica* **14** (1994) 127–134.

[BrKV03]   P. BRASS, G. KÁROLYI, P. VALTR: A Turán-type extremal theory of convex geometric graphs, in: *Discrete and Computational Geometry – The Goodman–Pollack Festschrift*, B. Aronov et al., eds., Springer 2003, 275–300.

[CaP92]   V. CAPOYLEAS, J. PACH: A Turán-type theorem on chords of a convex polygon, *J. Combinatorial Theory Ser. B* **56** (1992) 9–15.

[ErS46]   P. ERDŐS, A. STONE: On the structure of linear graphs, *Bull. Amer. Math. Soc.* **52** (1946) 1087–1091.

[Fl01]   T. FLEINER: The size of 3-cross-free families, *Combinatorica* **21** (2001) 445–448.

[Gy87]   A. Gyárfás: Problems from the world surrounding perfect graphs, *Zastos Mat.* **19** (1987) 413–441.

[Ha98]   H. Harborth: Empty triangles in drawings of the complete graph, *Discrete Math.* **191** (1998) 109–111.

[KaL78]   A.V. Karzanov, M.V. Lomonosov: Flow systems in undirected networks (in Russian), in: *Mathematical Programming*, O.I. Larichev, ed., Institute for System Studies, Moscow, 1978, 59–66.

[Ko88]   A.V. Kostochka: Upper bounds on the chromatic number of graphs (in Russian), *Trudy Inst. Mat. (Novosibirsk), Modeli i Metody Optim.* **10** (1988) 204–226.

[KoK97]   A.V. Kostochka, J. Kratochvíl: Covering and coloring polygon-circle graphs, *Discrete Math.* **163** (1997) 299–305.

[KoN98]   A.V. Kostochka, J. Nešetřil: Coloring relatives of intervals of the plane I: Chromatic number versus girth, *European J. Combinatorics* **19** (1998) 103–110.

[KuP96]   Y.S. Kupitz, M.A. Perles: Extremal theory for convex matchings in convex geometric graphs, *Discrete Comput. Geom.* **15** (1996) 195–220.

[MaT05]   A. Marcus, G. Tardos: Intersection reverse sequences and geometric applications, in: *Graph Drawing 2004*, J. Pach, ed., Springer *LNCS* **3383** (2005) 349–359.

[Mc00]   S. McGuinness: Colouring arcwise connected sets in the plane I, *Graphs Combinatorics* **16** (2000) 429–439.

[PaP*05]   J. Pach, R. Pinchasi, M. Sharir, G. Tóth: Topological graphs with no large grids, *Graphs Combinatorics*, to appear.

[PaP*02]   J. Pach, R. Pinchasi, G. Tardos, G. Tóth: Geometric graphs with no self-intersecting path of length three, in: *Graph Drawing 2002*, M.T. Goodrich et al., eds., Springer *LNCS* **2528** (2002) 295–311.

[PaRT03]   J. Pach, R. Radoičić, G. Tóth: Relaxing planarity for topological graphs, in: *JCDCG 2002* (Jap. Conf. Discrete Comput. Geom.), J. Akiyama et al., eds., Springer *LNCS* **2866** (2003) 221–232.

[PaSS96]   J. Pach, F. Shahrokhi, M. Szegedy: Applications of the crossing number, *Algorithmica* **16** (1996) 111–117.

[PaT05]   J. Pach, G. Tóth: Disjoint edges in topological graphs, in: Proc. Indonesian-Japan Joint Conference on Combinatorial Geometry and Graph Theory, (Bandung, Indonesia, 2003), Springer *LNCS* **3330** (2005) 133–140.

[**Pe94**]   P.A. Pevzner: Non-3-crossing families and multicommodity flows, in: *Selected Topics in Discrete Mathematics* (Moscow, 1972–1990), *Amer. Math. Soc. Transl. Ser. 2* **158**, Amer. Math. Soc., 1994, 201–206.

[**PiR03**]   R. Pinchasi, R. Radoičić: Topological graphs with no self-intersecting cycle of length 4, in: *SCG 03, Proc. 19th ACM Symp. Comp. Geom.*, ACM Press, 2003, 98–103.

[**Ta05**]   G. Tardos: Construction of locally plane graphs, *SIAM J. Discrete Math.*, to appear.

[**Va97**]   P. Valtr: Graph drawing with no k pairwise crossing edges, in: *Graph Drawing 1997*, Springer *LNCS* **1353** (1997) 205–218.

9.7 Ramsey-Type Problems

In classical Ramsey theory, one wants to find large monochromatic subgraphs in a complete graph whose edges are colored with several colors [GrRS90]. Most questions of this type can be generalized to complete geometric graphs, where the monochromatic subgraphs are required to satisfy certain geometric conditions.

Bialostocki and Dierker conjectured, and Károlyi, Pach, and Tóth [KáPT97] proved, that if the edges of a complete geometric graph with n vertices are colored by two colors, then one can find

(1) a noncrossing spanning tree all of whose edges are of the same color,

(2) $\lfloor \frac{n+1}{3} \rfloor$ pairwise disjoint edges of the same color.

The same statements had been known to be true for abstract graphs, where the geometric constraints are ignored.

Károlyi et al. [KáP*98] proved that for any two-coloring of the edges of a complete geometric graph with n vertices, there exists a noncrossing (simple) path of length $\Omega(n^{2/3})$ all of whose edges are of the same color.

Conjecture 1 *[KáPT97] For every $k \geq 2$ there exists a constant $c_k > 0$ such that if the edges of a complete geometric graph with n vertices are colored by k colors, one can always find a monochromatic noncrossing path of length at least $c_k n$.*

In the special case of $k = 2$, this conjecture has been verified for convex geometric graphs [KáP*98]. Here one can find a monochromatic noncrossing path of length $\lfloor \frac{n+1}{2} \rfloor$. This bound cannot be improved.

It is possible that the following stronger statement is also true: for every $k \geq 2$, there exists an integer n_k such that the vertices of every complete geometric graph whose edges are colored by k colors can be covered by at most n_k noncrossing monochromatic paths. The corresponding statement for abstract graphs was established by Gyárfás [Gy89].

It is not difficult to see [KáP*98] that every complete geometric graph with n vertices whose edges are colored by two colors contains noncrossing monochromatic cycles of length $3, 4, \ldots, \lfloor \sqrt{n/2} \rfloor$. On the other hand, there are examples without noncrossing monochromatic cycles of length at least $\sqrt{n} + 1$.

Problem 2 *(G. Károlyi et al. [KáP*98]) Let $k_2(n)$ denote the largest number k such that every complete geometric graph with n vertices whose edges are colored by two colors contains a noncrossing monochromatic cycle of length k. Determine $\lim_{n \to \infty} k_2(n)/\sqrt{n}$.*

There are many interesting problems for vertex-colored complete geo-metric graphs [KaK03]. It is well known that for any set of n red and n blue points in general position in the plane, one can find a noncrossing matching in which every edge connects two vertices of different colors. (To see this, consider a matching with this property whose total length is minimum.) What happens if we want to find a long noncrossing alternating path? Erdős conjectured that there always exists such a path of length $3n/2$, but this has been disproved be Abellanas et al. [AbG*03] and by Kynčl, Pach, and Tóth [KyPT05], who gave examples without noncrossing alternating paths of length roughly $4n/3$. It is obvious that one can always find a noncrossing alternating path of length n. Moreover, if the red and blue points are separated by a line, there also exists a noncrossing alternating path including all the points [AbG*97].

Problem 3   (*Erdős*) *Determine the largest number c such that for any n red points and n blue points in general position in the plane, one can always find a noncrossing alternating path of length at least cn.*

Another variant of the matching problem was studied by Dumitrescu and Steiger [DuS00]. Their goal was to prove that there always exists a large noncrossing matching in which every edge connects two points of the *same* color. Surprisingly, they found two-colored n-element point sets in general position in the plane in which every matching with the required property covers fewer than $(1-\varepsilon)n$ vertices, for some positive constant ε. The best known estimates for ε were found by Dumitrescu and Kaye [DuK01].

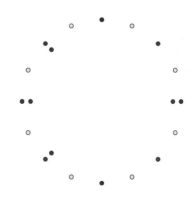

20 POINTS: ANY NONCROSSING
MONOCHROMATIC MATCHING LEAVES
FOUR POINTS UNMATCHED

Geometric graphs can also be regarded as families of segments. Larman et al. [LaM*94] discussed the following question.

Problem 4   *What is the largest integer $s(n)$ such that any family of n closed segments in general position in the plane has $s(n)$ members that are either pairwise disjoint or pairwise cross-ing?*

It is known that
$$n^{1/5} \leq s(n) \leq cn^{\log 4/\log 27}$$
for a suitable constant c [LaM*94],
[KáPT97]. It can be conjectured
that the answer to Problem 4 remains asymptotically the same if instead of families of segments we consider families of arbitrary convex
sets.

However, for families of axis-
parallel *rectangles* the corresponding function $s'(n)$ is at least
$$s'(n) \geq \Omega(\sqrt{n/\log n}).$$
The following conjecture is folklore.

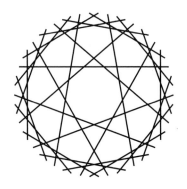

27 SEGMENTS WITH NO 5
DISJOINT OR CROSSING ONES

Conjecture 5   *There exists a constant $c > 0$ such that every family of n
rectangles in the plane whose sides are parallel to the axes
of the coordinate system has at least $c\sqrt{n}$ members that
are either pairwise disjoint or pairwise intersecting.*

It is possible that the answer to this question is positive even in the
following much stronger sense: in any system of n axis-parallel rectangles in
the plane, the product of the sizes of a largest subsystem of pairwise disjoint
rectangles and a largest subsystem of rectangles having a point in common
is at least cn for an absolute constant $c > 0$ (see [GyL85], [LaM*94]). It is
known that this product is $\Omega(n/\log n)$.

The last two problems are closely related to an old purely combinatorial
question of Erdős and Hajnal [ErH89] (see also [AlPS01], [Gy97]).

Problem 6   *[ErH89] Does there exist for every graph H a constant
$\varepsilon(H) > 0$ such that any graph G with n vertices that
does not contain an induced subgraph isomorphic to H
has either a complete subgraph or an empty subgraph of
size $n^{\varepsilon(H)}$?*

For instance, it is not known whether such an $\varepsilon(H)$ exists if H is a cycle
of length five. One can always guarantee the existence of a complete or an
empty subgraph with $e^{\varepsilon(H)\sqrt{\log n}}$ vertices, where $\varepsilon(H) > 0$ is a constant
[ErH89]. Erdős, Hajnal, and Pach [ErHP00] solved the bipartite version
of Problem 6: under the same assumptions one can always find either in
G or in its complement $\overline{G}$ a complete bipartite graph with at least $n^{\varepsilon(H)}$
points in its classes. This statement has many geometric applications. For
instance, it is easy to see [EhET76], [PaS01] that, as k tends to infinity,
almost no graph with k vertices can be obtained as (an induced subgraph
of) the intersection graph of a family $\mathcal{F}$ of arcwise connected sets in the

plane. Therefore, we immediately obtain that there exists a constant $\varepsilon > 0$ such that every family $\mathcal{F}$ of n arcwise connected sets in the plane has two subfamilies $\mathcal{F}_1, \mathcal{F}_2 \subseteq \mathcal{F}$ with at least n^ε members such that either every member of $\mathcal{F}_1$ intersects all members of $\mathcal{F}_2$ or no member of $\mathcal{F}_1$ intersects any member of $\mathcal{F}_2$. Moreover, if the elements of $\mathcal{F}$ are semialgebraic sets of a fixed degree, then this statement is true with εn in the place of n^ε [AlP*05].

It follows that every complete topological graph with n vertices whose edges are algebraic curves of degree at most d has two sets of edges, E_1 and E_2, each with at least $\varepsilon_d n$ elements, such that all of their endpoints are distinct and either every element of E_1 is disjoint from every element of E_2 or every element of E_1 crosses every element of E_2 (for a suitable constant $\varepsilon_d > 0$).

One may ask the following question:

Problem 7 (Pach) *Does there exist for every d a constant $\varepsilon_d > 0$ such that every complete topological graph with n vertices, whose edges are algebraic curves of bounded degree d, has at least $\varepsilon_d n$ pairwise disjoint or pairwise crossing edges with distinct endpoints?*

Pach, Solymosi, and Tóth [PaST03] proved that every *simple* topological complete graph (any pair of whose edges cross at most once) contains a complete subgraph of size at least constant times $\log^{1/8} n$, whose crossing pattern is the same as that of the edges of a *convex* or a so-called *twisted* complete graph of this size.

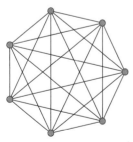
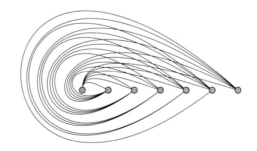

CONVEX DRAWING OF K_7 TWISTED DRAWING OF K_7

[**AbG*97**] M. Abellanas, J. García, G. Hernández, M. Noy, P. Ramos: Bipartite embeddings of trees in the plane, in: *Graph Drawing '96*, S. North, ed., Springer *LNCS* **1190** (1997) 1–10.

[AbG*03]   M. Abellanas, J. García, F. Hurtado, J. Tejel: Alternating paths (in Spanish), in: *Proc. X Encuentros de Geometría Computacional* (Sevilla, 2003), 7–12.

[AlP*05]   N. Alon, J. Pach, R. Pinchasi, M. Sharir, R. Radoičić: Crossing patterns of semi-algebraic sets, *J. Combinatorial Theory Ser. A*, to appear.

[AlPS01]   N. Alon, J. Pach, J. Solymosi: Ramsey-type theorems with forbidden subgraphs, *Combinatorica* **21** (2001) 155–170.

[DuK01]   A. Dumitrescu, R. Kaye: Matching colored points in the plane: Some new results, *Comput. Geom. Theory Appl.* **19** (2001) 69–85.

[DuS00]   A. Dumitrescu, W. Steiger: On a matching problem in the plane, *Discrete Math.* **211** (2000) 183–195.

[EhET76]   G. Ehrlich, S. Even, R.E. Tarjan: Intersection graphs of curves in the plane, *J. Combinatorial Theory Ser. B* **21** (1976) 8–20.

[ErH89]   P. Erdős, A. Hajnal: Ramsey-type theorems, *Discrete Applied Math.* **25** (1989) 37–52.

[ErHP00]   P. Erdős, A. Hajnal, J. Pach: A Ramsey-type theorem for bipartite graphs, *Geombinatorics* **10** (2000) 64–68.

[GrRS90]   R.L. Graham, B.L. Rothschild, J. Spencer: *Ramsey Theory*, 2nd ed., Wiley, New York, 1990.

[Gy97]   A. Gyárfás: Reflections on a problem of Erdős and Hajnal, in: *The Mathematics of Paul Erdős, Vol. 2*, R.L. Graham et al., eds., *Algorithms and Combinatorics* Ser. **14**, Springer-Verlag, 1997, 93–98.

[Gy89]   A. Gyárfás: Covering complete graphs by monochromatic paths, in: *Irregularities of Partitions*, G. Halász et al., eds., *Algorithms and Combinatorics* **8**, Springer-Verlag, 1989, 89–91.

[GyL85]   A. Gyárfás, J. Lehel: Covering and coloring problems for relatives of intervals, *Discrete Math.* **55** (1985) 167–180.

[KaK03]   A. Kaneko, M. Kano: Discrete geometry on red and blue points in the plane — A survey, in: *Discrete and Computational Geometry—The Goodman–Pollack Festschrift*, B. Aronov et al., eds., *Algorithms and Combinatorics* **25**, Springer-Verlag, 2003, 551–570.

[KáPT97]   G. Károlyi, J. Pach, G. Tóth: Ramsey-type results for geometric graphs. I, *Discrete Comput. Geom.* **18** (1997) 247–255.

[KáP*98]   G. Károlyi, J. Pach, G. Tóth, P. Valtr: Ramsey-type results for geometric graphs. II, *Discrete Comput. Geom.* **20** (1998) 375–388.

[KyPT05]   J. Kynčl, J. Pach, G. Tóth: Long alternating paths in bicolored point sets, in: *Graph Drawing 2004*, J. Pach, ed., Springer *LNCS* **3383** (2005) 340–348.

[LaM*94]   D.G. Larman, J. Matoušek, J. Pach, J. Törőcsik: A Ramsey-type result for convex sets, *Bull. London Math. Soc.* **26** (1994) 132–136.

[PaS01]   J. Pach, J. Solymosi: Crossing patterns of segments, *J. Combinatorial Theory Ser. A* **96** (2001) 316–325.

[PaST03]   J. Pach, J. Solymosi, G. Tóth: Unavoidable configurations in complete topological graphs, *Discrete Comput. Geom.* **30** (2003) 311–320.

9.8 Geometric Hypergraphs

If we want to generalize the results on geometric graphs (i.e., systems of segments between points in the plane) to systems of simplices in higher dimensions, we face unexpected difficulties. Even if we restrict our attention to systems of triangles induced by three-dimensional point sets in general position, it is not completely clear how "crossings" should be defined. If two segments cross, they do not share an endpoint. Should this remain true for triangles?

A *geometric r-hypergraph* H_r^d is a pair (V, E), where V is a set of points in general position in d-space, and E is a set of *closed* $(r-1)$-dimensional simplices induced by some r-tuples of V. The sets V and E are called the *vertex set* and the *(hyper)edge set* of H_r^d. Clearly, a geometric graph is a two-dimensional geometric two-hypergraph H_2^2.

A class $\mathcal{F}$ of geometric hypergraphs not permitted to be contained in the geometric hypergraphs under consideration is called a family of *forbidden subhypergraphs*. Given a class $\mathcal{F}$ of forbidden geometric hypergraphs, let $\text{ex}_r^d(\mathcal{F}, n)$ denote the maximum number of edges that a d-dimensional geometric r-hypergraph H_r^d of n vertices can have without containing a geometric subhypergraph belonging to $\mathcal{F}$.

A set of simplices is said to have a *nontrivial intersection* if their relative interiors have a point in common. A common point of the relative interiors of k simplices, all of whose vertices are *distinct*, is called a *crossing*. We talk about *crossing simplices* if such a point exists. It is possible that a set of simplices is *pairwise crossing* but not crossing.

Let $\mathcal{D}_k^r$ denote the class of all geometric r-hypergraphs consisting of k pairwise disjoint edges. Let $\mathcal{I}_k^r$ denote the class of all geometric r-hypergraphs consisting of k simplices, any two of which have a nontrivial intersection. Similarly, let $\mathcal{C}_k^r$ denote the class of all geometric r-hypergraphs consisting of k pairwise crossing edges.

Akiyama and Alon [AkA89] proved the following result. Let $V = V_1 \cup \ldots \cup V_d$ ($|V_1| = \ldots = |V_d| = n$) be a dn-element set in general position in d-space, and let E consist of all $(d-1)$-dimensional simplices having exactly one vertex in each V_i. Then E contains n disjoint simplices. This result can be applied to deduce that

$$\text{ex}_d^d(\mathcal{D}_k^d) \leq n^{d-(1/k)^{d-1}}.$$

Conjecture 1   *[AkA89] For any positive integers d and k, we have $\text{ex}_d^d(\mathcal{D}_k^d)$ $\leq O(n^{d-1})$. That is, the maximum number of $(d-1)$-dimensional closed simplices that can be selected from n vertices in d-space so that no k of them are pairwise disjoint is at most a constant times n^{d-1}.*

As we have seen in Section 9.5, this conjecture is true in the plane.

If for some $d \geq 3$, a d-dimensional geometric d-hypergraph of n vertices is not allowed to have $k \geq 3$ pairwise crossing edges, it is not difficult to show [DeP98] that it has $O(n^{d-(1/k)^{d-1}})$ edges.

Conjecture 2   *[DeP98] For any positive integers $d \geq 3$ and $k \geq 2$, we have $\mathrm{ex}_d^d(\mathcal{C}_k^d) \leq O(n^{d-1+\varepsilon})$, for every $\varepsilon > 0$. That is, the maximum number of $(d-1)$-dimensional closed simplices that can be selected from n vertices in d-space so that no k of them are pairwise crossing is at most a constant times $n^{d-1+\varepsilon}$.*

This statement is true for $k = 2$ with $\varepsilon = 0$.

It follows from known results in geometric graph theory [DeP98] that $\mathrm{ex}_d^d(\mathcal{I}_k^d)$, the maximum number of $(d-1)$-dimensional closed simplices that can be selected between n vertices in d-space so that no k of them have pairwise nontrivial intersection, is $O(n^{d-1})$ for $k = 2$ or 3 and $O(n^{d-1} \log n)$ for $k > 3$. For $d = 3$, $k \leq 3$, this result is asymptotically tight.

The following conjecture is due to G. Kalai (unpublished).

Conjecture 3   *Let $\mu(n)$ denote the maximum number of hyperedges that a three-dimensional geometric three-hypergraph of n vertices can have in which any pair of hyperedges are disjoint or share at most one vertex. Then we have*
$$\lim_{n \to \infty} \frac{\mu(n)}{n^2} = 0.$$

Károlyi and Solymosi [KáS02] showed that $\mu(n) \geq \Omega(n^{3/2})$.

Already in the plane there are several interesting unsolved questions concerning systems of triangles, that is, two-dimensional geometric three-hypergraphs with $n \geq 3$ vertices. By summing the interior angles of the triangles around each vertex we easily obtain that if no two triangles of such a system are allowed to overlap at a common vertex, the number of triangles cannot exceed $2n - 5$. This bound is tight for any triangulation of a point set whose convex hull is a triangle. In other words, in this case the "local planarity" condition (that there is no crossing at the vertices) has the same effect as the "global" one (that there is no crossing at all). What happens if no two triangles with a common vertex are allowed to share even an edge? By choosing half of the interior faces in a maximal triangulation, we find $n - 2$ triangles with the required property, but we do not have any upper bound better than $2n - O(1)$.

Problem 4   *(Brass) What is the maximum number of hyperedges in a two-dimensional geometric three-hypergraph with n vertices in which no two edges incident to a vertex have any other point in common?*

Similar questions can be asked for *convex geometric r-hypergraphs*, that is, under the assumption that the vertices must be in convex position [Br04]. In the plane this restriction allows us to deduce a simple Erdős-Stone type theorem with a properly defined variant of the chromatic number. This asymptotically solves all problems in which the forbidden configurations are not "*r*-chromatic." However, several interesting problems remain open even for forbidden configurations consisting of two triangles.

Problem 5   *(Brass [Br04]) What is the maximum number of hyperedges (triangles) in a two-dimensional convex geometric three-hypergraph with n vertices that contains neither ⨝ nor ⊠ ?*

[**AkA89**]   J. Akiyama, N. Alon: Disjoint simplices and geometric hypergraphs, in: *Combinatorial Mathematics*, G.S. Bloom et al., eds., *Annals New York Acad. Sci.* **555** (1989) 1–3.

[**Br04**]   P. Brass: Turán-type extremal problems of convex geometric hypergraphs,   in: *Towards a Theory of Geometric Graphs*, J. Pach, ed., *Contemporary Mathematics* **342**, AMS 2004, 25–33.

[**DeP98**]   T.K. Dey, J. Pach: Extremal problems for geometric hypergraphs, *Discrete Comput. Geom.* **19** (1998) 473–484.

[**KáS02**]   G. Károlyi, J. Solymosi: Almost disjoint triangles in 3-space, *Discrete Comput. Geom.* **28** (2002) 577–583.

10. Lattice Point Problems

10.1 Packing Lattice Points in Subspaces

One of the oldest and most extensively studied geometric questions concerning lattice points is the "no-three-in-line" problem raised by Dudeney [Du17] (pages 94 and 222) in the special case $n = 8$. What is the maximum number of points that can be selected from an $n \times n$ lattice square $\{1,\ldots,n\}^2$ such that no three of them are in a line? Clearly, $2n$ is an upper bound, since each of the n horizontal (or vertical) lattice lines can contain at most two selected points.

Problem 1   *(Dudeney [Du17]) Is it true that for every $n \geq 2$ one can select $2n$ points from the $n \times n$ lattice square such that no three of them are collinear?*

In a long sequence of papers [AdHK74], [HaJ*75], [CrH76], [Kl78], [Kl79], [An79], [HaOP89], [Fl92], [Fl98], many examples were constructed, up to $n = 52$, for which the bound $2n$ is attained. Most of these sets were found by computer search, and no general pattern has emerged. Erdős discovered that if n is a prime, then there are no three collinear elements among the integer points $(x,y) \in \{1,\ldots,n\}^2$ of the "modular parabola" $y \equiv x^2 \pmod{n}$; see the *Appendix* to Roth's paper [Ro51]. The best known general construction, due to Hall et al. [HaJ*75] has $\frac{3}{2}n - 3$ points whenever $n = 2p$ and p is a prime, and hence at least $(\frac{3}{2}-\varepsilon)n$ points, for any $\varepsilon > 0$ and every $n > n_0(\varepsilon)$ [HaJ*75]. This can be achieved by selecting the points on appropriate "hyperbolas" $xy \equiv k \pmod{p}$.

Based on (not completely convincing) probabilistic evidence, Guy and Kelly [GuK68] conjectured that for sufficiently large n, the maximum number of points that can be chosen from $\{1,\ldots,n\}^2$ without having three collinear points is

$$\left(\frac{2\pi^2}{3}\right)^{\frac{1}{3}} n \approx 1.87n.$$

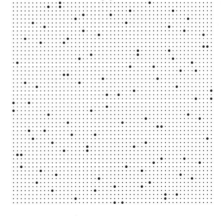

104 POINTS, NO 3 COLLINEAR
IN THE 52×52 LATTICE SQUARE

However, in view of the many examples for which the trivial upper bound is attained, it seems more likely that one can always select (roughly) $2n$ points meeting the requirements.

The modular analogue of this problem is much simpler. It is well known that if instead of the real lattice square $\{1, \ldots, p\}^2$, we consider the affine plane over the finite field $GF(p)$, then the largest sets with no three collinear elements are quadratic curves consisting of $p + 1$ points. Since real collinearity implies collinearity $\bmod p$, any such set also satisfies the conditions of the original problem, but its size $p + 1$ is much smaller than the conjectured optimum $2p$ in the real plane.

The similar "no-four-on-a-circle" problem is attributed to Erdős and Purdy ([Gu81] problem F3, p. 133). Thiele [Th95a], [Th95b] gave a construction of $\frac{n}{4}$ points selected from the $n \times n$ lattice square without four points on a circle, and an upper bound of $\frac{5}{2}n - \frac{3}{2}$. Here a line is also considered a (degenerate) circle.

Problem 2   (Erdős and Purdy) What is the largest number of points that can be selected from an $n \times n$ lattice square with no four on a circle?

Similar questions can be asked in higher dimensions. The number of points that can be selected from the d-dimensional lattice cube $\{1, \ldots, n\}^d$ with no $d + 1$ elements on the same hyperplane is at most dn, because the lattice cube can be partitioned into n hyperplanes. On the other hand, if n is a prime one can select n points with the required property. For example, every hyperplane intersects the "modular moment curve" $(t, t^2, \ldots, t^d) \pmod{n}$, $t = 1, \ldots, n$ in at most d points. For the analogous function for spheres in place of hyperplanes, Thiele [Th95b] gave a lower bound of $cn^{\frac{1}{d-1}}$, which is far from the trivial upper bound $(d + 1)n$.

Problem 3   What is the maximum number of points that can be selected from a d-dimensional $n \times \cdots \times n$ lattice cube with no $d + 1$ points in a hyperplane?

Problem 4   What is the maximum number of points that can be selected from a d-dimensional $n \times \cdots \times n$ lattice cube with no $d + 2$ points on a sphere or a hyperplane?

One can also restrict the number of points lying in lower-dimensional affine subspaces.

Problem 5   [BrK03] For given k and $d > k$, what is the maximum number of points that can be selected from a d-dimensional $n \times \cdots \times n$ lattice cube with no $k + 2$ points lying in a k-dimensional affine subspace?

Since the d-dimensional lattice cube $\{1, \ldots, n\}^d$ can be partitioned into n^{d-k} affine k-dimensional lattice cubes, we have the trivial upper bound $(k+1)n^{d-k}$, but few reasonable lower bounds are known. For $d = 3$ and $k = $

1, Pór and Wood [PóW05] proved that every line intersects the "modular paraboloid" $x_3 \equiv x_1^2 + x_2^2 \pmod{n}$ in at most two points, provided that n is a prime, $n \equiv 3 \pmod 4$. For higher dimensions, Brass and Knauer [BrK03] showed that if n is a prime, every line intersects the "modular moment surface" $(x_1, \ldots, x_d)$ with $x_1^1 + x_2^2 + \ldots + x_d^d \equiv q \pmod{n}$ in at most d points. Thus, there exists a set of n^{d-1} points such that each line contains at most d (rather than just two) points. Using a simple randomized selection process, we can obtain at least $n^{d-k-\varepsilon}$-element subsets of $\{1, \ldots, n\}^d$ such that every k-dimensional affine subspace contains at most a constant number $c(k, d, \varepsilon)$ of points. It would be interesting to improve these lower bounds and give a construction of at least $c(k, d) n^{d-k}$ points such that each k-dimensional subspace contains at most $k + 1$ of them.

Let $h_n(d)$ denote the maximum number of points that can be selected from $\{1, \ldots, n\}^d$ such that no n elements are collinear. The asymptotic behavior of this function when n is fixed and $d \to \infty$ was studied by Riddell [Ri71] and Erdős [Er73], who conjectured that $h_n(d) = o(n^d)$. This statement, which can be regarded as the density version of the Hales–Jewett theorem, was proved almost twenty years later by Furstenberg and Katznelson [FuK91]. From the other direction, Riddell [Ri71] gave a lower bound of $\Omega\left(\frac{n^{d+\frac{1}{2}}}{\sqrt{d}}\right)$ for $d > d_0(n)$. No nontrivial results seem to be known if instead of excluding "full lines" in the lattice cube, shorter collinear sequences are forbidden.

Problem 6   (Riddell [Ri71], Erdős [Er73]) What is the maximum number of points that can be selected from the lattice cube $\{1, 2, 3\}^d$ with no three points collinear?

The analogous questions for linear rather than affine subspaces appear to be more difficult. The maximum number of points that can be selected from the lattice cube $\{1, \ldots, n\}^d$ such that any k-dimensional linear subspace contains at most k points is asymptotically known only for the cases $k = 1$ and $k = d - 1$. For $k = 1$, this quantity coincides with the number of primitive lattice points in the cube and has been studied in great detail in number theory. In particular, we know that its order of magnitude is $\Theta(n^d)$, a positive fraction of the total number of lattice points. For $k = d - 1$, it was shown by Bárány, Harcos, et al. [BáH*01] that the maximum is $\Theta\left(n^{\frac{d}{d-1}}\right)$. This suggests that the maximum may be $\Theta\left(n^{\frac{(d-k)d}{d-1}}\right)$ for every k.

Problem 7   (Brass) Is it true that for any k and d, the maximum number of lattice points that can be selected from the d-dimensional $n \times \cdots \times n$ lattice cube such that each k-dimensional linear subspace contains at most k points is $\Theta\left(n^{\frac{(d-k)d}{d-1}}\right)$?

Körner [Kö95] studied "no-three-in-line" problems for metric lines with respect to the Hamming distance. For further problems and results of this kind, see the surveys of Lagarias [La95] and Gritzmann and Wills [GrW93].

[AdHK74] M.A. ADENA, D.A. HOLTON, P.A. KELLY: Some thoughts on the no-three-in-line problem, in: *Combinatorial Mathematics, Proc. 2nd Australian Conf.*, D.A. Holton, ed., *Lecture Notes Math.* **403**, Springer 1974, 6–17.

[An79] D.B. ANDERSON: Update on the no-three-in-line problem, *J. Combinatorial Theory Ser. A* **27** (1979) 365–366.

[BáH*01] I. BÁRÁNY, G. HARCOS, J. PACH, G. TARDOS: Covering lattice points by subspaces, *Period. Math. Hungar.* **43** (2001) 93–103.

[BrK03] P. BRASS, C. KNAUER: On counting point-hyperplane incidences, *Comput. Geom. Theory Appl.* **25** (2003) 13–20.

[CrH76] D. CRAGGS, R. HUGHES-JONES: On the no-three-in-line problem, *J. Combinatorial Theory Ser. A* **20** (1976) 363–364.

[Du17] H.E. DUDENEY: *Amusements in Mathematics*, Nelson, London 1917.

[Er73] P. ERDŐS: Problems and results on combinatorial number theory, in: *A Survey of Combinatorial Theory*, J.N. Srivastava et al., eds., North-Holland 1973, 117–138.

[Fl98] A. FLAMMENKAMP: Progress in the no-three-in-line problem. II, *J. Combinatorial Theory Ser. A* **81** (1998) 108–113.

[Fl92] A. FLAMMENKAMP: Progress in the no-three-in-line problem, *J. Combinatorial Theory Ser. A* **60** (1992) 305–311.

[FuK91] H. FURSTENBERG, Y. KATZNELSON: A density version of the Hales–Jewett theorem, *J. Anal. Math.* **57** (1991) 64–119.

[GrW93] P. GRITZMANN, J.M. WILLS: Lattice points, in: *Handbook of Convex Geometry, Vol.2*, P.M. Gruber et al., eds., Elsevier 1993, 765–797.

[Gu81] R.K. GUY: *Unsolved Problems in Number Theory*, Springer-Verlag, 1981.

[GuK68] R.K. GUY, P.A. KELLY: The no-three-in-line problem, *Canadian Math. Bulletin* **11** (1968) 527–531.

[HaJ*75] R.R. HALL, T.H. JACKSON, A. SUDBERY, K. WILD: Some advances in the no-three-in-line problem, *J. Combinatorial Theory Ser. A* **18** (1975) 336–341.

[HaOP89]   H. HARBORTH, P. OERTEL, T. PRELLBERG: No-three-in-line for seventeen and nineteen, *Discrete Math.* **73** (1989) 89–90.

[Kl79]   T. KLOVE: On the no-three-in-line problem. III, *J. Combinatorial Theory Ser. A* **26** (1979) 82–83.

[Kl78]   T. KLOVE: On the no-three-in-line problem. II, *J. Combinatorial Theory Ser. A* **24** (1978) 126–127.

[Kö95]   J. KÖRNER: On the extremal combinatorics of the Hamming space, *J. Combinatorial Theory Ser. A* **71** (1995) 112–126.

[La95]   J.C. LAGARIAS: Point lattices, in: *Handbook of Combinatorics, Vol. 1*, R. Graham et al., eds., Elsevier 1995, 919–966.

[PóW05]   A. PÓR, D.R. WOOD: No-three-in-line-in-3D, in: *Graph Drawing 2004*, J. Pach, ed., Springer *LNCS* **3383** (2005) 395–402.

[Ri71]   J. RIDDELL: A lattice point problem related to sets containing no l-term arithmetic progression, *Canadian Math. Bull.* **14** (1971) 535–538.

[Ro51]   K.F. ROTH: On a problem of Heilbronn, *J. London Math. Soc.* **26** (1951) 198–204.

[Th95a]   T. THIELE: The no-four-on-circle problem, *J. Combinatorial Theory Ser. A* **71** (1995) 332–334.

[Th95b]   T. THIELE: *Geometric Selection Problems and Hypergraphs*, Dissertation, FU Berlin 1995.

10.2 Covering Lattice Points by Subspaces

Some of the packing problems discussed in the previous section have dual
counterparts for covering. The intersection of a convex body with a lattice
is called a *convex set of lattice points*. Typically, we would like to cover a
convex set of lattice points by a minimal number of lines, hyperplanes, or
other subspaces.

Corzatt [Co85] conjectured that if in
the plane a convex set of lattice points can
be covered by n lines, then it can also be
covered by n lines having at most four different slopes. He gave arbitrarily large examples for which four directions are indeed
necessary in any covering of minimum size.
This is the lattice analogue of the "plank
theorem" mentioned in Section 3.4, where—
without the lattice restriction—one parallel
class was sufficient for a minimum covering.

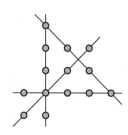

SET WHOSE MINIMAL COVER
REQUIRES FOUR DIRECTIONS

Conjecture 1 *(Corzatt [Co85]) If a convex set of lattice points in the
plane can be covered by n lines, then it can be covered by
n lines having at most four different slopes.*

This question can be generalized in at least two different ways.

Problem 2 *Does there exist for every dimension d a constant a_d such
that if a convex lattice set in $\mathbb{R}^d$ can be covered by n
hyperplanes then it can also be covered by n hyperplanes
belonging to at most a_d parallel classes?*

Problem 3 *(Brass) For every d, find the smallest constant b_d such
that if a convex lattice set in $\mathbb{R}^d$ can be covered by n
hyperplanes, then it can also be covered by $b_d n$ parallel
hyperplanes.*

The last problem is related to the *lattice width* of a convex body. Some
results concerning this parameter were found by K. Bezdek and Hausel
[BeH94] and Talata [Ta97].

Alon [Al91] studied coverings of the lattice cube $\{1, \dots, n\}^d$ in d-
dimensional space by lines spanned by a small subset of the lattice points.
He almost completely answered a question from Guy's problem book [Gu81]
(problem F3, p. 133), by showing that the minimum number of lattice points
with the property that the set of lines spanned by them together cover the

d-dimensional lattice cube is $\Omega\left(n^{\frac{d(d-1)}{2d-1}}\right)$ and $O\left(n^{\frac{d(d-1)}{2d-1}}\log n\right)$.

Problem 4 *(Guy [Gu81]) What is the minimum number of points that can be selected from the d-dimensional $n \times \cdots \times n$ lattice cube such that all elements of the lattice cube are covered by the lines spanned by them?*

Cheng and Huang [ChH83] discussed coverings of a special class of lattice octagons by lines.

One can also try to cover a set of lattice points by a polygonal path with as few bends as possible (with minimum "link length"). The d-dimensional lattice cube $\{1,\ldots,n\}^d$ can be covered by n^{d-1} segments. If we want to connect these segments to form a polygonal path, $2n^{d-1}$ vertices are certainly sufficient. Kranakis, Krizanc, and Meertens [KrKM94] constructed an orthogonal path with $\frac{3}{2}n^2 + n - 1$ vertices that covers the three-dimensional lattice cube $\{1,\ldots,n\}^3$. Collins [Co04] established the lower bound $\frac{4}{3}n^2 - O(n)$ for the minimum number of vertices in such a path. (There seems to be an irreparable error in the proof of the lower bound claimed in [CoM98].)

Conjecture 5 *[KrKM94] The minimum number of bends of a polygon that covers all points of the d-dimensional $n \times \cdots \times n$ lattice cube is $\left(1 + \frac{1}{d-1}\right)n^{d-1} + O(n^{d-2})$.*

Covering by linear subspaces instead of affine ones is more difficult. Bárány et al. [BáH*01] obtained asymptotically sharp estimates for the minimum number of $(d-1)$-dimensional linear subspaces necessary to cover the lattice points $C \cap \mathbb{Z}^d$ contained in a d-dimensional 0-symmetric convex body C, in terms of the successive minima of C. No results are known for the number of lower-dimensional subspaces, with the exception of the one-dimensional linear subspaces. In the latter case, the problem is again equivalent to a number-theoretic question discussed in [HuN96]: what is the number of primitive lattice vectors contained in C?

Problem 6 *What is the minimum number of k-dimensional linear subspaces necessary to cover the d-dimensional $n \times \cdots \times n$ lattice cube?*

[Al91] N. ALON: Economical coverings of sets of lattice points, *Geom. Funct. Anal.* **1** (1991) 225–230.

[BáH*01] I. BÁRÁNY, G. HARCOS, J. PACH, G. TARDOS: Covering lattice points by subspaces, *Period. Math. Hungar.* **43** (2001) 93–103.

[BeH94]   K. BEZDEK, T. HAUSEL: On the number of lattice hyper-
           planes which are needed to cover the lattice points of a convex
           body, in: *Intuitive geometry* (Szeged, 1991), K. Böröczky et
           al., eds., *Colloq. Math. Soc. János Bolyai* **63** (1994) 27–31.

[ChH83]   S.-S. CHENG, Y.-T. HUANG: Optimal intersections of plane
           lattice points with vertical and horizontal straight lines, *Math.
           Proc. Camb. Philos. Soc.* **94** (1983) 381–388.

[Co04]   M.J. COLLINS: Covering a set of points with a minimum
           number of turns, *Int. J. Comput. Geom. Appl.* **14** (2004)
           105–114.

[CoM98]   M.J. COLLINS, B.M.E. MORET: Improved lower bounds for
           the link length of rectilinear spanning paths in grids, *Informa-
           tion Proc. Letters* **68** (1998) 317–319.

[Co85]   C.E. CORZATT: Covering convex sets of lattice points with
           straight lines, *Congressus Numerantium* **50** (1985) 129–135.

[Gu81]   R.K. GUY: *Unsolved Problems in Number Theory*, Springer-
           Verlag, 1981.

[HuN96]   M.N. HUXLEY, W.G. NOWAK: Primitive lattice points in
           planar domains, *Acta Arith.* **76** (1996) 271–283.

[KrKM94]   E. KRANAKIS, D. KRIZANC, L. MEERTENS: Link length of
           rectilinear hamiltonian tours on grids, *Ars Combinatoria* **38**
           (1994) 177–192.

[Ta97]   I. TALATA: Covering the lattice points of a convex body with
           affine subspaces, in: *Intuitive Geometry*, I. Bárány et al., eds.,
           Bolyai Soc. Math. Stud. **6** (1997) 429–440.

10.3 Sets of Lattice Points Avoiding Other Regularities

Lattices are very "regular" point sets: they are periodic, and therefore they determine few distinct distances and few slopes, they span many similar triangles, etc. At least how many points have to be deleted to destroy these regularities? Conversely, at most how many points can be selected from an $n \times \cdots \times n$ lattice cube so that all distances or all slopes determined by them are distinct, no two triangles spanned by them are similar, or they do not contain all vertices of some other forbidden configuration? These questions are particularly interesting for those properties for which, without any restriction on the point set, lattice sections are (conjectured to be) extremal.

As we have seen in Section 5.3, the points of the $n \times n$ lattice square S_n determine only $O\left(\frac{n^2}{(\log n)^{\frac{1}{2}}}\right)$ distinct distances. Therefore, if all distances between the elements of a subset $X \subset S_n$ are distinct, then X consists of at most $O\left(\frac{n}{(\log n)^{\frac{1}{4}}}\right)$ points. Erdős and Guy [ErG70] pointed out that by a greedy algorithm, one can always find $\Omega\left(n^{\frac{2}{3}-\varepsilon}\right)$ points in S_n whose pairwise distances are all distinct. This was improved by Thiele [Thi95] and Lefmann [LeT95] to $\Omega\left(n^{\frac{2}{3}}\right)$, using random selection combined with the deletion method (see [AlS00]).

Problem 1   *What is the maximum number of points that can be selected from an $n \times n$ lattice square such that all distances determined by them are distinct?*

The lower bound remains valid in higher dimensions, but for $d \geq 3$ we do not have any upper bound better than the trivial estimate $O(n)$.

Problem 2   *What is the maximum number of points that can be selected from a d-dimensional $n \times \cdots \times n$ lattice cube such that all distances determined by them are distinct?*

Similarly, one can look for the maximum size of a subset X of the lattice cube with the property that all $\binom{|X|}{2}$ lines spanned by X have different slopes. (Consequently, no three elements of X are on a line.) Some explicit values in the plane were determined by computer search by Peile and Taylor [PeT00]. They showed that $\max |X| = n + 1$ for $n = 2, \ldots, 6$ and n for $n = 7, 8, 9$.

Problem 3   *What is the maximum number of points that can be selected from an $n \times n$ lattice square such that all lines spanned by their pairs have distinct slopes?*

Since all slopes determined by an $n \times n$ lattice square are rational, $\frac{a}{b}$ with $|a|, |b| < n$, there are only $O(n^2)$ available values. This immediately implies that the size of any subset X that determines $\binom{|X|}{2}$ distinct slopes is at most $O(n)$. Erdős, Graham, Ruzsa, and Taylor [ErG*92] improved this bound to $O(n^{\frac{4}{5}})$.

On the other hand, a greedy construction gives the lower bound $\Omega(n^{\frac{1}{2}})$, which was improved by Zhang [Zh93] and Thiele [Thi95] to $\Omega\left(\frac{n^{\frac{2}{3}}}{(\log n)^{\frac{1}{3}}}\right)$. Here, the best known construction can be obtained again by random selection combined with the deletion method. Note that modular constructions do not help here, because for any prime number p there are only p possible slopes $\bmod\, p$, in contrast to the cp^2 rational slopes for integers. Thus, the maximum size of a set with distinct slopes $\bmod\, p$ is $O(\sqrt{p})$.

9 POINTS IN THE LATTICE
WITH DISTINCT SLOPES

The last problem can be generalized to d-dimensional $n \times \cdots \times n$ lattice cubes for $d \geq 3$. Now, no two lines spanned by different point pairs are allowed to span parallel or identical lines. By straightforward extension of the methods applied in the plane, we obtain the lower bound $\Omega(n^{\frac{d}{3}})$ and the upper bound $O\left(n^{\frac{d^2}{2d+1}}\right)$ [Thi95].

The restriction that all slopes must be distinct is much stronger and more difficult to handle than the condition that all difference vectors are distinct. The transformation $(x_1, \ldots, x_d) \mapsto \sum_{i=1}^{d} x_i n^{i-1}$ maps any set $X \subset \{0, \ldots, n-1\}^d$ with distinct difference vectors to a set of integers $\hat{X} \subset \{0, \ldots, n^d - 1\}$ with distinct differences (which is essentially a so-called Sidon set). Thus, the maximum cardinality of such a set X is $\Theta(n^{\frac{d}{2}})$. In particular, we can select $\Theta(n)$ points with distinct difference vectors from the $n \times n$ lattice square in the plane, but only $O(n^{\frac{4}{5}})$ points with the property that all directions assumed by their difference vectors are distinct.

Instead of excluding four vertices that span a parallelogram (i.e., two pairs of vectors with identical differences), we may wish to avoid axis-parallel squares.

Problem 4   *What is the maximum number of points that can be selected from an $n \times n$ lattice square such that no four of them span an axis-parallel square?*

The maximum is known to be $o(n^2)$. An elegant quantitative argument was given by Solymosi [So04] (see also [So03]). A more general qualitative result had been established by Furstenberg [Fu81] (p. 152, Theorem 7.16). He showed that for any finite pattern $F \subset \mathbb{Z}^2$, the maximum cardinality of a subset of the $n \times n$ lattice square containing no homothetic copy of F is $o(n^2)$. On the other hand, Ajtai and Szemerédi [AjS74] constructed a set of $\Omega\left(n^{2-\frac{c}{\sqrt{\log n}}}\right)$ lattice points that does not even contain an axis-aligned right isosceles triangle (half a square). The actual bound is $\Omega(r_3(n)^2)$, where $r_3(n)$ is the maximum cardinality of a subset of $\{1, \ldots, n\}$ containing no arithmetic progression of length three. Given a sequence of numbers $a_1, \ldots, a_k \in \{1, \ldots, n\}$ that does not contain an arithmetic progression of length three, it is easy to see that the points $((a_j - a_i), (a_j + a_i))$ do not span any axis-aligned isosceles right triangle.

The minimum cardinality of a maximal square-free subset of the lattice square $\{1, \ldots, n\}^2$ (i.e., a subset to which no extra point can be added without creating an axis-parallel square) was studied by Abbott, Hanson, and Katchalski [AbH75], [AbK82]. They showed that this number is at least $\Omega\left(n^{\frac{4}{3}-\varepsilon}\right)$ and at most $O\left(n^{\frac{\log 3}{\log 2}}\right)$.

If we want to find a large subset of the lattice square $\{1, \ldots, n\}^2$ that does not contain the vertices of any axis-parallel rectangle (not only squares), the problem becomes equivalent to the following classical question on 0-1 matrices: What is the maximum number of 1s in an $n \times n$ 0-1 matrix containing no 2×2 submatrix, all of whose entries are 1? The answer is $\Theta\left(n^{\frac{3}{2}}\right)$; see [Fü83]. Many similar questions are surveyed article [BrKV03].

It is a well-known exercise in combinatorics to show that any set X of at least $2^d + 1$ points of $\mathbb{Z}^d$ has two elements whose midpoint belongs to $\mathbb{Z}^d$. Indeed, it follows by the pigeonhole principle that there exist $x_1, x_2 \in X$ such that their corresponding coordinates have the same parities; hence $\frac{1}{2}(x_1 + x_2)$ is an integer point. The set of 2^d points of $\{0, 1\}^d$ shows that this result is sharp. Harborth generalized this problem to k-tuples.

Problem 5   (Harborth [Ha73]) What is the minimum number $f(d, k)$ with the property that any set of $f(d, k)$ lattice points chosen from $\mathbb{Z}^d$ has k elements $x_1, \ldots, x_k$ whose centroid $\frac{1}{k}(x_1 + \cdots + x_k)$ is also a lattice point?

It depends only on the residues of the coordinates mod k whether the centroid of a point set belongs to $\mathbb{Z}^d$, so the problem can be rephrased as follows. What is the smallest number $f(d, k)$ such that any multiset of $f(d, k)$ points from $\mathbb{Z}_k^d$ has k elements that sum to zero?

According to the Erdős–Ginzburg–Ziv theorem [ErGZ61], $f(1, k) = 2k - 1$. We have $f(d, 2) = 2^d + 1$. Harborth [Ha73] established the general inequalities $(k - 1)2^d + 1 \leq f(d, k) \leq (k - 1)k^d + 1$. He also determined the exact values $f(d, 2^r) = (2^r - 1)2^d + 1$, $f(2, 3^r) = 4 \cdot 3^r - 3$, $f(2, 2^r 3^s) = 4 \cdot 2^r 3^s - 3$, and $f(3, 3) = 19$. All of these values except the last one agree with the lower bound in Harborth's inequality.

For the two-dimensional case, Kemnitz [Ke82], [Ke83] conjectured that $f(2, k) = 4k - 3$, i.e., that the lower bound in Harborth's inequality is always sharp. He proved this conjecture for numbers of the form $k = 2^a 3^b 5^c 7^d$. Various bounds for $f(2, k)$ were obtained in [AlD95], [Ró00], [Ga01], [Tha01]. Finally, Reiher [Re05] found a remarkably short proof of Kemnitz's conjecture.

For dimensions three and higher, the only known values are $f(3, 2) = 9$, $f(3, 3) = 19$ [Ha73], and $f(3, 4) = 41$ [Ke82], [Ke83]. Alon and Dubiner [AlD95] established the upper bound $f(d, k) \leq O\left((d \log d)^d k\right)$, in which the constants have been somewhat improved by Kubertin [Ku02]. Thus, for any fixed d the asymptotic value of $f(d, k)$ is known to be $O(k)$, but the factor depending on d might well be $O(c^d)$ rather than $O((d \log d)^d)$. The currently best lower bound for higher dimensions is $f(d, k) \geq 2^d \left(\frac{9}{8}\right)^{\lfloor \frac{d}{3} \rfloor} (k - 1) + 1$, where $k, d \geq 3$ and k is odd. It was found by Elsholtz [El04].

[AbH75]　H.L. Abbott, D. Hanson: On a combinatorial problem in geometry, *Discrete Math.* **12** (1975) 389–392.

[AbK82]　H.L. Abbott, M. Katchalski: On a set of lattice points not containing the vertices of a square, *European J. Combinatorics* **3** (1982) 191–193.

[AjS74]　M. Ajtai, E. Szemerédi: Sets of lattice points that form no squares, *Studia Sci. Math. Hungar.* **9** (1974/75) 9–11.

[AlD95]　N. Alon, M. Dubiner: A lattice point problem and additive number theory, *Combinatorica* **15** (1995) 301–309.

[AlS00]　N. Alon, J. Spencer: *The Probabilistic Method* (2nd ed.), Wiley 2000.

[BrKV03]　P. Brass, G. Károlyi, P. Valtr: A Turán-type extremal theory for convex geometric graphs, in: *Discrete and Computational Geometry — The Goodman-Pollack Festschrift*, B. Aronov et al., eds., Springer-Verlag 2003, 275–300.

[El04]　C. Elsholtz: Lower bounds for multidimensional zero sums, *Combinatorica* **24** (2004) 351–358.

[ErG70]　P. Erdős, R.K. Guy: Distinct distances between lattice points, *Elemente Math.* **25** (1970) 121–123.

[ErG*92]　P. Erdős, R.L. Graham, I.Z. Ruzsa, H. Taylor: Bounds for arrays of dots with distinct slopes or lengths, *Combinatorica* **12** (1992) 39–44.

[ErGZ61]   P. ERDŐS, A. GINZBURG, A. ZIV: Theorem in the additive
           number theory, *Bull. Res. Council Israel* **10F** (1961) 41–43.

[Fü83]   Z. FÜREDI: Graphs without quadrilaterals, *J. Combin. The-
           ory Ser. B* **34** (1983) 187–190.

[Fu81]   H. FURSTENBERG: *Recurrence in Ergodic Theory and Com-
           binatorial Number Theory*, Princeton University Press, 1981.

[Ga01]   W. GAO: Note on a zero-sum problem, *J. Combinatorial
           Theory Ser. A* **95** (2001) 387–389.

[Ha73]   H. HARBORTH: Ein Extremalproblem für Gitterpunkte, *J.
           Reine Angew. Math.* **262/263** (1973) 356–360.

[Ke83]   A. KEMNITZ: On a lattice point problem, *Ars Combinatoria*
           16-B (1983) 151–160.

[Ke82]   A. KEMNITZ: *Extremalprobleme für Gitterpunkte*, Disserta-
           tion, TU Braunschweig (1982).

[Ku02]   S. KUBERTIN: *Nullsummen in Z_p^d*, Diplomarbeit, TU Claus-
           thal 2002.

[LeT95]   H. LEFMANN, T. THIELE: Point sets with distinct distances,
           Combinatorica **15** (1995), 379–408.

[Me87]   N.S. MENDELSOHN: Packing a square lattice with a rectangle-
           free set of points, *Mathematics Magazine* **60** (1987) 229–233.

[PeT00]   R.E. PEILE, H. TAYLOR: Sets of points with pairwise distinct
           slopes, *Comput. Math. Appl.* **39** (2000) 109–115.

[Re05]   C. REIHER: On Kemnitz' conjecture concerning lattice-points
           in the plane, manuscript.

[Ró00]   L. RÓNYAI: On a conjecture of Kemnitz, *Combinatorica* **20**
           (2000) 569–573.

[So04]   J. SOLYMOSI: Note on a question of Erdős and Graham,
           Combin. Probab. Comput. **13** (2004) 263–267.

[So03]   J. SOLYMOSI: On a generalization of Roth's theorem, in: *Dis-
           crete and Computational Geometry – The Goodman–Pollack
           Festschrift*, B. Aronov et al., eds., Springer 2003, 825–827.

[Tha01]   R. THANGADURAI: On a conjecture of Kemnitz, *C. R. Math.
           Acad. Sci., Soc. R. Can.* **23** (2001) 39–45.

[Thi95]   T. THIELE: *Geometric selection problems and hypergraphs*,
           Dissertation, FU Berlin 1995.

[Zh93]   Z. ZHANG: A note on arrays of dots with distinct slopes,
           Combinatorica **13** (1993) 127–128.

10.4 Visibility Problems for Lattice Points

Two lattice points x, y are *visible* from each other if there is no other lattice point between them on the segment $\overline{xy}$. A lattice vector is said to be *primitive* if it is not the multiple of any other lattice vector, that is, if the greatest common divisor of its coordinates is one. Thus, the primitive lattice vectors are precisely those lattice points that are visible from 0. This concept has been studied and used in a number of quite distinct areas: in number theory (e.g., the number of primitive lattice vectors in a 0-symmetric convex body [HuN96]), in integer optimization, and even in theoretical physics [BaGW94], [BoCZ00], [Mo92].

The oldest geometric problem on visible lattice points is concerned with the size of "holes" in this set. Although the "global" density of the set of visible (primitive) points in the d-dimensional integer lattice is well known to be $\frac{1}{\zeta(d)}$ (where ζ is Riemann's zeta function), its distribution is rather irregular, and arbitrarily large "holes" occur. Erdős [Er58] proved that for any point (x, y) with $x, y \leq n$ there is a visible point within a distance of at most $O(\frac{\log n}{\log \log n})$. He also showed that for some of these points (x, y) the distance to the nearest visible point is at least $\Omega\left((\frac{\log n}{\log \log n})^{\frac{1}{2}}\right)$. The method of [Er58] generalizes to higher dimensions.

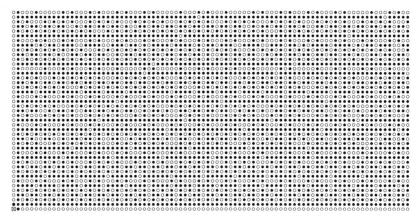

PRIMITIVE LATTICE POINTS; THE ORIGIN IS THE LOWER LEFT CORNER.

Problem 1   *(Erdős [Er58]) What is the smallest number $f_d(n)$ with the property that the distance between any point of the d-dimensional lattice cube $\{1, \ldots, n\}^d$ and the nearest primitive lattice point is at most $f_d(n)$?*

Herzog and Stewart [HeS71] studied the structure of various patterns of visible and invisible lattice points whose translates occur somewhere in the lattice. They characterized all possible patterns in every dimension, and hence completely answered some earlier questions of Winfee [Wi65] and Altgeld [Al66] on the connectivity of the lattice subgraphs of visible points.

Several papers deal with visibility from a set of points rather than a single point. Perhaps the most interesting open problem of this type is the following. What is the minimum number of lattice points that can be selected from $\{1, \ldots, n\}^d$ such that each point of this lattice cube is visible from at least one of them. Abbott established the lower bound $\Omega\left(\frac{\log n}{\log \log n}\right)$, which holds in all dimensions. Adhikari and Balasubramanian [AdB96] reduced the planar upper bound $O(\log n)$ given in [Ab74] to $O\left(\frac{\log n \log \log \log n}{\log \log n}\right)$. For $d \geq 3$, Adhikari and Chen [AdC99] obtained an $O\left(\frac{\log n}{\log \log n}\right)$ construction, matching the lower bound. Thus, only the case $d = 2$ is open.

Problem 2   *(Abbott [Ab74]) What is the smallest number of points that can be selected from the $n \times n$ lattice square such that each point of the lattice square is visible from at least one of the selected points?*

There are some results concerning the density of visible lattice points in the whole lattice, not restricted to a lattice square [Re66], [Ru66], [KrP94]. Rumsey [Ru66] proved that the density of the set of d-dimensional lattice points visible from *all* points of a fixed finite set X is $\prod_{p \text{ prime}}(1 - \frac{x_p}{p^d})$, where x_p is the number of mod p distinct points in X. Thus, for a fixed $k \geq 2$, there is no k-element set maximizing this density. Indeed, for any t we can select k points in such a way that they coincide modulo the first t primes. In this respect, the problem of maximizing the density of the set of points visible from *at least one* element of X is different: here there exist extremal sets X. A set of at most 2^d lattice points is extremal if and only if its points are pairwise visible. In particular, every subset of $\{0, 1\}^d$ is extremal. Kranakis and Pocchiola [KrP94] gave a more complicated criterion that characterizes all extremal sets in the d-dimensional lattice with up to 3^d points.

Another concept related to primitive lattice vectors is the number of distinct lines spanned by a set of lattice points. In a series of papers, Sheng [Sh77a], [Sh77b], [Sh78a], [Sh78b] showed that if we let points(D) denote the number of lattice points in D, and let lines(D) denote the number of distinct lines spanned by these lattice points, then for any d-dimensional convex body C we have

$$\lim_{r \to \infty} \frac{\text{lines}(rC)}{(\text{points}(rC))^d} = \frac{2^d - 1}{2^{d+1}\zeta(d)},$$

independently of C. He also obtained similar results, independent of C, for the numbers of k-point lines and $\geq k$-point lines, and found several examples of nonconvex sets C for which these results are false [Sh78b]. This is a surprising phenomenon that still requires explanation.

[Ab74]   H.L. ABBOTT: Some results in combinatorial geometry, *Discrete Math.* **9** (1974) 199–204.

[AdB96]   S.D. ADHIKARI, R. BALASUBRAMANIAN: On a question regarding visibility of lattice points, *Mathematika* **43** (1996) 155–158.

[AdC99]   S.D. ADHIKARI, Y. CHEN: On a question regarding visibility of lattice points II, *Acta Arith.* **89** (1999) 279–282.

[Al66]   J.P. ALTGELD: Solution to advanced problem 5263, *Amer. Math. Monthly* **73** (1966) 209–211.

[BaGW94]   M. BAAKE, U. GRIMM, D.H. WARRINGTON: Some remarks on the visible points of a lattice, *J. Physics A: Math. Gen.* **27** (1994) 2669–2674.

[BoCZ00]   F.P. BOCA, C. COBELI, A. ZAHARESCU: Distribution of lattice points visible from the origin, *Commun. Math. Phys.* **213** (2000) 433–470.

[Er58]   P. ERDŐS: On an elementary problem in number theory, *Canadian Math. Bulletin* **1** (1958) 5–8.

[HeS71]   F. HERZOG, B.M. STEWART: Patterns of visible and nonvisible lattice points, *Amer. Math. Monthly* **78** (1971) 487-496; Corrigendum: in same volume, p. 870.

[HuN96]   M.N. HUXLEY, W.G. NOWAK: Primitive lattice points in planar domains, *Acta Arith.* **76** (1996) 271–283.

[KrP94]   E. KRANAKIS, M. POCCHIOLA: Camera placement in integer lattices, *Discrete Comput. Geom.* **12** (1994) 91–104.

[Mo92]   R. MOSSERI: Visible points in a lattice, *J. Physics, A: Math. Gen.* **25** (1992) L25–L29.

[Re66]   D. REARICK: Mutually visible lattice points, *Norske Vid. Selsk. Forhdl.* **39** (1966) 41–45.

[Ru66]   H. RUMSEY: Sets of visible points, *Duke Math. J.* **33** (1966) 263–274.

[Sh78a]   T.K. SHENG: Lines containing q lattice points in $V \subset \mathbb{R}^n$, *Nanta Mathematica* **11** (1978) 71–77.

[Sh78b]   T.K. SHENG: Lines determined by lattice points in nonconvex regions, *Nanta Mathematica* **11** (1978) 78–83.

[**Sh77a**]   T.K. SHENG: Lines determined by lattice points in $\mathbb{R}^n$, *Nanta Mathematica* **10** (1977) 194–200.

[**Sh77b**]   T.K. SHENG: Lines determined by lattice points in $\mathbb{R}^2$, *Nanta Mathematica* **10** (1977) 77–81.

[**Wi65**]   A. WINFEE: Advanced problem 5263, *Amer. Math. Monthly* **72** (1965) 192–193.

11. Geometric Inequalities

11.1 Isoperimetric Inequalities for Polygons and Polytopes

The classical isoperimetric inequality states that among all (convex) sets with a given perimeter, the circular disk has the greatest area. This is a very old result that has been generalized in many directions. Apart from the fact that one does not need convexity here, essentially the same result holds in all scenarios in which the notions of "perimeter" and "area" can be naturally defined. The embedding space can also be varied: similar inequalities are true in higher dimensions (balls have maximum volume among all bodies with a given surface area), and problems of this type have also been considered in many spaces other than the Euclidean. Moreover, many analogous inequalities have been established concerning other measures besides the perimeter and the area. For the extensive literature on the many aspects of these problems see [ScA00], [BuZ88], [Tal93], [Lu93], [Fl93], [Ha57]. In this section, we concentrate on isoperimetric problems in discrete geometry, and avoid most questions that largely belong to convexity, differential geometry, or geometric measure theory.

Any inequality relating two geometric measures of a set, such as area, perimeter, diameter, width, inradius, or circumradius, is called an "isoperimetric" inequality. Many such inequalities are known for convex sets that are tight for balls, simplices, or degenerate bodies. Nevertheless, many interesting problems are still open, even for convex polygons in the plane.

For a fixed $n \geq 3$, it is easy to prove the analogue of the classical isoperimetric inequality for convex n-gons: among all n-gons of a given perimeter, the regular n-gon has the largest area. (See Siegel's Dido-type theorem [Si02], [Si01], and [Pa78], [BöB*86], [BeB89], for various generalizations.) The same holds for any other pair chosen out of the following parameters: area, perimeter, circumradius, inradius. Fixing the value of one of them, every other parameter attains its maximum or minimum for a regular n-gon. However, problems involving the diameter are harder for n-gons, because they depend on the divisibility properties of n. This was first noticed by Reinhardt [Re22], who determined the n-gons with a given diameter and maximal perimeter for all values of n that are not powers of two. He proved that if n has an odd divisor k, then every polygon that can be obtained from a regular Reuleaux k-gon by subdividing each of its boundary arcs into $\frac{n}{k}$ equal pieces, and taking the convex hull of the resulting vertices, is extremal. This has been rediscovered by many authors [Vi50], [Ha56], [Sc57], [Bi61], [LaT84], [Ta87], [Da97], but none of them has managed to settle the missing cases. The extremal quadrilateral is not a square: it is an equilateral triangle whose one side is subdivided. Vincze [Vi50] showed that the extremal octagon is not regular either.

Problem 1   *(Reinhardt [Re22]) What is the maximum perimeter of a convex 2^r-gon of unit diameter ($r \geq 3$)?*

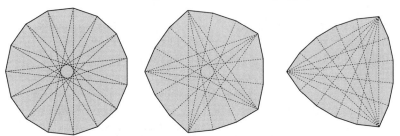

REGULAR 15-GON AND SUBDIVIDED REULEAUX 5- AND 3-GONS:
15-GONS OF MAXIMUM PERIMETER WITH A GIVEN DIAMETER

Bezdek and Fodor [BeF00] proved the exact analogue of Reinhardt's theorem, with the same extremal configurations and with the same missing cases $n = 2^r$, for the maximum width in place of the maximum perimeter.

We face similar difficulties in the diameter versus area scenario. Again, Reinhardt [Re22] answered the problem for many cases. He showed that for all odd values of n, the n-gon with a given diameter and the largest area is the regular n-gon, but for even values of $n \geq 6$ this is not the case. There are many extremal quadrilaterals beside the squares: any quadrilateral with two orthogonal diameters is extremal. The extremal hexagons and octagons were determined by Graham [Gr75] and Audet et al. [AuH*02], respectively: these are slightly perturbed regular pentagons and heptagons with one of their sides subdivided.

Problem 2   *(Reinhardt [Re22]) What is the maximum area of a convex n-gon of unit diameter when n is even and $n \geq 10$?*

In view of the situation for $n = 6, 8$, one cannot expect that the above maximum can be written in a nice closed form, but it might be possible to establish a structure theorem. Graham [Gr75] conjectured that for n even, the graph of diameters of the convex n-gon of the largest area with a given diameter is always an $(n-1)$-cycle with an additional vertex of degree one.

We can ask similar questions in the larger class of *simple* (i.e., non-self-intersecting) polygons. For some of these problems, the extremal simple polygons are automatically convex. This is the case, for example, in the diameter versus area scenario. For some other parameters the problem becomes trivial. For instance, the maximum perimeter of a simple n-gon of diameter one (more precisely, the supremum) is n. The following question for simple polygons seems to be new and interesting.

Problem 3   *(Brass) For $n \geq 5$ odd, what is the maximum perimeter of a simple n-gon contained in a disk of radius one?*

For even values of n, one can come arbitrary close to the trivial upper bound $2n$ by a simple n-gon whose sides go back and forth near a diameter of the disk, but for odd n this construction does not work.

For polytopes in three-dimensional space the situation is more complicated, since we have no good three-dimensional analogue for a regular polygon. There are many isoperimetric-type inequalities for three-dimensional convex polytopes (see [FeT72], [Fl93]), but most of them are sharp only in very few cases, namely for certain Platonic solids.

The search for a three-dimensional isoperimetric inequality for polytopes with n faces is an old project, perhaps started by Lhuilier in the eighteenth century. Since then the problem has been studied by Steiner, Lindelöf, Minkowski, Steinitz, and many others. There are some structural results on the extremal polytopes, such as the necessary condition found by Lindelöf [Li869], [Li899]: all extremal polytopes must have an inscribed sphere that touches each face of the polytope at its centroid. As Steinitz [St27], [St28] pointed out, in the class of simple polytopes of a given combinatorial type, this condition is also sufficient.

Recall that a three-dimensional polytope is called *simple* if each of its vertices belongs to precisely three edges (and to precisely three faces). Two polytopes in $\mathbb{R}^3$ have the same combinatorial type if there is a one-to-one correspondence between their vertex sets, edge sets, and face sets that preserves containment. The number of different combinatorial types of simple polytopes with a small number of faces is not too large. Therefore, for $n \leq 7$, it was possible to find all the extremal polytopes: the regular tetrahedron, a right triangular prism, the cube, and a right pentagonal prism (see [Go35]). The regular dodecahedron is also extremal, but we do not know any further examples.

The following general inequality for convex polytopes P with n faces was announced by Goldberg [Go35]. Goldberg's argument was completed by Florian [Fl56]. The first rigorous proof was given by L. Fejes Tóth [FeT48], [FeT64].

$$(\mathrm{Surf}(P))^3 \geq 54(n-2)\tan\left(\frac{n}{n-2}\frac{\pi}{6}\right)\left(4\left(\sin\left(\frac{n}{n-2}\frac{\pi}{6}\right)\right)^2 - 1\right)(\mathrm{Vol}(P))^2,$$

where equality holds only for regular tetrahedra, cubes, and dodecahedra.

Problem 4   *For $n \notin \{4,5,6,7,12\}$, what is the largest volume of a polytope with n faces and a given surface area?*

Of course, for large values of n these polytopes will converge to a ball, but the answers for small values of n may be interesting. Goldberg [Go35] made a conjecture on the structure of the extremal polytopes.

Conjecture 5 *(Goldberg [Go35]) All three-dimensional polytopes with a given face number and surface area and with maximum volume are simple.*

The isoperimetric problem is equally interesting for polytopes with a fixed number of vertices rather than faces. Surprisingly, the corresponding inequality is not known to be true [Fl93]. For its higher-dimensional analogues, see [BöB96].

Problem 6 *Is it true that any three-dimensional convex polytope P with n vertices satisfies the inequality*

$$\frac{(\operatorname{Surf}(P))^3}{(\operatorname{Vol}(P))^2} \geq \frac{27\sqrt{3}}{2}(n-2)\left(3\left(\tan\left(\frac{n}{n-2}\frac{\pi}{6}\right)\right)^2 - 1\right),$$

with equality only for the regular tetrahedron, octahedron, and icosahedron?

Concerning the last problem, it is an old and still open conjecture of Steiner that the icosahedron is extremal, at least within its own combinatorial type. L. Fejes Tóth [FeT50] (see also [FeT64], [Fl93]) formulated the following common generalization [FeT50], [Fl93] of the inequalities for given vertex numbers and face numbers.

Conjecture 7 *(L. Fejes Tóth) Any convex polytope P with v vertices, e edges, and f faces satisfies the inequality*

$$\frac{(\operatorname{Surf}(P))^3}{(\operatorname{Vol}(P))^2} \geq 9e\sin\left(\frac{\pi f}{e}\right)\left(\left(\tan\left(\frac{\pi f}{2e}\right)\right)^2\left(\tan(\frac{\pi v}{2e})\right)^2 - 1\right),$$

with equality only for the Platonic solids.

Similar problems can be raised with given face, edge, and vertex numbers for any other pair of parameters such as surface area, volume, inradius, and circumradius. In some cases it is known that certain Platonic solids are extremal [Fl93]. The cases in which the diameter is fixed are especially complicated, even for small vertex numbers. Among d-dimensional simplices of a given diameter, the regular simplex has the largest volume. Kind and Kleinschmidt [KiK76] determined the extremal polytope of the largest volume with a given diameter and $d+2$ vertices: it is the convex hull of two concentric orthogonal regular simplices of dimensions $\left\lceil\frac{d}{2}\right\rceil$ and $\left\lfloor\frac{d}{2}\right\rfloor$. The same problem for d-dimensional polytopes with $d+3$ vertices was settled by Klein and Wessler [KlW03]. It seems likely that for a given diameter and given vertex or face numbers, the extremal polytopes are never centrally symmetric.

Problem 8   *[KiK76] Find the three-dimensional polytopes of maximum volume with $n \geq 6$ vertices (faces) and a given diameter.*

Just as in the case of the "ordinary" isoperimetric inequality, most of the above problems are interesting mainly for small vertex or face numbers n. As n tends to infinity, the extremal polytopes always converge to a ball, the unique solution of the corresponding "continuous" isoperimetric problem.

Another interesting class of geometric objects for which isoperimetric questions have been asked are cell decompositions of space whose cells are not necessarily congruent but their size is somehow restricted. A classical example of such a question is the direct analogue of the isoperimetric inequality: the so-called honeycomb problem. What is the minimum average perimeter of a cell in a decomposition of the plane into convex cells of area one? It was discovered by L. Fejes Tóth [FeT72] (and by the bees [Da859]) that the regular hexagonal tiling is extremal. Allowing more general types of cells introduces more technical difficulties without changing the ultimate solution (see Hales [Ha01]). At least for convex cells, the same proof can be applied to any other pair of parameters, such as perimeter, area, inradius, circumradius. We can always restrict our attention to *normal* tilings [GrS87], for which the Euler formula gives the average edge number per cell, and then use the corresponding isoperimetric-type inequality together with Jensen's inequality to conclude that the hexagonal tiling is extremal.

However, this argument breaks down for all problems involving the diameter of the cells, since the regular polygons are not extremal in this respect. The case of area versus diameter is especially interesting and was studied by Lenz [Le56a], [Le56b]. The question was rediscovered by Heppes.

Problem 9   *(Lenz) What is the maximum average area of the cells in a decomposition of the plane into cells of diameter at most one?*

It is not hard to prove that the regular hexagonal tiling is optimal among all tilings combinatorially equivalent to it.

Almost nothing is known about three-dimensional cell structures. The best known open problem of this kind is the direct analogue of the usual isoperimetric inequality.

Problem 10   *What is the minimum average surface of the cells in a decomposition of three-dimensional space into convex cells of unit volume?*

Kelvin's conjecture, a long-standing open problem on the structure of the extremal cell decomposition with respect to the last problem, was finally disproved by Phelan et al. [WeP94], [Ph96], who constructed rather complicated cell decompositions consisting of several incongruent types of cells.

[AuH*02]   C. Audet, P. Hansen, F. Messine, J. Xiong: The largest small octagon, *J. Combinatorial Theory Ser. A* **98** (2002) 46–59.

[BeB89]   A. Bezdek, K. Bezdek: On a discrete Dido-type question, *Elemente Math.* **4** (1989) 92–100.

[BeF00]   A. Bezdek, F. Fodor: On convex polygons of maximal width, *Arch. Math.* **74** (2000) 75–80.

[Bi61]   H. Bieri: Zweiter Nachtrag zu Nr. 12, Bericht von H. Hadwiger, *Elemente Math.* **16** (1961) 105–106.

[BöB*86]   K. Böröczky, I. Bárány, E. Makai Jr., J. Pach: Maximal volume enclosed by plates and proof of the chessboard conjecture, *Discrete Math.* **60** (1986) 101–120.

[BöB96]   K. Böröczky Jr., K. Böröczky: Isoperimetric problems for polytopes with given number of vertices, *Mathematika*, **43** (1996) 237–254.

[BuZ88]   Y.D. Burago, V.A. Zalgaller: *Geometric Inequalities*, Springer-Verlag 1988.

[Da859]   C. Darwin: *On the Origin of Species by Means of Natural Selection, or the Preservation of Favoured Races in the Struggle for Life*, John Murray publishers, London, 1859.

[Da97]   B. Datta: A discrete isoperimetric problem, *Geometriae Dedicata* **64** (1997) 55–68.

[FeT72]   L. Fejes Tóth: *Lagerungen in der Ebene, auf der Kugel, und im Raum*, zweite Auflage, Springer-Verlag 1972.

[FeT64]   L. Fejes Tóth: *Regular Figures*, Pergamon Press 1964.

[FeT50]   L. Fejes Tóth: Extremum properties of the regular polyhedra, *Canadian J. Math.* **2** (1950) 22–31.

[FeT48]   L. Fejes Tóth: The Isepiphan Problem for n-hedra, *Amer. J. Math.* **70** (1948) 174–180.

[Fl93]   A. Florian: Extremum problems for convex discs and polyhedra, in: *Handbook of Convex Geometry, Vol. 1*, P.M. Gruber et al., eds., Elsevier 1993, 177–221.

[Fl56]   A. Florian: Eine Ungleichung über konvexe Polyeder, *Monatshefte Math.* **60** (1956) 130–156.

[Go35] M. GOLDBERG: The isoperimetric problem for polyhedra, *Tohoku Math. J.* (First Series) **40** (1935) 226–236.

[Gr75] R.L. GRAHAM: The largest small hexagon, *J. Combinatorial Theory Ser. A* **18** (1975) 165–170.

[GrS87] B. GRÜNBAUM, G.C. SHEPHARD: *Tilings and Patterns*, W.H. Freeman and Company 1987.

[Ha57] H. HADWIGER: *Vorlesungen über Inhalt, Oberfläche und Isoperimetrie*, Springer-Verlag 1957.

[Ha56] H. HADWIGER: Ungelöste Probleme 12, *Elemente Math.* **11** (1956) p. 86.

[Ha01] T.C. HALES: The honeycomb conjecture, *Discrete Comput. Geom.* **25** (2001) 1–22.

[KiK76] B. KIND, P. KLEINSCHMIDT: On the maximal volume of convex bodies with few vertices, *J. Combinatorial Theory Ser. A* **21** (1976) 124–128.

[KlW03] A. KLEIN, M. WESSLER: The largest small *n*-dimensional polytope with *n* + 3 vertices, *J. Combinatorial Theory Ser. A* **102** (2003) 401–409.

[LaT84] D.G. LARMAN, N.K. TAMVAKIS: The decomposition of the *n*-sphere and the boundaries of plane convex domains, in: *Convexity and Graph Theory*, M. Rosenfeld et al., eds., *Annals of Discrete Math.* **20** North-Holland, 1984, 209–214.

[Le56a] H. LENZ: Über die Bedeckung ebener Punktmengen durch solche kleineren Durchmessers, *Archiv Math.* **6** (1956) 34–40.

[Le56b] H. LENZ: Zerlegung ebener Bereiche in konvexe Zellen von möglichst kleinem Durchmesser, *Jahresbericht Deutsch. Math. Ver.* **58** (1956) 87–97.

[Li899] L. LINDELÖF: Recherches sur les polyèdres maxima, *Acta Soc. Sci. Fenn.* Helsingfors **24** (1898/1899).

[Li869] L. LINDELÖF: Propriétés générales des polyèdres etc., *Bull. Acad. Sci. St. Petersburg* **14** (1869) 258–269.

[Lu93] E. LUTWAK: Selected affine isoperimetric inequalities, in: *Handbook of Convex Geometry, Vol. 1*, P.M. Gruber et al., eds., Elsevier 1993, 151–176.

[Pa78] J. PACH: On an isoperimetric problem, *Studia Sci. Math. Hungar.* **13** (1978) 43–45.

[Ph96] R. PHELAN: Generalisations of the Kelvin problem and other minimal problems, *Forma* **11** (1996) 287–302.

[Re22]   K. REINHARDT: Extremale Polygone gegebenen Durch-
           messers, *Jahresbericht Deutsch. Math.-Ver.* **31** (1922) 251–
           270.

[Sc57]   J.J. SCHÄFER: Nachtrag zu Nr. 12, Bericht von H. Hadwiger,
           Elemente Math. **13** (1957) 85–86.

[ScA00]   P.R. SCOTT, P.W. AWYONG: Inequalities for convex sets,
           J. Ineq. Pure Appl. Math, **1** (2000) Art. 6.

[Si02]   A. SIEGEL: A Dido problem as modernized by Fejes Tóth,
           Discrete Comput. Geom. **27** (2002) 227–238.

[Si01]   A. SIEGEL: Some Dido-type inequalities, *Elemente Math.* **56**
           (2001) 1–4.

[St28]   E. STEINITZ: Über isoperimetrische Probleme bei konvexen
           Polyedern II, *J. Math.* **159** (1928) 133–143.

[St27]   E. STEINITZ: Über isoperimetrische Probleme bei konvexen
           Polyedern I, *J. Math.* **158** (1927) 129–153.

[Ta87]   N.K. TAMVAKIS: On the perimeter and area of the convex
           polygons of a given diameter, *Bull. Soc. Math. Gréce* (N. S.)
           28 (1987) 115–132.

[Tal93]   G. TALENTI: The standard isoperimetric theorem, in: *Hand-
           book of Convex Geometry, Vol. 1,* P.M. Gruber et al., eds.,
           Elsevier 1993, 73–123.

[Vi50]   S. VINCZE: On a geometrical extremum problem, *Acta Sci.
           Math. (Szeged)* **12A** (1950) 136–142.

[WeP94]   D. WEAIRE, R. PHELAN: A counter-example to Kelvin's
           conjecture on minimal surfaces. *Philos. Mag. Lett.* **69** (1994)
           107–110, also appeared in *Forma* **11** (1996) 209–213.

11.2 Heilbronn-Type Problems

How should one distribute n points in the unit square so that the area of the smallest triangle determined (spanned) by them is as large as possible? This classical question of H.A. Heilbronn can be interpreted as a problem of finding a set that is as far as possible from containing three collinear points.

Problem 1 (*Heilbronn*) *What is the smallest* $f^{\min\text{-area}}(n)$ *such that any set of n points in the unit square spans a triangle whose area is at most* $f^{\min\text{-area}}(n)$?

Since any triangulation of the convex hull of such a point set consists of at least $n - 2$ (possibly degenerate) triangles whose total area does not exceed one, we obtain the upper bound $f^{\min\text{-area}}(n) \leq \frac{1}{n-2} = O(\frac{1}{n})$. This was improved several times by Roth and Schmidt [Ro51], [Sc71], [Ro72a], [Ro72b], [Ro73], [Ro76]. The best known upper bound, $O\left(\frac{1}{n^{\frac{8}{7}-\epsilon}}\right)$ for every $\epsilon > 0$, was established by Komlós, Pintz, and Szemerédi [KoPS81]. Heilbronn conjectured that the true order of magnitude of this function is $O(\frac{1}{n^2})$, which can be attained by selecting $\Omega(n)$ points from the $n \times n$ lattice square with no three collinear elements and scaling it down to a unit square (compare to the no-three-in-line problem discussed in Section 10.1). In this set, the area of any triangle is at least $\frac{1}{2(n-1)^2}$, half of the area of a (scaled) lattice square. Surprisingly, Komlós et al. [KoPS82] disproved Heilbronn's conjecture by raising the lower bound to $\Omega(\frac{\log n}{n^2})$. Erdős [Er85] conjectured that the new bound is asymptotically tight.

Note that a random uniform choice of the points does not give a good distribution: the expected area of the smallest triangle is $\Theta(\frac{1}{n^3})$ [JiLV02], [GrJ03].

Heilbronn's problem for small values of n has been studied in several papers [Go72], [YaZZ92], [CoY02]. We have the following exact values:

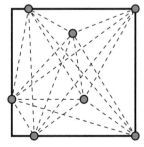

7 POINTS IN THE UNIT SQUARE
ALL TRIANGLES HAVE LARGE AREA

n	3	4	5	6
$f^{\min\text{-area}}(n)$	$\frac{1}{2}$	$\frac{1}{2}$	$\frac{1}{3\sqrt{3}}$	$\frac{1}{8}$

Instead of selecting the points from a unit square, they can be distributed in regions of other shapes, for example, in triangles [YaZZ94]. The question arises [DrYZ95], [YaZ95], which shape of unit area maximizes the value of the smallest area? One can also try to maximize the ratio between the areas of the smallest triangle and the largest triangle [Du95].

A surprising fact that illustrates the difficulty of Heilbronn's question is that the answer completely changes if instead of all triangles (convex hulls of triples) spanned by a point set, we consider the convex hulls of all four-tuples. Schmidt [Sc71] constructed n-element point sets in the plane such that the area of the convex hull of any four-tuple determined by them is at least $\Omega(n^{-\frac{3}{2}})$. In general, for any $k \geq 4$, Bertram-Kretzberg, Hofmeister, and Lefmann [BeHL00] found n-element point sets in which the area of the convex hull of any k-tuple is at least $\Omega\left(n^{-1-\frac{1}{k-2}}\right)$. On the other hand, for $k \geq 4$ we have no better upper bound than the trivial estimate $O(\frac{1}{n})$.

Problem 2   *What is the smallest number $f_k^{\min\text{-area}}(n)$ for which any set of n points in the unit square contains k elements whose convex hull has area at most $f_k^{\min\text{-area}}(n)$?*

Barequet [Ba01a] has generalized Heilbronn's problem to higher dimensions. One can select $\Theta(n)$ points from the d-dimensional $n \times n \times \cdots \times n$-lattice cube (say, the points $(t, t^2, \ldots, t^d) \bmod n$, $t = 1, 2, \ldots, n$, on the "modular" moment curve, where n is a prime) such that no $d+1$ elements lie in a hyperplane. The volume of any full-dimensional simplex spanned by them is at least $\frac{1}{d!}$. Thus, for a fixed d, by scaling we obtain n points in the unit cube such that the volume of every simplex is $\Omega(\frac{1}{n^d})$. This lower bound was improved to $\Omega(\frac{\log n}{n^d})$ by Lefmann [Le03]. Lefmann's bound is not constructive: adapting the approach in [KoPS82], we can randomly and uniformly select a set with slightly more than n points, and define the hypergraph of "bad" simplices whose volumes are small. It turns out that with large probability this hypergraph contains an independent set of size n. An algorithmic approach to construct "explicit" examples, at least in three-dimensional space, was suggested by Lefmann and N. Schmitt [LeS02].

Since every set of n points in $\mathbb{R}^d$ can be triangulated using $\Omega(n)$ simplices, it follows that the volume of at least one of them is at most $O(\frac{1}{n})$. For odd dimensions, slightly better bounds were found by Brass [Br05]; e.g., $O(n^{-\frac{6}{5}})$ in $\mathbb{R}^3$. It is not clear whether the above upper bound can be improved by finding larger triangulations of point sets. Does every d-dimensional point set have a large triangulation? Projection arguments fail to lift the two-dimensional upper bounds to higher dimensions, because the projection of a simplex is not a triangle.

Problem 3   *What is the smallest number $f_d^{\min\text{-vol}}(n)$ such that any set of n points in the d-dimensional unit cube has $d+1$ elements that span a simplex of volume at most $f_d^{\min\text{-vol}}(n)$?*

The problem becomes simpler if one allows points to be collinear, but considers only the area of the smallest nondegenerate triangle. This variant goes back to Erdős, Purdy, and Straus [ErPS82], [St78], who proved that among all sets X of n points, not all on a line, the ratio of the area of the largest triangle to the area of the smallest nondegenerate triangle spanned by X is at least $\lceil \frac{n}{2} \rceil - 1$. Equality holds here when X consists of two roughly equal sets of equidistant points lying on two parallel lines. However, the proof in [ErPS82] works only for $n \geq 32$. The largest value of n for which this statement is known to be false is five: the regular pentagon is a counterexample. It may be interesting to decide whether there are any larger counterexamples. Komlós, Pintz, and Szemerédi [KoPS84] studied the same question under the restriction that at most $f(n) \ll n^{\frac{1}{8}}$ points of X are on a line. The analogous problems in higher dimensions are open.

Problem 4   *Is it true that for any set X of $n > n_0(d)$ points in $\mathbb{R}^d$ that do not lie in a hyperplane, the ratio of the largest volume to the smallest positive volume of a nondegenerate simplex spanned by X is at least $\lceil \frac{n}{d} \rceil - 1$?*

Rothschild and Straus [RoS85] solved the similar problem in which one compares the smallest positive volume of a simplex to the volume of the convex hull of all points.

Heilbronn's triangle problem is the prototype of a large class of problems of the following type. Define a measure μ on k-tuples and find a set of n points that maximizes the minimum measure of its k-element subsets. Of course, even for $k = 3$ there are many natural measures μ, but not all of them are equally interesting. For example, in choosing μ to be the perimeter of the triangle spanned by a triple, which is a very "local" measure, the problem becomes similar to packing problems. On the other hand, the area of a triangle can be small even if its vertices are far apart. The inradius and width should behave similarly to the area.

Conjecture 5   *(Motzkin, Schmidt) Let $f^{\text{width}}(n)$ be the smallest number such that any set of n points in the unit square $[0,1]^2$ has a triple that can be covered by a strip of width $f^{\text{width}}(n)$. Then we have $f^{\text{width}}(n) = o(\frac{1}{n})$.*

Using analytic techniques, Beck [Be90] proved this conjecture in the special case that the set is uniformly distributed in the sense that it contains precisely one point from each small square square of a $\sqrt{n} \times \sqrt{n}$ chessboard. With this restriction one can always find a triple that can be covered by a strip of width $O\left(\frac{1}{n^{1.116}}\right)$.

If the measure μ is the minimum (not necessarily strictly positive) angle of the triangle spanned by a triple, the asymptotic behavior of the solution is simple: one can always find a triple of measure $\Theta(\frac{1}{n})$, and it can be conjectured that the extremal configuration is a regular n-gon.

Conjecture 6   *Any set of n points in the plane has three elements that determine an angle not greater than $\frac{\pi}{n}$.*

Barequet [Ba01b] pointed out that the following conjecture of L. Fejes Tóth is related to the dual version of Heilbronn's problem and the Motzkin–Schmidt conjecture mentioned above.

Conjecture 7   *(L. Fejes Tóth [FeT87]) Given n great circles on the unit sphere in $\mathbb{R}^3$, no three of which pass through the same point, the area of the smallest cell determined by their arrangement is $o(\frac{1}{n^2})$.*

Some initial results in this direction were obtained by Ismailescu [Is03].

Erdős and Szemerédi [Er84] raised an interesting variant of Heilbronn's problem involving two different measures that probably cannot be simultaneously maximized. In every set X of n points selected from the unit square, the smallest distance $d_{\min}$ is $O(n^{-\frac{1}{2}})$, which is asymptotically tight, for example, for a regular triangular lattice. At the same time, the smallest (possibly zero) angle spanned by X is at most $O(n^{-1})$. Is it true that the order of magnitude of the product of the smallest distance and the smallest angle determined by n points in the plane is smaller than $O(n^{-\frac{3}{2}})$, the product of the above two upper bounds? The best known lower bound, $\Omega(n^{-2})$, is realized by the regular n-gon.

Conjecture 8   *(Erdős and Szemerédi [Er84]) The product of the smallest distance and the smallest angle determined by n points in the unit square is $o(n^{-\frac{3}{2}})$.*

Blumenthal [Bl39] studied the largest angle $\alpha = \text{max\_angle}(n) \in [0, \pi]$ such that any set of n points in the plane has three elements whose angle is at least α. The first values of this function are $\text{max\_angle}(3) = \frac{1}{3}\pi$, $\text{max\_angle}(4) = \frac{1}{2}\pi$, $\text{max\_angle}(5) = \frac{3}{5}\pi$, $\text{max\_angle}(6) = \text{max\_angle}(7) = \text{max\_angle}(8) = \frac{2}{3}\pi$. The problem had been addressed in a series of papers [Sz41], [ErS61], [Se92], [Se93], before it was completely solved by Sendov [Se95]. As in the case of Heilbronn's triangle problem, the most obvious conjecture turned out to be wrong. The correct values are $\text{max\_angle}(n) = \left(1 - \frac{2}{2k+1}\right)\pi$ for $2^k < n \leq \frac{5}{4}2^k$ and $\text{max\_angle}(n) = \left(1 - \frac{1}{k+1}\right)\pi$ for $\frac{5}{4}2^k < n \leq 2^{k+1}$ for $k \geq 2$.

The same problem was discussed but only partially solved in higher dimensions [ErF83].

Problem 9   *(Erdős and Szekeres [ErS61]) What is the largest angle $\alpha = \text{max\_angle}_d(n)$ such that any set of n points in d-*

> dimensional space contains a triple that spans an angle
> not smaller than α?

It is known [Cr61], [Sc63], [Gr63] that any set of six points in three-dimensional space determines an angle larger than or equal to $\frac{\pi}{2}$. This is best possible: there is a set of five points such that all angles determined by them are acute ($< \frac{\pi}{2}$). On the other hand, the eight vertices of a cube do not span any angle larger than $\frac{\pi}{2}$. Danzer and Grünbaum [DaG62] proved that any set in $\mathbb{R}^d$ that does not determine an angle larger than $\frac{\pi}{2}$ consists of at most 2^d points. This bound is attained for the vertex set of a cube. We know that the d-dimensional function $\max\_\mathrm{angle}_d(n)$ satisfies $\max\_\mathrm{angle}_d(d+1) = \frac{\pi}{3}$ and $\max\_\mathrm{angle}_d(2^d) = \frac{\pi}{2} < \max\_\mathrm{angle}_d(2^d + 1)$. Erdős and Füredi [ErF83] showed that for any $\alpha > \frac{\pi}{3}$ there exists a constant $c_\alpha > 1$ with the property that one can find at least c_α^d points in $\mathbb{R}^d$ such that all angles determined by them are smaller than α.

A parametric version of this problem was raised by Conway et al. [CoC*79], who studied the minimum numbers $f(n,\alpha), g(n,\alpha)$ of triples among n points in the plane that determine an angle less than α or larger than α, respectively. They managed to show that the limits $\lim\limits_{n\to\infty} \frac{f(n,\alpha)}{n^3}$ and $\lim\limits_{n\to\infty} \frac{g(n,\alpha)}{n^3}$ exist for every α and determined these limits for certain values of α.

[Ba01a]   G. BAREQUET: A lower bound for Heilbronn's triangle problem in d dimensions, *SIAM J. Discrete Math.* **14** (2001) 230–236.

[Ba01b]   G. BAREQUET: A duality between small-face problems in arrangements of lines and Heilbronn-type problems, *Discrete Math.* **237** (2001) 1–12.

[Be90]   J. BECK: Almost collinear triples among N points on the plane, in: *A Tribute to Paul Erdős*, A. Baker et al., eds., Cambridge Univ. Press 1990, 39–57.

[BeHL00] C. BERTRAM-KRETZBERG, T. HOFMEISTER, H. LEFMANN: An algorithm for Heilbronn's problem, *SIAM J. Comput.* **30** (2000) 383–390.

[Bl39]   L.M. BLUMENTHAL: Metric methods in determinant theory, *Amer. J. Math.* **61** (1939) 912–922.

[Br05]   P. BRASS: An upper bound for the d-dimensional analogue of Heilbronn's triangle problem, to appear in *SIAM J. Discrete Math.*

[CoY02]   F. COMELLAS, J.L.A. YEBRA: New lower bounds for Heilbronn numbers, *Electron. J. Combin.* **9** (2002) Research Paper 6.

[CoC*79]   J.H. CONWAY, H.T. CROFT, P. ERDŐS, M.J.T. GUY: On the distribution of values of angles determined by coplanar points, *J. London Math. Soc.* 2. Ser. **19** (1979) 137–143.

[Cr61]   H.T. CROFT: On 6-point configurations on 3-space, *J. Lond. Math. Soc.* **36** (1961) 289–306.

[DaG62]   L. DANZER, B. GRÜNBAUM: Über zwei Probleme bezüglich konvexer Körper von P. Erdős und V.L. Klee, *Math. Zeitschrift* **79** (1962) 95–99.

[DrYZ95]   A.W.M. DRESS, L. YANG, Z. ZENG: Heilbronn problem for six points in a planar convex body, in: *Minimax and Applications*, D.-Z. Dhu et al., eds., *Nonconvex Optim. Appl. Ser.* Vol. 4, Kluwer Acad. Publ. 1995, 173–190.

[Du95]   J.B. DU: A lower bound for Heilbronn's problem (Chinese. English summary) *Hunan Jiaoyu Xueyuan Xuebao (Ziran Kexue)* **13** (1995) 41–47.

[Er85]   P. ERDŐS: Problems and results in combinatorial geometry, in: *Discrete Geometry and Convexity*, J.E. Goodman et al., eds., *Annals New York Acad. Sci.* **440** (1985): 1–11.

[Er84]   P. ERDŐS: Some old and new problems in combinatorial geometry, in: *Convexity and Graph Theory*, M. Rosenfeld et al., eds., *Annals Discrete Math.* **20** (1984) 129–136.

[ErF83]   P. ERDŐS, Z. FÜREDI: The greatest angle among n points in the d-dimensional Euclidean space, *Annals Discrete Math.* **17** (1983) 275–283.

[ErPS82]   P. ERDŐS, G. PURDY, E.G. STRAUS: On a problem in combinatorial geometry, *Discrete Math.* **40** (1982) 45–52.

[ErS61]   P. ERDŐS, G. SZEKERES: On some extremum problems in elementary geometry, *Ann. Univ. Sci. Budapest. Rolando Eötvös, Sect. Math.* **3–4** (1961) 53–62.

[FeT87]   L. FEJES TÓTH: On spherical tilings generated by great circles, *Geometriae Dedicata* **23** (1987) 67–71.

[Go72]   M. GOLDBERG: Maximizing the smallest triangle made by N points in a square, *Math. Magazine* **45** (1972) 135–144.

[GrJ03]   G. GRIMMETT, S. JANSON: On smallest triangles, *Random Structures Algorithms* **23** (2003) 206–223.

[Gr63]   B. GRÜNBAUM: Strictly antipodal sets, *Israel J. Math.* **1** (1963) 5–10.

[Is03]   D. ISMAILESCU: Slicing the pie, *Discrete Comput. Geom.* **30** (2003) 263–276.

[JiLV02]   T. JIANG, M. LI, P. VITÁNYI: The average-case area of Heilbronn-type triangles, *Random Structures Algorithms* **20** (2002) 206–219.

[KoPS84]   J. KOMLÓS, J. PINTZ, E. SZEMERÉDI: On a problem of Erdős and Straus, in: *Topics in Classical Number Theory, Vol. II*, G. Halász, ed., *Colloq. Math. Soc. János Bolyai* **34** North-Holland 1984, 927–960.

[KoPS82]   J. KOMLÓS, J. PINTZ, E. SZEMERÉDI: A lower bound for Heilbronn's problem, *J. London Math. Soc.* **25** (1982) 13–24.

[KoPS81]   J. KOMLÓS, J. PINTZ, E. SZEMERÉDI: On Heilbronn's triangle problem, *J. London Math. Soc.* 2. Ser. **24** (1981) 385–396.

[Le03]   H. LEFMANN: On Heilbronn's problem in higher dimension, *Combinatorica* **23** (2003) 669–680.

[LeS02]   H. LEFMANN, N. SCHMITT: A deterministic polynomial-time algorithm for Heilbronn's problem in three dimensions, *SIAM J. Computing* **31** (2002) 1926–1947.

[Ro76]   K.F. ROTH: Developments in Heilbronn's triangle problem, *Advances Math.* **22** (1976) 364–385.

[Ro73]   K.F. ROTH: Estimation of the area of the smallest triangle obtained by selecting three out of n points in a disc of unit area, in: *Analytic Number Theory*, H.G. Diamond, ed., *Proc. Sympos. Pure Math.* **24** Amer. Math. Soc., 1973, 251–262.

[Ro72a]   K.F. ROTH: On a problem of Heilbronn II, *Proc. London Math. Soc.* **25** (1972) 193–212.

[Ro72b]   K.F. ROTH: On a problem of Heilbronn III, *Proc. London Math. Soc.* **25** (1972) 543–549.

[Ro51]   K.F. ROTH: On a problem of Heilbronn, *J. London Math. Soc.* **26** (1951) 198–204.

[RoS85]   B.L. ROTHSCHILD, E.G. STRAUS: On triangulations of the convex hull of n points, *Combinatorica* **5** (1985) 167–179.

[Sc71]   W.M. SCHMIDT: On a problem of Heilbronn, *J. London Math. Soc.* 2. Ser. **4** (1971/72) 545–550.

[Sc63]   K. SCHÜTTE: Minimale Durchmesser endlicher Punktmengen mit vorgeschriebenem Mindestabstand, *Math. Ann.* **150** (1963) 91–98.

[**Se95**]   BL. SENDOV: Minimax of the angles in a planar configuration
            of points, *Acta Math. Hungar.* **69** (1995) 27–46.

[**Se93**]   BL. SENDOV: Angles in a plane configuration of points *C. R.
            Acad. Bulg. Sci.* **46** (1993) 27–30.

[**Se92**]   BL. SENDOV: On a conjecture of P. Erdős and G. Szekeres,
            C. R. Acad. Bulg. Sci. **45** (1992) 17–20.

[**St78**]   E.G. STRAUS: Some extremal problems in combinatorial geo-
            metry, in: *Combinatorial Mathematics* (Proc. Conf. Canberra
            1977), D.A. Holton et al., eds., Springer *Lect. Notes Math.*
            686 (1978) 308–312.

[**Sz41**]   G. SZEKERES: On an extremum problem in the plane, *Amer.
            J. Math.* **63** (1941) 208–210.

[**YaZ95**]   L. YANG, Z. ZENG: Heilbronn problem for seven points in a
            planar convex body, in: *Minimax and Applications*, D.-Z. Dhu
            et al., eds., *Nonconvex Optim. Appl. Ser.* Vol. 4, Kluwer
            Acad. Publ. 1995, 173–190.

[**YaZZ94**] L. YANG, J.Z. ZHANG, Z. ZENG: On the Heilbronn numbers
            of triangular regions, (in Chinese, English summary), *Acta
            Math. Sinica* **37** (1994) 678–689.

[**YaZZ92**] L. YANG, J.Z. ZHANG, Z. ZENG: A conjecture on the first
            several Heilbronn numbers and a computation, (in Chinese, see
            MR93i:51045) *Chinese Ann. Math. Ser. A* **13** (1992) 503–515.

11.3 Circumscribed and Inscribed Convex Sets

Fix a family $\mathcal{C}$ of sets, for example, n-gons, ellipses, or rectangles in the plane and choose a measure to quantify the "size" of a set. What is the largest number x such that every convex set of size one contains a member of $\mathcal{C}$ whose size is at least x? What is the smallest number X such that every convex set of size one can be covered by a member of $\mathcal{C}$ whose size is at most X? For many choices of the set family $\mathcal{C}$ and the measure, these questions on large *inscribed* and small *circumscribed* sets are quite interesting [BrL01]. The questions for large inscribed sets are in a way "dual" to the Heilbronn-type problems discussed in the previous section: there we tried to maximize the smallest subset, here we have to minimize the largest one.

A prototype of such statements is the following theorem of Sas [Sa39], whose special case $n = 3$ had been discovered earlier by Blaschke [Bl17]. For any integer $n \geq 3$, every plane convex set of area one contains an inscribed n-gon whose area is at least as large as the area of the largest n-gon inscribed in an ellipse of area one.

The corresponding problem of circumscribing n-gons is open for every $n > 3$. For triangles, Gross [Gr18] and Eggleston [Eg53] (see also [Ch73], [ChL71]) proved that any convex set C is contained in a triangle whose area is at most twice the area of C. Equality holds here if and only if C is a parallelogram. Moreover, one can even prescribe the direction of one side of the circumscribing triangle [Ho51].

Problem 1   For $n > 3$, what is the smallest number $c_n^{\text{outer area}}$ such that every plane convex set C has a circumscribing n-gon P_n with $\operatorname{area}(P_n) \leq c_n^{\text{outer area}} \operatorname{area}(C)$?

The largest inscribed triangle of a convex set behaves similarly to the diameter, which can be interpreted as the largest inscribed 2-gon. The "isodiametric" theorem (which follows, for example, from the isoperimetric inequality) states that in the class of all plane convex sets of diameter at most one, the circle of unit diameter has the largest area. Its analogue for inscribed triangles is equivalent to Blaschke's [Bl17] result mentioned above: in the class of all plane convex sets C that do not contain an inscribed triangle of area larger than one, ellipses have the maximum area. As in Section 11.1, we may restrict our attention to the case that C is a polygon. Determine the maximum area of a convex k-gon C_k that does not contain an inscribed triangle of area larger than one. This problem was studied and solved for $k \leq 6$ by Fleischer et al. [FlM*92]. The extremal quadrilaterals and pentagons turned out to be regular, but the extremal hexagon is not.

The corresponding questions for the perimeter were answered by Eggleston [Eg53] for triangles, and by Schneider [Sc71] for all n. In the class of
all convex sets of a given perimeter, the circle maximizes the length of the
smallest circumscribing n-gon and minimizes the length of the largest inscribed n-gon. In the case of circumscribed n-gons the circle is the only
extremal set. However, for inscribed n-gons, the uniqueness of the extremal
sets is still unclear for odd $n \geq 23$ and for even $n \geq 44$.

For the diameter instead of the perimeter, the analogous question for
inscribed n-gons is trivial: every plane convex body contains an inscribed
n-gon with the same diameter. Finding the minimum diameter of an n-gon
circumscribed about a convex set of diameter one is an open problem for
every $n \geq 4$. For $n = 3$, Gale's work [Ga53] implies that every C permits
a circumscribing triangle P_3 with $\mathrm{diam}(P_3) \leq \sqrt{3}\,\mathrm{diam}(C)$, which is sharp
for the circle.

Problem 2   *What is the smallest number c_n^{outer} such that every plane
              convex set C permits a circumscribing n-gon P_n with*
              $$\mathrm{diam}(P_n) \leq c_n^{\mathrm{outer}}\,\mathrm{diam}(C)?$$

In measuring the size of a set by its inradius, it is obvious that every
convex set C permits a circumscribing triangle (which possibly degenerates
to a parallel strip) with the same inradius. But what is the largest inradius
of an inscribed n-gon? If C is a circle, then this number is $\cos(\pi/n)$ times
the radius of C. Thus, any convex set C has an inscribed n-gon P_n with
$\mathrm{inradius}(P_n) \geq \cos(\pi/n)\,\mathrm{inradius}(C)$, and this bound cannot be improved.

If we replace the inradius by the circumradius, the resulting questions
remain easy. Every plane convex set C contains a triangle Δ with circumradius $\mathrm{radius}(\Delta) = \mathrm{radius}(C)$, and is contained in an n-gon P_n with
$\mathrm{radius}(P_n) \leq \frac{1}{\cos\frac{\pi}{n}}\,\mathrm{radius}(C)$.

For the scenario in which sets are measured by their widths, it is again
obvious that for every C and every $\varepsilon > 0$ there is a triangle P_3 with $C \subset P_3$
and $\mathrm{width}(P_3) \leq (1 + \varepsilon)\,\mathrm{width}(C)$. The dual problem is open.

Problem 3   *What is the largest number c_n^{inner} such that every plane
              convex set C has an inscribed n-gon P_n with $\mathrm{width}(P_n) \geq
              c_n^{\mathrm{inner}}\,\mathrm{width}(C)?$*

For $n = 3$, Gritzmann and Lassak [GrL89] conjectured that regular pentagons are extremal.

Some of the questions concerning inscribed polygons of a convex set
C remain interesting if we restrict our attention to convex polygons C and
insist that the inscribed polygon P_n must be a "subpolygon" of C; i.e., its
vertices must be selected from the vertex set of C. Note that one can always

find an inscribed polygon of maximum area that is a subpolygon of C in this sense, but the analogous statement is false for many other measures.

It is not hard to see that among all $(n+1)$-gons circumscribed about a unit disk, the regular one minimizes the inradius of the largest sub-n-gon. If we measure a convex polygon containing the origin O by the radius of its largest inscribed disk centered at O, we obtain a similar result. Bárány and Heppes [BáH94] and Brass [Br97] proved that any convex polygon with vertex set X that contains a unit disk around 0 has a subpolygon with vertex set $Y \subseteq X$, $|Y| \leq n$, containing a disk of radius $\frac{\cos(2\pi/(n+1))}{\cos(\pi/(n+1))}$ around O. Equality holds for a regular $(n+1)$-gon. In higher dimensions, several analogous results (quantitative Steinitz theorems) were established by Bárány, Katchalski, and Pach [BáKP82].

A similar question for the width rather than the inradius was raised by Tsintsifas [Ts76].

Set of points of width 1
Three-point subset of width $\frac{\sqrt{5}-1}{2}$

Problem 4 *(Tsintsifas [Ts76]) What is the largest number $c_n^{\overset{\text{width}}{\text{sub}}}$ such that every convex polygon with vertex set X has a sub-n-gon with vertex set $Y \subseteq X$ whose width satisfies* $\text{width}(\text{conv}(Y)) \geq c_n^{\overset{\text{width}}{\text{sub}}} \text{width}(\text{conv}(X))$?

For subtriangles ($n = 3$), the regular pentagon is conjectured to be extremal again.

As mentioned at the beginning of this section, similar questions can be asked for other types of inscribed and circumscribing objects, including ellipses (the so-called Löwner–John ellipses) [BeFT99], [La03], affinely regular hexagons [La02], axially symmetric sets [La02], and rectangles [Ra52], [Ko57], [La93a]. Every plane convex set C contains a rectangle whose area is at least half the area of C and is contained in a rectangle whose area is at most twice the area of C. These bounds are best possible.

A completely different way to approximate a plane convex body C_1 by an inscribed closed set $C_2 \subset C_1$ was proposed by Fekete. If there is a point $x \in C_1 \setminus C_2$ from which C_2 can be seen spanning a "small" angle, then C_2 is considered a "bad" approximation for C_1. We define for $C_2 \neq C_1$ the *aspect ratio* by

$$\mu^{\overset{\text{aspect}}{\text{ratio}}}(C_1, C_2) := \inf_{x \in C_1 \setminus C_2} \sup_{y_1, y_2 \in C_2} |\sphericalangle y_1 x y_2|.$$

For $C_2 = C_1$, the aspect ratio is defined to be π. We want to determine a convex n-gon $P_n \subset C$ maximizing $\mu^{\overset{\text{aspect}}{\text{ratio}}}(C, P_n)$. It is important to insist in this definition that $x \notin C_2$. Otherwise, the quality of the approximation at x is determined by the interior angle of C_1 at x if x lies on the boundary of C_2, and is equal to $\pi/2$ if x lies in the interior of C_2. With respect to the *aspect ratio*, the best approximation of a convex polygon C_1 is always realized by a subpolygon $C_2 \subset C_1$, and the worst aspect ratio is given by an edge $\overline{y_1 y_2}$ of this subpolygon and a vertex x of the polygon cut off by that edge. It is easy to determine the best approximation by triangles: every convex set C contains a triangle $P_3 \subset C$ such that $\mu^{\overset{\text{aspect}}{\text{ratio}}}(C, P_3) \geq \frac{\pi}{2}$, and the square does not admit any better approximation. One just takes the smallest circle containing C and selects at most three of the touching points that contain the center of the circle in their convex hull. The best approximation by n-gons for $n \geq 4$ is still not determined, although there is a simple lower bound of $\frac{n-2}{n}\pi$.

Problem 5   *(Fekete) Is it true that for every convex set C there is an n-gon $P_n \subset C$ with $\mu^{\overset{\text{aspect}}{\text{ratio}}}(C, P_n) \geq \frac{n-1}{n+1}\pi$?*

The above questions can also be asked in higher dimensions. For most of the direct analogues of the questions concerning polygons (i.e., for problems on the size of the largest inscribed or smallest circumscribed polyhedra with a given number of vertices or faces) we have only asymptotic results. One of the few exceptions is Macbeath's theorem [Ma51], which is a generalization of Sas's result in the plane. For any d and $n \geq d+1$, among all d-dimensional convex bodies of volume one, the only extremal sets that minimize the volume of the largest inscribed polytopes with n vertices are the ellipsoids. There are many results concerning the simultaneous approximation of a convex body by a pair of inscribed and circumscribed sets that are homothetic to each other. For instance, the results on Löwner–John ellipsoids, the analogous statements for parallelotopes [La93b], [La91], [ChS67], and the quantitative Steinitz theorems [BáKP82] belong to this category. In all of these examples, the size of a set is measured by its volume [Sc73], and in the case of homothetic pairs we use the scale factor.

Problem 6   *[BáKP82], [KiMY92] Given $d \geq 3$, what is the largest number c_d such that any set X in $\mathbb{R}^d$ whose convex hull contains a unit ball centered at the origin has a $2d$-element subset Y whose convex hull contains a ball of radius c_d centered at the origin?*

As we have seen before, the corresponding problem in the plane has been solved, and the vertex set of a regular pentagon is extremal [BáH94], [Br97].

Gritzmann and Lassak [GrL89] studied what happens if we want to select a k-element subset in $\mathbb{R}^d$ whose width is as large as possible.

Problem 7   Let $k \geq d + 1$. What is the largest number $c_d(k)$ such that every set X in $\mathbb{R}^d$ has a k-element subset Y with width$(Y) \geq c_d(k)$ width(X)?

[BáH94]   I. BÁRÁNY, A. HEPPES: On the exact constant in the quantitative Steinitz theorem in the plane, *Discrete Comput. Geom.* **12** (1994) 387–398.

[BáKP82]   I. BÁRÁNY, M. KATCHALSKI, J. PACH: Quantitative Helly-type theorems, *Proc. Amer. Math. Soc.* **86** (1982) 109–114.

[BeFT99]   A. BEZDEK, F. FODOR, I. TALATA: Applications of inscribed affine regular polygons in convex discs, in: *Proc. Int. Sci. Conf. Mathematics, Vol. 2* (Žilina, Slovakia, 1998), V. Bálint, ed., EDIS, Žilina University Publisher, (1999) 19–27.

[Bl17]   W. BLASCHKE: Über affine Geometrie III: Eine Minimumeigenschaft der Ellipse, *Ber. Ver. Sächs. Akad. Wiss. Leipzig, Math.-Nat. Klasse* **69** (1917) 3–12.

[Br97]   P. BRASS: On the quantitative Steinitz theorem in the plane, *Discrete Comput. Geom.* **17** (1997) 111–117.

[BrL01]   P. BRASS, M. LASSAK: Problems on approximation by triangles, *Geombinatorics* **10** (2001) 103–115.

[Ch73]   G.D. CHAKERIAN: Minimum area of circumscribed polygons, *Elemente Math.* **28** (1973) 108–111.

[ChL71]   G.D. CHAKERIAN, L.H. LANGE: Geometric extremum problems, *Math. Mag.* **44** (1971) 57–69.

[ChS67]   G.D. CHAKERIAN, S.K. STEIN: Some intersection properties of convex bodies, *Proc. Amer. Math. Soc.* **18** (1967) 109–112.

[Eg53]   H.G. EGGLESTON: On triangles circumscribing plane convex sets, *J. London Math. Soc.* **28** (1953) 36–46.

[FlM*92]   R. FLEISCHER, K. MEHLHORN, G. ROTE, E. WELZL, C. YAP: Simultaneous inner and outer approximation of shapes, *Algorithmica* **8** (1992) 365–389.

[Ga53]   D. GALE: On inscribing n-dimensional sets in a regular n-simplex, *Proc. Amer. Math. Soc.* **4** (1953) 222–225.

[GrL89]   P. GRITZMANN, M. LASSAK: Estimates for the minimal width of polytopes incribed in convex bodies, *Discrete Comput. Geom.* **4** (1989) 627–635.

[Gr18]   W. GROSS: Über affine Geometrie XIII: Eine Minimumeigenschaft der Ellipse und des Ellipsoids, *Ber. Ver. Sächs. Akad. Wiss. Leipzig, Math.-Nat. Klasse* **70** (1918) 38–54.

[Ho51]   J. Hodges: An extremal problem in geometry, *J. London Math. Soc.* **26** (1951) 311–312.

[KiMY92]   D. Kirkpatrick, B. Mishra, C. Yap: Quantitative Steinitz's theorems with applications to multifingered grasping, *Discrete Comput. Geom.* **7** (1992) 295–318.

[Ko57]   A. Kosiński: A proof of the Auerbach–Banach–Mazur–Ulam theorem on convex bodies, *Colloq. Math.* **17** (1957) 216–218.

[La03]   M. Lassak: Affine-regular hexagons of extreme areas inscribed in a centrally symmetric convex body, *Adv. Geom.* **3** (2003) 45–51.

[La02]   M. Lassak: Approximation of convex bodies by axially symmetric bodies, *Proc. Amer. Math. Soc.* **130** (2002) 3075–3084 (electronic), Erratum in **131** (2003) p. 2301 (electronic).

[La93a]   M. Lassak: Approximation of convex bodies by rectangles, *Geometriae Dedicata* **47** (1993) 111–117.

[La93b]   M. Lassak: Estimation of the volume of parallelotopes contained in convex bodies, *Bull. Polish Acad. Sci. Math.* **41** (1993) 349–353.

[La91]   M. Lassak: Approximation of convex bodies by parallelotopes, *Bull. Polish Acad. Sci. Math.* **39** (1991) 219–223.

[Ma51]   A. Macbeath: An extremal property of the hypersphere, *Proc. Cambridge Philosophical Soc.* **47** (1951) 245–247.

[Ra52]   K. Radziszewski: Sur une problème extrémal relatif aux figures inscrites et circonscrites aux figures convexes, *Ann. Univ. Mariae Curie-Sklodowska, Sect. A* **6** (1952) 5–18.

[Sa39]   E. Sas: Über eine Extremaleigenschaft der Ellipse, *Compositio Math.* **6** (1939) 468–470.

[Sc73]   R. Schneider: A characteristic extremal property of simplices, *Proc. Amer. Math. Soc.* **40** (1973) 247–249.

[Sc71]   R. Schneider: Zwei Extremalaufgaben für konvexe Bereiche, *Acta Math. Acad. Sci. Hungar.* **22** (1971/72) 379–383.

[Ts76]   G. Tsintsifas: The strip (Problem 5973), *Amer. Math. Monthly* **83** (1976) p. 142.

11.4 Universal Covers

Perhaps the oldest question on universal covers was asked by Lebesgue: what is the smallest area of a set C that can be used to cover by a rigid motion any set of diameter one in the plane? Another famous problem of this type is L. Moser's "worm problem": what is the smallest area of a set that can be used to cover any curve of length one? (What is the smallest blanket mother worm can use to cover her little baby?) Both problems have stimulated a lot of research, and both are still unsolved. Before discussing them in detail we formulate a more general question.

A *universal cover* for a given class $\mathcal{X}$ of objects is a set C that contains a congruent copy of every element $X \in \mathcal{X}$. Given any parameter for measuring the "size" of a set, we may wish to find a smallest universal cover belonging to a fixed family of sets $\mathcal{Y}$. Thus, for any choice of this parameter, for any families $\mathcal{X}$ and $\mathcal{Y}$ of objects that we want to cover and by which we want to cover, and for any type of motions we are allowed to use (congruence, translation), we obtain a different question.

Problem 1   *(Lebesgue, 1914; see [Pá20]) What is the smallest area of a convex set in the plane that contains a congruent copy of every planar set of unit diameter?*

Lebesgue's problem was first studied by Pál [Pá20], who found that $0.8257 \leq \text{area}(min\ cover) \leq 0.8454$. The upper bound has been successively reduced to 0.844 by various constructions [Sp36], [Eg63], [Ha75], [Du80], [Ha81], [Ha92b], but the progress since Sprague's work [Sp36] has been almost infinitesimal. Pál's original lower bound is the area of the smallest convex set that contains a unit equilateral triangle and a unit circle. It had not been improved for more than seventy years, when Elekes [El94] increased it to 0.8271, the area of the smallest convex set that contains all regular 3^k-gons of unit diameter $(k = 1, 2, \ldots)$. The lower bound was further increased to 0.832 by Brass and Sharifi [BrS05]; this number is the area of the smallest convex set that contains a circle, an equilateral triangle, and a regular pentagon of unit diameter. In this way, one could continue to obtain better lower bounds by using additional sets of diameter one that must be contained in a universal cover. However, since each additional set allows another independent motion, the parameter space becomes too large.

In these results, the covering set was always assumed to be convex, but the same questions are quite as natural without this assumption, and they may lead to different answers. Duff [Du80] constructed a nonconvex universal cover C smaller than any known convex cover. But the lower bounds also decrease without the assumption of convexity of the universal cover.

No work appears to have been done on the following natural variant of Lebesgue's question.

Problem 2   *What is the smallest diameter of a set in the plane that contains a congruent copy of every set of unit diameter?*

Here the smallest universal cover can be chosen to be convex. The unit disk and unit equilateral triangle together force a lower bound of $\frac{1}{2} + \frac{1}{\sqrt{3}} = 1.077$.

The family of all sets of area one in the plane does not admit a universal cover of finite area. Indeed, it is easy to see that for covering all $r \times \frac{1}{r}$ rectangles, we already need a set of infinite area. However, there exists a bounded universal cover for all sets of unit perimeter, because the diameter of any such set is at most $\frac{1}{2}$. A slight modification of this problem, which can be regarded as the variant of Moser's worm problem for "closed worms," is the following.

Problem 3   *What is the smallest area of a set in the plane that contains a congruent copy of every closed curve of length at most one?*

It is not clear whether the answer to the last question would change if we required the covering set to be a topological disk (see [Ko86] and [We00] for related results). A closed curve can be replaced by the boundary of its convex hull without increasing its length, so under the above assumption we may reformulate the last question as follows. What is the minimum area of a universal cover C for all plane convex sets of unit perimeter? Wetzel [We73] proved that for convex C this number is in the interval $[0.155, 0.159]$. It would also be interesting to find the minimum perimeter of a universal cover for this class. It may be easier to solve these problems for "closed worms" than to answer Moser's original question for "open" ones.

Problem 4   *(L. Moser) What is the smallest area of a set in the plane that contains a congruent copy of every open curve of length at most one?*

Wetzel [We73] gave a lower bound of 0.2194 for convex covers. An elegant unpublished argument of Meir shows that a closed semidisk of radius $\frac{1}{2}$ is a universal cover for all curves of length one. Its area is $\frac{\pi}{8} \approx 0.3927$. For convex covers, this upper bound has been reduced in several steps [We73], [PoG73], [GeP74], [Ha92a], [NoPL92]. The best known upper bound, 0.2753, is due to Norwood and Poole [NoP03]. It was further improved to 0.246 by Hansen [Ha92a], using a nonconvex cover.

One can also make other restrictions on the family $\mathcal{X}$ of sets to be covered, or on the family $\mathcal{Y}$ of sets that can be used as covers. For any triangle

Δ, Wetzel [We70], [We72] determined the smallest triangle Δ' similar to Δ that is a universal cover for all closed curves of length one: the perimeter of the incircle of Δ' is 1. For any integer $n \geq 4$, Chakerian and Logothetti [ChL88] found the smallest regular n-gon that covers every planar set of diameter one. Füredi and Wetzel [FüW00] and Kovalev [Ko83] determined the (unique) universal convex covers of smallest area for all triangles of unit perimeter and unit diameter, respectively. Universal covers of finite sets were studied by Wastun [Wa88].

The oldest result on universal covers using covering sets of a specific type was established by Jung [Ju01], [Ju10], [BlW41]. It states that any d-dimensional set of unit diameter is contained in a ball of diameter $\sqrt{\frac{2d}{d+1}}$, circumscribed about a regular simplex of edge length one. A ball of this size is often called a *Jung ball*. The analogous problem, asking for the smallest regular simplex that covers all sets of unit diameter, was solved by Gale [Ga53]: any d-dimensional set of unit diameter is contained in a regular simplex of edge length $\sqrt{\frac{d(d+1)}{2}}$. This is the simplex circumscribed about a ball of unit diameter, so Gale's result is also best possible. For the case of cubes, it is easy to see that every set of unit diameter is contained in a unit cube, and one can even choose this cube to be axis-aligned. The analogous result for regular octahedra whose opposite faces are at distance one was stated by Gale [Ga53] without a proof. A stronger statement occurs as a lemma in the proofs of the three-dimensional Borsuk partition conjecture [HeR56], [He57], [Gr63], [Gr57]. The same problem for higher-dimensional crosspolytopes remains open.

Problem 5   *Is it true that a d-dimensional crosspolytope circumscribed about a ball of unit diameter contains a congruent copy of every set of unit diameter?*

Makai (personal communication) pointed out that it follows from a heuristic counting argument comparing "degrees of freedoms," based on a result of Makeev [Ma82], that the answer to the last question is expected to be negative in dimensions d larger than five.

Consider a d-dimensional regular simplex of edge length one. For each edge e of the simplex, take the slab bounded by the hyperplanes that are perpendicular to e and pass through the endpoints of e. Let U_d be the intersection of these $\binom{d+1}{2}$ slabs. In the plane, U_2 is a regular hexagon of side length $\frac{1}{\sqrt{3}}$ and, according to Pál's theorem [Pá20] that can be proved by a straightforward continuity argument, it is a universal cover for all sets of unit diameter. In three-dimensional space, Makeev [Ma97], Hausel, Makai, and Szűcs [HaM*97], [HaM*00], and G. Kuperberg [Ku99] proved independently and at about the same time that U_3 contains a congruent copy of every set of unit diameter. Actually, U_3 is a rhombo-dodecahedron, which is the dual of a cubo-octahedron, the convex hull of the midpoints of the edges of a cube.

Conjecture 6   *(Makeev [Ma94]) For any $d > 3$, the set U_d contains a congruent copy of every d-dimensional set of unit diameter.*

There are several papers about covering "worms" by balls [RuS48], [Ni71], [We71], [ChK73] and by hypercubes [ScW72], [ChK73]. It is an immediate consequence of the triangle inequality that every open curve of unit length is contained in a ball of radius $\frac{1}{2}$ around the curve's midpoint. For closed curves one needs to do a bit more work to show that every closed curve of unit length is contained in a ball of radius $\frac{1}{4}$. This result, which holds in every dimension, is obviously best possible. It was perhaps first proved by Segre [Se34] and has been rediscovered several times [RuS48], [Ni71], [We71], [ChK73]. For d-dimensional cubes in the place of balls, Schaer and Wetzel [ScW72] showed that every open curve of unit length is contained in a cube of diagonal one. This result is again best possible.

Without restricting the type of the covering set, very little is known about these problems in higher dimensions. For the minimum volume of a universal cover for all open curves of unit length, some bounds have been obtained by Lindström [Li97] and by Håstad et al. [HåLW01]. Weissbach [We87] obtained a universal cover for all sets of unit diameter in $\mathbb{R}^d$, by taking the intersection of $d+2$ parallel slabs of width one containing a ball of unit diameter. Eggleston [Eg63] proved that for $d \geq 3$, there exist inclusion-minimal closed bounded universal covers of arbitrarily large diameter for all d-dimensional sets of unit diameter (i.e., no proper closed convex subset of them is a universal cover). For $d = 2$, he and Grünbaum [Gr63] announced that the diameter of any such set is smaller than three, and gave an explicit construction for an inclusion-minimal closed bounded universal cover. Some lower bounds on various parameters of d-dimensional centrally symmetric covers were given in [We98].

All of the above questions can also be raised for translative covers, that is, for sets that contain a translate of every element of a given class of sets $\mathcal{X}$. These variants are usually easier, and many of them have been completely solved. The oldest theorem of this type was again found by Pál [Pá21]: the minimum area of a universal translative cover for all unit segments (in fact, for all open curves of unit length) in the plane is $\frac{1}{\sqrt{3}}$, attained by an equilateral triangle of height one. The direct analogue of Lebesgue's original problem is the following.

Problem 7   *What is the smallest area of a set in the plane that contains a translate of every planar set of unit diameter?*

K. Bezdek and Connelly [BeC97] proved that the perimeter of a translative cover for all sets of unit diameter is minimized by a disk of diameter

$\frac{2}{\sqrt{3}}$. In any dimension, the universal translative cover of minimum mean width for all sets of unit diameter is a ball (the Jung ball) [BeC98]. In the plane, every set of constant width $\frac{1}{2}$ is a universal translative cover for all sets of unit perimeter [BeC89], and every such set minimizes the perimeter of a universal translative cover for the same class of sets. Furthermore, any convex set whose minimum width is at least $\frac{1}{\sqrt{3}}$ is a universal translative cover for the same class, and this bound is also tight [Be90].

Problem 8   *(Wetzel [We73]) What is the smallest area of a set in the plane that contains a translate of every set of unit perimeter?*

For convex covers, Wetzel proved the lower bound 0.155. The best known construction, due to Mallée [Mal94], gives the upper bound 0.164.

We finish this section with an unusual covering problem. A *carpenter's ruler* is divided into segments of different lengths that are hinged where the pieces meet, which makes it possible to fold it. Hopcroft, Joseph, and Whitesides posed the problem of determining the smallest one-dimensional case (shortest interval) into which a given ruler fits when folded. This problem, which is known to be computationally hard, motivated the following question.

Problem 9   *(Călinescu and Dumitrescu [CăD05]) What is the smallest area of a plane convex set of unit diameter that contains a folded copy of any ruler whose pieces are of length at most one?*

It was shown in [CăD05] that this number c satisfies

$$\frac{3}{8} \le c \le \frac{\pi}{3} - \frac{\sqrt{3}}{4}.$$

Note that the number of pieces the ruler is divided into, and hence the total length of the ruler, plays no role in this question. In a folded position of the ruler the pieces may cross each other.

[Be90]   K. BEZDEK: On certain translation covers, *Note Mat.* **10** (1990) 279–286.

[BeC98]   K. BEZDEK, R. CONNELLY: The minimum mean width translation cover for sets of diameter one, *Beiträge Algebra Geom.* **39** (1998) 473–479.

[BeC97]   K. BEZDEK, R. CONNELLY: Minimal translation covers for sets of diameter 1, *Period. Math. Hungar.* **34** (1997) 23–27.

[BeC89]   K. Bezdek, R. Connelly: Covering curves by translates of
            a convex set, *Amer. Math. Monthly* **96** (1989) 789–806.

[BlW41]   L.M. Blumenthal, G.E. Wahlin: On the spherical surface
            of smallest radius enclosing a bounded subset of n-dimensional
            space, *Bull. Amer. Math. Soc.* **47** (1941) 771–777.

[BrS05]   P. Brass, M. Sharifi: A lower bound for Lebesgue's uni-
            versal cover problem, *Internat. J. Comput. Geom. Appl.*, to
            appear.

[CăD05]   G. Călinescu, A. Dumitrescu: The carpenter's ruler fold-
            ing problem, in: *Current Trends in Combinatorial and Com-
            putational Geometry*, J.E. Goodman et al., eds., Cambridge
            Univ. Press, to appear.

[ChK73]   G.D. Chakerian, M.S. Klamkin: Minimal covers for closed
            curves, *Math. Mag.* **46** (1973) 55–61.

[ChL88]   G.D. Chakerian, D. Logothetti: Minimal regular poly-
            gons serving as universal covers in $\mathbb{R}^2$, *Geometriae Dedicata*
            26 (1988) 281–297.

[Du80]   G.F.D. Duff: A smaller universal cover for sets of unit
            diameter, *C. R. Math. Rep. Acad. Sci. Canada* **2** (1980/81)
            37–42.

[Eg63]   H.G. Eggleston: Minimal universal covers in E^n, *Israel J.
            Math.* **1** (1963) 149–155.

[El94]   G. Elekes: Generalized breadths, Cantor-type arrange-
            ments and the least area UCC, *Discrete Comput. Geom.* **12**
            (1994) 439–449.

[FüW00]   Z. Füredi, J.E. Wetzel: The smallest convex cover for tri-
            angles of perimeter two, *Geometriae Dedicata* **81** (2000) 285–
            293.

[Ga53]   D. Gale: On inscribing n-dimensional sets in a regular n-
            simplex, *Proc. Amer. Math. Soc.* **4** (1953) 222–225.

[GeP74]   J. Gerriets, G. Poole: Convex regions which cover arcs
            of constant length, *Amer. Math. Monthly* **81** (1974) 36–41.

[Gr63]   B. Grünbaum: Borsuk's problem and related questions, in:
            Convexity, V. Klee, ed., *Proc. Sympos. Pure Math.* **7**, Amer.
            Math. Soc., 1963, 271–284.

[Gr57]   B. Grünbaum: A simple proof of Borsuk's conjecture in
            three dimensions, *Proc. Cambridge Philos. Soc.*, **53** (1957)
            776–778.

[Ha92a]   H.C. Hansen: The worm problem (in Danish), *Normat* **40**
            (1992) 119–123, p. 143.

[**Ha92b**]   H.C. Hansen: Small universal covers for sets of unit diameter, *Geometriae Dedicata* **42** (1992) 205–213.

[**Ha81**]   H.C. Hansen: Towards the minimal universal cover (in Danish), *Normat* **29** (1981) 115–119, p. 148.

[**Ha75**]   H.C. Hansen: A small universal cover of figures of unit diameter, *Geometriae Dedicata* **4** (1975) 165–172.

[**HaM*00**]   T. Hausel, E. Makai Jr., A. Szűcs: Inscribing cubes and covering by rhombic dodecahedra via equivariant topology, *Mathematika* **47** (2000) 371–397.

[**HaM*97**]   T. Hausel, E. Makai Jr., A. Szűcs: Polyhedra inscribed and circumscribed to convex bodies, in: *Proc. Third Internat. Workshop on Diff. Geom. and its Appls. and First German-Romanian Seminar on Geom.* (Sibiu, Romania, 1997), *General Math.* **5** (1997) 183–190.

[**HåLW01**]   J. Håstad, S. Linusson, J. Wästlund: A smaller sleeping bag for a baby snake, *Discrete Comput. Geom.* **26** (2001) 173–181.

[**He57**]   A. Heppes: On the partitioning of three-dimensional point sets into sets of smaller diameter (in Hungarian), *Magyar Tud. Akad. Mat. Fiz. Oszt. Közl.* **7** (1957) 413–416.

[**HeR56**]   A. Heppes, P. Révész: Zum Borsukschen Zerteilungsproblem, *Acta Math. Sci. Hungar.* **7** (1956) 159–162.

[**Ju10**]   H.W.E. Jung: Über den kleinsten Kreis, der eine ebene Figur einschließt, *J. Reine Angew. Math.* **137** (1910) 310–313.

[**Ju01**]   H.W.E. Jung: Über die kleinste Kugel, die eine räumliche Figur einschließt, *J. Reine Angew. Math.* **123** (1901) 241–257.

[**Ko86**]   M.D. Kovalev: The smallest Lebesgue covering exists (in Russian), *Mat. Zametki* **40** (1986) 401–406, translation in *Math. Notes* **40** (1986) 736–739.

[**Ko83**]   M.D. Kovalev: A minimal convex covering for triangles (in Russian), *Ukrain. Geom. Sb.* **26** (1983) 63–68.

[**Ku99**]   G. Kuperberg: Circumscribing constant-width bodies with polytopes, *New York J. Math.* **5** (1999) 91-100.

[**Li97**]   B. Lindström: A sleeping bag for a baby snake, *Math. Gaz.* **81** (1997) 451–452.

[**Ma97**]   V.V. Makeev: On affine images of a rhombo-dodecahedron circumscribed about a three-dimensional convex body (in Russian), *Zap. Nauchn. Sem. S.-Peterburg. Otdel. Mat. Inst. Stek-*

lov (POMI) **246** (1997), *Geom. i Topol.* **2**, 191-195, 200; translation in *J. Math. Sci. (New York)* **100** (2000) 2307–2309.

[Ma94]   V.V. MAKEEV: Inscribed and circumscribed polygons of a convex body (in Russian), *Mat. Zametki* **55** (1994) 128–130; translation in *Math. Notes* **55** (1994) 423–425.

[Ma82]   V.V. MAKEEV: Dimension restrictions in problems of combinatorial geometry (in Russian), *Sibirsk. Mat. Zh.* **23** (1982), 197–201, p. 222.

[Mal94]   H. MALLÉE: Translationsdeckel für geschlossene Kurven, *Arch. Math. (Basel)* **62** (1994) 569–576.

[Ni71]   J.C.C. NITSCHE: The smallest sphere containing a rectifiable curve, *Amer. Math. Monthly* **78** (1971) 881–882.

[NoP03]   R. NORWOOD, G. POOLE: An improved upper bound for Leo Moser's worm problem, *Discrete Comput. Geom.* **29** (2003) 409–417.

[NoPL92] R. NORWOOD, G. POOLE, M. LAIDACKER: The worm problem of Leo Moser, *Discrete Comput. Geom.* **7** (1992) 153–162.

[Pá20]   J. PÁL: Über ein elementares Variationsproblem, *Math.-fys. Medd., Danske Vid. Selsk.* **3** (1920) no. 2, 35 p.

[Pá21]   J. PÁL: Ein Minimumproblem für Ovale, *Math. Annalen* **83** (1921) 311–319.

[PoG73]   G. POOLE, J. GERRIETS: Minimum covers for arcs of constant length, *Bull. Amer. Math. Soc.* **79** (1973) 462–463.

[RuS48]   H. RUTISHAUSER, H. SAMELSON: Sur le rayon d'une sphère dont la surface contient une courbe fermée, *C. R. Acad. Sci. Paris* **227** (1948) 755–757.

[ScW72]   J. SCHAER, J.E. WETZEL: Boxes for curves of constant length, *Israel J. Math.* **12** (1972) 257–265.

[Se34]   B. SEGRE: Sui circoli geodetici de una superficie a curvatura totale constante che contengono nell'interno un linea assegnata, *Boll. Un. Mat. Ital.* **13** (1934) 279–283.

[Sp36]   R. SPRAGUE: Über ein elementares Variationsproblem, *Mat. Tidsskr. Ser. B* (1936) 96–98.

[Wa88]   C.G. WASTUN: Universal covers of finite sets, *J. Geometry* **32** (1988) 192–201.

[We00]   B. WEISSBACH: On a covering problem in the plane, *Beiträge Algebra Geom.* **41** (2000) 425–426.

[We98]   B. WEISSBACH: Quermaßintegrale zentralsymmetrischer Deckel, *Geometriae Dedicata* **69** (1998) 113–120.

[**We87**]   B. WEISSBACH: Polyhedral covers, in: *Intuitive Geometry* (Siófok, 1985), K. Böröczky et al., eds., *Colloq. Math. Soc. János Bolyai* **48** North-Holland, 1987, 639–646.

[**We73**]   J.E. WETZEL: Sectorial covers for curves of constant length, *Canadian Math. Bull.* **16** (1973) 367–375.

[**We72**]   J.E. WETZEL: On Moser's problem of accommodating closed curves in triangles, *Elemente Math.* **27** (1972) 35–36.

[**We71**]   J.E. WETZEL: Covering balls for curves of constant length, *Enseignement Math.* (2) **17** (1971) 275–277.

[**We70**]   J.E. WETZEL: Triangular covers for closed curves of constant length, *Elemente Math.* **25** (1970) 78–82.

11.5 Approximation Problems

The approximation of planar convex sets by polygons is a classical subject that was already studied by Blaschke [Bl17] in 1917. There are many important results on the best approximation by n-gons and, in particular, on the asymptotic behavior of the error term as $n \to \infty$. Nevertheless, many elementary problems are unsolved. For various aspects of the subject see the surveys of Gruber [Gr93], Alt and Guibas [AlG99], and Brass and Lassak [BrL01].

To specify an approximation problem, one has to choose a class $\mathcal{X}$ of objects to be approximated, a class $\mathcal{Y}$ of sets by which we approximate, and some "distance function" $d(X,Y)$ between the elements of $X \in \mathcal{X}$ and $Y \in \mathcal{Y}$ for measuring the quality of the approximation. Occasionally we impose special restrictions on the approximating objects $Y \in \mathcal{Y}$. For example, in the case of inner and outer approximation, they should be contained in or should contain the set $X \in \mathcal{X}$, respectively.

In the classical versions of inner (outer) approximation of plane convex sets X by inscribed (circumscribing) convex n-gons Y, we measure the quality of the approximation by area$(X \setminus Y)$ (respectively area$(Y \setminus X)$). As we have seen in Section 11.3, from the point of view of inner approximation by convex n-gons, the worst possible plane convex bodies of unit area are ellipses. This is not true for outer approximation. For example, from the point of view of outer approximation by (circumscribing) triangles, the worst possible convex bodies of unit area are not ellipses, but parallelograms.

It is natural to extend these measures to any convex n-gon Y, not necessarily contained in or containing X, and try to minimize the distance between X and Y with respect to the *symmetric difference metric*

$$d^{\mathrm{diff}}_{}{}^{\mathrm{symm}} (X,Y) := \mathrm{area}(X \setminus Y) + \mathrm{area}(Y \setminus X).$$

This metric is often applied in the theory of packing and covering.

The problem of finding the optimal approximation of X by the best "unrestricted" n-gon P_n (whose vertices can be "anywhere" in the plane) is open.

Problem 1   *What is the smallest number δ_n such that for any planar convex set C with area$(C) = 1$, there is a convex n-gon P_n with $d^{\mathrm{diff}}{}^{\mathrm{symm}} (C, P_n) \leq \delta_n$? Is it true that ellipses are extremal?*

For any $\alpha, \beta \in (\frac{1}{2}, \infty)$ with $\frac{1}{\alpha} + \frac{1}{\beta} = 2$, one can consider the "asymmetric" difference metric

$$d^{\alpha,\beta}(X,Y) = \alpha \operatorname{area}(X \setminus Y) + \beta \operatorname{area}(Y \setminus X).$$

Let $\delta_n^{\alpha,\beta}$ denote the smallest number satisfying the condition in Problem 1 with respect to the metric $d^{\alpha,\beta}$. Clearly, we have $\delta_n^{1,1} = \delta_n$. According to G. Fejes Tóth (personal communication), if a sequence of pairs (α_k, β_k) tends to (α, β), then $\delta_n^{\alpha_k, \beta_k} \to \delta_n^{\alpha,\beta}$ holds for any fixed n, as $k \to \infty$. Moreover, under proper norming, the extremal configurations also change smoothly. In particular, for $n = 3$, the ellipse continuosly changes into a parallelogram as α moves from ∞ toward $\frac{1}{2}$. In view of the fact that approximation with respect to the symmetric difference metric is "half way" between inner and outer approximations, it would be quite surprising if the answer to the second question in Problem 1 were yes.

A particularly useful property of the symmetric difference metric was established by Eggleston [Eg57]. If P_n is an optimal n-gon in the sense that the distance $d^{\text{diff}}_{\text{symm}}(C, P_n)$ is as small as possible, then we have

$$d^{\text{diff}}_{\text{symm}}(C, P_n) \leq \frac{1}{2} d^{\text{diff}}_{\text{symm}}(C, P_{n-1}) + \frac{1}{2} d^{\text{diff}}_{\text{symm}}(C, P_{n+1}),$$

for every $n \geq 4$. In other words, the sequence $d^{\text{diff}}_{\text{symm}}(C, P_n)$ $(n = 3, 4, \ldots)$ is convex. For the largest inscribed and smallest circumscribing n-gons the same assertion had been discovered earlier by Dowker [Do44]. G. Fejes Tóth [FeT77] proved that this statement also holds with respect to the weighted metrics $d^{\alpha,\beta}(X,Y)$ defined above. This nice property is not shared by most other metrics. For example, it fails for the Hausdorff metric.

The Hausdorff distance of two nonempty sets $X, Y \subset \mathbb{R}^d$ is defined as

$$d^{\text{dorff}}_{\text{Haus-}}(X,Y) := \max\left(\sup_{x \in X} \inf_{y \in Y} d(x,y) \, , \, \sup_{y \in Y} \inf_{x \in X} d(x,y) \right).$$

The Hausdorff metric is defined on the family of all nonempty compact sets, a much larger class than the family of convex sets. Since "real-life" geometric objects are usually not convex, the Hausdorff distance has been the preferred distance measure in computational geometry. One does not even have to use the Euclidean metric: given any metric on the underlying set, the Hausdorff distance defines a metric on the family of all nonempty compact sets. Some problems become computationally simpler if one uses a polyhedral metric like L_∞ as the underlying metric.

However, it is much harder to estimate the error term for Hausdorff approximation, especially if one wants to obtain more than just an asymptotic statement. In various settings, the questions on the optimal error bounds are open.

Problem 2   *(Popov) What is the smallest δ_n such that for any planar convex set C with unit perimeter there is a n-gon P_n with $d^{\text{Haus-}}_{\text{dorff}}(C, P_n) \leq \delta_n$? Is it true that the only extremal sets are the regular $(n + 1)$-gons?*

Problem 3   *(Popov) What is the smallest number δ_n^{inner} with the property that for any planar convex set C with unit perimeter there is an inscribed n-gon P_n with $d^{\text{Haus-}}_{\text{dorff}}(C, P_n) \leq \delta_n^{\text{inner}}$? Is it true that regular $(n + 1)$-gons are the only extremal sets?*

Problem 4   *(Popov) What is the smallest number δ_n^{outer} with the property that for any planar convex set C with unit perimeter there is a circumscribing n-gon P_n with $d^{\text{Haus-}}_{\text{dorff}}(C, P_n) \leq \delta_n^{\text{outer}}$? Is it true that regular $(n + 1)$-gons are the only extremal sets?*

Popov [Po68], [Po70] conjectured that in each of the last three problems, the convex sets hardest to approximate are the regular $(n + 1)$-gons. He proved that

$$\delta_n \approx \frac{\pi}{4n^2}, \quad \delta_n^{\text{inner}} \approx \frac{\pi}{2n^2}, \quad \delta_n^{\text{outer}} \approx \frac{\pi}{2n^2}.$$

Regarding Problems 3 and 4, it was shown by Georgiev [Ge84] and Ivanov [Iv73] that the conjecture holds in the class of all $(n + 1)$-gons. Kenderov [Ke73] studied a similar problem in which the quality of the approximation is measured by Hausdorff distance, where the underlying metric is the L_1-metric instead of the Euclidean one. In this setting, the convex set that is hardest to approximate by triangles is a correctly aligned square.

A subpolygon of a polygon P is a polygon whose vertices are chosen from the vertex set of P, in the original order. Thus, a subpolygon of a convex polygon P is always contained in P, and the approximation of convex polygons by subpolygons is a special case of inner approximation. For inner approximation of convex polygons P with respect to the symmetric difference metric, the best approximating n-gons can always be chosen to be subpolygons. However, this is not necessarily the case for inner approximation with respect to the Hausdorff metric. In this setting, typically some of the vertices of the best n-gons do not belong to the vertex set of P. It is a convenient feature of approximation by subpolygons that to find an optimal solution, it is sufficient to consider subsets of a finite set (the vertex set of P), instead of free placement of the vertices. Brass [Br00], [BrL01] made a strong Popov-type conjecture concerning a large group of subpolygon approximation problems.

Conjecture 5 *(Brass [Br00]) Let $\mathcal{P}$ be any class of polygons that is closed under taking subpolygons. If $\mathcal{P}$ has an element that is hardest to approximate by its sub-n-gons with respect to the Hausdorff metric, then one can also find an $(n+1)$-gon with this property.*

The class $\mathcal{P}$ could consist of, e.g., all polygons of perimeter at most one, or diameter at most one, or we can impose any other "normalizing" restriction that prevents the approximation quality becoming arbitrarily bad by blowing up a polygon. It is known that in any class meeting the above requirements, among the elements hardest to approximate by sub-n-gons there is always an m-gon with $m \equiv 1 \bmod n$ [Br00].

Observe that it is sufficient to prove the above conjecture for classes $\mathcal{P}$ consisting of all subpolygons of a given polygon (namely, all subpolygons of an element of the class that is hardest to approximate).

There is another class of interesting geometric approximation problems combining inner and outer approximations that are related to the *Banach–Mazur distance*. Given two real normed spaces U, V of the same finite dimension, let

$$\delta^{\text{Banach}}_{\text{Mazur}}(U, V) := \inf \left\{ \|\phi\| \, \|\phi^{-1}\| \mid \phi \colon U \to V \text{ is a linear bijection} \right\}.$$

This quantity is not a metric, but its logarithm

$$d^{\text{Banach}}_{\text{Mazur}}(U, V) := \ln(\delta^{\text{Banach}}_{\text{Mazur}}(U, V))$$

is, and it plays a central role in geometric functional analysis. In some sense, the Banach–Mazur distance measures the distance between the unit balls of two normed spaces, that is, between affine equivalence classes of centrally symmetric convex bodies. An alternative definition is the following. For any two d-dimensional convex bodies C_1 and C_2, let

$$\delta^{\text{Banach}}_{\text{Mazur}}(C_1, C_2) := \inf \left\{ \lambda \; \middle| \; \begin{array}{l} \text{there is an affine map } a \text{ and a homo-} \\ \text{thety } h_\lambda \text{ of ratio } \lambda > 0 \text{ such that} \\ a(C_1) \subset C_2 \subset h_\lambda(a(C_1)) \end{array} \right\}.$$

The restriction to affine equivalence classes is a natural step in functional analysis, since these classes correspond to the isometry classes of normed spaces. However, it seems somewhat artificial in a geometric context, where it is more convenient to use the following definition:

$$\delta^{\text{BM-hom}}(C_1, C_2) := \inf \left\{ \frac{\lambda_2}{\lambda_1} \; \middle| \; \begin{array}{l} \text{there are homotheties } h_{\lambda_1} \text{ and } h_{\lambda_2} \\ \text{of ratios } \lambda_1, \lambda_2 > 0 \text{ such that} \\ h_{\lambda_1}(C_1) \subset C_2 \subset h_{\lambda_2}(C_1) \end{array} \right\}.$$

This function compares homothety equivalence classes of convex bodies. It is not a metric, but as in the case of the original Banach–Mazur distance, its logarithm $\ln \delta^{\text{BM-hom}}$ is.

Problem 6   *What is the smallest γ_n such that for every planar convex set C there is an n-gon P_n with $\delta^{\text{BM-hom}}(C, P_n) \leq \gamma_n$?*

For $n = 3$ this problem was studied in [BáB*93], [FlM*92], [La92], but even this case has not been settled. The best known lower and upper bounds for γ_3 were given by Fleischer, Mehlhorn, Rote, Welzl, and Yap [FlM*92]: $2.118 \leq \gamma_3 \leq 2.25$. It is conjectured that in this case the extremal set is the regular pentagon.

In some variants of the Banach–Mazur metric it is required that the centroids of the homothetic copies of C_1 coincide with the centroid of C_2, or at least with each other [BrL01], [Gr63], [La92]. Similar questions on approximation by centrally or axially symmetric sets, rectangles, affinely regular hexagons, and sets belonging to various other special classes have also been considered [Gr63], [La93], [La98], [La02], [La03].

Stromquist [St81] proved that the Banach–Mazur diameter of the family of centrally symmetric plane convex bodies is $\frac{3}{2}$. Actually, he proved a stronger statement: the convex body

$$C_1 = \left\{ (x, y) \ \middle| \ |y| \leq 1, \ \ x^2 + y^2 \leq 2, \ \ \frac{x^2}{2} + y^2 \leq \frac{4}{3} \right\}$$

has the property that its distance $\delta^{\text{Banach}}_{\text{Mazur}}(C_1, C_2)$ to any other centrally symmetric plane convex body C_2 is less than or equal to $\sqrt{\frac{3}{2}}$. The Banach–Mazur distance between the regular hexagon and the square is $\frac{3}{2}$. Lassak [La89] showed that $\delta^{\text{Banach}}_{\text{Mazur}}(C_1, C_2) \leq 1 + \sqrt{2}$ for any convex body C_1 and any centrally symmetric convex body C_2 in the plane. This bound is attained when C_1 is a triangle and C_2 is a parallelogram.

[AlG99]   H. ALT, L.J. GUIBAS: Discrete geometric shapes: matching, interpolation and approximation, in: *Handbook of Computational Geometry*, J.-R. Sachs et al., eds., Elsevier 1999, 121–153.

[BáB*93]   V. BÁLINT, A. BÁLINTOVÁ, M. BRANICKÁ, P. GREŠÁK, I. HRINKO, P. NOVOTNÝ, M. STACHO: Translative covering by homothetic copies, *Geometriae Dedicata* **46** (1993) 173–180.

[Bl17]   W. BLASCHKE: Über affine Geometrie III: Eine Minimumeigenschaft der Ellipse, *Ber. Ver. Sächs. Akad. Wiss. Leipzig, Math.-Naturw. Klasse* **69** (1917) 3–12.

[**Br00**]   P. BRASS: On the approximation of polygons by subpoly-
               gons, in: *Euro-CG 2000, Proc. European Workshop Comput.
               Geom.* (Eilat, 2000), 59–61.

[**BrL01**]   P. BRASS, M. LASSAK: Problems on approximation by tri-
               angles, *Geombinatorics* **10** (2001) 103–115.

[**Do44**]   C.H. DOWKER: On minimum circumscribed polygons, *Bull.
               Amer. Math. Soc.* **50** (1944) 120–122.

[**Eg57**]   H.G. EGGLESTON: Approximation of plane convex curves I:
               Dowker-type theorems, *Proc. London Math. Soc.* **7** (1957)
               351–377.

[**FeT77**]   G. FEJES TÓTH: On a Dowker-type theorem of Eggleston,
               Acta Math. Acad. Sci. Hungar. **29** (1977) 131–148.

[**FlM*92**]   R. FLEISCHER, K. MEHLHORN, G. ROTE, E. WELZL,
               C. YAP: Simultaneous inner and outer approximation of
               shapes, *Algorithmica* **8** (1992) 365–389.

[**Ge84**]   P. GEORGIEV: Approximation of convex n-gons by $(n-1)$-
               gons (in Bulgarian, English summary, see Zbl 545.41045), in:
               *Mathematics and Education in Math., Proc. 13th Spring Conf.
               Bulg. Math. Soc.* (Sunny Beach, Bulgaria, 1984), 289–303.

[**Gr93**]   P.M. GRUBER: Aspects of the Approximation of Convex
               Bodies, in: *Handbook of Convex Geometry, Vol. A*, P.M. Gru-
               ber et al., eds., Elsevier 1993, 319–345.

[**Gr63**]   B. GRÜNBAUM: Measures of symmetry of convex sets, in:
               Convexity, V. Klee, ed., *Proc. Sympos. Pure Math.* **7** Amer.
               Math. Soc. 1963, 233–270.

[**Iv73**]   R. IVANOV: Approximation of convex n-polygons by means of
               inscribed $(n-1)$-polygons (in Bulgarian, English summary, see
               Zbl 262.52002), in: *Mathematics and Education in Mathemat-
               ics, Proc. 2nd Spring Conf. Bulg. Math. Soc.* (Vidin, Bulgaria,
               1973), 113–122.

[**Ke73**]   P. KENDEROV: On an optimal property of the square (in
               Russian, see Zbl 329.52010), *C. R. Acad. Bulg. Sci.* **26** (1973)
               1143–1146.

[**La03**]   M. LASSAK: Affine-regular hexagons of extreme areas in-
               scribed in a centrally symmetric convex body, *Adv. Geom.* **3**
               (2003) 45–51.

[**La02**]   M. LASSAK: Approximation of convex bodies by axially sym-
               metric bodies, *Proc. Amer. Math. Soc.* **130** (2002) 3075–3084
               (electronic), Erratum in **131** (2003) p. 2301 (electronic).

[La98]   M. Lassak: Approximation of convex bodies by centrally symmetric bodies, *Geometriae Dedicata* **72** (1998) 63–68.

[La93]   M. Lassak: Approximation of convex bodies by rectangles, *Geometriae Dedicata* **47** (1993) 111–117.

[La92]   M. Lassak: Approximation of convex bodies by triangles, *Proc. Amer. Math. Soc.* **115** (1992) 207–210.

[La89]   M. Lassak: Approximation of plane convex bodies by centrally symmetric bodies, *J. London Math. Soc. 2. Ser.* **40** (1989) 369–377.

[Po70]   V.A. Popov: Approximation of convex sets (in Bulgarian, with Russian, English summary), *B'lgar. Akad. Nauk. Otdel. Mat. Fiz. Nauk. Izv. Mat. Inst.* **11** (1970) 67–80.

[Po68]   V.A. Popov: Approximation of convex bodies (in Russian, see Zbl 215.50601), *C. R. Acad. Bulg. Sci.* **21** (1968) 993–995.

[St81]   W. Stromquist: The maximum distance between two-dimensional Banach spaces, *Math. Scand.* **48** (1981) 205–225.

Author Index

Subject Index